Die Binnengewässer

Einzeldarstellungen aus der Limnologie und ihren Nachbargebieten

Begründet von August Thienemann

Unter Mitwirkung von Fachgenossen herausgegeben von

Dr. H.-J. Elster

Professor für Limnologie, Direktor des
Limnologischen Instituts der Universität
Freiburg (Walter-Schlienz-Institut),
Konstanz/Bodensee

Dr. W. Ohle

Professor für Limnologie,
Universität Kiel, Max-Planck-Institut
für Limnologie,
Plön/Holstein

Band XXV

1971

E. Schweizerbart'sche Verlagsbuchhandlung
(Nägele u. Obermiller) Stuttgart

Biology of Brackish Water

by

Dr. Adolf Remane
Professor of Zoology, University of Kiel

Dr. Carl Schlieper
Professor of Marine Zoology, University of Kiel

Second Revised Edition

With 165 Figures and 50 Tables in the text and on 4 Folders

1971

E. Schweizerbart'sche Verlagsbuchhandlung
(Nägele u. Obermiller) Stuttgart

Wiley Interscience Division
John Wiley & Sons, Inc. New York-Toronto-Sydney

Second Edition of »Die Biologie des Brackwassers«
Die Binnengewässer, Vol. XXII

First Edition

© E. Schweizerbart'sche Verlagsbuchhandlung
(Nägele u. Obermiller) Stuttgart 1958

Second Edition

© E Schweizerbart'sche Verlagsbuchhandlung
(Nägele u. Obermiller) Stuttgart 1971

Printed by: ČGP Delo, Ljubljana
Printed in Yugoslavia

Library of Congress Catalog Card Number: 70-134683
ISBN 3 510 40034-8
ISBN 0 471 71640-5

Published in the United States, Canada, Australia by Wiley Interscience Division: John Wiley & Sons,
Inc., New York, Toronto, Sydney
in all other countries: E. Schweizerbart'sche Verlagsbuchhandlung (Nägele u. Obermiller) Stuttgart

Preface

This volume on the "Biology of Brackish Water" represents more than just the English translation of the German "Biologie des Brackwassers" published in 1958; this English edition has been completely revised and covers a much broader field of knowledge. Both the authors, ecologist REMANE and physiologist SCHLIEPER, have worked on the aspects of the brackish water for more than forty years. In 1934, at the meeting of the German Zoological Society, REMANE gave his first extensive lecture on this subject referring to his own results on the ecological singularities of the biotope of brackish water. In 1932, SCHLIEPER treated the life conditions of inhabitants of brackish water from a physiological viewpoint in a lecture at the International Limnological Congress in Amsterdam. His first experimental research on the influence of low salt concentrations in brackish water on marine invertebrates originates from the year 1929. REMANE as well as SCHLIEPER have been constantly publishing on these subjects since that time. Both authors have their laboratories in Kiel (however, not in the same institute) on the shores of the Baltic. Therefore, their extensive investigations were started at this brackish sea, the ecological problems of which have also been well treated with different emphases by Finish and Scandinavian biologists. It is certain that these publications have supported and stimulated the research of brackish waters throughout the world.

The biologists of the Netherlands especially have done important work in this field of science. The desalinisation of the Zujder Zee and the Ijselmeer have been excellent and impressive objects for studies. The Roumanian and Russian investigations of the Pontokaspian Seas are also very important, since observations on the behaviour of many interesting endemic species of brackish water there as well as the analysis of numerous species immigrated from other regions into brackish waters with changed chemical composition have been possible. The increasing world acknowledgement of the importance of this kind of research and the general interest in the biotope between oceans and fresh waters have been stressed by international meetings in Venice (1958) and on Jekyll-Island (1964) which were devoted to special problems of brackish waters and estuaries, respectively.

This new treatise provides a survey of aspects of the biology of brackish water considered essential by the authors.

The editors are convinced that this English edition will further stimulate the international research on brackish water and enlarge the general scientific interest in this field.

The Authors, the Editors and the Publisher

Contents

Part I

Ecology of brackish water

By Adolf Remane

Part II

Physiology of brackish water
(Physiological features of life in brackish water)
By Carl Schlieper

To Part I and II

Part I

Ecology of brackish water

by

Adolf Remane

Kiel

With 81 Figures and 7 Tables in the text and on 3 Folders

Introduction

Studies of brackish water as a habitat were neglected for a long time compared with those of the sea and fresh water. It was MÖBIUS in his work on the fauna of Kiel Bay who directed attention to the specific problems of this zone of transition. The most important result to emerge was the recognition that brackish water is not merely a region of transition between sea and fresh water; it displays a number of distinct phenomena which warrant a special study of this area. In the last decades the number of publications relating to brackish water has increased enormously. The present compilation can no longer represent a handbook surveying the whole field of investigations; it is confined to a presentation of the important ecological features peculiar to the region of brackish water and the problems they pose. Due to my personal knowledge of the brackish waters of Central Europe and, in particular, of the Baltic, these regions have been specially emphasized. This may be justified by the fact that, thanks to the work of Scandinavian, Finnish, German and Dutch biologists, they are the best known areas of brackish water. I wish to thank my collaborators Dr. PETER AX, Dr. SEBASTIAN GERLACH, Dr. OTTO KINNE and Dr. ROLF SIEWING for their varied help, Professor THIENEMANN for help with sources of literature, Prof. H. CASPERS, Prof. Dr. SEGERSTRÅLE, Dr. h. c. CHR. BROCKMANN, Dr. S. JAECKEL jun. for information, and the publishers for their courtesy during publication of the book.

My researches into the biology of brackish-water organisms have been carried out with the support of the Deutsche Forschungsgemeinschaft, the Akademie der Wissenschaften und Literatur and the Deutsche Kommission für Meeresforschung.

I. Salinity as a factor in the distribution of animals and definitions of brackish water

A. Boundaries and subdivisions of brackish water

Salinity is a crucial factor which exerts a peculiar influence on living organisms. In establishing the range of salinity which permits life, we find on the one hand fresh water with a minimal salt content, nearly approaching zero, on the other hand the region of saturated salt water (about 263—267$^0/_{00}$ salt content). Several species occur in such high salinity as is specially found in saline waters. In the red Lake Bulack near the Caspian Sea, supposedly with 285$^0/_{00}$ S. Suworow (1909, p. 676) found in addition to flagellates (*Monas dundalii* probably *Dunaliella*) und Cyanophyceae (*Lyngbya*), the larvea of two species of Diptera (5 mm and 3 mm), a species of Oligochaete, copepods, ("*Canthocamptus*" sp. probably *Cletocamptus*) and a Rotifer ("*Diaschiza*" sp.). Caspers (1952, p. 253) found in the evaporation basins of the saline Anchialo on the Black Sea, at about the same salinity, some Diptera larvae of the species *Haliella caspersi*. The brine shrimp *Artemia salina* and the fly larvae of the genus *Ephydra* exceed the upper limit of 200$^0/_{00}$ S. Microorganisms are even more resistant. Rippel-Baldes (1952, p. 95) writes: ,,Viele Bakterien z. B. Sporenbildner des Bodens, sind halotolerant und vertragen, bei mannigfacher Abstufung hinsichtlich der einzelnen Arten, bis zu 30% Kochsalz in der Nährlösung. Sehr verbreitet sind dann außer Leuchtbakterien andere halophile Formen, z. B. *Beggiatoa*-Arten, ferner eine Reihe von Formen, die nur bei hohen Kochsalzmengen gedeihen. Hierher gehören namentlich zahlreiche rotgefärbte Arten, die oft auf Salzfischen auffällig zur Entwicklung kommen. Russisches, aus Salzseen gewonnenes Salz enthält 100000 bis 200000 lebende Keime in 1 Gramm, darunter 7000—120000 des rotgefärbten *Micrococcus roseus*, der sich in konzentrierten Kochsalzbrühen entwickelt." [Many bacteria, e. g. spore-forming ones from the soil are halotolerant and tolerate up to 30% NaCl in the nutrient solution — bearing in mind that different species display varying degrees of tolerance. Apart from luminous bacteria there are other, widely distributed halophilic froms such as species of *Beggiatoa* as well as a number of forms which will only thrive in the presence of large amounts of NaCl. Among these, in particular, are numerous red species which are often conspicuous on salt fishes. Russian salt, obtained from salt lakes, contains 100 000—200 000 living germs in 1 gramme; among these there are 70000—120000 of the red-coloured *Micrococcus roseus* which develops in concentrated brine.]

J. Ruinen (1938) placed natural salt samples from salt-waters, brine and salt works from different continents into culture media. A good number of Flagellata from various orders grew, even in cultures with a high salt concentration. Thus in 3$^0/_{00}$ S up to saturation there flourished: *Monosiga brevicollis, Tetramitus cosmo-*

politus, T. ovoideus, Dunaliella salina, D. euchlora, Asteromonas gracilis, at 100⁰/₀₀ S to saturation: *Bodo caudatus, Jolyella bunebungensis*; at 160⁰/₀₀ to saturation: *Phyllomitus yorkeensis, Amphimonas rostratus, Pleurostomum flagellatum, Petalomonas* spec., species of *Dunaliella* etc.

At an even earlier date (1925) during his hydrobiological studies of the brackish waters of Kujawi LIEBETANZ had placed soil samples from saline waters into NaCl culture solutions of 0.5⁰/₀₀—20⁰/₀₀ and at 200⁰/₀₀ S he was still able to grow the following organisms: the diatom *Fragillaria virescens* var. *halophila*, the flagellates *Oicomonas salina* and *Dunaliella salina*, the fungus *Torula salina*, the amoeba *Amoeba salina*.

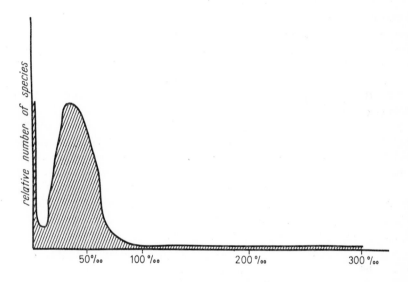

Fig. 1. Areas of salt concentration colonized by living organisms. The number of species at different salinities is indicated.

The salinity spectrum (fig. 1) of living organisms ranges from zero to about 270⁰/₀₀ (300⁰/₀₀). It is customary to subdivide such a spectrum — after NAUMANN — into an oligotype, a mesotype and a polytype. In relation to many factors, such as temperature, pH, the majority of organisms will settle in the mesotype of an environmental factor and the number of species decreases as they approach the extremes. If such a classification were made for salt content (such as oligotype 0—100, mesotype 100—200, polytype 200—300) the result would show, surprisingly, that over 99% of all organisms belong to the oligotype, since both sea water and fresh water are found within its lower range. These represent two optima for living organisms and between them lies the peculiar medium, brackish water.

A definition of this brackish water is essential. But it must not be forgotten that definitions are merely scientific aids and their practical usefulness as well as convention are important considerations.

There is a difference of opinion whether the term brackish water should be restricted to fresh water mixed with sea water or whether inland saline waters of a medium salinity should be included. REDEKE (1935) defines: "Brackwasser ist ein Gemisch von Süßwasser und Meerwasser s. str." [brackish water is a mixture of fresh water and sea water s. str.], thus excluding the inland waters of low salinity. But since the organisms from these inland waters are nearly all identical with those occurring in brackish sea water at the sea shore, such a division is impracticable for the biologist. Therefore I designate all waters of a medium salinity as brackish waters.

It is more difficult to demarcate brackish water from sea and fresh water since all gradations occur and any boundary is artificial. The delimitations which have been made vary considerably. MÖBIUS and HEINCKE (1883, p. 284) put the boundary between brackish water and sea at $7.5^0/_{00}$. VAN OYE, REDEKE, E. DAHL and many other authors at $30^0/_{00}$ since at this point marked biological differences appear as compared with true sea water ($> 30^0/_{00}$), while other authors (EKMAN 1953, HILTERMANN 1949, REMANE 1934, SCHLIENZ 1927) favour a delimitation at $17—18^0/_{00}$. The boundary between brackish and fresh water is given at $0.1^0/_{00}$ (JOHANSEN 1918), as 0.184 (REDEKE), 0.5 (SCHLIENZ, EKMAN) or about $3^0/_{00}$ (REMANE). SEGERSTRÅLE (1959) gives a historical survey.

The reasons for these differences are to be found in the fact that one does not simply wish to establish arbitrary boundaries within the continuity of a hydrographic factor, but tries to find natural limits, that is those at which a change takes place in the biological composition or the biological behaviour of the organisms. VÄLIKANGAS, in particular, has emphasized this requirement. But such natural limits will only slowly be unravelled by painstaking, individual work.

It is certain that biologically the salinity range from sea to fresh water is not continuous, but capable of subdivision into distinct stages. Attempts to divide this range have been going on for a long time. HEIDEN distinguished seven grades of salinity regions on the basis of their diatom flora.

REDEKE's classification has been widely accepted

g/l Cl		$^0/_{00}$ S
< 0.1 fresh water		$<— 0.21$
0.1— 1.0 oligohaline		0.21— 1.84
1.0—10.0 mesohaline	brackish water	1.84—18
10.0—17.0 polyhaline		18 —30
17.0 sea water		>30

REDEKE takes the chlorine content, not the salt content, as his unit of measurement. The total salt content — at least in sea water — stands in a fixed relation to the chlorine content, as expressed in KNUD's formula $S = 0.030 + 1.806$ Cl. However, addition of river water may upset this relation between salt content and chlorine content as it often contains Ca, Mg and SO_3 in different percentages. Thus, in this case, salt content is higher than that calculated from the chlorine content. As regards marine brackish water, the deviation from KNUDSEN's formula is slight, according to RINGER (1907) barely over $0.1^0/_{00}$. The situation is different for continental brackish waters which arise through evaporation of river water or by the addition of dissolved salts from salt deposits. In these cases the differing salt composition may lead to a considerable shift in chlorine and salt relations (see p. 147).

Already REDEKE (1922, p. 332) calls the mesohaline zone "das Gebiet der autochthonen Brackwasserformen" and "das Brackwassergebiet s. str." [the region of authochthonous brackish-water forms and the brackish-water region s. str.]. VÄLIKANGAS (1926, p. 215) divides this region into two subzones — the meio- or β mesohaline 2—8°/$_{oo}$ and the pleio- or mesohaline subzone 8—16.5^o/$_{oo}$. REDEKE makes an analogous division, but he calls the lower one 1.8—10°/$_{oo}$ α and the upper one 10—18°/$_{oo}$ β mesohaline.

HILTERMANN (1949, 1966) while retaining these demarcations, goes on to further subdivisions and introduces specific technical terms. He calls sea water below 30°/$_{oo}$ brachyhaline sea water and subdivides brackish water which begins at about 18°/$_{oo}$ into pliohaline (18—10°/$_{oo}$), mesohaline (10—5°/$_{oo}$), miohaline (5—3°/$_{oo}$) and oligohaline (3—0.5^o/$_{oo}$) brackish water; from 0.5^o/$_{oo}$ there is the fresh water region (see fig. 2). There is a further subdivision into the infrahaline region with 0.5—0.2^o/$_{oo}$ S. LÖFFLER (1961) uses the term hyperlimnic for the region of 0.05—0.1^o/$_{oo}$ S, the region of fresh water poor in ions. JÄRNEFELDT (1940b, p. 95) and HALME (1944, p. 40) call the region of 0.05—0.5^o/$_{oo}$ limnohaline region.

EKMAN (1953, p. 117) provides a special classification:

fresh water	0—0.5^o/$_{oo}$	salinity
oligohaline brackish water	0.5—3°/$_{oo}$	salinity
mesohaline brackish water	3—10°/$_{oo}$	salinity
polyhaline brackish water	10—17 (20?)^o/$_{oo}$	salinity
oligohaline sea water	17 (20?)—30°/$_{oo}$	salinity
mesohaline sea water	30—34 (?)^o/$_{oo}$	salinity
polyhaline sea water	34 (?)^o/$_{oo}$	salinity

In this context the terms oligohaline, mesohaline, polyhaline do not refer directly to a specific salinity zone, but are used to characterize the zones both in brackish and sea water.

The terms oligohaline, mesohaline and polyhaline have now been generally accepted for specific salinity areas within the whole region. It will be difficult to accept them with a different meaning and occurring twice within the salinity spectrum.

At a "Symposium on the Classification of Brackish Waters" 1958 the following classification was suggested ("The Venice System")

zone	salinity ^{o}/$_{oo}$
hyperhaline	$> \pm 40$
euhaline	± 40—± 30
mixohaline	$(\pm 40)\ \pm 30$—± 0.5
mixoeuhaline	$> \pm 30$ but $<$ adjacent euhaline sea
(mixo-)polyhaline	± 30—± 18
(mixo-)mesohaline	± 18—± 5
(mixo-)oligohaline	± 5—± 0.5
limnetic (fresh water)	$< \pm 0.5$

Both mesohaline and oligohaline zones are further subdivided into α and β regions, as follows:

α mesohaline (= pliomesohaline) .... ± 18—± 10
β mesohaline (= miomesohaline) .... ± 10—± 5
and α oligohaline (= pliooligohaline) .... ± 5—± 3
β oligohaline (= miooligohaline) .... ± 3—± 0.5

The name "poikilohaline" (DAHL) has been proposed for zones with marked fluctuations in salinity: mixohaline in this system refers to water in which the salt water is supplied by the sea. It is advisable to confine the term mixohaline to the total area of mixing of sea and fresh water from ± 30—$40^0/_{00}$ and $0.5^0/_{00}$ and to reserve the term brackish water (eau saumâtre) for the biologically specific region of $18^0/_{00}$—$5^0/_{00}$. It is for the limnologists to decide whether the boundary towards fresh water should be fixed at $0.5^0/_{00}$. PRICE and GUNTER (1964) prefer $0.15^0/_{00}$, LÖFFLER (1956) $1^0/_{00}$. I am using the terms euhaline for 30—$40^0/_{00}$, polyhaline (= brachyhaline) for 30 to $18^0/_{00}$, pliohaline for 8—$18^0/_{00}$, miohaline for 8—$3^0/_{00}$, oligohaline for 3—$0.5^0/_{00}$. Mesohaline equals pliohaline + miohaline. HILTERMANN (1966) suggests that the mesohalinikum should be subdivided into three; this may prove to be justifiable, but the designation of mesohaline for the intermediate zone between plio- and miohaline may lead to misunderstanding. Interhaline would be a better term.

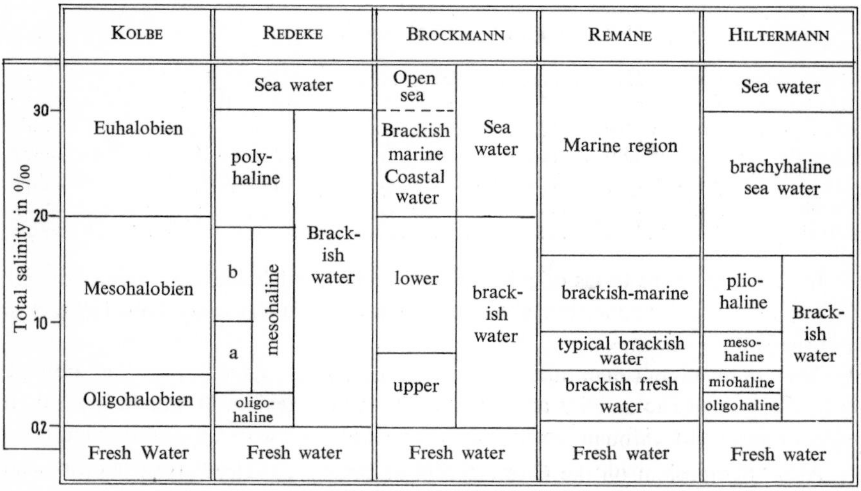

	KOLBE	REDEKE		BROCKMANN		REMANE		HILTERMANN	
30 —	Euhalobien	Sea water		Open sea	Sea water	Marine region		Sea water	
		poly-haline		Brackish marine Coastal water				brachyhaline sea water	
20 —	Mesohalobien	b	Brack-ish water	lower	brack-ish water	brackish-marine	plio-haline		Brack-ish water
10 —		mesohaline						meso-haline	
		a				typical brackish water			
	Oligohalobien	oligo-haline		upper		brackish fresh water	miohaline		
							oligohaline		
0,2 —	Fresh Water	Fresh water		Fresh water		Fresh water		Fresh water	

Total salinity in $^0/_{00}$

Fig. 2. Various classifications of the intermediate region between the sea and fresh water. After ROTTGARDT 1952.

The biologist needs to classify not only the areas of water, but also the organisms as regards their reaction to salt content. Frequently the terms denoting the aquatic regions (oligohaline, mesohaline, polyhaline) have simply been transferred to the animals; one talked of oligohaline plant and animal species. But owing to the peculiar, specific reactions of living organisms to the salinity spectrum this may lead to misunderstanding. An animal living in an oligohaline water may be a "halophilous" fresh-water animal, or else an extremely euryhaline marine animal or a specific oligohaline brackish-water animal. The same species would have been given a different name, had it extended into neighbouring areas. Thus the eelgrass *Zostera marina* would be designated a mesohaline species in an area of Finland, but as polyhaline in the Belt Sea and the Baltic. Therefore it is preferable to confine the terms oligohaline, mesohaline etc. only to the water or waters and name organisms according to their main habitats: sea or fully saline = halobionts, fresh water = limnobionts, brackish water = Hyphalmyrobien. Any of these groups may occur in a narrow or wide range of salinity, that is, it may be stenohaline or euryhaline. The names for the aquatic regions might be used — as appended to marine or limnic — to characterize the boundary area of a species; such as marine-mesohaline for a marine species extending from the sea into the mesohalinikum where the limits of its range are; limnic-mesohaline for a species penetrating from fresh water into the mesohalinikum, without advancing any further (see p. 64). For species restricted to narrow salinity ranges KOLBE's terms might be applied: Euhalobien, Mesohalobien, Oligohalobien (see p. 87).

WILLER has used the terms polyhaline, mesohaline, oligohaline in a different sense. WILLER examined lagoons which were connected with the sea. In these cases there always is a region with strongly marine influence (polyhaline), a middle (mesohaline) and a region of fresh-water influence (oligohaline). The absolute salinity values of these regions obviously differ from lagoon to lagoon, as the salinity of the adjoining sea and the inflow vary; for these reasons the terms are only relative. LILLELUND (1955) rightly, substitutes polythalass, mesothalass and oligothalass for WILLERS terms.

B. The origin of brackish water

If the term "brackish water" is taken in its widest sense as mentioned before, then brackish water will arise anywhere where a balanced inflow of salts and fresh water produces an intermediate salinity of the water. In this the salts, or corresponding zones respectively, are the more conservative elements because in enclosed bodies of water only small amounts enter the cycle of living matter. Organisms utilize nitrates and phosphates, which are commonly present in relatively small amounts, and take calcium and silica compounds for skeleton formation. A considerable proportion of these substances is returned to the water on mineralisation. The salts are able to undergo a mainly internal circulation in the water. Fresh water, however, is always subject to an external circulation by virtue of its physical factors. Water is constantly lost through evaporation and any brackish water in which this loss is not continously made good by addition of fresh water from outside will have its salinity rising rapidly and turn from brackish water into salt water of high concentration.

The supply of the two components, salt and fresh water, may take place in several different ways.

Supply of salts into brackish water comes about in three ways.

1) Sea water penetrates into the body of water. This is the commonest way in which brackish water arises. The numerous brackish waters at river mouths, lagoons (Haffe, Etangs) and shore pools are such examples. Normally sea water enters through some connection above ground, but under certain conditions a subterranean infiltration of sea water may take place into shore pools or lagoons, though these bodies of water are cut off above ground.

A well-known example of such a lagoon is the relict lake Mogilnoje (see DER-
JUGIN 1925) on the Island Kildin in the Barents Sea. A shingle bar cuts it off completely from the sea, but at a depth of several metres (probably 5—6 to 12 m) sea
water seeps through the bar; thus a deep layer, rich in salts is produced beneath
a brackish surface layer. For these marine brackish waters the striking constancy in the chemical composition of the sea water is significant.

2) Fresh water invades geological salt deposits, e. g. rock salt, the salt is dissolved; the salt water then appears in salt springs and forms salt pools, less frequently
salt lakes which contain brackish water, but they may greatly exceed sea water in
their salinity. In Central Europe the salt waters of Oldesloe (THIENEMANN), West-
phalia (SCHMIDT), Halle, Stassfurt, Lothringia (FLORENTIN) and others are known.
Since the geological salt deposits may show considerable deviation from sea water
in the relative proportions of salts, the same holds true for these saline brackish-
waters.

R. W. KOLBE (1927) reports on the origin of such saline waters near Sperenberg
south of Berlin. In the gypsum area deep borings were carried out 1867—1871.
In spite of the bore hole being subsequently closed, strongly saline water from the
depth percolated out; this finally reached the Krummensee and adjoining ditches
and canals, transforming them into salt or brackish waters. This condition remained
more or less stationary.

3) Salts occurring free in the soil or rocks are dissolved by fresh water and removed by rivers. In normal fresh water the concentration of these salts remains

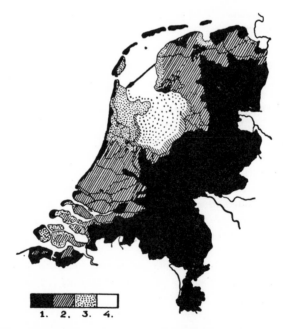

Fig. 3. Distribution of brackish waters in the Netherlands. 1, limnetic region — 2, predominantly oligohaline waters — 3, predominantly mesohaline waters — 4, poly- and
euhaline region. — No zones are marked in the estuaries. After REDEKE.

so low that, by definition, it is still fresh water. But in arid regions evaporation may be so considerable that most of the river water will evaporate. There remain relict waters without outlet and with increasing salinity owing to a constant supply of salts and the evaporation of water. This often results in brackish lakes or even highly concentrated salt pools. In these, too, the kind and relative amounts of salts may vary according to local conditions (see p. 146).

4) Industrial effluents may contribute to a higher salinity in some stretches of rivers (Werra, Wipper).

Influx of fresh water into brackish water can also take place in a variety of ways.

1) Direct supply of precipitation (rain water) is important in humid climates as loss of water by evaporation can be compensated for. In the Baltic, for example, annual precipitation about equals evaporation. Water derived directly from precipitation is presumably the main contributing factor for fresh-water supply only in some brackish rock pools and in the brackish "interstices" of the dense stands of algae (*Enteromorpha* zone) in the upper eulittoral of the sea. No doubt it plays an important part in the temporary small collections of brackish water which form behind the undercut slope on sandy shores, in the so-called "storm beaches". Even in marine Watt flats heavy rainfall at low tide may lead to the formation of limited layers of brackish water in the sandy soils; these may well have biological effects and harbour some brackish-water organisms.[1] H. MEYER and E. SCHULZ found that after heavy rainfall at a locality in the eulittoral of the North Sea the interstitial water of the upper layers of sand showed a reduction in salinity from $46.3^0/_{00}$ in June to $9.44^0/_{00}$ in August. The low salinity values in the tropical zones of the doldrums are evidence that heavy downpours have an influence even on the surface salinity of the ocean.

2) Fresh ground water in the soil makes subterranean contact with marine ground water, resulting in a zone of subterranean brackish water. Conditions for this are suitable on the sea-shore wherever water can percolate into the soil. This occurs anywhere on sea-shore with sandy or gravel ground; the same applies for the loose, more or less humous soils of the salt meadows. This influx of fresh ground water is clearly visible on the shore where exposed impermeable beds prevent fresh water from sinking in so that at low water it issues in runnels on the sea-shore.

3) Rivers, brooks and runnels of melt water above ground carry fresh water into sea and brackish water. This "evident" supply of fresh water and its important role for the brackish-water zone at river mouths have long been known. Melt waters are of special importance in polar regions where frozen soils prevent water from sinking in, as well as in the fjords of mountainous rocky coasts in regions of plentiful snow. In these areas melt waters can bring about a considerable freshening of the surface layers.

4) In addition to melt water the effects of ice have to be considered. As regards calving of glaciers and icebergs, their effect on salt content is slight as water is only

---

[1] The biologically destructive effect of excessive rainfall on the organisms of the marine eulittoral, by reducing salinity at low water, will not be discussed in detail. I only mention that in the Great Coral Reefs of Australia such situations were responsible for vast numbers of animals dying.

slowly given off. But the effect of an ice cover on brackish water may be more far reaching as this allows further extension of river mouths into the sea under a cover of ice. In this way ice exerts an indirect effect by producing a considerable freshening of the surface marine layers during a period of freeze-over (see H. LUTHER 1951).

To sum up, the following diagram illustrates the origin of brackish water:

Increasing salinity	decreasing salinity
invasion of sea water	outflow of water from rivers above ground
supply of salts from geological salt deposits	direct water of precipitation
influx of dissolved salts from soil	infiltration by fresh ground water (subterranean)
evaporation	(supply of ice from glaciers)
industrial effluents	

An intermediate salinity of brackish water results from a certain equilibrium between supply of salts, evaporation and influx of fresh water. This balance is of necessity a labile one since nearly all the factors depend on the weathei — storm floods, precipitation, intensity of sunlight etc.

C. Specific hydrographic features of brackish waters

Brackish waters differ from sea and fresh water not only in their salinity; hydrographic conditions also set them apart both from the sea and fresh water. These special features are of great biological significance.

Primarily, these are salinity stratification and fluctuations in salt content; secondarily the considerable oxygen depletion and the accumulation of hydrogen sulphide in the deeper layers. All these result from the way in which brackish waters originate.

a) Salinity stratification. At the same temperature water with higher salt concentration is heavier than that poor in salts; thus, if areas of different salt content are in contact with one another the water rich in salts will sink to the bottom. Mixing of water by turbulence and slowly acting diffusion militate against the isolation or bodies of water; but they frequently lead to the formation of new intermediate zones of medium salinity so that several layers of differing salt content may come to lie one above the other (fig. 4). Apart from the discontinuity layer due to temperature which is equally found in fresh water and sea, there exist in brackish water one or more discontinuity layers, often of great intensity, which are caused by salt content (haline stratification). Haline stratification is by no means absent from the open sea, but is far more pronounced in brackish water. While in the ocean salinity differences of $1^0/_{00}$ for every 100 m represent values above the average, salinity gradients in brackish water may rise to 10—70 times that amount (see fig. 5); locally they may be even greater. Within the brackish waters this stratification shows regional differences. It is specially high 1) in brackish — water seas, in the areas of contact and mixing of sea and brackish water. Thus in the approaches to the Baltic — that is in the Belt Sea, Sound and partly in the Kattegat, salinity stratification may at times reach enormous proportions (fig. 5), but it is variable. 2) at river mouths and in fjords in which the morphology of the ground and possibly the absence of tides result in a slight degree of mixing of the different types of water. There is often a sharp contrast between an upper layer, either poor in salts or fresh, and a counter-current on the floor which is rich in salts. 3) in small bodies of saline water and some lagoons in which fresh water enters on the surface in small rills, while salt

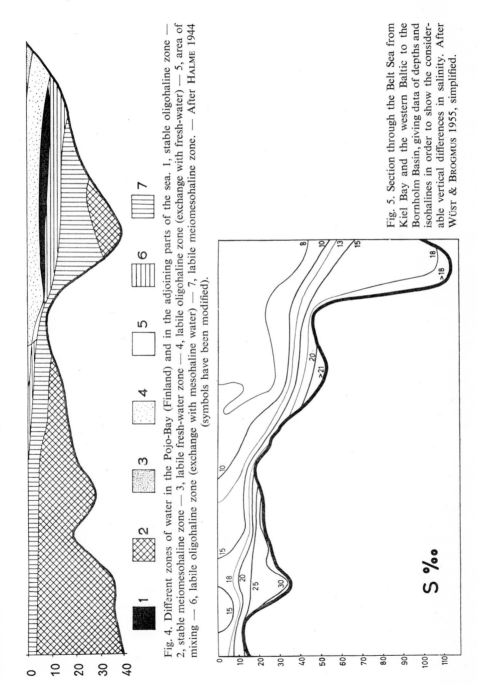

Fig. 4. Different zones of water in the Pojo-Bay (Finland) and in the adjoining parts of the sea. 1, stable oligohaline zone — 2, stable meiomesohaline zone — 3, labile fresh-water zone — 4, labile oligohaline zone (exchange with fresh-water) — 5, area of mixing — 6, labile oligohaline zone (exchange with mesohaline water) — 7, labile meiomesohaline zone. — After HALME 1944 (symbols have been modified).

Fig. 5. Section through the Belt Sea from Kiel Bay and the western Baltic to the Bornholm Basin, giving data of depths and isohalines in order to show the considerable vertical differences in salinity. After WÜST & BROGMUS 1955, simplified.

water comes in "subterraneously" as seepage water. The relict lake Mogilnoje has a surface layer of about 0—5 m with a salt content of 0—4°/$_{00}$, a middle zone (5 to 8 m) with 5—28°/$_{00}$ and a bottom layer (from 12—13 m) with 30—32°/$_{00}$ (DERJUGIN 1925). In a shore pool of Kiel Bay inwards from the dike (dam east of Stein) I found the surface water at 0.3°/$_{00}$, the water at 1 m depth at 3°/$_{00}$ S. 4) in the interstitial water of the floor in the coastal area on the Kniepsand (Amrum, west coast of Schleswig-Holstein) E. SCHULZ and H. MEYER (1939, p. 325) found in the Farbstreifenwatt in the surface grey layer (3 mm) 46.3°/$_{00}$ S, in the green layer below it (5 mm) 33.4°/$_{00}$ S, in the red layer (1 mm) 29.8°/$_{00}$ S, in the black layer (5—10 mm) 23°/$_{00}$, and in the ground water 7.4°/$_{00}$ S. Here salinity stratification was inverted; this was due to the concentration of salts at the surface, to evaporation and the inhibition of exchange through circulation.

In the areas of transition between brackish seas and the sea proper haline stratification is very labile; in the Kiel Bay, for example, it is mostly degraded in winter, but builds up again in spring. It may even be built up and disappear again several times in the course of a year. The effect of strong winds is an important factor in the degrading or prevention of haline stratification. It is particularly effective in shallow waters. In estuaries tides counteract the formation of a stable stratification.

b) Salinity fluctuations. The supply of fresh and salt water into brackish waters is never constant, but is influenced by precipitation, effects of wind and, partly, by tides. This variability in the supply combined with evaporation which also depends on weather and climate, account for salinity changes in the same locality which may be extensive. In the ocean salinity fluctuations in the same place are locally only small and fresh water exhibits no such changes; by contrast, brackish water represents a special region in which changes may at times be enormous and varied. They vary considerably from region to region. They are most pronounced where inflow and outflow vary to a large extent: in the connecting channels of the sea and at the entrance to brackish-water seas that is in the western Baltic, the Belt Sea, the eastern Kattegat, the Black Sea near the Bosphorus and the Sea of Marmara. Furthermore there are great fluctuations in the salt content in the region of river mouths with changing volume of water, particularly if the mouth is in the tidal area. In lagoons as well in which periods of rain bring a strong influx of fresh water and times of drought cause sea water to be sucked in, this may sometimes result in an interchange between fresh water and sea water (Lake Chilka).

Similar conditions obtain in some central areas of fjords. Great fluctuations also occur in small bodies of saline water where even heavy rainfall can produce such an effect and in particular in those small sheets of water flooded by sea water at intervals (spring tides) (fig. 77).

Since fluctuations everywhere depend on the weather (precipitation, atmospheric pressure, wind) they are primarily irregular. According to seasonal distribution — such as heavy precipitation in winter and little evaporation, more frequent outflow — there is an additional yearly rhythm of change; in tidal areas the inflow at spring tides produces a further tidal rhythm.

However, wide areas of brackish water display only small fluctuations. Even in the transitional areas mentioned above — with alternating inflow and autflow — there is a tendency to form specific bodies of water with medium salinities; they are demarcated

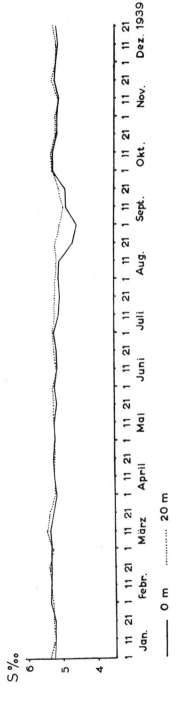

Fig. 6. Salinity in the Finngrundet on the East coast of Sweden (Öregrund-Archipelago) at 0 m (solid line) and at 20 m depth (dotted line) to show the constancy of the salt content in the Baltic. Drawn from data by WAERN (1952, p. 15).

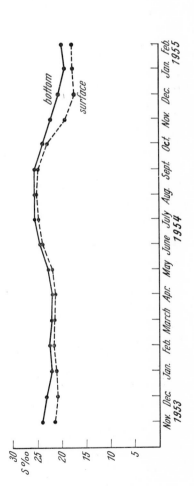

Fig. 7. Relatively constant salt content in a brackish lake of arid zones (Lac Quarun, Fayum, Egypt). After NAGUIB.

from one another by certain frontiers (discontinuity layers) (see WATTENBERG 1949, WYRTKI). In the Belt Sea region, for example, there is Baltic water, Belt Sea water, Kattegat water. These masses of water may be shifted to and fro when the position of inflow and outflow changes and their otherwise oblique lines of demarcation are displaced more steeply. Planktonic organisms are able to follow these displacements and in this way remain in a medium with much reduced changes while the geographical region is subject to extreme fluctuations. Similar intermediate masses of brackish water occur at river mouths with plentiful intermingling of water, e. g. in the Elbe (see CASPERS 1955). These regions may represent homogeneous habitats though the constituent parts of water are continuously being changed.

The salt content in the large basins of brackish-water seas is more uniform. In these there is an upper layer that is almost homohaline and which, in the Baltic, extends to a depth of 40—80 m (fig. 6). In shallow marginal zones (Bodden, shores) that lie at some distance from the mouths of rivers the bottom fauna inhabits an area of brackish water whose salt content varies little. According to DIETRICH (1950) the annual salinity fluctuation in the main part of the Baltic Basin as well as the Bottensee and Bottenwiek is of the order of less than $0.5^0/_{00}$; in the Belt Sea it rises to $3—5^0/_{00}$, in the Kattegat to over $7^0/_{00}$, in parts of the Norwegian fjords to $10^0/_{00}$, but in the central and western North Sea it drops again to below $0.5^0/_{00}$. Even the deep water of these seas is largely homohaline so long as it is some distance removed from the areas of inflow. This is particularly the case in the Black Sea and the Caspian Sea. In the deep layer of the Black Sea, beginning at 200—150 m depth, there is a constant salinity of $21—22^0/_{00}$, but at great depths it hardly rises above $22^0/_{00}$. In the central and southern regions of the Caspian Sea salinity is always $12.6—12.9^0/_{00}$ from the surface to the bottom (900 m) (ZENKEVITCH 1963). Shallow inland saline waters may be largely homohaline, such as water of salt springs and some brackish lakes in arid regions. Lake Moeris (Lac Quarun) in the Fajum area of Egypt has almost identical salinity at the surface and the bottom and the annual fluctuations are slight (NAGUIB, fig. 7).

Brackish waters differ from the open sea and from fresh water in their salinity fluctuations; within the brackish waters themselves these show considerable variation. Large areas of brackish water approach conditions of sea and fresh water in this respect; in some parts their salinity fluctuations are less marked than those found in shallow marine areas. Fluctuations in the Wattensea of the North Sea are greater than in the surface layer of the Baltic.

The secondary special features of brackish water are oxygen deficiency and abundance of H_2S in the lower layers; both are found much more pronouncedly than in non-brackish waters of comparable morphological structure. Both depend on the existence of haline stratification; they are present only when this stratification is permanent or persists over long periods.

Haline stratification which occurs in one or more discontinuity layers is an additional feature, besides the usual thermal stratification. It prevents vertical exchange of water masses. Even slight differences in salinity are sufficient to bring it about. RUTTNER (1937) established the fact that in the Traunsee, a fresh-water lake in Austria, the influx of industrial effluents produced an increase in the concentration of the deep water by 95 mg per litre, that is by 46%. This was sufficient to prevent complete circulation. Furthermore, stable discontinuity layers may lead to the accumulation of detritus and planktonic organisms so that transparency of the water above them decreases (WATTENBERG 1938). If horizontal exchange in the water is slight, e. g. in basins without or with little communication with the sea, the permanent water in the deep suffers a depletion of oxygen. At the same time remains of organisms sink from the upper layers so that this environment is a favourable field for the activities of anaerobic bacteria or for obligate anaerobes. Production of H₂S is due partly to the decomposition of sulphates by microorganisms *(Microspira)*, partly to the process of decay of organisms which have sunk to the bottom. The H₂S content, in its turn, creates favourable conditions for the sulphur bacteria *(Chromatium, Beggiatoa* etc.) which thrive here. The division into a normal surface layer rich in living organisms and an azoic depth lacking oxygen and rich in H₂S is well known from the Black Sea. This zone may be very rich in bacteria. In the Black Sea its biomass (mg/m³) is usually greater than in the surface layer. The minimum occurs in the region of the discontinuity layer (KRISS 1958).

The situation is less pronounced in the Baltic, partly because the basin is shallower, partly because the wider Belts allow a better horizontal supply of water for the deep layers. But even in the Belt Sea there is, occasionally, a situation resembling that in the Black Sea. With marked saline stratification oxygen disappears from the basins of the Belt Sea (Eckernförder Bay and others) which are more or less land locked; the bottom animals and the fish in the nets die and the floor is covered with the grey felt of the sulphur bacterium *Beggiatoa mirabilis*; amongst them may be some nematods and hypotrichous ciliates. These "infectious" or "dead grounds" only occur sporadically. But such deep zones are normal for brackish lagoons which have little communication with the sea, e. g. in the relict lake Mogilnoje (DERJUGIN 1925) as well as in isolated pools on the sea shore.

Fig. 8. Change of haline stratification depending on water temperature in Lake Varna. The arrows indicate the course of salt water penetrating from the sea (A) in summer (when haline stratification occurs) and (B) in winter (when haline stratification disappears). After VALKANOV 1937 from CASPERS 1951.

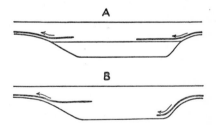

If the zone rich in H₂S extends upwards into the illuminated regions, flourishing zones of purple bacteria may form. They depend on light for their assimilation. A layer of pink coloured water may develop in this zone caused by mass development of *Chromatium* species. As these organisms oxidize H₂S DERJUGIN speaks of a bacterial film in Lake Mogilnoje protecting the organisms above it from extinction. In fact, the transport by storms, of layers rich in H₂S, to higher levels may kill all the animals there, including fishes, in the waters affected. This has been observed in the lagoons of the Black Sea (CASPERS 1951, PASPALEW 1941).

Similar phenomena are known to occur in fresh water, but under natural conditions they are of far less consequence. This becomes evident from a comparison of waters of similar types.

The lower layers of water also become enriched with CO_2. The temperature gradient often shows a striking minimum near the haline discontinuity layer (fig. 64c). Brackish waters thus represent certain adverse conditions for the development of organisms, both in comparison with sea and with fresh water. They are not bound up with a certain salinity as such, but depend on the intensity and kind of exchange of water masses (salinity fluctuations, haline stratification) and on the morphology of the waters (stagnant deep zone with O_2 depletion and abundance of H_2S).

Strong haline stratification need not necessarily lead to O_2 depletion and accumulation of H_2S. MUNRO FOX (1926) records from Lake Timsah in the area of Suez Canal an upper layer of almost fresh water, containing *Ceratophyllum*, while the deep zone (about 0.35—1.70m) is highly saline and inhabited by a marine fauna with sponges, ascidians (*Polyclinum saturninum*). The whole lake is shallow and the intense sunlight permits good illumination down to the bottom, allowing diatoms and others, to flourish. This flora evidently supplies the O_2 fot the deep.

II. Special ecological features of brackish-water organisms

A. Poverty of species

It has been known for a long time that brackish waters are poor in species and the number of species living in brackish water is much smaller than in marine regions with similar habitats and much smaller than in fresh water. There are very few species in brackish water. This is true for brackish seas, continental brackish lakes, estuaries etc. A special example of this is given in table 1 which illustrates the reduction of marine species from the North Sea and from the Kattegat, in the direction of the Baltic. Table 2, after S. JAECKEL (1952) provides a survey of the molluscs in the North Sea-Baltic area; fig. 10, after JOHANSEN refers to the molluscs in the Randersfjord. The reduction in the number of algae from the North Sea to the Baltic can be seen from table 3 after C. HOFFMANN (1933).

A general picture of the relations between number of species and salinity is represented in a curve fig. 9; it has been compiled from individual records. It displays the following special features. The lowest number of species is not halfway between fresh water and marine salinity, but is displaced close to fresh water. It is at about 5—7⁰/₀₀. This "asymmetric" position of the minimum is due to different behavior of fresh-water and marine animals. While the number of fresh-water species decreases rapidly even at a slight increase in salinity (see p. 62) the reduction of marine species takes longer. In the middle range of salinity at 17—18⁰/₀₀ it is about half that found at marine salinity, but reduction seems to be more rapid at about 10⁰/₀₀ S.

This salinity-species curve is based on many single records. An attempt has been made to eliminate the effects of other factors found in nature and to represent the relation of total number of species to salinity alone. Naturally, the course of the graph may be altered by local salinity fluctuations, temperature and O_2 content of the individual body of water. But for all thalassine brackish waters and some inland ones the minimum number of species close to fresh water can be demonstrated. It

Table 1. Number of species of marine and brackish-water animals (excluding parasites) in the region North Sea-Kattegat, Belt Sea, Baltic.

	I	II	III	IV	Öres.
Foraminifera	about 80	47	? 5	0	
Porifera	64	18	1	0	
Hydroid polyps	82	34	7	3	35
Scyphozoa	20	6	2	1	
Anthozoa		5	3	0	13
Ctenophora	4	3	1	1	
Bryozoa	about 90	35	5	2	
Kamptozoa	3	2	0	0	
Rotatoria	about 30	about 50			
Nematoda	about 350	about 300			
Tardigrada	5	3	2	0	
Polychaeta	about 250	about 100	22	4	111
Hirudinea	10	1	6	0	
Phyllopoda	6	5	5	5	
Ostracoda	about 100	50	20		
Cop. Harpactic.	about 320	126	27	20	
Cirripedia	10—12	5	1	1	10
Mysidea	15	7	5	4	15
Amphipoda	147	55	20	9	76
Isopoda	35	13	8	6	(about 25)
Cumacea	about 25	5—6	1—2	0	17
Halacaridae	43	24	10	2	
Decapoda	about 50	12	6	2	32
Pantopoda	10	5	0	0	5
Polyplacophora	about 7	1	(1)	0	
Prosobranchia	114	26	13	1	33
Opisthobranchia	about 70	23	6	3	
Bivalvia	92	32	11	4	58
Echinodermata	39	10	2	0	30
Ascididae	24	7	1	0	
Copelata	4	2	2	2	
Teleostei	120	69	41	20	

I = North Sea from Holland to the Dogger Bank, Kap Skagen and Kattegat above the 100 m isobath, II = German Belt Sea (Kiel Bay and Bay of Mecklenburg), III = southern and central Baltic, IV = northern Baltic (Aland Sea, Gulf of Finland and Gulf of Bothnia), Öres. = Öresund. Number of species in Öresund after BRATTSTRÖM.

In a determination of the correlation of number of species : salinity the numbers of species listed above cannot simply be correlated with the average salinity of the areas in question as this number is also affected by other factors in these areas. Thus I (North Sea — Kattegat) is many times larger than II (German Belt Sea), it has typical rocky coasts which are absent from II; it is deeper, the Belt Sea does not go below 35 m; while I has a permanent bottom in the deep this is lacking in II; II has a very high salinity fluctuation; the occasional ice cover in winter periodically reduces the bottom fauna and the same effects are produced by oxygen deficiency and the formation of H_2S in the troughs during the summer. Due to the action of all these factors the number of species in I will be smaller than could be expected from a direct comparison of salinities. Against that the number of species in II is increased as some are carried by the inflow of highly saline water. Many of these species are unable to live permanently in the salinity of the Belt Sea, but they appear in the list of species; in addition a number of species of II and III only occur in the deep, richly saline water, that is, in a salinity which exceeds the average of the region. The graph in fig. 9 has been constructed while taking all these secondary effects into account. It may be that the drop in the curve between 33 and 18 should have been drawn a little more steeply.

Table 2. Number of species of marine molluscs in the North Sea and the Baltic, after
S. Jaeckel (1952).

	North Sea	Kattegat	SW part of the Kattegat	Belts	Öresund (northern part)	Kiel Bay	Bay of Mecklenburg	Outer Baltic	Central Baltic	Gulf of Finland and Gulf of Bothnia
Solenogastres	8	1								
Scaphopoda	11	1	1	1	1					
Lamellibranchia	189	92	56	42	61	32	21	11	5	4
Polyplacophora	11	6	6	5	5	1		(1)		
Prosobranchia	210	101	58	40	69	26	18	13	3	1
Opisthobranchia	141	61	43	28	41	23	11	6	6	3
Pterpoda	3	1	1			1				
Cephalopoda	32	14	7	5	8	4	1			

Table 3. Reduction in number of algal species, after C. Hoffmann (1933 b)

	North Sea	Baltic			
		part I	part II	part III	part IV
Salinity in ⁰/₀₀	33	20—10	10—7	7—5	5—3
Red Algae	181	94	43	18	8
Brown Algae	124	75	48	23	10
Green Algae	101	76	68	45	28
Total	406	245	159	86	46

The parts of the Baltic are: part I Belt Sea to Darsser Ridge. Part II Sea of Arcona, Bornholmsee and the southern part of the Gotland Sea excluding the Gulf of Riga. Part III Northern Gotland Sea including Aland Sea and Schärenmeer the Baltic between, the numerous small island and the Gulf of Finland. Part IV Botten Sea and Bottenwiek.

should be noted that different groups of animals vary in their reactions to decreasing salinity; some will react more strongly than depicted in the curve, others to a lesser degree. There are marine groups highly sensitive to any dilution of sea water and their species curve drops quite rapidly (Radiolaria, Cephalopoda, Octocorallia, Madreporia etc., see p. 88); others are more tolerant of salt. The following general differences appear to be of ecological significance: a) On reduction of salinity the marine macrofauna decreases more rapidly than the microfauna. This is clearly shown among the Crustacea, taking Decapoda, Amphipoda, Isopoda and Cumacea as representatives of the macrofauna, and Ostracoda and Copepoda for the microfauna. Reduction of Echinodermata, Tunicata, Porifera (macrofauna) is above average; as far as can be assessed from available data it is below average in the Turbellaria (excluding Polyclada!) and Nematoda; it is undoubtedly far below average in the Ciliata, marine Rotifera, Gastrotricha. b) Reduction of species in groups forming a calcareous skeleton is greater than in their relations lacking such a skeleton. This is particularly marked within certain groups, e. g. the relatively strong reduction of Polychaeta and Bryozoa forming calcareous shells (see p. 39). c) Groups which have invaded the saline area from fresh water and have developed distinct species in brackish waters and in the sea display the usual reduction of species where the brackish region starts; but there is no minimum of species in brackish water or else it is only slightly indicated. Among the Rotifera and Oligochaeta a number of species are restricted to the polyhaline or even euhaline region, but their absence from the mesohaline region is compensated for by specific brackish-water species, hence there is no minimum. The same applies to dipterous larvae (Nematocera); very low numbers appear to occur in Chaetonotoidea (Gastrotricha) and in marine flowering plants. d) For some groups there is a complete gap in the mesohalinikum, that is they exist in high and in low salinities, but not in intermediate ones. The Hirudineae have ten species in the Kattegat, one marine one in the Belt Sea; there is a gap from 15 to about 10⁰/₀₀ and then one finds the fresh-water Hirudineae. The Porifera (sponges) have a gap in their occurrence at about 15—5(7)⁰/₀₀ in the North Sea-Baltic region. This gap is considerable in water mites (*Hydrachnella*, excluding Halacaridae) and is marked in the Cyprididae among the Ostracoda as regards the marine regions proper (in small bodies of saline water Cyprididae occur in the whole of mesohaline and polyhaline areas, for instance species of *Heterocypris*, *Cypridopsis* and *Eucypris*). This gap in brackish water displayed by certain groups is only known from the North Sea-Baltic area with some degree of accuracy (cf. water mites p. 78). Among plants such a gap seems to exist in the monocotyledonous family of Hydrocharitacea, between the submerged marine genera (*Halophila*, *Thalassia*) and the fresh-water ones (*Elodea*, *Vallisneria* and others).

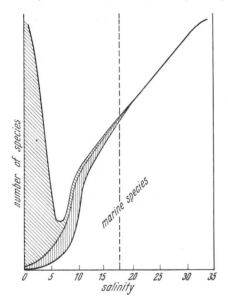

Fig. 9. Number of species in relation to salinity. The graph has been based on numerous single records. Obliquely hatched: Proportion of fresh-water species. Vertical hatching: Proportion of specific brackish-water species. Light: marine species. Black (at base): holeuryhaline species. In each case the number of species corresponds to the vertical extent of the respective area. After Remane 1934.

The graph constructed by me (1934, fig 1; here fig. 9) for the number of species in the salinity range between 33 and 0⁰/₀₀ was modified by E. Dahl 1956 (see fig. 12). He particularly stresses the considerable reduction in number of species in the range 34—30⁰/₀₀ S, said to include 75% of the original species. In my opinion the differences are due to the following facts: My graph is a reconstruction based on a comparison of numerous localities since, in any individual case not only salinity, but other factors are variable and have a bearing on the reduction of species. Dahl starts from a special case, the reduction of species from the Skagerrak to the Öresund. Here we find a very steep gradient in depth, from a few 100 m to the shallow areas of 20—40 m depth. Together with a north-south orientation and the opening of the deep zone towards the northern seas, this produces a temperature gradient and a gradient of temperature changes as well. Only extensive comparative studies will show how many of the species restricted to deeper water in the Skagerrak are limited by low salinity, how many by increased and fluctuating temperature, how many are bound to the stable deep floors, and avoid the bottoms of the shallow areas with their changing structure. In bringing out a relationship between the species-curve Skagerrak—Kattegat—Sund and the salt content it is necessary — for the sake of accuracy — to deduct all arctic boreal species whose southern boundary lies in the North Sea (about Cape Skagen to Scotland) from the number of

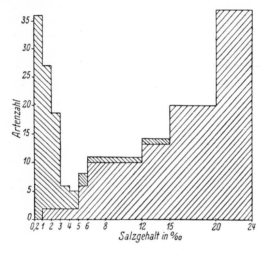

Fig. 10. Distribution of numbers of molluscan species in areas of differing salinities in the Randersfjord. The proportion of freshwater molluscs and of marine species is shown by different kinds of hatching. After JOHANSEN 1918 from REMANE 1934.

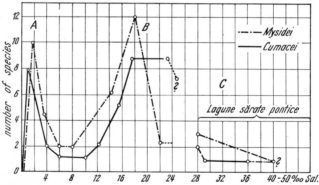

Fig. 11. Course of graph: number of species — salinity for Mysideae and Cumaceae in the region of the Black Sea. A, the unusual peak in low-saline areas. The peak B only occurs in waters of higher salinity (over $18^0/_{00}$) in lagoons of the Black Sea. Using the number of species from the Sea of Marmara to the Mediterranean this peak is likely to disappear since about 45 species of Mysideae and 51 species of Cumaceae inhabit the Mediterranean. C, number of species in salt lagoons. After BAČESCU 1954.

species occurring in the richly saline areas of the Skagerrak—Kattegat. Temperature is probably a limiting factor for those species (see ELOFSON 1941, p. 454). Further-more the marine species penetrating into the Kattegat meet a zone of very intensive salinity fluctuations towards the south. Baltic water of low salinity which leaves the Baltic during periods of outflow just reaches the Swedish coast as a result of the currents being deviated to the right and this leads to considerable, chiefly non-periodic fluctuations of the surface water. In the area chosen by DAHL for his comparisons several factors combine to a cumulative effect on the reduction of species; thus the curve conditioned by salinity only should take a less steep course.

In addition DAHL's graph is based on four groups of the macrofauna (Echino-dermata, Amphipoda, Isopoda, Cumacea). It has been pointed out (REMANE 1941, p. Ia, 6) that reduction of species is more marked in the macrofauna than in the microfauna. As shown in table 1, all the groups selected by DAHL are above average in their behaviour. A curve for Nematoda, Ciliata, marine Rotifera, Copepoda, Harpacticoidea and Ostracoda would be very much flatter. For comparison I present the species gradient for Ostracoda after ELOFSON (1943) which he obtained from investigations in the same area Skagerrak—Sund—Baltic. The reduction in species is far more gradual than in DAHL's graph. My curve was meant to take the average of all groups into account. As seen in table 3, it is in better agreement with the be-haviour of plants. In addition LEVRING's data (1940) can be quoted; he gives the number of algal species in Scotland as 350, for the west coast of Norway, extending into the arctic region as 480, for the east coast of Sweden as 350 and for Blekinge on the Baltic coast of Sweden (about 6.8—7.2$^0/_{00}$ S) as 120. From full-strength salt areas to 7$^0/_{00}$ S there is a reduction to about 25—30% of the original number of species; in this case reduction of species is even less than in my graph fig. 9.

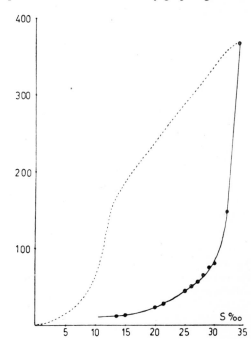

Fig. 12. Course of graph salinity—number of species after E. DAHL. Constructed from 4 groups of the macrofauna (Echinodermata, Am-phipoda, Cumaceae, Isopoda) in the region of the West coast of Sweden. Dotted line: graph compi-led by REMANE. After DAHL 1956.

However, the graph for reduction of species from 34—30⁰/₀₀ has probably been drawn a little too flat by me; the exact course cannot be determined as we do not know the extent to which the occurrence of deep sea forms is affected by temperature or salinity. In my view the reduction by 75% as assumed by DAHL ("that in the interval between 34 and 30⁰/₀₀ a reduction by 75% has already taken place" 1956) appears far too high if plants and animals, macroorganisms and microorganisms (diatoms!) are taken into account. In my opinion it is well below 20%. Investigations in marine areas with a somewhat lowered salt content such as obtains e. g. in the Pacific on the coast of Panama and Columbia will help to clarify this question.

The list of animal species from the Mediterranean and the Black Sea as given by ZENKEVITCH (1963) also provides support for a greater drop of the species-curve from fully salt water (35—38⁰/₀₀) to half-salinity (18—19⁰/₀₀). According to this list there are 5,244 species in the Mediterranean and 1,145 in the Black Sea. The number of species drops to 21.8%. These differences in the number of species are not only due to salt content. Animals colonize the Black Sea only to a depth of 100—150 m while they settle in the Mediterranean down to its greatest depth. Plants are more suitable for comparison on account of their dependence on light which restricts them to rather shallow zones in both seas. The number of their species is 423 in the Mediterranean and 270 in the Black Sea, that is 63.7%. In addition the lower temperatures of the Black Sea act as a barrier for some mediterranean animals. There is an undoubted anomaly in species reduction in the Black Sea where several groups show an increase of species in the oligohaline or miomesohaline region respectively. As an illustration I reproduce the graph on the occurrence of Mysidea and Cumacea in the salinity spectrum as drawn by BAČESCU 1954 (fig. 12). Both groups have an optimum at A in the range of 1—4⁰/₀₀. This peak is brought about by the crowding together into this area of Tertiary brackish-water species of the Pontocaspian brackish sea after the transgression of more saline water through the Bosphorus (for details see p. 144). The drop above 18—23⁰/₀₀ S, the peak at B in fig. 11 is caused purely by position, since the Sea of Marmara with its salinity fluctuations and restriction of biotope presents an unfavourable line of approach for the penetration of many mediterranean forms; (in addition the species are probably not fully known). In the Mediterranean there are more Mysidea and Cumacea than in the whole of the Pontocaspian region.

Another deviation from the ideal curve fig. 9 occurs if the comparison is confined to the salinity gradient of saline waters alone, that is salt pools and salt lakes of inland areas. Here, too, there is an obvious reduction of species from the direction of fresh water, even if we confine the comparison to the fauna of fresh-water pools. After a sudden drop there is a continuous decrease in numbers of species up to high salinity values, without any marked peak in the region of marine salinity, that is at 30—40⁰/₀₀. This at least applies to the macrofauna as well as to Rotifera, Ostracoda, Calanoidea. HUSTEDT (1925, p. 100) reports on the diatoms of such areas that following a minimum number of species close to fresh water the number of brackish-water species rose with increasing salinity, but only to a certain point: „Sobald nämlich etwa die normale Konzentration des Meerwassers erreicht ist, tritt das Gegenteil ein, die Zahl der Arten nimmt ab." [As soon as about normal concentration of sea water has ben reached the oppposite takes place and the number of species declines.] Judging from KAHL's study (1928) of Ciliata in the salt water of Oldesloe these display similar behaviour so that at least there are some groups in this environment with a species curve resembling that of the marine region.

It still remains an open question to what extent the salt content itself or the changes in salinity are responsible for the reduction of species and the poverty of species in brackish water. Undoubtedly any change in an ecological factor, especially a large and irregular one, may in itself be a powerful factor in deterioriation.

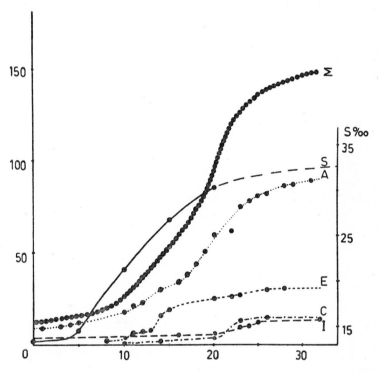

Fig. 13. The bathymetric distribution of the marine Amphipoda (A), Echinodermata (E), Cumaceae (C) and Isopoda (I) in the Öresund. The graph Σ represents the total number of species of the four groups, the graph S the average salinity in relation to depth, after daily measurements from the light ship Lappegrund during the years 1923—1938. After E. DAHL 1956.

This usually applies in those cases where deflections, at least to one side, approach the extreme values of tolerance. BACCI (1954b) and DAHL (1956) stress the importance of salinity fluctuations.

Within the brackish water there are regions of greatly varying poikilohalinity. As has been shown before (p. 14) the salt areas of the large brackish-water seas (Baltic, Black Sea etc.) are almost homohaline, especially in their extensive surface areas. The difference in poikilohalinity between the Belt Sea and the Baltic is many times greater than that between the Baltic, fresh water and sea. But the smallest number of species does not occur in the region of greatest poikilohalinity; on the contrary, there is a rich fauna and flora of predominantly marine character in this part (see REMANE 1934). The lowest number of species is found in the zone of transition to the homohaline brackish water of the Baltic and not in the main area of poikilohalinity. The same situation obtains in the Black Sea. According to OSTROUMOW (1894) 151 species of molluscs live in the poikilohaline area in the northern mouth of the Bosphorus, but only 91 in the practically homohaline region of the Baltic proper. The reduction of species from the deep regions of high and constant salinity to the shallower areas of lower and fluctuating salt content — so typical of Kattegat, Belt Sea and Öresund (see fig. 13) — cannot be interpreted as outlined above. In the deep basins of the western Baltic (Arkona Basin, Bornholm Basin) the deep water shows greater salinity fluctuations than the surface layer and yet the fauna of the deep is richer in species (molluscs, Crustacea, Polychaeta).

The great importance of the salinity factor has been demonstrated by the increase of salinity in the Baltic over the last decades. It rose by $0.5^0/_{00}$ in the surface waters of the Finnish coasts. This slight increase resulted in many marine species extending their area into the Baltic or into its bays: e. g. the medusae *Aurelia aurita, Cyanea capillata, Melicertum octocostatum,* the Copelata *Fritillaria borealis,* the Copepod *Temora longicornis, Pseudocalanus elongatus, Centropages hamatus,* the Amphipod *Calliopius laeviusculus,* the Cirriped *Balanus improvisus,* the bivalves *Astarte borealis* and *Cyprina islandica,* several fishes, especially the cod *(Gadus callarius).* Conversely the area occupied by the relict crustacean *Limnocalamus grimaldii,* coming from fresh water, has been narrowed down (see LINDQUIST 1960, MANKOWSKI 1961, 1962, SEGERSTRÅLE 1965 a).

Ax and Ax made an experimental investigation into the problem whether a minimum number of species can be demonstrated with constant salinity. In cultures of ciliates from brackish water and from fresh water 52 species developed at $15^0/_{00}$, 39 at $8^0/_{00}$, 29 at $5^0/_{00}$, 33 at $3^0/_{00}$. Undoubtedly the reduction of species in the range of $15—5^0/_{00}$ salinity is due to the salt content itself and not to salinity changes. It has been suggested that the rapid formation and disappearance of brackish waters, that is their short duration, may be responsible for the paucity of species (DAHL).

It is difficult to assess the significance of the historical factor in respect of paucity or abundance of species. The large, continuous marine basins undoubtedly provide better opportunities for the development of different types of organization; while regions subdivided into small areas offer, by virtue of their isolation, a good chance for differentiation into many closely related species. They will therefore contain fewer classes and orders, but not necessarily fewer species. These conditions are very similar for fresh and for brackish water. Both are split into innumerable individual bodies of water; no doubt the amount of brackish water exceeds that of fresh water since the volume of the largest fresh-water lakes is considerably less than that of brackish-water seas (Baltic, Black Sea, Caspian Sea). The number of bodies of fresh water is certainly larger though we must not underrate the large number of brackish lagoons and brackish-water pools and lakes in arid regions. As regards geological continuity the majority of fresh water and brackish waters are of short duration since displacement of inflow and silting-up constantly produce changes. Only a few have been in the same position over periods of geological time. Among these are brackish-water seas such as the Black and Caspian Seas which have been brackish-water seas since mid-Tertiary times. Conditions for species differentiation in fresh water and in brackish water closely resemble each other. Differences are to be found not so much in the behaviour of the brackish-water fauna which is fairly normal in respect of the original medium, of the sea, than in the peculiar features of the fresh-water fauna. Why should this extreme environment with its scarcity of ions and constant changes of individual biotopes have undergone a much richer differentiation than the brackish water resembling it? How is it that these fresh-water species, as inhabitants of an extreme environment, have become stenohaline to such an extent that even more recent species and genera which at one time invaded fresh water from the sea over brackish water are intolerant of salt to a high degree? The behaviour of fresh-water organisms is the main factor contributing to the lowest number of species in brackish water (see p. 62) and accounts for this to be situated close to fresh water.

B. Changes in form and way of life in brackish water

It is well known that marine animals and plants occurring in brackish water differ in size and often in from form other members of their species found in highly saline areas. This is true of plants as well as animals, though the extent varies considerably in individual groups.

1. Changes in size (reduction in brackish water, depauperization) of marine organisms

Any scientist working on brackish water is well acquainted with the fact that many marine animals and plants attain a smaller size in brackish water than in the sea. One may be inclined to regard the small forms in brackish water as stunted as frequently happens in boundary areas. A closer inspection, however, reveals the special features of such a reduction in brackish water. The reduction in size in brackish water differs from that in the usual area of stunted forms in so far as it is a gradual process extending over hundreds of kilometres; even within the area of reduction many species, notwithstanding their small size, form highly vital populations rich in individuals (see REMANE 1934). Reduction in size in brackish water is therefore comparable to that due to BERGMANN's law which states that there is an increase in size within species for populations of colder climates as against those of warmer regions.

Both BERGMANN's law and the one regarding reduction of size in brackish water are rules, that is they are manifest in many species; exceptions are rare — at least in certain groups of animals and plants.

In the following pages an attempt will be made to describe the reduction in size in brackish water more precisely and to establish the differences between various species and groups. It is essential to compare as many regions and salinity gradients as possible since the size itself is influenced by a number of factors which in individual cases may override the dependence of size on salinity. This applies particularly to nutrition. Thus it is not surprising to find many deviations in local populations. JAECKEL jun. (1950) draws a curve for reduction in size of *Mytilus edulis* from the Schlei, a Förde of Kiel Bay, which reveals an anomalous situation since the mussels are larger at 8—9⁰/₀₀ than those at 10—11⁰/₀₀ (fig. 14). JAECKEL emphasizes that these relatively larger mussels come from the Strait of Lindaunis which is rich in

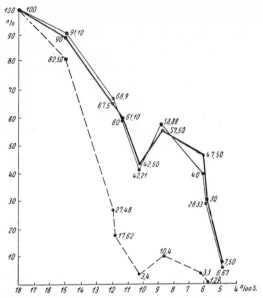

Fig. 14. Reduction in size of *Mytilus edulis* in the Schlei, a brackish Förde of Kiel Bay. Figures denote % values compared with measurements of *Mytilus* in Kiel Bay (S = 18⁰/₀₀). Solid dark line: length; solid light line: width; broken line: weight. After S. JAECKEL jun. 1950.

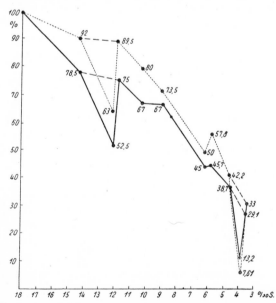

Fig. 15. Reduction in size of *Mya arenaria* in the Schlei. Solid line: length; dotted line: width; broken line: presumed course of graph if more material were available. After S. JAECKEL jun. 1950.

currents. A rapid exchange of water may favour growth, especially in sessile animals, as it facilitates a renewal of nutrients and the removal of excretory material. In the Kiel Canal where the conditions in the biotope remain uniform while salinity decreases, *Mytilus* shows a perfectly uniform reduction in size. Among fishes the Riesenströmlinge (Giant herring) of the Baltic are a group which does not conform to reduction in size. In general herrings (Strömlinge) show reduction in size with the exception of the isolated Giant herrings of the east coast of Sweden. According to the investigations of HESSLE (1925) this is due to a change in diet taking place in some of the Strömlinge — whether they are a biological race or due to external factors. While herrings and their small forms from the Baltic (Strömlinge) are plankton feeders and remain such, a number of individuals change their diet with age; according to HESSLE they feed on *Gammarus*, Isopoda *(Idotea, Mesidotea)* and even small fishes. This change to a predatory way of life results in increased growth in size. It is not known whether the transition itself has been caused genetically or by environmental factors.

Even when all other factors affecting growth and size have been taken into account the fact remains that in many species salinity exerts a decisive influence, either directly or indirectly.

Molluscs. Reduction in size is most marked in many molluscs, especially bivalves. Fig. 16 illustrates the behaviour of maximal shell length in the most common species.[2]

Fig. 16 illustrates the fact that large species, especially *Mya arenaria* and *Mytilus edulis* show a steady and enormous reduction in size. Their length at about $5^0/_{00}$ is approximately $\frac{1}{3}$ of that at about $30^0/_{00}$ salinity. To my knowledge these are the largest differences in size in the animal kingdom, occurring in regional populations within a species. Reduction in size in *Cardium edule* is also considerable, but it is minimal for *Macoma baltica* which in this respect occupies a special position. Only few data are available for the other bivalves, but *Aloides (Corbula) gibba* and *Syndosmya (Abra) alba* also display a reduction in size.

Reduction in size is clearly recognisable in a number of Gastropod species; once again it is most marked in the large forms. The whelk *Buccinum undulatum* reaches about 11 cm height in the Dutch North Sea (BENTHEM JUTTING 1933), in the Kiel Bay 6—7 cm (JAECKEL jun. 1952), *Neptunea antiqua* in Holland about 20 cm, in the Kiel Bay 10 cm, *Aporrhais pes pelecani* in Holland 5 cm, in the Kiel Bay 1.8 cm. Differences are less pronounced in the Littorineae: *L. obtusata* in Holland about

---

[2] Data for the maximum are naturally less favourable than data for a mean value since the maximum is more variable and does not permit a calculation of the coefficient of variation etc. But the marine molluscs are animals which grow continuously and age determination by means of annual rings is only possible in special cases (most likely for *Cardium edule*). Any collection therefore contains animals of different ages and the mean value is also influenced by the lower limit to which smaller animals have been included. This setting of the lower limit is, of course, subjective and so the mean value may be subjectively influenced. But if large quantities of material from many localities are available — and this is the case for the molluscs in question — the curve for reduction in size can be given with a high degree of accuracy. Both the general course of the curves as well as specific differences in fig. 16 seem to me to be established. See p. 37 for differences between individuals of the same age.

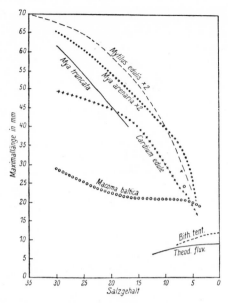

Fig. 16. Reduction of maximal length of some molluscs with diminishing salinity. Individual values exceeding those of the normal maximum in the population are inserted as isolated circles or crosses. For *Mytilus edulis* and *Mya arenaria* the measurements have been divided by two before they were entered on the graph. The graphs are based on data from the literature and own measurements. Between salinities 30 and 17⁰/₀₀, parts of the course of the graphs are uncertain. After REMANE 1934.

1.5 cm, in Kiel Bay up to 1 cm, *L. littorea* in Holland up to 3.6 cm, in Kiel Bay up to 2.4 cm in the shallower parts, up to 3.3 cm in the deeper ones (JAECKEL jun. 1952).

Fishes. Another group which shows a distinct reduction in size in brackish water are the teleosts though their blood concentration behaves differently from that of the other marine animals. MÖBIUS and HEINCKE wrote in 1883, p. 285, about the fishes of the Baltic: ,,Die Brackwasserrassen sind kleiner, ihr Rumpf ist höher, die Bewaffnung des Körpers mit Stacheln and ähnlichen Hautbildungen ist schwächer." [The brackish-water races are smaller, the body is higher, an armour of spines and other dermal processes is less developed.] For fishes it is also difficult to establish a maximum accurately as both growth and the size they attain depend to a large extent on the density of population, that is on intraspecific competition; this has been demonstrated particularly for flatfish. Consequently periods of high population density may result in the production of smaller forms, periods of reduced density in larger ones. A reduction of density through fishing may lead to more vigorous growth and higher maximal values in the remaining animals. Thus when plaice *(Pleuronectes platessa)* was removed from the dense stand of the North Sea and transferred into the Belt Sea, with less competition and abundant nutrients, it exhibited good growth there (POULSEN 1938) (fig. 20).

Temperature also exerts its influence on fishes so that the factors determining the final size are as manifold as for molluscs. A wealth of material is available on differences between fishes of the same age and on the course of their growth (see p.35).

Undoubtedly salt content, among others, is an important factor in determining size. It may exert its influence in a variety of ways: by reducing metabolic performance, retarding development, causing earlier onset of sexual maturity etc. Here are a few typical examples of reduction in size:

KÄNDLER (1944) made a thorough study of the turbot *Scophthalmus (Rhombus) maximus*. In the North Sea its maximal length is about 1 m, in the Baltic Basin 50 cm, „beide Geschlechter werden also im Nordseegebiet rund doppelt so groß und dementsprechend achtmal so schwer wie in der eigentlichen Ostsee" (KÄNDLER). [thus both sexes are about twice the size and consequently eight times the weight in the North Sea as compared with the Baltic proper.]

SCHNAKENBECK (1938) quotes the following maximal lengths for plaice (*Pleuronectes platessa*): Northern North Sea 77 cm, southern North Sea 73 cm, western Baltic 50 cm, eastern Baltic 40 cm; here, too, the maximal length drops to nearly half.

According to POULSEN (1938a) four-year old sand dab *Limanda (Pleuronectes) limanda* attain 24 cm in the Kattegat, 22 cm in the Belt Sea and 19 cm in the Baltic.

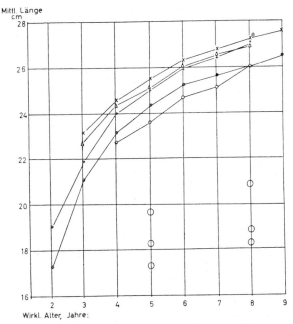

Fig. 17. Growth curves of the herring. The solid lines show different populations from the North Sea. From BÜCKMANN 1938. Curves with × signify: southwestern North Sea with entrance to Canal 1934—1937 after TESCH; curve with triangle, as above, after LE GALL; with + region of Dogger Bank 1934—1935; with ● spent herrings from the Belgian coast 1930—1931; with ○ year 1921 in the region of Eastern England. The individual circles show values for the length of 5—8 year old herrings in the Baltic after ANDERSSON 1938; the upper circles refer to herrings from the Botten Sea, the middle and bottom ones to herrings from the Central Baltic (Nynäshamn). The herrings from the low-saline Botten Sea are larger than those from the central Baltic.

Clupeidae also show a reduction in size. In the southern North Sea the average length for herrings is up to about 28 cm, the Baltic form (Strömling) according to Hessle (1925) 20.5 cm in the Bornholm Basin, 18 cm in the central Baltic between the Aland Islands and Gotland. However, both the herring and the cod display an increase in size in the Bottensee as shown in the graphs after Andersson (1938) (figs. 17, 18). Thus in addition to the Giant herring (p. 27) the herring exhibits another reversion of the reduction in size. Andersson relates these anomalies to nutritive conditions. The upper layers of the central Baltic are distinctly poor in plankton, while there is an abundance of plankton in the Bottensee, making nutritional conditions for the herring, and through it for the cod, more favourable. Lower temperature may be adduced as a subsidiary factor in the sense of Bergmann's law; this explains an increase in size towards the North by means of the temperature factor.

The majority of the other marine fishes show a moderate reduction in size in brackish water. It also applies to small species. As can be seen from table 4 Hass (1937) found that *Pomatoschistus* (*Gobius*) *minutus* has a marked reduction in size in the coastal populations when salinity diminishes. Even within a limited area of brackish water, e. g. the Schlei (Kiel Bay) of about 40 km length such a relation may be evident in fishes. Hass (1936) found the following lengths in shallow-water populations[3] of *Gobius microps* (at 0.3—1 m depth).

Table 4.

	S	Average value cm	Range of variation	n
Kieler Bucht: Bottsand	14	3.8	2.1—5.0	169
Schlei: Gunneby	10	3.0	2.0—4.6	100
Schlei: Große Breite	5	2.5	1.3—3.6	219

In some species of fishes no reduction in size takes place. Thus e. g., according to Hass (1937) the populations of *Gobius niger* display a striking constancy in size: Naples (maximum of 12 cm), Oslofjord 12 cm, Gullmarfjord 14 cm, Limfjord 12 cm, Kieler Förde 12 cm, Kiel Canal 13 cm. The population from the mesohaline brackish water of the Kiel Canal is not smaller than those from Naples or the Oslofjord.

According to Hass *Myoxocephalus* (*Cottus*) *scorpio* shows similar behaviour as can be seen from his table 3. It even reaches particularly high values in a population in the brackish water of the southern Baltic (Samland); this probably represents a further case in which growth in size is favoured by local conditions.

Brattström (1941) has produced valuable and detailed data for the echinoderms which show varying behaviour. It must be borne in mind that nearly all the echinoderms of the North Sea-Baltic region have already reached their boundary in the Belt Sea and the Sound, that is at 15—20°/₀₀ S (excluding *Ophiura albida* and *Asterias rubens*). The echinoderms extend over a much smaller salinity range than the molluscs and fishes. Within this range distinct reduction in size is found in *Psammechinus mili-*

---

[3] In deeper water populations both of *Gobius minutus* and of *G. microps* behave differently. They are e. g. smaller than the corresponding populations from shallow water (temperature factor? nutritional factor?) (see Hass 1936, 1937).

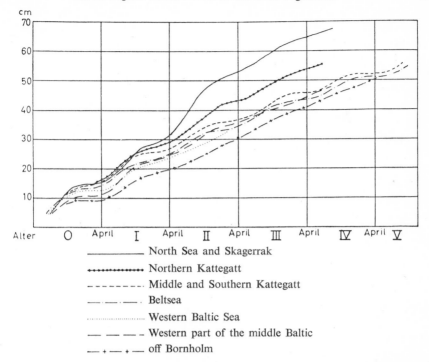

—————— North Sea and Skagerrak
••••••••••••• Northern Kattegatt
– – – – – – – Middle and Southern Kattegatt
— · — · — · Beltsea
·················· Western Baltic Sea
— — — – Western part of the middle Baltic
— + — + — off Bornholm

Fig. 18. Growth curves of cod *(Gadus morrhua)* as an example of average reduction in size from the North Sea and Skagerrak into the southern Baltic. After POULSON 1938. But according to ANDERSSON 1938 5-year-old cod in the low-saline Botten Sea attain 67 cm, i. e. nearly the same values as in the North Sea.

aris, Echinocyamus pusillus (16—8 to 9 mm), *Echinocardium cordatum* (60—44 mm), *Astropecten irregularis, Solaster papposus, Henricia sanguinulenta* etc. Reduction in size in brittle stars is far less than for sea urchins and starfish. It is still recognisable in *Ophiura texturata* and *O. albida,* but has not been demonstrated for *Amphiura chiajei, Ophiopholis aculeata, Ophiura robusta;* BRATTSTRÖM found strikingly large specimens of *Ophiothrix fragilis* and the sea urchin *Strongylocentrolus dröbachiensis* in the Öresund, that is near their limit; but the average size in this area was smaller than in the more saline regions of the Kattegat. The sea cucumber *Cucumaria elongata* and the dendrochirote holothurians as a whole has particularly large specimens in the Öresund. It seems that in this case the specially favourable nutritional conditions alone were effective in this transitional zone.

The behaviour of the common starfish *Asterias rubens* has not been fully elucidated; populations of different size and varying development of a calcareous skeleton occur in the same area, partly in ecological isolation. BRATTSTRÖM (1941, p. 168) records from the Öresund: A. Small forms, radius maximum 10—15 cm. 1) Relatively small, soft and vividly purple *Asterias,* chiefly on soft floors, but also on sand. 2) *Asterias* relatively poor in lime, lacquer red, red, yellowish-red, brown, black or greenish, on harder floors, partly among algae; also at greater depths. B. Large forms. Radius over 15 to a maximum of 26 cm. 3) Reddish *Asterias,* poor in lime, at different localities, often at a depth of about 28—30 m. 4) *Asterias* with a strongly developed skeleton and coarse spines. Light yellow, grey, greyish-white or brown. At isolated stations. At least types 1, 2 and 4 (at 25—30 m depth) also occur in Kiel Bay. We fully agree with BRATTSTRÖM's remark: „Eine systematische Untersuchung dieses Seesterns mit biologischen und zoogeographischen Beobachtungen kombiniert, würde wahrscheinlich auch sehr interessante Resultate ergeben." [A systematic study of this starfish, combined with biological and zoogeographical observations is likely to yield very interesting results] (1941, p. 168).

Finally reduction in size correlated with salinity occurs in Cnidaria. Among pelagic medusae *Aurelia aurita* is a striking example; in the eastern Baltic it attains about half the dimensions of specimens from the Belt Sea. No exact data are available for Hydromedusae; any reduction which occurs is slight. Reduction is found in sessile species of many colonial forms; but reference to this will be made later as it is connected with the different growth processes of such a colony. Among solitary polyps I know of no reduction, either for Lucernariae or for hydroid polyps.

PAX writes about Actinia: „Brackwasseraktinien sind kleiner als ihre Artgenossen in salzreichem Wasser". [Brackish-water Actinia are smaller than members of their species living in highly saline water.] Thus a race of *Actinia equina* living on the Bulgarian coast of the Black Sea never attains the average size of animals from the North Sea. No doubt *Metridium senile* shows only a slight reduction in size and the same applies for *Tealina felina* from the North Sea to the Belt Sea, taking the coastal form var. *coriacea* as the only valid base for comparison. Reduction in size seems more marked only in *Halcampa duodecimcirrata* which rarely attains 2 cm in Kiel Bay.

Nor is reduction in brackish water uncommon among plants. It appears to have been established for them in the first place. AGARDH wrote in 1817: „Mare Balticum id habet insigne, ut Algas Oceani mirifice contrahet, et sibimet ipsis dissimiles reddat" and ARESCHOUG (1847) calls the algae of the Baltic „formas diminutas s. contractas, gracilescentes et vulgo steriles" (quoted after WAERN 1952, p. 8). T. LEVRING (1940) describes the behaviour of the algae as follows: „Für die ganze Ostseeflora ist es auch sehr bezeichnend, daß die Mehrzahl der Algen als mehr oder weniger stark reduzierte Formen vorkommen. Die Grundursache dazu ist natürlich auch der geringe Salzgehalt." „Von den vorkommenden Arten, die sich auch in anderen Meeren mit höherem Salzgehalt finden, gibt es einige, die im Gebiet ganz normal entwickelt zu sein scheinen. Es sind dies *Monostroma latissimum, Enteromorpha*-Arten, die *Prasiola*-Arten, *Urospora penicilliformis, Cladophora crystallina, Pylaiella rupincola, Sphacelaria racemosa, Stictyosiphon tortilis, Hildenbrandia prototypus, Ceramium diaphanum* und *strictum, Polysiphonia violacea* (die drei letzten sind jedenfalls oft sehr gut entwickelt, können aber als etwas reduzierte Formen auftreten). Die Chlorophyceen sind sogar oft besser und reichlicher entwickelt. Ebenso verhält es sich mit *Pylaiella,* die in der Ostsee in besonders großen Mengen vorkommt. Daß gewisse Formen in der Ostsee zu einer solchen Entwicklung gelangen, kann natürlich auch mit Konkurrenzfragen zusammenhängen." „Bei der Mehrzahl der Arten äußert sich die Reduktion darin, daß die Fäden oder die Thallusflächen etwas dünner sind, die ganze Pflanze etwas kleiner geworden ist. Sonst haben sie ein fast normales Aussehen." [It is typical of the whole Baltic flora that the majority of the algae occur as more or less strongly reduced forms. The chief reason for this is, of course, low salinity. Among the species which are also found in other seas of higher salinity some appear to show quite normal development. These are *Monostroma latissimum,* species of *Enteromorpha* and *Prasiola, Urospora penicilliformis, Cladophora crystallina, Pylaiella rupincola, Sphacelaria racemosa, Stictyosiphon tortilis, Hildenbrandia prototypus, Ceramium diaphanum* and *strictum, Polysiphonia violacea* (the last three are often very well developed, but may also occur as somewhat reduced forms). The Chlorophyceae are frequently better and more richly developed. The same is true for *Pylaiella* which is particularly plentiful in the Baltic. It may well be due to problems of competition that some forms attain such development in the Baltic. In the majority of forms reduction is brought about by the filaments or sheets of the thalus being somewhat thinner and the whole plant a little smaller. Otherwise their appearance is almost normal.]

The overall result is that the algae do not exhibit the extreme type of reduction shown by many molluscs and fishes, but a medium type. Only in the boundary zone *Fucus* (dwarf *Fucus*) shows extreme reduction. The outermost outposts of *Laminaria* in the western Baltic are also only a few decimetres long. It is difficult to give exact measurements for comparison since these algae vary enormously in their growth forms (see HAYREN, LEVRING, WAERN).

All the groups discussed so far have one thing in common: reduction in brackish water occurred in a high percentage of the species and, in part, attained a high degree. In contrast there are a number of classes and orders in which such a reduction is either entirely lacking or else can only be detected to a slight degree in a few representatives. Nearly the whole of the microfauna belongs here. Absence of reduction has been specially emphasized for the Foraminifera (RHUMBLER, ROTTGARDT). Reduction in size has been recorded for some cases of microfauna. LE CALVEZ and LE CALVEZ (1951) report it for Ammonia (Rotalia) and Nonionidae in mediterranean etangs of the south coast of France. But here it occurs under extreme conditions. G. HARTMANN (1966) found reduction of size in brackish water among some Ostracoda, e. g. *Hemicytherura* and *Xestoleberis* as well as in *Cyprideis torosa* (personal communication). But it is unknown for Rotifera, Gastrotricha, Nematoda, Copepoda, Cladocera and Halacaridae. There appears to be a correlation between absolute body size and an occurrence of reduction as this is almost exclusively found in the macrofauna and, within it, especially in large forms.

Reduction of body size is remarkably slight in higher Crustacea (Malacostraca) and, as far as is known, in Polychaeta, including the large forms. Among Malacostraca a slight reduction in size occurs in *Mysis mixta* according to APSTEIN and in *Crangon crangon* (MAUCHER 1961). I cannot find any data about reduction in Amphopida, Isopoda and Tanaidacea, only the marine Cumacea of the Black Sea are designated as "nains" (BAČESCU 1949). For the shore crab *Carcinides maenas as* well as for *Pachygrapsus* and *Eriphia* PORA even records an increase in size in the Black Sea as compared with the Mediterranean.

The microflora can only be briefly discussed. HUSTEDT (1925) and KOLBE (1927, 1932) mention differences in size and form in diatoms from regions of different salinities. Based on investigations in the saline waters of Oldesloe (A. THIENEMANN 1925) HUSTEDT mentions: 1) *Diploneis interrupta,* average length 0.466 mm at 25.8⁰/₀₀ NaCl, 0.444 mm at 10.2⁰/₀₀ NaCl; 2) *Navicula elegans,* at roughly similar

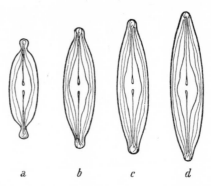

Fig. 19. Dependence of the size of shell and its shape on the salinity of the locality in the diatom *Caloneis amphisbaena*. a, forma *typica* from fresh water; b, c, transitional forms from low-saline localities (1300—1900 mg Cl/l); d, var. *aequata* from the Krummen Lake (about 9000 mg Cl/l). After KOLBE 1932.

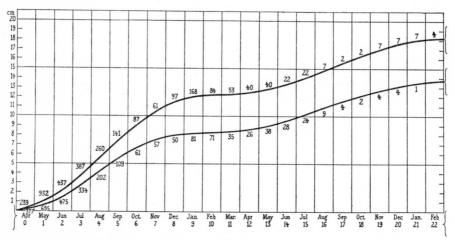

Fig. 20. Growth of plaice *(Pleuronectes platessa)* transplanted from the North Sea into the Belt Sea, compared with the growth of plaice indigenous in the Belt Sea. After BLEGVAD 1940.

differences in NaCl 0.913 and 0.773 mm resp.; 3) *Nav. peregrina* 0.9445 mm and 0.900 mm. The ranges of variation widely cut across each other, the differences are merely averages and smaller than for typical reduction. HUSTEDT stresses ,,daß fast alle Brackwasserformen, die ich in Oldesloe gefunden habe, in stärker konzentriertem Meerwasser auch größere Dimensionen erreichen" [that nearly all brackish-water forms I have found in Oldesloe attain larger dimensions in more concentrated sea water]. In some species extending into fresh water a reduction of salinity produces a shortening of the apical axis, frequently accompanied by special, button-like constriction at the poles (fig. 19). This is true for *Caloneis amphisbaena, Anomoeoneis sphaerophora, Navicula hungarica*. The fresh-water form of *Fragilaria construens* has the central part of its shell inflated.

Reduction in size with lowered salinity may have different causes which will have to be individually and experimentally investigated. It is even possible that several factors may be affecting one species.

1) The performance of certain organs may be reduced by low salinity. This has been demonstrated by SCHLIEPER (see part II) in numerous experiments on *Mytilus*. The rate at which the cilia beat is reduced and this reduces the rate of transport of food.

2) Additional work in brackish water requires an increased O_2-demand.

3) Growth is reduced in brackish water. BRANDT (1897) observed a generation of *Mytilus edulis* of the same age in the North Sea and the Baltic and found less growth in lowered salt content. This may also be due to reduction in nutrition as mentioned under 1), but other factors may lead to reduced growth. It is interesting that plaice *(Pleuronectes platessa)* which had been transferred to the Belt Sea as young fish from the overcrowded stand in the North Sea showed on an average a higher rate of growth than the indigenous plaice of the Belt Sea (fig. 20). The growth rate here is not determined solely by the environment during development.

As shown in the graphs (fig. 21) reduced growth at diminished salinity is a decisive factor in fishes.

4) Length of life is shorter in brackish water than in the sea so that the final size reached in brackish water is less.

5) Sexual maturity is reached earlier. In species in which sexual maturity terminates or retards growth, an earlier onset of sexual maturity may result in reduction of final size; if sexual maturity is delayed size may increase above normal. BRADSHAW (1957) found such "delayed maturation" in the Foraminifer *Ammonia beccarii tepida*. Strangely enough it took place when salinity was either abnormally low or else abnormally high and it led to slight increase in size. The onset of sexual maturity may also depend on nutrition, e. g. be determined by a proliferation of diatoms (L. SCHÜTZ 1960).

It must be stressed once more that reduction of size in whole populations may be brought about by environmental factors other than salinity, especially nutrition. Nutritional differences may be caused by vastly differing kinds of situations. Increased population density reduces the amount of food available to the individual animal and leads to a reduction of growth rate.

6) Time for taking-in food is shortened. This happens to all animals of the tidal zone which feed only while covered with water. This period is shorter the higher in the tidal region the animals occur; thus, in *Mytilus* for example a considerable reduction in size is found in the upper tidal region. The same effect can be produced if the water is rich in inorganic detritus or if there is an overabundance of diatoms which have sunk to the bottom. Filter feeders and ciliary feeders will then stop ingestion of food. Two examples will serve to illustrate the complexity of the situation:

SEGERSTRÅLE (1960) examined three populations of the bivalve *Macoma baltica* in the Finnish waters of the Baltic, from 3, 20 and 33 m depth. The results are: "Duration of life increases with depth; in the deepest locality, the *Macoma* population lives, on an average, about 25 years as against a mere (sic) of 7—8 years in the 3 m locality. This prolongation of life span with depth is accompanied by retardation of growth rate. The two features are attributed to the lower temperature and poorer feeding conditions at greater depths. The final size attained decreased with depth. This discrepancy would appear to be mainly due to the deterioration of the nutritive conditions towards the deep parts of the area studied. There may be striking variations in growth rate (size) within one and the same year-class."

2. Changes in size in fresh-water organisms

Do fresh-water organisms which penetrate more or less deeply into brackish water behave like the marine ones as the water gets less salt? The answer is negative. No case is known which would correspond to the genuine reduction of the larger marine molluscs or fishes. Yet reduction is not entirely lacking. The fresh-water Prosobranchia and apparently Pulmonata as well which live in brackish water show a reduction in size which, however, is far smaller than the typical one of marine animals; it corresponds more to a normal reduction in size at the boundaries of an area. KINNE (as yet unpublished) found in his experiments that *Limnaea stagnalis* grows less in salt than in fresh water. Sexually mature animals which had been reared from eggs showed at first retarded growth at 3⁰/₀₀ S, but later they caught up so that after 185 days they were no smaller than the other specimens growing in fresh water. But at 6⁰/₀₀ S they remained smaller (21 against 25 mm).

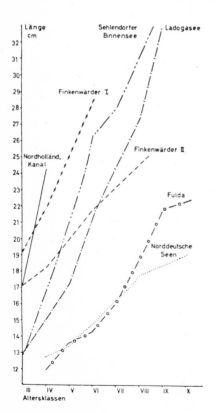

Fig. 21. Growth in length of some populations of the perch. The populations from brackish water (Sehlendorfer Inland Lake; North Holland Canal) display good growth. After W. SEEMANN 1960.

But the situation is different for bivalves. According to A. C. JOHANSEN and S. JAECKEL jun. *Unio pictorum* grows better in weak brackish water than in the fresh water of the water meadows. This statement is based on a comparison of growth rings of the shells; it does not necessarily prove that salinity promotes growth, but it is unlikely to have an inhibitory effect.

Similarly, no inhibition of growth is known for fresh-water fishes invading brackish water; frequently brackish-water populations have been considered to have grown particularly well (see fig. 21, 22).

Some fresh-water plants display a reduction in size. *Lemna minor,* in cultures of clones grown in weak salt solutions showed, at times, an initial increase of cell size, reproduction and length of roots, that is a stimulation; but later length of roots and leaf size decreased sharply and the cells became smaller (REMANE and LÄSSIG 1960).

A number of cases are known with certainty in which reduction of size proceeds continuously from sea to fresh water without showing a minimum in brackish water. Some diatoms behaving in this way have already been mentioned. A clear and continuous reduction in size from sea to fresh water is evident in the sticklebacks *Gasterosteus aculeatus* and *Pungitius pungitius.* The smelt *(Osmerus eperlanus)* is much smaller in fresh-water populations than in those from brackish water (MARRE 1931). After the cutting-off of the Zuiderzee and its freshening the smelt decreased in size to such an extent that they became useless for direct human consumption. But they became an important food for the pike-perch (BEAUFORT 1953, p. 258 and p. 355). However, as with the Giant herring (p. 27) there is a diffference in diet. The large brackish-water smelt ate mainly marine Mysidae *(Neomysis* and *Meso-podopsis),* the fresh-water smelt fed on pelagic Cladocera (HAVINGA 1941). There is a marked difference in size in all Salmonidae which possess forms remaining in fresh water (e. g. brook trout) as well as those migrating into the sea. Invariably the form living in the sea for part of its life is considerably larger. It has been possible to demonstrate for salmon and trout that transition from fresh water into sea or brackish water produces a rapid increase in growth (HENKING 1929, P. F. MEYER 1932, WILLER and QUEDNAU 1931, 1934). Trout of the same stand which remain in fresh water turn into river trout, transferred into the sea into large sea trout *(Salmo fario trutta).* This does not take effect in the first juvenile stages which are normally spent in fresh water; on the contrary, in this case sea water has rather an inhibitory effect. It was not possible to hatch salmon eggs in sea water. Later a prolonged stay in fresh water has a retarding effect on the growth of the salmon, but migration into the sea releases a rapid "marine growth" which continues for two years, irrespective of the age at which immigration into the sea has taken place (WILLER and QUEDNAU 1934).

The relationships between salinity of the medium, body size and growth are very complex. The following facts are important: A marked reduction occurs only in a part of the marine species; it is most pronounced for some molluscs *(Mytilus edulis, Mya arenaria)* and in fishes *(Rhombus maximus).* This reduction is essentially brought about by a diminished rate of growth in media of low salinity; only occasionally there is an additional factor in some molluscs *(Cardium),* that is shorter life span in low-salinity areas. There are differences in the behaviour of species inhabiting fresh and brackish water respectively. In some there is a reduction in size from fresh to brackish water, though to a small extent (Gastropoda), others exhibit an increase in size in that direction (fishes, some diatoms) so that these groups have a continuous series of reductions from the sea into fresh water. The microfauna and -flora are not affected by this reduction — or at least to a lesser degree than the macrofauna and -flora. The diatoms appear to be an exception.

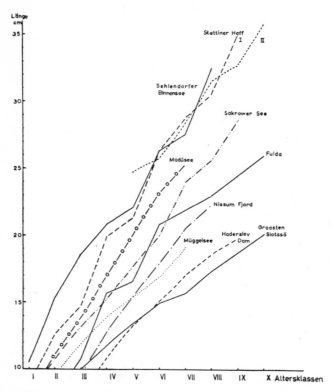

Fig. 22. Growth in length of roach *(Leuciscus rutilus)*. The populations from brackish water (Sehlendorfer Inland Lake; Stettiner Haff) show particularly good growth. After W. SEEMANN 1960.

It may be worth examining briefly what happens to the size of marine animals when salinity exceeds that of normal sea water. Fox (1926) has studied this problem in the course of his work on highly saline waters in the Suez Canal region, especially in the Bitter Lake with about 50⁰/₀₀ S. Apart from a few deformities in Foramini-fera Fox (1926) found "in other groups of animals inhabiting the Bitter Lakes no deformity nor diminution in size of individuals was observed as compared with specimens from the outside sea". In a few cases such as the fish *Mugil cephalus* and the ascidian *Phallusia nigra* increase in size could be established in highly saline water.

3. Deviation of form and structure of organisms in brackish water

Brackish-water populations may differ from their marine or fresh-water relations not only by their size, but also in their morphological features. Such changes in species can only briefly be mentioned here. It is highly desirable that these phenomena should be investigated in detail.

a) Reduction of calcareous skeletons

Already Möbius and Heincke (1883) stressed the fact that brackish-water races of fishes have a body armature with fewer spines or similar dermal processes. This reduction of calcareous parts continues into fresh water. The best known example is provided by the sticklebacks. The fresh-water form of *Gasterosteus aculeatus* (f. *hologymna* or f. *gymnura* resp.) lacks lateral plates completely — or nearly so — while in salt water the plates are more or less well developed (f. *trachura* or f. *semiarmata* resp.) Analogous differences, though less pronounced, are found in the tenspined stickleback *(Pungitius pungitius)*. There is a distinct reduction of lime in molluscs. According to an account by Levander (1899) the large shells of *Macoma baltica* in Finnish waters were paper thin. G. Afanashev (1938) has shown that the Black Sea "molluscs have a lighter shell than those of the fully saline seas. The ratio of the weight of the shells to that of the body for the Black Sea bivalves varies from 0.95 to 4.5% (average 1.8%), while for molluscs of fully saline seas it varies from 1.25 to 10.8% (average 3.5%)" (Zenkevitch 1963, p. 375). Some species of Foraminifera can produce shells without lime in the brackish water near the limits of their occurrence. Le Calvez and Le Calvez (1951) found a reduction in the thickness of shells and in sculpturing in Foraminifera of the brackish etangs in southern France. Similarly G. Hartmann (1966) found a reduction of the sculpturing of the shell in brackish water in *Hemicytherura reticulata,* an Ostracod living in the Patagonian fjords. In part II Schlieper discusses well-studied examples of reduction in lime.

There is another aspect of this reduction of lime; groups of animals in which only some of the species produce calcareous shells in brackish and fresh water show a regression of forms producing such shells. Rhumbler (see Remane 1934) points out the % increase of Foraminifera with sandy shells from the open sea (about 20.5%) to the brackish water of Kiel Bay (61.5% of species); among Cnidaria only forms without calcareous shells occur in brackish water. F. Borg (1936) emphasizes the fact that there is regression of Bryozoa forming calcareous shells; only species without lime occur in fresh water. In his studies of *Membranipora crustulenta* the same author reports that within this species calcification of the zooecium walls as compared with highly saline water (Borg 1931). The distribution of animals with calcareous shells is not solely determined by salinity. Higher temperatures favour deposition of lime.

b) Meristic reductions and anomalies

The number of vertebrae in fishes bears a well known relation to temperature, but in marine teleosts there is, in addition, an average reduction with diminishing salinity (see Hass 1937).

Reduction in number of vertebrae and in size — in the direction of fresh water do not go hand in hand. In the region of the Zuiderzee the smelt *(Osmerus)* has more vertebrae in its fresh-water populations than in those from brackish water (HAVINGA 1928). After the water had become fresh which brought about a reduction in size, the number of vertebrae increased insignificantly (BEAUFORT 1953, p. 338). According to HASS (1938) *Gobius microps* behaves contrary to the rule, the number of vertebrae increases with diminishing salinity (Kiel Bay 30.2, Schlei, Grosse Breite 30.82). Annelids show a reduction in the number of setae. HAGEN (1954) found a reduction of setae in different directions among the Oligochaeta: from sea to brackish water in *Amphichaeta sannio,* from sea to fresh water in *Rhizodrilus pilosus, Enchy-traeoides arenarius* and *Paranais litoralis,* from fresh water to brackish water in *Nais barbata, Stylaria lacustris* and *Chaetogaster diaphanus* and from fresh water to the sea in *Fridericia bulbosa.* The Polychaete *Fabricia sabella* has only 8—9 uncini on the thorax in the brackish water of the Kiel Canal at 7—10⁰/₀₀ S, while the typical animals, of comparable body length, from Heligoland have 9—12 uncinate setae on the thorax (BANSE 1956).

There is a striking accumulation of anomalies in brackish water. Such a phe-nomenon is not uncommon in the boundary areas of species, but it seems to me to reach a particularly high degree in the brackish-water region. HASS (1936) found in *Gobius microps* within Kiel Bay (Schlei) that fusion of vertebrae increased from 4.5% to 16.3% as salinity declined from $17^0/_{00}$ to $5^0/_{00}$.

Anomalies in brackish water have also been reported for Rotifera (RENTZ) and Oligochaeta (HAGEN 1954). In his experiments with the brackish-water polyp *Cordy-lophora* KINNE (1956) found many anomalies in cultures of high salinity. Older co-lonies have a tendency to malformations at high salinities ($24^0/_{00}$, $30^0/_{00}$). The growth zone at the base of the head produces more and more hydranth tissues and less and less stolon and caulome material; the coordination of the processes of growth and differentiation is disturbed. Anomalous hydranths arise. Several heads arranged one behind the other form ,,einachsige Hydranthenaggregate" [uniaxial aggregates of hydranths]. Or else heads are arranged at angles to each other: ,,mehrachsige Hydranthenaggregate" [aggregates of hydranths with several axes]. Sometimes large bodies of coenenchyma are produced from which single hydranths protrude. Occa-sionally even spherical complexes of hydranths form without stalks or stolons which — after detachment from the original stock — lie freely on the bottom of the experimental dish (KINNE 1956, p. 634).

c) Other deviations

The brackish-water forms of some molluscs show marked differences in appearance from the forms of highly saline regions and have frequently been described as special varieties. This phenomenon is most conspicuous in *Cardium edule* and *Mya arenaria*. The cockle *Cardium edule* occurs in brackish waters not only as small specimens, but with a reduced number of ribs (19—21 against 24—28). EISMA (1965) has demon-strated a far-reaching correlation between salinity and number of ribs. This makes it possible to establish salinity to an accuracy of $1.3^0/_{00}$ Cl when the number of ribs is over 20.9 on an average. In *Mya arenaria* width is less reduced than length at diminishing salinity so that the shells become more rounded.

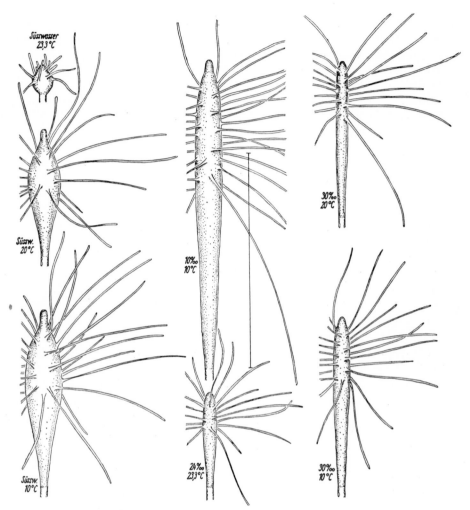

Fig. 23. Differences in size, shape and number of tentacles of the hydroid polyp *Cordylophora caspia* at different salinities and different temperatures. After O. KINNE.

HOOP (1940) reports the following differences from a histological comparison of the typically euryhaline bivalves from the North Sea (32⁰/₀₀ salinity), Kiel Bay (about 18⁰/₀₀ S) and the mouth of the Schwentine (<10⁰/₀₀). For animals of equal size the epithelia of the mantle become thinner; in *Mya arenaria* and *Cardium edule* from Kiel Bay into the brackish water of the mouth of the Schwentine, in *Macoma baltica* only in the small animals from the boundary of their occurrence, *Mytilus edulis* remains constant with regard to this feature. Furthermore the shell secreting glands of the skin are less well developed in the mesohaline region.

At reduced salinity the Ostracoda often produce animals with a hump-like arching of the shell, usually together with normal animal (G. HARTMANN 1964, SCHÄFER 1953). Thus in water of low salinity (from 5⁰/₀₀?) *Cyprideis litoralis* produces the form *torosa* which has often been considered a species of its own. Similar variants occur in the Cytheridea which have become completely or predominantly inhabitants of fresh water. In *Limnocythere inopinata* the brackish-water form usually has no hump (f. *incisa*), the normal fresh-water form shows a varying, but generally distinct formation of a hump. In other purely fresh-water Cytheridea the humps are a constant feature *(Limnocythere stationis, Cytherissa lacustris)*, also in a form of *C. torosa* from the Issyk-Kiel *(C. t. pedaschenkoi)*.

KINNE's (1956) experimental and ecological studies of the brackish-water polyp *Cordylophora* have demonstrated what varied morphological changes may be produced by changing salinity (fig. 23). Within a few days or weeks the shape of the polyps' heads undergoes a reversible adaptation to the salinity and temperature conditions provided. On either side of $16.7^0/_{00}$ S, at $20°C$, the length of the heads decreases; at $10°C$ and $10^0/_{00}$ S it is greatest. The same holds true for the number and length of tentacles; the width of the head is largest in fresh water and declines continuously over $2^0/_{00}$, $5^0/_{00}$, $10^0/_{00}$, $16.7^0/_{00}$, $24^0/_{00}$ to $30^0/_{00}$. "Innerhalb des $S^0/_{00}$ Normalbereichs $(1—24^0/_{00})$ sind die Hydranthen bei $10—18°C$ am längsten und tragen die meisten und längsten Tentakel" [Within the normal range of salinity $(1—24^0/_{00})$ the hydranths are longest at $10—18°C$ and they bear the largest number and longest tentacles.] In addition to the shape of the head, the growth form of the colony[*] changes with the salt content of the surrounding water. Fresh-water colonies consist of stolons for 75—90% of their total length, they lie flat on their substratum. At certain intervals short, mainly unbranched stem polyps bud from the stolons ("Rasenkolonie") [turflike colonies]. With rising salinity the proportion of stolons is reduced in favour of the caulome: The stalks of the stem polyps elongate and lateral polyps sprout. The branching system grows higher and higher above the anchoring stolons into a „Bäumchenkolonie" [arborescent colony] (p. 632—633). E. ARNDT (1965) found a somewhat different behaviour. The length of the hydranths was greatest at $3.6^0/_{00}$ (temperature? certainly above $10°C$).

Among algae morphological changes can assume such proportions that in brackish water the species habitually adopt a completely different appearance. Some Red Algae (*Polyides rotundus, Ahnfeldtia plicata, Plumaria elegans, Delesseria sanguinea, Phycodrys sinuosa, Polysiphonia urceolata*) have altered so much in the Baltic that only anatomical and comparative studies will establish their species relationship (LEVRING 1940). Above all thalloid species become more or less pronouncedly filamentous (see p. 33). According to LEVRING these morphological deviations occur particularly in the reduced algae. But in the Baltic other factors have also changed. Many algae are found at greater depths and with it, in a different "light climate"; they also form detached stands of different growth forms — therefore detailed ecological and experimental investigations will be necessary in order to establish the part played by salinity in these changes.

4. Reproductive changes in brackish water

Over a hundred years ago botanists noticed that the algae of the Baltic display sexual reproduction only to a slight degree or it may be absent altogether ("vulgo sterile" ARESCHOUG 1847). Reproduction is predominantly or wholly vegetative.

Sexual reproduction, like size, may be influenced by many factors. Apparently detached growth of algae, as often happens in the Baltic, leads to a reduction in the formation of gametes. This is shown by the Sargasso weeds floating in the Atlantic Ocean; therefore it is not conditioned by brackish water. On the boundaries of distribution areas as well as in extreme environments many organisms have parthenogenetic forms replacing the bisexual ones (see Crustacea). It does appear, however, that the accumulation of sterile forms in brackish water cannot be traced to these general factors alone, and thus it deserves a brief discussion.

According to LEVRING, HAYREN, and WAERN those algae which show a marked reduction in size are the ones which are mainly sterile over wide areas of the Baltic. Some Red Algae (*Callithamnion furcellaria, Polysiphonia nigrescens, Rhodomela subfusca*) of the brackish water of the Baltic are found solely or predominantly as tetrasporous plants, not as cystocarpous ones though the species thrive vegetatively. *Fucus* shows locally a slight development of reproductive organs (deep water *F. vesiculosus, F. serratus, f. arctica*). LEVRING (1940, p. 143) found *Phyllophora membranifolia* only sterile near Blekinge on the Baltic coast of Sweden, though it was well developed vegetatively. Closely related species may show differing behaviour. Thus LEVRING found *Phyllophora Brodiaei* near Blekinge fertile in parts; similarly *Polysiphonia violacea* formed both tetraspores and cystocarps in that area. Even the marine flowering plant *Zostera marina* becomes sterile near its limits of occurrence on the Finnish coast; it still flowers, but does not produce any fruit (H. LUTHER 1951, p. 129).

It is interesting that genera found mainly in fresh water are predominantly or exclusively sterile in brackish water. As a result it has been impossible to determine to which species some of these stands belong. This happens for *Vaucheria, Spirogyra, Oedogonium, Bulbochaete, Zygnema* and others. We owe to W. HÖHNK (1952—1956) (see fig. 24) more detailed investigations on differences in the reproduction of fungi at varying regions. Here, too, the sexual phase is usually more sensitive than the phase of sporulation (see fig. 24).

Disturbances in reproduction are also exhibited by mosses which penetrate into brackish water. In submerged aquatic mosses (*Fontinalis* and others) neither reproductive organs nor sporogonia were found, but antheridia were observed in *Calliergon megalophyllum* at a geoamphibiotic locality (H. LUTHER 1951, p. 128). Among Characeae of the Finnish coast the same author found only antheridia in *Nitellopsis obtusa* and parthenogenetic oogonia only in *Chara canescens*.

Fresh-water flowering plants growing in brackish water also have reproductive deviations. According to H. LUTHER no flowers were formed in brackish water by *Elodea canadensis, Hydrocharis morsus-ranae, Lemna trisulca, L. minor, Utricularia minor (Littorella uniflora)*.[5] In a rock pool *Lemna minor* resorbed the flowering primordia before they opened. Other plants flowered, but did not set fruit: *Potamogeton nitens* — which may be a sterile species hybrid —, *Myriophyllum alterniflorum, Utricularia neglecta;* in other species sparse formation of fruits took place.

The situation is quite different in the animal kingdom. While vegetative reproduction takes place or is possible in most aquatic plants, it occurs among Me-

---

[5] This species only occurs submerged.

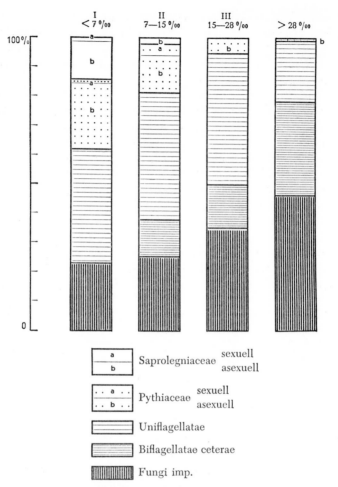

Fig. 24. % distribution of various lower fungi at different degrees of salinity. For Saprolegniaceae and Pythiaceae the relative proportions of sexual and asexual phases in the total number of fungal records have been indicated. After Höhnk 1956.

tazoa only in a number of sessile species and a few worms, such as the Oligochaetae families of Naididae and Aeolosmatidae and Turbellaria e. g. *Microstomum;* (it is commonest in Cnidaria and Naididae). But in none of these groups it represents the exclusive way of reproduction of a species or population. There are no purely vegetatively growing stands as commonly found in plants.

Therefore a "brackish-water sterility" in animals — especially those without vegetative reproduction — can only mean that some species invade a region, presumably chiefly by transport of larvae, in which they can no longer attain full sexual maturity. H. Meyer (1935) and Remane (1941) report such sterile stands of *Asterias rubens* and *Lucernaria quadricornis* in Kiel Bay; here they appear to be restricted to periods in which the environment is unfavourable as in other years they were found to be sexually mature in this area.

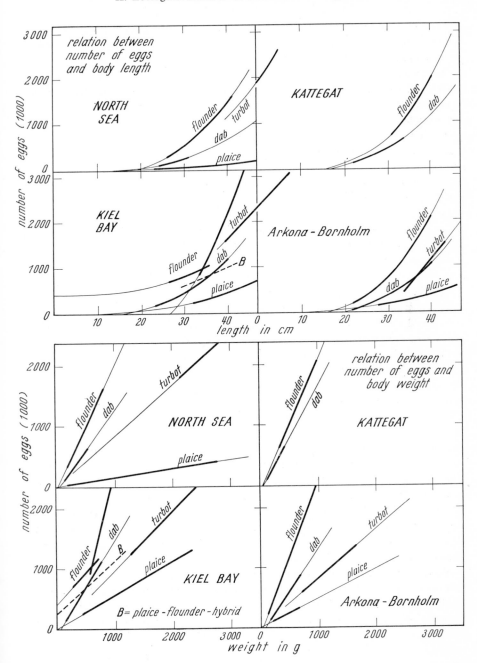

Fig. 25. Relationships between number of eggs, body weight, body length of plaice *(Pleuronectes platessa)*, flounder *(P. flesus)*, sand dab *(P. limanda)* and turbot *(Rhombus maximus)* in North Sea, Kattegat, Belt Sea and western Baltic.
Areas of observation are marked by a thickening of regression lines. After KÄNDLER and PIRWITZ 1957.

Even in animals a number of cases are known in which reproduction has been somewhat restricted in brackish water or with diminishing salinity. A detailed study will undoubtedly greatly increase the number of such cases.

In the brackish-water polyp *Cordylophora caspia* the number and size of gonophores and eggs diminish in fresh water (KINNE 1956). According to MARX and HENSCHEL (1941) flatfish are unable to be fertilized in brackish water, *Pl. limanda*, for example, from $15^0/_{00}$ S downwards in experiments.

Fresh-water animals also reduce part of their reproduction on transition into brackish water. The egg capsules of the snail *Theodoxus fluviatilis* are smaller in brackish than in fresh water and contain only 55—80 eggs instead of over 100 (BONDESEN 1941). According to SEIFERT (1938) *Ephydatia fluviatilis* forms no gemmulae in brackish water which means a reduction of vegetative reproduction.

A delay or narrowing-down of the reproductive period within the annual cycle seems to be widespread. A comparison of population from the North Sea to the brackish water of the Baltic shows a delay of "spawning times" both for fishes and echinoderms (*Asterias,* see part II) and Cnidaria. The common jellyfish *Aurelia aurita,* for instance attains sexual maturity in June/July in the Belt Sea, but in August/September in the southeastern Baltic.

A reduction in reproductive rate clearly means a diminished chance of survival. One would therefore expect to find some mechanism of compensation which in plants would be represented by vegetative reproduction. The loss may be partly made good by lack of competition in brackish water which is poor in species, chiefly a reduced rate of destruction due to lack of consumers. But there are other factors favouring reproduction. In several fishes sexual maturity of the Baltic populations starts at an earlier age than for those from the North Sea. According to KÄNDLER (1941) some species in the Baltic are sexually mature 1—2 years earlier than those in the North Sea. This applies e. g. for the economically important *Rhombus maximus,* the herring and others. KÄNDLER considers this a compensation for increased difficulties in reproductive conditions in brackish water. Recently KÄNDLER and PIRWITZ published important studies on the fertility of flatfish in the North Sea-Baltic region (1957) (see fig. 25). In the fresh-water snail *Theodoxus fluviatilis* whose reduction in cocoon size and number of eggs have been referred to, often several eggs in a cocoon will turn into larvae in brackish water, while in fresh water normally only one egg will develop and the remaining ones decompose to give nutritional material (BONDESEN 1941).

Further investigations on the relationship between salt content and reproduction promise to yield interesting results. SCHÜTZ stresses that in some animals the reproductive period is set off by the onset of a proliferation of diatoms; differences in the beginning of this proliferation, as a nutritional factor, may cause differences in the onset of reproduction for some animals.

5. Special ecological features of brackish-water inhabitants

a) Occurrence at greater depths in brackish water

As long ago as 1875 MEYER and MÖBIUS emphasized that many molluscs living in the sublittoral of the Kiel Bay inhabit the tidal belt of the English coast. In these two areas the species live in regions of different depths; in weakly saline areas their occurrence is displaced towards the depth — either on an average or at least in one of the boundaries. This phenomenon has been termed "brackish-water submergence" (REMANE 1955). It is widespread, but not universal. Like many other, similar ecological phenomena it only has the character of a rule.

Submergence can take place in a number of ways, according to a displacement into deeper regions of the lower limit of occurrence, its upper limit or both. For the benthic fauna of the Baltic the important submergence is one which is predominantly or solely confined to the lower limit (basal submergence). Through this process the area of the deep is considerably extended towards the brackish water of the Baltic, in parts over 100 m! Such a submergence occurs chiefly in the species of the *Macoma-baltica-community* PETERSEN which in the North Sea and in other richly saline seas lives in the eulittoral zones and the adjoining areas of shallow water down to 10—15 m depth. In the Baltic the following species penetrate into depths of over 100 m: the common mussel *Mytilus edulis, Macoma baltica, Hydrobia ulvae,* the Polychaeta *Pygospio elegans* and *Scoloplos armiger* (see BRANDT 1890, DEMEL and MULICKI 1954). The upper limit of the bivalves hardly changes throughout their range of distribution.

For plants basal submergence hardly comes into question since light — quite independently of salinity — is primarily the limiting factor for depth. This necessitates emergence rather than submergence of the lower limit in brackish seas where light penetration is, on an average, reduced. And yet analogous phenomena are found in algae, though to a slight degree. The bladder wrack *Fucus vesiculosus* is a typical plant of the eulittoral of the sea forming a narrow belt on rocks. According to WAERN (1952, p. 167) it extends over a zone of 0—4.7 m depth in the Gullmarfjord on the Swedish coast which is no longer fully marine in character; in the Alandssee on the east coast of Sweden it stretches from 0.5—11.5 m depth so that the displacement in depth of the lower limit has led to a considerable widening of the zone. Further north the area is narrowed down by the lowering of the upper limit to 4—10 m.

A biological explanation of this basal submergence is not always easy to find. For the bivalves mentioned above the most likely explanation is the fact that there is no biological competition in the Baltic with its paucity of species; of special significance is the absence of echinoderms and snails which feed on bivalves. The situation for *Fucus vesiculosus* is complex. WAERN stresses particularly that in the Gullmarfjord he saw no *Fucus vesiculosus* below 4.7 m even in such places where there was no inhibiting competition from other algae; from the fact that below 4 m the specimens were stunted — in marked contrast to the deep form of *Fucus* from the brackish water of the Baltic — he concludes that the existence of various biotopes must account for these differences.

The distribution of several brackish-water inhabitants, especially those in shallow zones with deposition of detritus, superficially resembles a basal submergence. They are found — though only sporadically — in the upper eulittoral zone in the North Sea, that is only in the upper region of the tidal belt. Their occurrence in the sublittoral in the Baltic — if only at a few metres in depth in most cases — and the fact of their reaching the water boundary makes their distribution in depth resemble a weak basal submergence. This, however, does not apply to their distribution within the Baltic and the Belt Sea, but only to a comparison of these seas with their occurrence in the North Sea. To these belong, above all, representatives of the *Cyprideis litoralis-Manayunkia aestuarina* community; that is in addition to the species from which the name of the community is derived there are *Protohydra,* some Ostracoda such as *Cytherois fischeri, Loxoconsha elliptica.* Similar behaviour is exhibited by species of the brackish-water phytal zone such as *Littorina saxatilis* and *Idotea viridis.* The first one is represented in the brackish-water phytal zone by a different race from that in the upper eulittoral of the rocky sea coasts. In all these cases one might speak of an emergence of brackish-water animals and euryhaline ones towards the sea rather than of a submergence of animals from sea to brackish water. Various possible explanations of this phenomenon suggest themselves. In part there are true brackish-water species which find suitable conditions in more saline seas only in the uppermost layer, often with diminished salinity; in part biotic competition may play its role for euryhaline animals, as the extreme upper regions of the tidal area have little competition; in part only the upper mud area of tidal seas, with its vegetation of lower plants, offers conditions of a biotope resembling those of the shallow coasts of brackish-water bays.

Far more widespread than basal submergence is the displacement in depth of the upper limits (upper submergence) or else a total submergence. At least half the marine animals penetrating into the Kattegat, the Belt Sea and partly into the Baltic show a more or less marked displacement of upper limits so that many animals living near the shore line or even in the tidal areas in the North Sea and the Atlantic are only found in the sublittoral zone, either in the Belt Sea or in the Baltic. REMANE (1955) gives a survey of these animals and their gradual and regional submergence. A personal communication by BRATTSTRÖM from Bergen (Norway) may be added: "Wenn man die Öresundfauna kennt, wirkt es oft überraschend, zu sehen, wie viele Arten hier in der Litoralzone leben, die im Öresund in den tieferen Schichten auftreten. *Chaetopterus* kommt z. B. in Norwegen auch litoral vor, ebenso *Echiurus.* Im Sund leben sie nur in der Tiefe. *Purpura* sinkt vom Litoral hier bis zu 15—20 m im Sund, und *Modiola modiolus,* die hier litoral ist, kommt im Sunde erst unter 20 m vor." [For someone knowing the fauna of the Öresund it is often surprising to find how many species occur here in the littoral zone which live in the lower layers in the Öresund. *Chaetopterus* e. g. occurs in Norway in the littoral as well as *Echiurus.* In the Sund they only live in the deep. *Purpura* sinks from the littoral here to 15—20 m in the Sund and *Modiolus modiolus* littoral here, only occurs below 20 m in the Sund.]

Among algae several forms typical of the eulittoral or the tidal zones displace their upper limit into the sublittoral, e. g. *Fucus serratus* and the *Laminaria* species do so already in the Belt Sea.

Conditions of salinity account for this typical displacement of the upper limit. Species approaching their lower salinity limit find suitable conditions for their existence only in the deeper layers with a higher salt content. They cannot inhabit the zones of shallow water because the general level of salinity is too low or because it periodically falls to a lethal level.

It must be stressed that not all species submerge. Many littoral species remain in that region up to their limit, e. g. the Polyclade *Stylochoplana maculata* while the related form *Leptoplana tremellaris* shows submergence. Among Cirripedia there is no submergence in *Balanus balanoides,* while *Balanus crenatus* shows it to a moderate degree and *Verruca stroemii* to a great extent. A comparison of the tidal zone of the North Sea coast and Kiel Bay (see REMANE 1955) revealed submergence in 64% of Polychaeta, 50% of molluscs, higher Crustacea (Malacostraca) 39%, and 16% for Copepoda Harpacticoidea. Here, too, the microfauna seems to display this phenomenon to a lesser extent than the macrofauna. It is interesting that nearly a whole community shows change in vertical position from the North Sea to the Baltic. It is the Halammohydra-coenosis inhabiting medium and coarse sands. In the North Sea it lives in the undercut sandy beaches of the tidal zone, in Kiel Bay at 6—15 m depth. The larger groups which lack submergence are mainly those families and orders which have invaded the sea from fresh water or from the land, though they may develop a number of distinct species in the sea. To these belong the Oligochaeta, Rotifera, Halacaridea.

Not all cases of submergence are due to salt content. Among the colonies of the tidal area the so-called Bathyporeia-coenosis of the pure fine sand is somewhat further displaced seawards than the colonies of the mud floor. But it still remains close to the water surface. The reason for this is the disturbing effect of irregular water cover as happens so extensively in the shore zone of brackish seas. Inhabitants of permeable sandy bottoms are forced to a greater extent to move their life zone deeper than those from water-holding mud floors. But this submergence is only a matter of a few centimetres! In addition a cover of ice on the northern coasts of the Baltic may bring about some submergence as against the southern Baltic coasts which are rather more free from ice.

b) Changes of biotope on transition from sea to brackish water

The characteristic bivalves of the *Macoma-baltica* community: *Mya arenaria, Macoma baltica, Cardium edule* as well as *Mya truncata* are typical inhabitants of fine sand in the sea, whereas in brackish water they occur predominantly on soft substrata; their main biotope in brackish water thus differs from that in the sea. In some species e. g. *Cardium edule* and *Macoma baltica* it is more an extension of a biotope, while in others it represents a change. This is true, e. g. for the heart urchin *Echinocardium cordatum,* the worm *Lagis koreni* and in particular for the whelk *Buccinum undatum* and the bivalve *Mya truncata*; these live mainly on soft mud in the Belt Sea; in the Baltic *Mya arenaria* also lives on soft mud even at lower salinities. This may be a combined effect of salinity and competition. The diminishing salt content displaces the species from the upper water layers, but there occurs a soft substratum in the more saline depth; tolerance of the substratum and reduced competition enable the species to change to soft substrata.

There are other reasons for an extension and a displacement of biotopes for some inhabitants of rocky coasts which settle on hard substrata. In the Belt Sea and the Baltic these are mainly inhabitants of the phytal zone such as the calcareous sponge *Sycon,* tube worms e. g. *Pomatoceros,* Ascidia etc. Evidently the disappear-

ance of the rocky floor induces a transition to the phytal zone which provides
the required substratum, either predominantly or exclusively. It is strange that
some sessile animals which are widespread on substrata show the tendency of living
epizoically on other animals in their boundary area e. g. the polyp *Clava multicornis*
on the snail *Theodoxus* (STAMMER 1928), the Bryozoon *Alcyonidium polyoum* on the
Gastropoda *Littorina* and *Zippora*. Quite striking is the transition of the Gastropod
Littorina saxatilis from the uppermost area of the rocky eulittoral into the beds
of *Ruppia* etc. in shallow brackish waters. The transition is not continuous; there
are two races of differing biological behaviour which e. g. in the Belt Sea still occur
almost side by side, both in their typical habitat. To a lesser degree *L. obtusata* and
Littorina litorea — which is extremely eurytopic in less saline water — carry out a
transfer into the stands of *Zostera*.

c) Differences of biotope in the range brackish water — fresh water

Corresponding to the submergence sea — brackish water one might expect to find
an emergence of fresh-water organisms invading the sea from fresh water. This
expectation is fulfilled in broad outline. As will be shown later (p. 78) the currents
of immigration from the fresh-water region into the sea move chiefly along the coast;
only a few descendants of the fresh-water fauna penetrate to depths over 20 m in
the sea (the Ostracod *Candona neglecta,* the Oligochaete *Limnodrilus heterochaetus,*
some Chironomus larvae). But as this phenomenon persists into highly saline regions
the salt content alone cannot simply be considered the decisive factor. Undoubtedly
a number of factors will have to be taken into account; among them the low salt
content may be significant for some fresh-water species e. g. Rotifera such as *Testudinella patina, Epiphanes senta* and others.

It is a well known fact that fresh-water animals and plants closely related to
marine species prefer far more streaming and moving fresh water in lakes than
they do in the sea or else they are more cold-water stenotherm in fresh water than
in the sea. The Red Algae *Hildebrandtia* and *Lemanea* and — slightly less pronounced — the Brown Alga *Pleurocladia lacustris* are examples from the group
of algae; striking examples among animals are *Cordylophora,* the fresh-water Nemer-
tinae (*Prostoma clepsinoides* and others), the Cytheridea among the Ostracoda
which occur in small bays and pools in the brackish water itself, while living in
larger lakes in fresh water; furthermore the Neritidae with *Theodoxus* and the Hydro-
biinae among the Gastropoda; the Gobiidae among the fishes etc. Among the
microfauna Rotifera such as *Pleurotrocha reinhardti* should be mentioned. There
are several possible explanations for this apparent rule, each of which may be valid
for certain species. Running water may provide more ions per unit time than stagnant water of similar composition. It may supply more O_2 in case the species has
a higher O_2 demand in fresh water. A uniform ecological phenomenon may well
originate from the combined effects of very varied forms of physiological reactions.
Only comparative and experimental investigations will clarify this.

Rotifera of the genus *Notholca* serve as example for a thermal restriction of
life in fresh water; in fresh water they are mostly typical spring forms; in the sea
they also occur in warmish shore pools, in part with the same species. In experiments

Notholca tolerated higher temperatures in fresh water as well (FOCKE 1961). Thermal restriction may also play its part in some organisms which prefer running water and lakes in fresh water.

A remarkable fact must be added here — which has been confirmed by numerous examples — that many outposts of marine groups live subterraneously in fresh water, above all in fissure water and interstitial water of gravelly sands, e. g. the Polychaete *Troglochaetus*, the Isopoda *Microcerberus*, *Microcharon*, the Amphipoda *Ingolfiella*, *Bogidiella* etc. To these a number of Turbellaria may be added which live in the interstitial system of river sands, e. g. Otoplanida (*Coelogynopora*, *Otoplana*), *Vejdovskya* and others (see P. AX).

III. Brackish water as an area of colonization of fresh-water and marine organisms and specific habitat of brackish-water organisms

A. Types of brackish-water inhabitants

Four types of organisms settle in the brackish-water region:

1) Holeuryhaline species of animals and plants, that is organisms which, as species, inhabit the whole salinity range from sea water into fresh water and frequently penetrate into salinities of over 35 to 40⁰/₀₀.

2) Euryhaline-fresh-water species penetrating from fresh water more or less extensively into the salinity ranges of the brackish water.

3) Euryhaline-marine species which reach the brackish water from the sea.

4) Genuine brackish-water organisms (VÄLIKANGAS) confined to brackish water.

In each of these four groups there are subdivisions which display the characteristic features of the group to a greater or lesser extent.

In order to allocate individual species to these groups it is necessary to know the salinity range within which they can exist.

There seem to be great difficulties in establishing the boundary region of a species as required for such a classification. The ecologist will determine the limits on the basis of records from nature together with the salinity measured at these localities. In this way he obtains the observed limits of occurrence within the salinity range, but these need not be the limits of existence of the species.

The area of existence comprises the region in which a species maintains itself by reproduction; beyond the limits individuals may survive in some stages — often the old ones, more rarely the juvenile ones — but can no longer preserve the species permanently.

The working-out of the ecological boundaries of existence is by no means as simple as it appears at first sight. A number of records are available as supplied by faunistic studies; often the salinity at the site is reported simultaneously, but more often the average salinity values and, partly, the amplitude of fluctuations are known from hydrographic investigations made independently of the faunistic ones.

In an attempt to deduce the ecological tolerance of the species from the manifold records the following facts must be considered, especially for establishing the salinity range.

a) The constant inflow of fresh water into brackish water by means of tributaries carries a continuous stream of fresh-water animals into brackish water. Many animals can live in this water for some time; but the influx takes place all the time and frequently — often large — numbers of fresh-water organisms are found in areas where they are unable to exist permanently. This applies in the first place to planktonic organisms, but inhabitants of the phytal zone (snails and insect larvae) can be carried downstream on floating vegetation. Conversely, animals dwelling near the mouth of rivers can be carried upstream by the highly saline countercurrent along the bottom. This may lead to their being found in places where the salinity is too low for their survival. Records in the vicinity of river mouths have to be treated with caution.

b) Relationships in areas of changing salt content are complex. Changing environmental factors have in themselves a deleterious effect; that applies particularly to irregular changes (cf. the effect of irregular precipitation on the vegetation of arid regions!). On transition into regions of fluctuating salinities organisms will tend to reach their limit of occurrence sooner than might be expected from the average salinity (see fig. 26). This is shown particularly well in H. LUTHER's (1951) investigations of higher plants in brackish water. SPÄRCK (1936) found that compared with the Baltic, fresh-water species in the Ringköpingfjord penetrated less into brackish water and marine species did not enter low-salinity areas so much. Both groups are more clearly separated in the Fjord than in the Baltic. The reason lies in the fact that in many cases the boundary of existence is determined by extreme values (minima or maxima of any factor) rather than by the average value. The destructive effects of occasional salt transgressions into fresh-water regions or

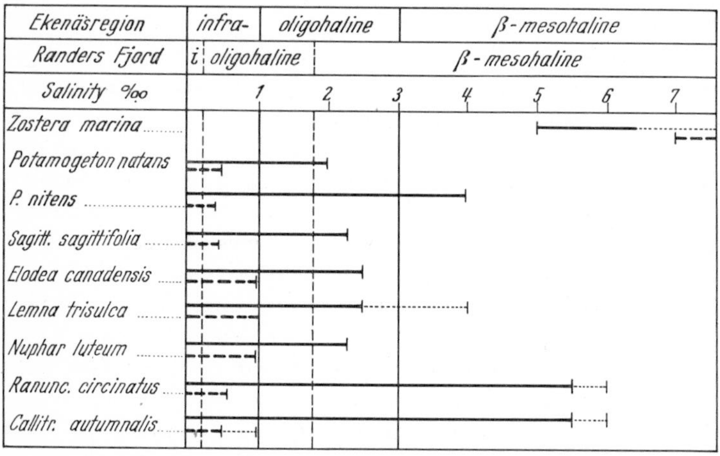

Fig. 26. Differences in salinity boundaries of certain plants in Finland (Ekenäs region) with relatively constant salinity, and in the Randersfjord with greatly fluctuating salinity. Area of occurrence in Finland: solid line; in Randersfjord: broken line. After H. LUTHER (1951).

occasional abnormal reduction of salinity in brackish or marine areas are known. There is an impressive description of the catastrophic results of heavy rain on the coral banks of the Barrier Reef at low water (see PORTIER).

SMITH (1955) found that *Nereis diversicolor* had a much higher salinity limit in Finland (Tvärminne) at about $4^0/_{00}$ S than in the Isefjord where the species nearly reaches fresh water with the regulatory performance remaining the same. Among the factors accounting for this SMITH refers to extreme situations in Finland "critically low salinities occurring in spring while temperatures are still very low".

Under extreme conditions many organisms are capable of adopting resistant stages with reduced vitality and in this way they eliminate the destructive effects of temporary extreme factors. This also applies to the salt content. OTTO (1936) found in his studies of the fauna of the *Enteromorpha* zone that Nematoda, f. i. get into a state of immobility at high salinities; when conditions return to normal they get back to an active life. W. SCHULZ found a similar reversible salinity rigor, accompanied by immobility and shrinking in *Enchytraeus albidus* which can tolerate a sojourn in $100^0/_{00}$ S in that condition. ZENKEVITCH (1938) observed this immobility due do salinity in many animals; in *Fabricia sabella*, *Nereis diversicolor* and *Balanus improvisus* he even speaks of anabiosis at $60^0/_{00}$, $80^0/_{00}$ and $100^0/_{00}$ S.

From these cases we come to organisms which normally spend some time as resistant resting stages such as hydroid polyps with their dermant regressed states and numerous other sessile animals (some Tunicata, Scyphozoa, Bryozoa). Groups with quick successions of generations often produce resting eggs (Cladocera, Rotifera) even in the sea or they form cysts (Protozoa). All such organisms can survive in a medium which provides the necessary conditions for existence only for the span of their development. In the same brackish water there may be species of very diverse salinity requirements, the one in a period of lower salinity, the other at higher salt content. In such cases the limits of tolerance cannot be deduced from the extreme values nor from the average salinity value.

c) The frequent salinity stratification results in the fact that in the same body of water different aquatic regions of very different salinities lie closely one above the other and therefore the faunas are also vastly different. A catch with the net may bring these together in one haul. According to DERJUGIN (1925, p. 100) in the relict Lake Mogilnoje the same plankton net will produce marine medusae (*Cyanea*, *Rathkea*) and fresh-water Crustacea (*Cyclops*, *Diaptomus*, *Daphnia*). The reed *Phragmites communis* grows at the shore of the North Sea at almost marine salinity, but only in those places where the roots find a layer of soil of low salinity.

This provides an indication that in establishing limits of tolerance attention must be paid in the first place to regions with a gradual gradient and small fluctuatious in salt content, viz. the Baltic, parts of the Black Sea, Kiel Canal etc. In these many species show the same limits in different regions, but for many others the limiting value of salinity varies in the more distant areas. The reasons for this may be manifold:

d) The limiting effect of salinity is influenced by other environmental factors, acting either as stimulation or as inhibition. This is specially true for temperature. It has long been known that in warmer regions marine and brackish-water animals penetrate farther into fresh water than in the North Sea and the Baltic. This statement was originally based on the obvious observation that many groups (families, orders) have fresh-water representatives in the tropics which are purely marine in the North (Heteronemertinae, Selachia, Cirripedia, Polychaeta of the family Nereidae etc.). But a comparison of the boundaries in the Baltic and the Black Sea shows that the same species can often tolerate lower salinities in the South (*Branchiostoma*, Polychaeta such as *Glycera*, *Melinna*, *Sabellaria* and others). Experiments revealed that the effects of temperature on salinity tolerance are varied (see KINNE 1964). Some apecies are more tolerant of lower salinity at lower temperatures and of higher salinity at higher temperatures. To this low low and high high type the following animals belong: *Gammarus duebeni*, *Rithropanopeus harrisi*, *Cordylophora caspia*, *Crangon crangon*, *Cyprinodon macularius*. The other combination low/high at which lower salinity was better tolerated at higher temperatures was found in *Palaemonetes varians*, *Leander serratus*, *Penaeus duorerum*, *P. aztecus* etc.

There also exists a relation between Ca-content and salinity. Higher Ca-content facilitates a toleration of lower salinity. For details see part II.

e) In certain regions the limit may be a mere coincidence with a salinity value, the true reasons being changes of biotope or biotic factors. The Isopod *Idotea granulosa* and the Amphipod *Calliopus laeviusculus* attain their lower salinity limit at $14—15^0/_{00}$ in the Belt Sea (Kiel Bay), but only at $5—6^0/_{00}$ in the northern part of the Baltic, on the Finnish coast. Both species inhabit the phytal zone in open moving water. In Kiel Bay the brackish waters with $< 14^0/_{00}$ are only lagoons and Förden lacking these biological factors. Similarly many species of the interstitial fauna of the sand penetrate much farther into regions of low salinity in the northern parts of the Baltic than they do in Kiel Bay where the corresponding sands are missing etc. Any ecologist will take these factors into account in his assessment of the salinity range of species.

It is difficult to study the inhibiting effect of one species on another which is called competition. Competition undoubtedly exists but it can rarely be measured accurately. But this factor probably exerts less influence in brackish water than in truly marine or fresh-water regions. Brackish water is poor in species and thus both fresh-water and marine organisms encounter regions with reduced numbers of species; this does not preclude competition — which can be effected by a single competing species — but will on the whole reduce it. At the mouth of the Schwentine in the Kieler Förde *Laomedea loveni* is restricted in its normal area by the fact that *Mytilus edulis* is strongly developed and grows over the summer generation of *Laomedea* while in the autumn its recovery is prevented by *Balanus crenatus* (SCHÜTZ 1964). SEGERSTRÅLE (1965) explains the gaps in the settling of the bivalve *Macoma baltica* by mass colonization carried out by the Amphipod *Pontoporeia affinis*. KINNE (1953) showed that in the area of contact between *Gammarus duebeni* and *G. zaddachi* the males of *zaddachi* mount females of *duebeni* and thus prevent reproduction.

f) The regional differences in salinity limits which have been considered so far have been brought about by external factors. The behaviour of the species was taken to be constant. But it is highly probable that in such an important medium as salinity many species consist of populations exhibiting different reactions. The individuals of any species are not identical in their reactions; the majority of species consist of an immense number of biotypes and often a number of subspecies. Varying behaviour of populations towards salt content may arise: 1) by adaptive adjustment of organisms to the special salinity conditions of the area in which they grow up = adaptive modification. Such "nongenetic adaptations" may be irreversible (see KINNE 1962 for *Cyprinodon macularius*); 2) by "genetic adaptation", by natural selection of those biotypes which are fittest for any particular range of salinity. Evidence for the existence of such genetically different populations can only be obtained from breeding experiments. But it seems likely that many of the special races such as "relicts" in fresh water are genetically different from the original marine form: e. g. the fish *Myoxocephalus (Cottus) quadricornis,* the Crustacea *Mesidotea entomon* and *Pontoporeia affinis.* Detailed results are available for *Gasterosteus aculeatus* (see p. 62).

From the situation just described it follows that limits of tolerance for any species cannot be demarcated with salinity values in parts per mille; limiting values include a range of fluctuations of 1—2⁰/₀₀ in lower and 3—5⁰/₀₀ at higher values. In many cases boundaries will just have to be placed in the zones of oligohalinikum, miomesohalinikum etc.

Experimental studies suggest themselves as a means by which salinity limits might be established more easily and accurately. There is no doubt that experimental ecology is needed on a much larger scale, but it is equally certain that the ecological interpretation of experimental results is not easy. Most experiments on resistance were carried out with adult animals (see part II); that is adults were placed into media of varying salinity and the limits of their undisturbed vitality determined. If this physiological resistance agrees with the ecological tolerance this may be taken as explaining the situation. But this is frequently not the case; the experimentally determined range may be wider or narrower than the natural habitat. In the first case occurrence may have been narrowed down by competition or some other factors which prevent the occupation of an area though its salinity range is physiologically possible. But the adult stage which has been tested in the experiment may be more resistant than the juvenile one or the eggs. For this large numbers of examples can be quoted. The adult crab *Carcinides maenas* can survive up to 4⁰/₀₀ S, but in the North Sea its eggs will only develop in the range of 28—40⁰/₀₀ (BROEK-HUYSEN 1956); in Kiel Bay the same species still produces eggs and larvae at 15⁰/₀₀. The snail *Littorina littorea* shows similar behaviour. According to HAYES (1929) its eggs require a minimum salinity of 20⁰/₀₀, but in Kiel Bay eggs and larvae are viable even at 15⁰/₀₀. The adult snails have a wider salinity range. This also applies for the Gastropod *Nucella (Purpura) lapillus,* species of *Mytilus, Ostrea* and *Balanus* (see KINNE 1964b). SMITH (1964) found a narrower range for the development of eggs and larvae in the extremely euryhaline Polychaete *Neanthes (Nereis) diversicolor;* even for the population on the Finnish coast optimal development takes place at about 13⁰/₀₀.

Diminished salinity tolerance of the eggs or juvenile stages is also widespread in euryhaline fresh-water fishes. The sea trout *Salmo fario trutta* develops its eggs only up to about 8—10⁰/₀₀ S while the adult penetrates far into the sea. According to NELLEN (1965) many fresh-water euryhaline fishes of the Baltic only spawn in areas of low salinity of $< 5^0/_{00}$ while the adults penetrate as far as the mesohalinikum. The young stages of the sea trout *Salmo fario trutta* for example are more sensitive to salt than the adult.

The problem under discussion requires breeding experiments in different salinity gradients over several generations. Many such studies are available for unicellular organisms, but only a few for Metazoa (see KINNE: *Gammarus duebeni* 1953, *Cordylophora caspia* 1956a; FOCKE (1961) for species of *Notholca* and others). Bacterial cultures produce remarkable results. By a gradual increase in salinity it was possible to get strains with a low salinity tolerance to grow in media of $200^0/_{00}$ salinity. STATHER (1930) succeeded in doing this with strains resembling those growing on salt media. KLUYVER and BAARS (1932) started from two strains of vibrio, one halophilous, the other intolerant of salt; by a gradual change of the medium they were able to reverse the requirements of the two (quoted from W. L. FLANNERY 1956). This change in reaction corresponds with the sad experience we make with many pathogens which gradually acquire resistance to drugs. This is evidently due to a specific ability of bacteria to produce mutants which have altered the physiology of their metabolism and are being selected in the new medium. But generalizations should not be based on these extreme cases. Among bacteria SPRUIT and PIJPER (1952) found isolated bacteria from red salt incapable of growing in media of less than $160^0/_{00}$ S. All observations in waters in which the salt content has been more or less gradually changed (Zuiderzee, see BEAUFORT; Ringköpingfjord, see SPÄRCK 1936; Sperenberger salt water, see KOLBE 1927; Finnish waters, see SEGERSTRÅLE 1951, PURASJOKI 1953) revealed an immediate and drastic reaction of the organisms, including unicellular ones; so that for the majority of species salinity tolerance is a fixed quantity the change of which requires long, almost geological periods of time.

Frequently resistance established by experiment is less than ecological tolerance. The following reasons may account for such a difference: 1) The organisms investigated have already been adapted by their present medium, that is the original range of reactivity of the juvenile stage has been narrowed down in agreement with their real environment. The resistance must be considered a modification or dauermodification in the genetic sense. If marine Nudibranchia (Aeolididae, Dorididae) of the Belt Sea die quickly when transferred from $15^0/_{00}$ into $25^0/_{00}$ S this probably represents such a case as the same species are fully viable in the North Sea at $30^0/_{00}$; their stands in the Belt Sea are often supplemented by larvae transported from more saline water. 2) The species consists of a number of physiological races or subspecies. The experimenter has used one of these races for his experiments, thus obtaining the reactions of a race, but not of the species. This certainly happens very frequently. HÖHNCK (1956) clearly stresses in his studies of brackish-water mycology III (p. 103): ,,Verschiedene Isolierungen der gleichen Art von verschieden salzigen Habitaten reagieren unterschiedlich gegenüber einer gegebenen Salzgehaltsstufenfolge. Nicht die Art hat eine genotypisch festgelegte Reaktionsweise gegen Salzwerte, sondern die Standortrasse. Darum lassen sich auch Beispiele dafür finden, daß mehrere Isolierungen der gleichen morphologischen Art von gleich oder ähnlich salzigen Habitaten auch gleich oder ähnlich empfindlich sind gegen bestimmte Salzwerte." [Different isolates from the same morphological species from habitats of varying salinities show different reactions to a given series of salinity gradients. It is not the species which has a genotypically fixed way of reacting to salinity values, but the local race. Therefore examples can be found that several isolates of the same morphological species from habitats of the same or similar salinity are equally or similarly sensitive to certain salinity values.] The situation resembles that of the temperature factor. Here, too, many species subdivide into individual races with differing temperature optima and different tolerance of temperature fluctuations. This splitting-up into individual races with varying range of physiological reactions will be found chiefly in euryhaline species and less in stenohaline ones.

In several instances salinity tolerance in the experiment exceeded that found in nature. In nature, the fresh-water plant *Lemna minor* has its limit at about $3^0/_{00}$, in cultures it grew well and even reproduced at $14^0/_{00}$ S. But from $10^0/_{00}$ S the plants were smaller, practically without roots (REMANE and LÄSSIG). Such results can easily be understood. In cultures the inhibiting effects of the environment (competition etc.) have been eliminated so that even organisms with low vitality and little reproduction may thrive. In nature they would no longer be viable. An increased salinity tolerance has also been found in many Protozoa. In nature *Paramecium caudatum* hardly exceeds $3^0/_{00}$ S, in the experiment, with a slow increase of salt concentration, it was grown up to $10^0/_{00}$ (HAYES 1930) or even to $13.4^0/_{00}$ S (FRISCH 1940). In fresh water *Frontonia leucus* survives up to about $2^0/_{00}$, in cultures OBERTHÜR succeeded in keeping them up to $8^0/_{00}$.

In many cases it is doubtful whether a variable species is in reality a uniform species or whether it is made up of closely related, but different species. In that case the evaluation of salinity tolerance will change. From the former euryhaline *Cardium edule* a brackish-water species is now being split off, *C. lamarcki* which used to be counted as *f. rustica* of *C. edule* (MUUS 1967).

There is no doubt that further studies will change the picture of salinity tolerance and salinity limit for many species. However, there are numerous observations available which allow the following general picture of the behaviour of organisms to be drawn.

B. Holeuryhaline animals and plants

Species of animals and plants which exist in full sea water, brackish water and fresh water are called holeuryhaline. As salinity exerts a powerful influence on the distribution of animals and plants the number of holeuryhaline species is relatively low.

Among Protozoa several species are recorded for sea and fresh water, the Heliozoon *Actinophrys sol* and the Ciliata *Cristigera setosa*, *Pleuronema coronatum*, *Mesodinium pulex*, *Lionotus fasciola*, *Loxophyllum helus* and others (see KAHL, Ax and Ax).

There are no holeuryhaline species of Porifera and Cnidaria. Among Turbellaria *Gyratrix hermaphroditus* is widespread in fresh water, but occurs in the sea as well, e. g. in the North Sea, in the Red Sea up to $44^0/_{00}$ S and in inland salt springs (species in Germany up to over $48^0/_{00}$). A similar behaviour is displayed by *Breslauilla*

relicta and *Macrostomum appendiculatum* which P. Ax counts among the holeuryhaline animals (fig. 27). The Rotifera *Notholca bipalium, N. squamula, N. acuminata,
Proales reinhardti, Colurella colurus, C. adriatica* (see p. 67) behave in a similar
manner.

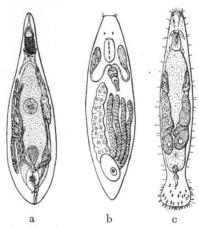

a   b   c

Fig. 27. Holeuryhaline Turbellaria. a, *Gyratrix hermaphroditus;* b, *Breslauilla relicta;* c,
Macrostomum appendiculatum. Enlarged.

The Oligochaete *Enchytraeus albidus* inhabits the margin of the shore and moist
biotopes rather than the aquatic region proper; it occurs both on the sea shore and
in fresh-water areas. It is fully viable in the range of $0—35^0/_{00}$ (Krizenecky 1916,
W. Schulz 1954); above $35^0/_{00}$ it gradually loses its mobility and shrinks at high concentration ($100^0/_{00}$) but it can survive a short sojourn in high concentrations ($100—
200^0/_{00}$). *Lumbricillus lineatus* has a similar distribution, but it occurs also in the
aquatic region proper and mainly in marine biotopes. *Nais elinguis* and *Mariona
subterranea* approach holeuryhalinity.

There is a surprisingly high number of Diptera larvae which are able to tolerate
extreme salinities. Remmert (1955) mentions as holeuryhaline: *Symplecta stictica,
Dasyhelia flaviventris, Culicoides circumscriptus, C. nubeculosus, Smittia edwardsi,
Nemotelus notatus, Stratiomyia furcata, Syntormon pallipes, Hydrophorus praecox,
Scatella subguttata, Sc. stagnalis, Lispa tentaculata.* According to Remmert the fol-
lowing show wide salinity ranges in their occurrence: *Aëdes dorsalis* $0—102^0/_{00}$,
A. salinus 0—52, *Aëdes mariae* 0—60, *Odontomya viridula* 0—28, *Caenia palustris,
C. fumosa* 0—30, *Ephydra riparia* 0—210, *E. scholtzi* 0—124, *E. micans* 0—29, *E.
micellaria* 0—60. The upper limit for some species will be displaced after further
studies of hyperhaline waters.

All fishes which migrate from the sea into fresh water for spawning (anadromous
fishes) or from fresh water into the sea (catadromous fishes, Anguilla) are holeuryhaline. Only the edults are capable of adaptation to varying salinity. But some
fishes are able to reproduce in fresh water and in the sea. To these belong the stickleback *Gasterosteus aculeatus* and a few Cyprinodontidae e. g. *Cyprinodon macularius*
(see Kinne 1963), *Fundulus heteroclitus* and probably *Aphanius fasciatus.*

Taking into account all known records there is a larger number of species which display a habitat range from full-strength sea water into fresh water, but have their main area in one of the three chief regions of sea, brackish water or fresh water; only extreme individual colonies comprise the whole salinity range. The species have been called facultatively holeuryhaline animals (REMANE 1941). They join on to the strongly euryhaline marine, brackish-water and fresh-water animals and can often be distinguished from them only with difficulty.

As an example I quote the Ostracod *Cyprideis torosa litoralis* which settles in masses on the floor of brackish bays and pools. So far it has always been considered a true brackish-water animal. But there are isolated records from fresh water, from North Germany, (Grubersee, Trammer See, see KLIE 1938) and from Africa (Lake Rudolf, KLIE 1939) as well as from thermal waters of Iceland (KLIE 1928). In those waters there may be some salinity but that is hardly the case in the Trasimenian Lake (HARTMANN). *Cyprideis* has also been found in highly saline lagoons of the Mediterranean (at $60^0/_{00}$ S in the saltmarsh on the Isle St. Marguerite near Cannes, HARTMANN 1957 and at $35^0/_{00}$ S in the Salt Lake Pomorie Bulgaria, CASPERS 1951 b). The Rotifera *Brachionus plicatilis* and *Hexarthra* (*Pedalia*) *oxyure* show similar behaviour. Both are normally indicator forms of brackish water, but they occur in inland salines at over 40 to about $70^0/_{00}$ S and there are isolated records from fresh water (*Br. plicatilis* in Lake Rudolf, Africa, see BEADLE 1943.) The Polychaete *Neanthes* (*Nereis*) *diversicolor* is an extremely euryhaline inhabitant of the sea. A. L. SMITH (1956) found it in the fresh water of the Tamar estuary, but its survival here is made possible by the supply of larvae from a more saline region. A closely related species or subspecies *N. limnicola* which is viviparous actually lives in fresh water. The sessile Polychaete *Merceriella enigmatica* was found in brackish water in the South of France by PETIT and RULLIER (1952).

The facultative holeuryhalinity of the shrimp *Gammarus duebeni* has been studied in detail. In nature it is predominantly an inhabitant of brackish water, especially of pools and lakes on the shore with extreme conditions. In fresh water it has only been recorded in a few waters in Ireland, but these populations evidently form a special physiological race (see REID 1939, BEADLE and CRAGG 1940, BEADLE 1943). Normal brackish-water *G. duebeni* do not tolerate fresh water for more than a few months and no eggs form in the ovaries (REID 1939). KINNE (1953) was only able to breed the species at a salinity of at least $1—2^0/_{00}$.

G. duebeni reacts differently to a high salt content. In nature it is absent from the open sea, but occurs in fully saline rock pools (FORSMAN 1951). In experiments FORSMAN found that the animals remain in "good vigour" at $50^0/_{00}$ S, at $81^0/_{00}$ they got into a reversible immobility and died after 24 hours' sojourn in $85^0/_{00}$ S. In nature the normal limit of the species towards sea water is not determined by salt content, but apparently by competition from the other species of *Gammarus* (see SEGERSTRÅLE 1950, KINNE 1952). KINNE discovered that *G. zaddachi* has a higher reproductive rate and its males seize the females of *G. duebeni* for pairing (praecopula), but no progeny is produced. Thus in areas of contact *G. zaddachi* depresses the reproductive rate of *G. duebeni* by appropriating the females (KINNE 1). But *G. duebeni* has a higher resistance to O_2 deficiency (FORSMAN 1951), hi er temperatures (KINNE 1952) and diminished salinity than the adjoining species of *Gammarus*; it can therefore maintain itself in habitats with extreme conditions. Possibly the brackish-water prawn *Palaemonetes varians* behaves in exactly the same way, but so far no records are available for high salt concentrations. Therefore it is listed on p. 114.

Among species which nearly attain holeuryhalinity from the sea the following should be mentioned: the planktonic Cladocera *Podon polyphemoides* LEUCKART. This species lives in the salinity range $1^0/_{00}$ to $> 35^0/_{00}$; but it has been found in fresh water at one locality in the Ödelsee in Västernorrland (Sweden) (LILLJEBORG 1901). The marine Nematode *Theristus setosus* BÜTSCHLI has been reported from the fresh water of Lake Ochrida, the marine Nematoda *Polygastrophora octobulba* and *Viscosia papillata* from Lake Nicaragua (GERLACH 1958). Judging from experimental results the Ciliate *Frontonia marina* might be included. It lives in the sea and inland saline places but has not yet been found in fresh water where *F. leucas* occurs. OBERTHÜR (1937) was able to culture *F. marina* in a range of $0—125^0/_{00}$ salinity! No doubt there are several facultatively or regionally holeuryhaline species among Copepoda; *Diaptomus* (*Arctodiaptomus*) *salinus* is found in salt waters up to $70^0/_{00}$ S in arid regions (BEADLE 1943), but occurs in fresh water as well e. g. in Germany (HERBST 1951). The habitat of *Cyclops viridis* may cover nearly the same salinity range in North Africa (BEADLE) while with us it is only a strongly euryhaline fresh-water animal; *Nitocra lacustris* shows similar behaviour and is brackish-fresh water in Europe. LANG (1948) lists *Canuella perplexa* as holeuryhaline and as possibly facultatively holeuryhaline species *Mesochra lilljeborgi, Metis holothuriae, Paradactylopodia brevicornis, Pseudonychocampus proximus, Tachidius discipes*.

The planktonic Rotifer *Keratella quadrata* displays peculiar behaviour. It is very common in fresh water and also found in brackish water up to about $6—7^0/_{00}$ S. But far beyond this limit large populations occurred in the North Sea (LEVANDER 1911). There is an analogous case among Crustacea in *Chydorus sphaericus*.

The reasons for facultative holeuryhalinity may be varied. In some cases only adults will survive and no reproduction takes place (*Neanthes diversicolor, Gammarus duebeni*), in other instances there are relict lakes of marine origin. The term fresh water is often applied to the whole body of water. But salt springs may issue into lakes (Trammer See near Plön) and in arid regions lakes may reach a certain degree of salinity (Lake Tanganjika); this may be so at least in shallow bays so that the environment of the species is not pure fresh water.

Among plants the Cyanophyceae have a high percentage of holeuryhaline species. „Der Salzgehalt scheint jedoch für die Verteilung der Cyanophyceen von geringerer Bedeutung zu sein, als wenn es sich um höhere Algen handelt. Dies geht u. a. daraus hervor, daß eine große Anzahl von Arten (39) sowohl an der schwedischen Westküste als auch in der Ostsee vorkommen, sowie daraus, daß eine erhebliche Anzahl (24 Arten) sowohl im Meer als im Süßwasser leben können." (LINDSTEDT 1943, p. 110) [Salinity appears to be of less importance in the distribution of Cyanophyceae than in the case of higher algae. This becomes evident from the fact (among others) that a large number of species (39) occur both on the west coast of Sweden and the Baltic and that a considerable number (24 species) are able to live both in the sea and in fresh water]. The author mentions the following species as occurring both in the sea and fresh water: *Gloeothecea rupestris, Nodularia spumigena, N. Har-*

veyana, Anabaena variabilis, A. torulosa, species of *Spirulina, Oscillatoria limosa, A. brevis, Phormidium fragile,: Lingbya aestuarii, L. lutea* and others.

In marked contrast to this are the diatoms which are rich in species. KOLBE (1932, p. 224) mentions only *Bacillaria paradoxa* as a species able to live both in fresh water and the sea. HUSTEDT (1925) refers to *Nitzschia frustulum* as a species thriving both in fresh water and in brine of 90°/00 S.

Among algae the Chlorophyceae are more tolerant of salt than the other groups; some species of Enteromorpha *(E. intestinalis)* invade from the sea into fresh water; some species of *Cladophora* approach euryhalinity from fresh water. According to BUDDE (1931) *Rhizoclonium riparium, Schizomeris Leibleinii* and *Ulothrix tenerrima* can tolerate large salinity fluctuations.

Among flowering plants I can only think of the genus *Ruppia* which extends from fresh water into high salinities and may be holeuryhaline.

As regards the holeuryhaline species there is the problem whether individuals are capable of tolerating extreme salinities or whether the species consists of a number of subspecies with limited tolerance. Both these possibilities have been realized. To the first type belongs, e. g. the fish *Cyprinodon macularius* (KINNE and KINNE 1962). The embryos develop in 0—70°/00 S with an extension of the incubation period in the hyperhaline region. All animals were derived from one pair (from Salton Sea, California). *Fundulus heteroclitus* tolerates a transfer from sea water into fresh water and the converse (BLACK 1951). The Rotifera *Notholca bipalium, N. squamula* and *N. actiminata* also belong to this type. FOCKE succeeded in transferring animals of these species gradually, within 6—10 hours, from sea water into pure fresh water. Here they continued to live and reproduced. A clone of *Notholca acuminata* marina was under observation in fresh water for 7 months, that is for about 100 generations. Similarly specimens of *Notholca acuminata* from fresh water (f. *lacustris*) survived in sea water for many generations. Sudden transfer between sea and fresh water caused damage, but a direct transfer from 10°/00 into 2°/00 S did no noticeable harm, while direct increase of salinity from 10°/00 to 30°/00 S was not tolerated. A gradual transfer into 50—60°/00 S led to reversible disturbances and increased mortality.

Ax and Ax obtained similar results with their culture of Ciliata in differing salt content. The holeuryhaline species *Loxophyllum setigerum, Uronema marinum, Prorodon discolor marinus, Prorodon teres, Paramecium calkini, Cyclidium elongatum* and *Frontonia marina* developed in all regions from fresh water to the fully marine one after direct transfer from brackish water (about 15°/00 S). *Loxophyllum setigerum* and *Uronema marinum* even tolerated 60°/00 S; OBERTHÜR was able to keep *Frontonia marina* in the range of 0—125.5°/00 S though in nature the species does not occur in fresh water. But Ciliata also have populations which react differently. The extreme behaviour of *Frontonia marina* with a tolerance of 0—125°/00 S is shown by a population from the salt spring of Artern (Central Germany) with 44°/00 S; a population of *F. marina* from the Bottsand (Kiel Bay) at about 15°/00 S only tolerated the range of 0—40°/00. The investigations by Ax and Ax reveal a better development in water whose salinity is that of the place of origin (15°/00). The existence of genetically different races with different salinity ranges has been demonstrated for several holeuryhaline species. The Turbellarian *Gyratrix herm-*

aphroditus occurs in Finland in three distinct chromosomal forms, a diploid fresh-water race 2 n = 4, a tetraploid brackish-water race which tolerates 11.9⁰/₀₀ S and a triploid form which appears to reproduce only parthenogenetically (M. REUTER 1961). Genetically controlled formation of races has also been established for the stickleback *Gasterosteus aculeatus* (MÜNZING 1959, 1962, 1963). This fish, widespread in the Palearctic and Nearctic regions, has two genetically different main forms: 1) a mostly marine migratory form with full development of the lateral bony plates = *trachurus*. It migrates into fresh water or brackish water for spawning; 2) a fresh-water form in which the lateral plates are reduced or absent = *leiurus* (= *gymnura* or *hologymna* resp.). The distribution of both types is complicated. In areas of contact they hybridize and produce mixed populations with an intermediate number of plates (= *semiarmata*). The *trachurus* race will still spawn in the sea, e. g. in Kiel Bay (Wendtdorfer Bucht) at about 15⁰/₀₀ S. In addition to the hybrid form *semiar-matus* a genetically independent *semiarmatus* form has been found in Lake Isnik in Anatolia. The eggs of *leiurus* also develop in brackish water (1/3 sea water) (HEUTS 1947). In a northern population (Belgium) salt water causes a delay, in a southern (Naples) a speeding-up of development.

Genetic differences are to be expected where forms of different reactivity live in close proximity such as the smelt *Osmerus eperlandus,* the bivalve *Dreissena poly-morpha* in the Caspian Sea (f. *marina* at about 7⁰/₀₀ S, f. *fluviatilis* up to about 2⁰/₀₀) *Hydra vulgaris* and others.

A survey of extremely salt tolerant species which have been mentioned above brings out the following features:

1) The majority of holeuryhaline organisms not only tolerate the salinity range fresh water — sea water, but penetrate far into hyperhaline regions, often into 60—200⁰/₀₀ S. The same phenomenon occurs in an attenuated manner in euryhaline halobionts.

2) A large number of holeuryhaline animals are phylogenetically derived from fresh-water animals, they are of fresh-water origin. That is true of Rotifera, Oligo-chaeta, Diptera larvae, some Copepoda. The Turbellaria *Gyratrix hermaphroditus* and *Breslauilla relicta* are of marine origin. This accounts for the fact that a much higher percentage of inhabitants of hyperhaline waters are of fresh-water origin than marine animals (see p. 175).

3) Ecologically many holeuryhaline organisms belong to the colonizers of moist-terrestrial biotopes, e. g. the Oligochaeta, especially the Enchytraeidae, the Diptera larvae of the genera *Ephydra, Scatella,* the Cyanophyceae. This semi-terrestrial way of life appears to create conditions favourable for acquiring holeuryhalinity. BEYER (1939) made a detailed physiological investigation of *Ephydra* larvae. Their internal medium shows complete independence of salinity fluctuations.

C. Behaviour of fresh-water animals and plants (limnobionts) towards brackish water

As mentioned before, the peculiar poverty of species in brackish water between 3 and 8⁰/₀₀ S is due to the high degree of susceptibility of most fresh-water organisms to slight salinity increases in the water: It leads to a steep drop in number of species on transition from fresh-water to brackish water of low salt content.

A discussion of the penetration of a whole specific fauna and flora into a different ecological region may be undertaken from various points of view. The first is the factual-statistical approach. It registers the different extent to which limnobionts have invaded brackish or salt water and classifies them into various groups according to their degree of eury — or stenohalinity. This classification takes only those species into account which have their centre of occurrence in fresh water. Any immigration into a new habitat very soon leads to the formation of new species, genera etc. which are adapted to the new area. Then fresh-water species become specific brackish-water species or even marine species of fresh-water origin. They are then placed among brackish-water or marine organisms. But new problems arise when the plasticity as well as the ecological development and origin of these immigrants are considered. It is well known in phylogeny that various phyletic lines retain a different plasticity over long geological periods. The following are independent of one another and behave differently:

1) The ability to form species. If this alone is effective the phylum divides into a large number of species with great uniformity of organization.

2) Plasticity. It leads to a transformation of organization, frequently the formation of new organs; expressed in the language of systematics, the branch rapidly acquires new genera and families.

3) Adaptability to new biotopes and new living conditions. Any new environment that an animal phylum enters offers a wealth of special possibilities for existence, of "niches" as they are called in modern phylogeny. The ability to fill all these "niches" in a short time varies considerably from group to group. Adaptation may be linked with the formation of new organs (see production of suckers by animals in running water). But this need not be the case.

Considered ecologically, this adaptability is in its turn dependent on conditions of the original environment. Limnobionts for example in two totally different areas are preadapted for the colonization of brackish water and sea. Firstly in large lakes and river mouths. They provide conditions of the large open water (pelagic zone) and of intensive water movement; this implies a restriction of aerial respiration at the water surface for air-breathing small aquatic animals. These factors have an even greater impact in the sea. Secondly preadaptation for colonizing brackish water is provided for the inhabitants of small pools which dry up. Before drying-up is complete it leads to a concentration of dissolved substances and to large temperature fluctuations. It is no coincidence that brackish waters, especially in arid regions, frequently contain representatives of genera and groups which in fresh water inhabit and prefer temporary bodies of water e. g. Cladocera of the genus *Moina,* the Anostraca, Cyclopidae, Ostracoda of the genera *Heterocypris, Eucypris* etc. There follow the inhabitants of moist soil, that is the semi-terrestrial forms which also possess the ability to colonize brackish and saline zones.

Ecologically the dispersal of groups coming from fresh water into the sea is neither unbounded nor random. A current of littoral-continental colonization along the shallow coastal waters and a region of thalassic colonization may be distinguished (REMANE 1952).

Tolerance of brackish water varies considerably in individual groups of animals and plants. A survey of the various groups will be given. The classification is based on the degrees of salinity which have been reached, but it must be pointed out that the limiting values are only rough approximations.

1) Stenohaline limnobionts = fresh-water organisms in the area of 0 to about $0.5^0/_{00}$ S. More accurate investigations will undoubtedly require further subdivision. Organisms from the range $0—0.1^0/_{00}$ S might be called anhaline limnobionts.

2) Euryhaline limnobionts of the 1 st degree = oligohaline fresh-water organisms. They survive from fresh water into the oligohalinikum, that is the salinity range of $0.5—3^0/_{00}$. This corresponds to the hypohaline animals of LINDBERG (1948).

3) Euryhaline limnobionts of the 2nd degree = meiomesohaline fresh-water organisms. Occurring from fresh water into the salinity region $3^0/_{00}—8^0/_{00}$.

4) Euryhaline limnobionts of the 3rd degree. Occurring from fresh water into salinity areas of over $8^0/_{00}$, pleiomesohaline and polyhaline fresh-water organisms.

Protozoa. The taxonomy of the naked Amoebae and colourless Flagellata is incomplete and accurate observations are so rare that no comprehensive survey can as yet be made.

But the shelled groups are well known and a sharp contrast is immediately evident: Foraminifera in the sea and Testacea (Thecamoebae) in fresh water. The well known genera of the Testacea, *Arcella, Difflugia, Centropyxis, Nebalia, Eugypha* are common in fresh water on plants and on the floor; this wealth of forms disappears in oligohaline water. There are isolated records of *Difflugia* in mesohaline water; VALKANOV even reports *Difflugia constricta* at $20^0/_{00}$ S, but these are more likely to be chance establishments than true sites of colonies. Only *Cyphoderia margaritacea* (EHRBG.) — which should be called *Campsascus vulgaris* VALKANOV — is euryhaline 3rd degree since, like a species of *Cochliopodium* it transgresses the $8^0/_{00}$ limit (REMANE 1950). In cultures Thecamoebae remain viable up to $5^0/_{00}$ S (ZUELZER 1910, FINLAY). Forms whose relationship with the typical Testacea is not established have been omitted such as the Gromiidae *Platoum, Pleurophrys* and others occurring in the sea. HOOGENRAAD and DE GROOT (1940) in the Fauna van Nederland mention only six species among 119 Thecamoebae which invade brackish water.

Altogether well over $90^0/_{00}$ of the Testacea species are stenohaline or perhaps euryhaline limnobionts of the 1 st degree; more pronounced euryhalinity is a rare exception; no specific species are produced in brackish water.

The Heliozoa are not strictly confined to fresh water, the well known *Actinophrys sol* occurs in fresh water and in the sea ($20^0/_{00}$ S); it may be holeuryhaline.

Our data about Ciliata are still incomplete. In a large number of species they inhabit the entire range from fresh water into the sea. Many genera are represented in the whole of the region (*Stentor, Lacrymaria, Coleps, Condylostoma, Frontonia, Vorticella, Carchesium* etc.), but there are only very few species which these habitats have in common. The fresh-water species are stenohaline or slightly euryhaline. KAHL (1933) stresses the fact that even at $3^0/_{00}$ S fresh-water, marine and true brackish-water Ciliata are fairly evenly mixed, ,,bei aber nur wenig stärkerer Konzentration eine typische Salzwasserfauna auftrat" [but with a salt concentration only a little higher a typical salt-water fauna was found]. Even the species of epizoic Ciliata on *Gammarus* and others are different in brackish water and the sea from those in fresh water (see PRECHT). Only an epizoon of fishes *(Trichodina domerguei)* is also found on sticklebacks in the sea (17⁰/₀₀ S); it is euryhaline of the 3 rd degree.

Undoubtedly a transition from fresh water into the sea and the reverse have frequently taken place among Ciliata. The numerous genera which they have in common are proof of this, but the direction of migration can only rarely be determined.

The same is true of Porifera and Cnidaria. A small branch of each (Spongillidae an Hydridae) has successfully invaded fresh water and split into species and genera. Most of these species are confined to fresh water and are not found in a salinity above 3⁰/₀₀ S. Only one species from each group is euryhaline of the 2nd degree an extends to waters of about 6—7⁰/₀₀ S. The sponge *Ephydatia xuviatili₃* (L.) is found on marine algae *(Furcellaria)* in the Greifswalder Botten; the polyp *Pelmatohydra oligactis* settles not only on *Potamogeton,* but also on *Fucus* in mesohaline water. According to investigations by GRESENS and PALMHERT the brackish-water form of *Pelmatohydra* appears to be a physiological race. SEIFERT reports on *Ephydatia:* „Kulturversuche ergaben, daß regenerierende Stücke unter entsprechenden Bedingungen innerhalb von 12 Stunden in reines Süßwasser zurückgebracht werden können und hier normal weiterwachsen" [Culture experiments showed that regenerating pieces could, under appropriate conditions, be returned to pure fresh water within 12 hours and they continued to grow normally.] The fresh-water sponges of the family Lubomirskiidae from Lake Baikal, Lake Ochrid and the Caspian Sea have two species of the genus *Metschnikowia* which in the Caspian Sea extend into the mesohaline region.

To my knowledge the small polyp *Microhydra* and its medusa *Craspedacusta* have only been found in pure fresh water.

The Turbellaria form a group inhabiting the sea, brackish water and fresh water with a large number of species an organizational types. Individual groups have produced many species an genera in fresh water. Among the Triclads these are the Paludicola with *Planaria, Dendrocoelum, Polycelis* and others, among the Rhabdocoela the Dalyellidae (Dalyelloida) and the Typhlophanidae (Typhloplano-idea). They provide the bulk of fresh-water Turbellaria. The Macrostomida with the genera *Macrostomum* an *Microstomum* have a distinct, but slighter development of their own in fresh water; the two genera also occur in the sea; the same holds for the Lecithoepitheliata with the Prorhynchidae. The Catelunida are almost exclusively fresh-water forms. In all the other groups there are, at the outset, isolated pioneers represented in fresh water *(Pseudosyrtis fluviatilis, Plagiostomum lemani* and others).

On transition into brackish water these fresh-water Turbellaria show a varied behaviour. The Triclads provide three euryhaline species of the 2nd degree: *Planaria torva, P. lugubris* and *Dendrocoelum lacteum.* They are commonly found in the phytal zone of brackish water, as well as on *Fucus* and *Furcellaria* and extend to regions of about 7—8⁰/₀₀ S. Near Greifswald *Planaria lugubris* invades 4⁰/₀₀ S (SEIFERT 1938); among species occurring in the area where fresh and brackish water mingle MEIXNER (1938) lists *Polycelis nigra.* FORSMAN (1956) records *Polycelis* from *Fucus* near Kalmar on the east coast of Sweden.

Bdellocephala punctata, widespread in North German lakes, and the more or less rheophile species *Planaria gonocephala, P. alpina, Polycelis cornuta* appear to be absent from brackish water. About half the Triclads occurring in the fresh-water lakes of northern Europe invade brackish water over $3^0/_{00}$ S. But this advance has not resulted in the formation of distinct species in brackish water and the sea; presumably these Triclads are old inhabitants of the sea.

The small Turbellaria of the fresh water are on the whole far less euryhaline. The majority of Catelunida are intolerant of brackish water, only *Stenostomum leucops* is euryhaline and appears to reach $3^0/_{00}$ S. (E. SCHULZ has found an apparently marine species in the marine sands of Kiel Bay. This species has not yet been published.) The Dalyellidae, common in fresh water, are almost completely absent from even low-saline water ($3^0/_{00}$ S). A. LUTHER (1955 p. 88) reports: ,,Manche Süßwasserarten vertragen schwaches B r a c k w a s s e r (bis etwa $1^0/_{00}$): *Gieysztoria virgulifera, pavimentata triquetra* (Buchten der Ostsee), *cuspidata* (Schärentümpel, Südfrankreich), andere etwa bis $5—6^0/_{00}$: *Microdalyellia fusca* und *G. maritima* (Ostsee), *G. subsalsa* (Ufer des Mittelmeeres), in schwach brackigem Wasser fand GRAFF *M. mohicana*" [Some fresh-water species tolerate weak brackish water (up to about $1^0/_{00}$): *Gieysztoria virgulifera, pavimentata triquetra* (bays of the Baltic), *cuspidata* (rock pools on skerries, South of France), others up to about $5—6^0/_{00}$: *Microdalyellia fusca* and *G. maritima* (Baltic), *G. subsalsa* (Mediterranean coast) in slightly brackish water GRAFF found *M. mohicana*]. Of the 118 fresh-water species listed by LUTHER only three exceed the salinity limit ot $1^0/_{00}$, that is only $4^0/_{00}$ of fresh-water species. In spite of this pronounced halophobia the Dalyellidae have developed some lines of distinct species in brackish water and the sea. A. LUTHER (1955, p. 88) reports: ,,An salzhaltiges Wasser sind vermutlich *Gieysztoria knipovici* (Kaspisches Meer) und *bergi* (Aralsee) gebunden. Das ist auch der Fall mit den beiden in der Ostsee lebenden *Halammovortex*-Arten, vom denen *H. macropharynx* noch in der Kieler Bucht vorkommt. (Die amerikanische *H. lewisi* lebt an der atlantischen Küste). Ähnlich scheinen sich die *Axiola*-Arten zu verhalten: *remanei* im Frischen Haff und in der Kieler Bucht, *luetjohanni* im Nord-Ostsee-Kanal; *Jensenia angulata* und *Beauchampiella oculifera* schließlich sind rein m a r i n e, im O z e a n l e b e n d e T i e r e. Wir finden somit unter den Dalyelliden alle Stufen der Anpassung vom Leben in rein süßem Wasser bis dem im Weltmeere".[7] [Presumably *Gieysztoria knipovici* (Caspian Sea) and *bergi* (Lake Aral) are limited to salt water. This also applies to the two species of *Halammovortex* from the Baltic; of these *H. macropharynx* still occurs in Kiel Bay. (The American *H. lewisi* occurs on the Atlantic coast). The species of *Axiola* appear to behave in a similar manner: *remanei* in the Frischen Haff and Kiel Bay, *luetjohanni* in the Kiel Canal; *Jensenia angulata* and *Beauchampiella oculifera* are purely m a r i n e a n i m a l s l i v i n g i n t h e o p e n s e a. Thus among the Dalyellidae all degrees of adaptation are found from life in pure fresh water to that in the ocean.]

---

[7] It may be that the marine genera are the original forms s t i l l living in the sea. KARLING (1956) describes a new marine genus *Alexlutheria* and discusses the possibility that the fresh-water forms might be derived from these marine ones by various ecological routes. In that case *Halammovortex* would be an ancestor of fresh-water forms *(Gieysztoria)* and not a descendant which had returned to the sea.

The same holds true for the Typhloplanidae. The numerous fresh-water species of this family are absent from the low-salinity regions. Among the euryhaline species which invade the brackish-fresh-water region MEIXNER lists *Castrada perspicua, C. stagnorum, C. lanceola, C. hoffmanni, C. intermedia, Mesostoma lingua, Bothromesostoma personatum, Phaenocora typhlops.* Only the last species seems to transgress the $3^0/_{00}$ boundary and produces a form *subsalina*. Among the marine and brackish-water species of the Typhloplanidae there are some which owe their origin to previous immigration from fresh water. Among these are *Castrada subsalsa* which LUTHER discovered in the brackish waters of Finland and *Opistomum immigrans* which Ax (1956) described from the meiomesohaline brackish water of the Etang de Canet (South of France).

It is difficult to assess the Macrostomidae. Many species, only recently identified, live in fresh water, brackish water and in the sea. Distinct species of *Macrostomum* were found by LUTHER (1947) in the brackish water of Finland and by Ax (1951) in other brackish waters; it appears that the fresh-water species hardly penetrate into salt water; only *M. appendiculatum* is holeuryhaline (p. 58). It is not clear in how far the individual species occurring in salt water are of fresh-water origin.

Prorhynchus stagnalis (*Lecithoepithelialia*) penetrates locally into brackish waters, it is euryhaline (? of the 1st degree).

On the whole the fresh-water Turbellaria are strongly intolerant of salt, the Triclads are most commonly euryhaline, the Catenulida and Typhloplanida least of all.

Rotifera. The Rotifera are represented in the fresh-water benthos by a large number of species not yet fully ascertained; in Europe it is hardly less than a thousand. Again there is a steep drop in brackish water, but there are a considerable number of euryhaline species of 2nd and 3rd degrees. At $3^0/_{00}$ the oligohaline brackish water is rich in fresh-water species, but this region has not been extensively studied. ALTHAUS (1957) stresses in her studies of the saline waters of Central Germany, „daß bei einem Gesamtsalzgehalt von 2.5—$3^0/_{00}$ die Rotatorienfauna einen besonderen Charakter annimmt. Euryhaline Süßwasserarten, die wenigstens lokal $8^0/_{00}$ überschreiten, sind z. B. *Philodina roseola, Mniobia magna* und *M. symbiotica* unter den Bdelloidea, *Collotheca ornata* und *Ptygura* spec. unter den sessilen Rotatorien, *Testudinella patina, Cephalodella gibba, Epiphanes senta* und *Eosphora ehrenbergi* unter den vagilen Monogononta." [At a total salt content of 2.5—$3^0/_{00}$ the Rotifera fauna assumes a special character. Euryhaline fresh-water species which exceed $8^0/_{00}$ at least locally are e. g. *Philodina roseola, Mniobia magna* and *M. symbiotica* among the Bdelloidea, *Collotheca ornata* and *Ptygura* spec. among the sessile Rotifera, *Testudinella patina, Cephalodella gibba, Epiphanes senta* and *Eosphora ehrenbergi* among the motile Monogononta.]

The high degree of euryhalinity displayed by the Rotifera is even more marked in the plankton than in the bottom fauna. By far the majority of all pelagic fresh-water rotifers occurring in lakes also inhabit oligohaline water of 0—$3^0/_{00}$. Here we find species of *Asplanchna*, of *Brachionis, Filinia, Keratella, Hexarthra* (*Pedalia*) also *H. mira, Kellicotia* = *Notholca longispina, Chromogaster, Conochilus, Collotheca, Polyarthra, Synchaeta, Ploesoma hudsoni* and others. The differences between oligohalinikum and fresh water are exceedingly small. Naturally the rotifers of bog and acid waters are absent from brackish water and a few other species seem to be lacking (e. g. *Synchaeta oblonga* etc.); but about 70—80% of the pelagic fresh-water Rotifera inhabit this region and determine the picture of the Rotifera plankton. Even the plankton of the 3—5⁰/₀₀ range is characterized by fresh-water species several of which extend to about the 8⁰/₀₀ boundary. Such euryhaline plank-tonic rotifers of the 2nd degree are e. g. *Brachionus angularis, Br. capsuliflorus, Filinia longiseta.* To these must be added the facultatively *Keratella quadrata.*

More striking is the rich development of new species in brackish water and the sea; they have undergone differentiation after their immigration from the sea. Almost invariably their organization resembles that of the original fresh-water forms so closely that they are retained in the same genus.

Full-strenght seas have been reached by the Bdelloidea e. g. by the epizoic *Zelinki-ella synaptae* (on *Synapta* in the Mediterranean) and another genus outwardly resem-bling *Rotaria,* but without wheel-organ which I have found in the sandy areas of the North Sea, the Atlantic and the Red Sea. The Monogononta have many migration currents from fresh water into the sea, having sent at least 25 lines of immigration from fresh water into the sea (e. g. the genera *Proales, Pleurotrocha, Cephalodella, Lindia, Encentrum, Wigrella, Wierzejskiella Aspelta, Erignatha, Lecane* etc.).

A similar situation obtains in the Gastrotricha; one of their orders (Chaetonoto-idea) has entered fresh water from the sea and there split into many species. Of these fresh-water Chaetonotoidea many have remained true fresh-water forms, e. g. the genera *Polymerurus, Neogossea, Dasydytes*; some Chaetonotidae are euryhaline e. g. *Lepidoderma squammatum* and *Heterolepiderma ocellatum.* Several lines, how-ever, have returned to the sea and differentiated into species and, in part, special subgenera *(Halichaetonotus).* Chaetonotoidae of fresh-water origin are relatively numerous as far as full-strength seas (Mediterranean see WILCKE 1954), especially in the sandy regions. These are inhabited by species of genera which are also repre-sented in fresh water *Aspidiophorus, Chaetonotus, Heterolepidoderma, Dichaetura, Ichthydium.* There are at least 7 lines of immigration (REMANE 1950).

In spite of their uniform morphology the ubiquitous N e m a t o d a have pro-duced a well differentiated fauna of the sea, the brackish water and fresh water; in addition there are many terrestrial species in soils soaked with fresh water. Ac-cording to a personal communication by S. GERLACH the Dorylaiminae, Trilobinae, Plectinae, Cylindrolaiminae, Anguillulidae, have undergone a special development in fresh water. Euryhaline fresh-water species which invade the oligohaline brackish-water region are *Dorylaimus stagnalis, Trilobus gracilis, Chromadorita leuckarti, Punctodora ratzeburgensis.* „Weitere Arten aus terrestrischen bzw. limnischen Gat-tungen werden in der Zone des oberen Eulitorals und des Supralitorals ebenso wie im Küstengrundwasser gefunden, z. B. in der *Enteromorpha*-Zone und im Cyano-phyceensand. Zum Teil handelt es sich dabei um Einzelfunde solcher Formen, die aus den benachbarten nicht salzigen Lebensräumen dorthin eingedrungen oder verschlagen sind (viele *Dorylaimus-, Plectus-, Monhystera-, Rhabditis-, Cephalobus-* und *Diplogaster*-Arten, dazu Anguillulinidae). Andere kommen so regelmäßig vor, daß sie zum festen Bestand solcher brackigen Lebensräume an der Grenze Land-Meer gehören, etwa *Dorylaimus carteri, D. eurydorys, Plectus granulatus, Acrobeles ciliatus* und *Diplogaster armatus.*

Alle oben genannten Gruppen limnischer Nematoden mit Ausnahme der kleinen Gruppe der Cylindrolaiminae haben ins Meer einwandernd eigene Arten gebildet, allerdings in mäßiger Zahl. Hierher gehören *Syringolaimus striatocaudatus* (bis 25 m Tiefe), *Dolicholaimus marioni* (bis 21 m), *Dorylaimus marinus* (Meeresalgen bei Lorrient), *D. maritimus* (Grönland). Eine besondere Gruppe schließlich bilden die Arten aus terrestrischen Gattungen, die bisher nur an der Küste gefunden worden sind, nicht aber im Binnenland, z. B. *Dorylaimus balticus, Tripyla cornuta, Mononchus spectabilis, M. rotundicaudatus, M. schulzi, Cephalobus strandi-cornutus, Rhabditis marina, Rh. ocypodis, Odontopharynx longicauda.* Alle diese Formen bleiben in ihrem Vorkommen auf die Randzone des Supralitorals beschränkt; sie dringen in die eigentlichen marinen Lebensräume nicht vor."

[Further species of terrestrial or fresh-water genera are found in the zone of the upper eulittoral and the supralittoral as well as in the coastal ground water for instance in the *Enteromorpha*-zone and in the Cyanophyceae sand. These are partly isolated records of such forms which have penetrated there from adjoining, non-saline habitats or have been carried there (many species of *Dorylaimus, Plectus, Monhystera, Rhabditis, Cephalobus* and *Diplogaster* as well as Anguillinidae). Others occur so regularly that they belong to the fixed stand of such brackish habitats at the boundary of land and sea, such as *Dorylaimus carteri, D. eurydorys, Plectus granulatus, Acrobeles ciliatus* and *Diplogaster armatus*.

With the exception of the small group of Cylindrolaiminae all the above groups of fresh-water Nematoda have formed a moderate number of distinct species on entering the sea. To these belong *Syringolaimus striatocaudatus* (up to 25 m depth), *Dolicholaimus marioni* (up to 21 m), *Dorylaimus marinus* (marine algae near Lorrient), *D. maritimus* (Greenland). A special group is represented by the species from terrestrial genera which so far have only been found on the coast, not inland e. g. *Dorylaimus balticus, Tripyla cornuta, Mononchus spectabilis, M. rotundicaudatus, M. schulzi, Cephalobus strandi-cornutus, Rhabditis marina, Rh. ocypodis, Odontopharynx longicauda.* All these forms are restricted to the marginal zone of the supralittoral; they do not penetrate into the marine habitat proper.]

There is a somewhat confusing situation in the Oligochaeta. Until recently it was thought that this important group with relatively large animals was well known as regards the species occurring in fresh water and the sea, at least in Europe. The investigations of salines of Oldesloe (Schleswig-Holstein) by THIENEMANN (1925) have resulted in a new genus, further species and genera were discovered by KNÖLLNER (1935) in Kiel Bay. More recent studies by TH. BÜLOW (1955—1957), NIELSEN and CHRISTENSEN (1959) revealed that many new forms exist in the brackish regions of the sea. HAGEN (1951) examined the areas of transition from fresh water to sea water in Kiel Bay and found some Oligochaeta in fresh water which had so far been recorded from brackish water only. Assessment of the salinity boundaries is further complicated by the fact that in brackish and sea water the Oligochaeta especially settle in the strip of shore of the eulittoral and supralittoral zones, that is the algae which have been thrown up, *Zostera* and masses of reed, grounds rich in detritus etc. These areas have great salinity fluctuations and the salt content has rarely been measured; so that in many cases no accurate data are available about mean salinity nor about the amplitude which is sure to be very large.

The Oligochaeta are a distinctly fresh-water and fresh-water-terrestrial group. The whole of their rich development took place in these regions. The immigration of a remote ancestor of the Oligochaeta must have occurred a very long time ago. It is difficult to decide whether the Oligochaeta entered the subterranean region directly from the sea — as the Polychaeta e. g. of the genus *Ophelia* and *Neanthes* are doing at the present time — and colonized fresh water secondarily, or whether they were fresh water animals first and only became terrestrial later.

Undoubtedly at the present time the Oligochaeta have a remarkably high number of euryhaline fresh-water species and in addition new species and genera have formed in brackish water and in the sea. This process can be clearly seen in several families.

The peculiar, purely aquatic Aeolosomatidae appear to send only one species *Aelosoma* spec. into pleiomesohaline brackish water. Here it occurs particularly in the coastal ground water while in fresh water it is found among plants. The species of the fresh-water ground water of the shore *(Potamodrilus fluviatilis, Rheomorpha = Aeolosoma neiswestnovae)* which are found in these regions in North Germany have not yet been found in the brackish coastal ground water.

The Naididae are rich in species; a number of them are euryhaline-fresh water and reach the mesohalinikum, in parts even the pliomesohalinikum. To these belong *Stylaria lacustris, Chaetogaster diastrophus, Nais variabilis, N. barbata, Amphichaeta leydigi;* according to KORN *N. elinguis* penetrates from fresh water into the poly-halinikum and may be holeuryhaline. In several cases specific species have developed in the sea. *Amphychaeta sannio* is a brackish-water species; the genus *Paranais* contains predominantly marine species *(P. litoralis, P. frici, P. botniensis)*; of these *P. litoralis* even invades fully marine areas. *Uncinais*, with one species *uncinata* is a marine genus of its own.

The Tubificidae contain many euryhaline fresh-water animals and distinct species in brackish water and the sea. Euryhaline limnobionts of the 3rd degree which penetrate into regions of over $8^0/_{00}$ and in some cases reach 15—$16^0/_{00}$ S are *Tubifex tubifex, Limnodrilus hoffmeisteri, L. claparedianus, Ilyodrilus bavaricus*; common and specific brackish-water species are *Tubifex costatus, T. nerthus*. Most of the species of the genus *Peloscolex* are brackish-water or marine ones *(P. benedeni, P. heterochaetus, P. canadensis, P. gabriellae)* as well as those of the genus *Monopylophorus (M. rubroniveus = Rhizodrilus pilosus)*. Six specific genera live in the marine region *(Aktedrilus, Spiridion, Thalassodrilus, Clitellio, Phallodrilus* and *Heterodrilus)*, the last two in the Mediterranean. I found *Tubifex* on the coast of the Red Sea in over $40^0/_{00}$ S (see BRINKHURST 1964).

The classification of the Enchytraeidae which is most difficult has recently been clarified by NIELSEN and CHRISTENSEN (1959). In addition to a number of euryhaline species two holeuryhaline ones have developed, the famous *Enchytraeus albidus* (see p. 58) and the predominantly marine *Lumbricillus (Pachydrilus) lineatus*. Two genera have produced a number of distinct species in the sea where the majority of species occur. These are the genus *Marionina* to which many species of *Michaelsena (M. subterranea, M. postchitellochaeta)*, as well as *Fridericia pseudoargentea* and *Enchytraeoides spiculus* have been allocated, and the genus *Lumbricillus*. The genus *Cernosvitoviella* contains one fresh-water and one marine species of the coastal ground water *(C. immotus)*. *Grania* is a special genus of the Enchytraeidae which inhabits the sea.

It is strange that the terrestrial Lumbricidae are so intolerant of salt; only *Eiseniella tetraedra* occurs in brackish soil. In the tropics several Megascolecidea *(Pontodrilus, Microscolex)* live on the sea-shore. The Lumbriculidae are also confined to the fresh-water region.

Altogether the Oligochaeta have invaded mesohaline brackish water more than twenty times, the fresh-water euryhaline species are included. Only the Tubificidae and Enchytraeidae have invaded the sea where both families have developed special halobiont species, the Tubificidae have even formed specific genera.

Among the Hirudicinea *Glossosiphonia complanata* and, in the Black Sea, probably *Hirudo medicinalis* are euryhaline of the 2nd degree, the fish leech *Piscicola geometra* even to the 3rd degree. From the family of fish leeches a purely marine branch of leeches has produced several distinct genera living as ectoparasites mainly on fishes.

Fresh-water molluscs. Thanks to the activity of many collectors extensive data are available on the distribution of this group in brackish water (see JOHANSEN 1918, JAECKEL 1950, LINDBERG 1948).

On the whole the fresh-water bivalves are rather intolerant of brackish water, only a few reach the $5^0/_{00}$ boundary, none transgress the $8^0/_{00}$ limit. No distinct brackish-water species of fresh-water origin have been formed.

Fresh-water Bivalves: Limits of salinity in $^0/_{00}$.

	JOHANSEN Randersfjord	JAECKEL jun. Schlei	JAECKEL jun. open Baltic	SCHLESCH	Others
Dreissensia polymorpha...		4.7 (5.6)	4	3	
Unio crassus				3	
U. tumidus	2—3			3	3
U. pictorum	2—3	1		3	
Anodonta piscinalis	2—3	3.8	2—3		5
Anodonta cellensis		1		3	
Pseudanod. minima				2	
Sphaerium corneum	2—3	3	3	3	
Sphaerium rivicola				2	
Sphaerium solidum				2	
Musculum lacustre	2—3			3	
Pisidium amnicum	0.5			0.5	< 1
P. cinereum	3			3	1.5—3.2
P. henslowanum	1—1.5			1.5	
P. subtruncatum	1			1	1.5
P. nitidum	1	4—5	1.5	1.5	1.5
P. obtusale	0.1—0.5			0.5	
P. milium	0.2—0.5			0.5	

18 species are listed in the table. But since there are 28 species in the fauna of northern and Central Europe a large proportion is fresh-water stenohaline *(Margaritana,* many species of *Pisidium)*. Even among the species listed in the table *Pisidium milium, P. obtusale, P. amnicum* and *Anodonta cellensis* still remain within

the 1⁰/₀₀ limit, a few others transgress it only locally *(Sphaerium rivicola, Sph. solida, Pseudanodonta, Pisidium henslowanum)*, so that they, too, are probably fresh-water stenohaline; this group would then comprise 70% of the whole number of species. 6 species would be euryhaline of the 1st degree, one species *(Dreissensia)* typically euryhaline of the 2nd degree and locally two others *(Anodonta piscinalis, Pisidium nitidum)*.

The small Pisidae appear to be far more sensitive to salt than the large Unionidae. BEADLE (1943) reports a *Pisidium* spec. indet. from salt pools in North Africa. In the tropics, especially of the Old World, as well as in the Caspian Sea species of *Corbicula* appear to invade the Mesohalinikum, but *C. consobrina* disappeared from Lake Quarun in Egypt when its salinity rose (at present polyhaline, BOULOS).

No invasion of the sea with subsequent development of new species in brackish water or the sea has been established for bivalves.

A more marked degree of euryhalinity is found in Gastropoda.

The table shows that all Prosobranchia of the North Sea and Baltic area are euryhaline, the majority of the 1st degree; the species of *Bithynia* and locally (Schlei) *Valvata cristata* (JAECKEL jun. 1952) are euryhaline of the 2nd degree. *Theodoxus fluviatilis* is euryhaline of the 3rd degree; it belongs to a family widespread in the sea and has produced a special form (f. *baltica*) in the Baltic. Locally (Zuiderzee) *Bithynia tentaculata* exceeds the 8⁰/₀₀ boundary.

Fresh-water Prosobrachii: Limits of salinity in ⁰/₀₀

	Randersfjord	Schlei	Baltic	SCHLESCH	Others
Amnicola steini	0.5			0.5	
Lithoglyphus naticoides ..				3	Zuidersee
Bithynia leachi	4—5 (7.4)	5—6	2	2	
B. tentaculata	3—4 (5)	6	6—7	7	12 Zuidersee
					7 Lindberg
Viviparus viviparus	1—3			3	
V. fasciatus ....:...	0.5				
Valvata piscinalis	2—3	2—3	about 2	about 3	2.56 Lindberg
V. pulchella		2—3			
V. cristata	(about 2)	5	about 1	1	1
Theodoxus fluviatilis	7—12 (15)	15.5	14—18	13	

The Pulmonata have numerous euryhaline species — contrary to expectation — as the families in question have had a long period of differentiation in fresh water and it is difficult for small air-breathing aquatic animals to colonize the sea.

Some evidently stenohaline limnobionts of the fresh water around the Baltic and North Sea are absent from this list *(Limnaea glabra, Aplexa hypnorum, Planorbis laevis, Pl. vorticulus* etc), but they do not add up to 30% of the total number (as against 70% in the bivalves). About 7 species are euryhaline limnobionts of the 2nd degree, of the 3rd degree are *Limnaea ovata* (= *Radix baltica)* and locally *L. palustris*. In the sea the fresh-water Gastropoda settle on *Fucus* (see SEGERSTRÅLE 1928, FORSMAN 1956) and occur at depths of several meters (*L. ovata* at 6 m according to SEIFERT).

Fresh-water Pulmonata: Limits of salinity in $^0/_{00}$.

	Randersfjord	JAECKEL		SCHLESCH	Others
		Schlei	Baltic		
Ancylus fluviatilis		? about 4			
Acroloxus lacustris ...	0.2—0.6			3	
Planorbis corneus				3	
Pl. planorbis	0.6 (? 3)	4	4—5	3	4.9 (LINDBERG)
Pl. carinatus				3	
Pl. vortex	1	3	2—3	5	5.4 (LINDBERG)
Pl. contortus	1	3	2—3	5	7 (LINDBERG)
Pl. spirorbis				2	5.7 (LINDBERG)
Pl. albus	0.5	1	3	2	3.5 (LINDBERG)
Pl. acronicus					3.3 (LINDBERG)
Pl. crista				2	1.5 (LINDBERG)
Pl. complanatus				2	
Segmentina nitida		3			
Physa fontinalis	1—1.5	6	5—6	6	6 (FORSMAN)
Limnaea stagnalis	1.5	6	7	7	7 (LINDBERG)
L. anricularia	2	6	1.5	1.5	
L. ovata (limosa)	6	13.7	10—11	10	
L. palustris	3 (? 4)	8	6	5	7 (LINDBERG)
L. truncatula	0.3	3		0.5	
Myxas glutinosa				3	

The Pulmonata form no new distinct species and genera in the sea; *Limnaea ovata* from the sea has been described as specific form (forma *baltica*), but the differences are slight. *Planorbis eichwaldii* from the Caspian Sea might be assessed as a true brackish-water species.

The various orders of Crustacea show varied behaviour regarding the colonization of fresh-water, brackish-water and marine habitats. The Phyllopoda and Anostraca are predominantly fresh-water inhabitants, the Copepoda and Ostracoda have many species in all three aquatic regions where they are of ecological importance; the Malacostraca and Cirripedia are chiefly marine.

The ancient forms among the Phyllopoda that is as the Notostraca with *Triops* and *Lepidurus* are confined to fresh water. Among the Onychiura the primitive Conchostraca are also predominantly fresh-water animals, but some species occur in saline continental waters of arid regions. LÖFFLER (1961) in his list of species reports *Eocyzicus obliquus* up to about 4$^0/_{00}$ and is inclined to count *E. orientalis* and possibly *Leptestheria dahalacensis* among the exclusive inhabitants of saline inland waters. The neotenic descendants of the Conchostraca, the Cladocera, have not only acquired a generally wide ecological plasticity, but also produced several euryhaline fresh-water species and a number of migration lines in salt water of higher concentration.

But the majority of the Cladocera are restricted to fresh-water areas below 1$^0/_{00}$ so that they also display a rapid drop in species beyond that zone. Of about 86 species from Central Europe barely 15 species have been recorded from the oligohaline brackish-water area of the Baltic. In the plankton for example *Leptodora, Diaphanosoma brachyurum*, species of *Bosmina, Daphnia longispina* and *hyalina* resp., *D. pulex, D. magna, Ceriodaphnia quadrangula, Chydorus sphaericus* occur in parts up to about 4—5⁰/₀₀; for more benthic species there are local records of *Sida crystallina, Simocephalus exspinosus, S. vetulus, Scapholeberis mucronata, Ceriodaphnia setosa, Alona affinis, A. rectangulus, Monospilus etripas* (6,5⁰/₀₀) *Chydorus sphaericus* (see LEVANDER 1901, S. SCHWARZ 1960, THUST 1964). Occasionally dense colonies of Cladocera are found in the mesohaline region (*Daphnia pulex* in shallow brackish water). *Chydorus sphaericus* even occurred among *Zostera nana* in the North Sea on the western side of the Island Pellworm. STEUER (1942) records *Diaphanosoma excisum* and *Moina dubia* from Lake Edku in the delta of the Nile (average salinity 5,4⁰/₀₀). But these isolated outposts do not alter the fact that the rich fresh-water fauna of the Cladocera is of no importance in the marine oligihalinikum. There are some exceptions in the Baltic: *Bosmina obtusirostris* has a var. *maritima*, a brackish-water race which occurs practically everywhere in the Baltic and is an important member of the plankton.

There are more Cladocera in brackish pools of arid regions than in the marine regions themselves (including the lagoons). SARS (1903), GAUTHIER (1928), BEADLE (1943) report a number of Cladocera from saline waters in Central Asia and North Africa. In a list of Crustacea LÖFFLER (1961, p. 355) mentions a number of species which partly exist in high salinities: *Daphnia atkinsoni* 19,7⁰/₀₀, *D. tibetana* 12,8⁰/₀₀, *Moina brachiata* 36.4⁰/₀₀, *M. hutchinsoni* 39⁰/₀₀, *M. macrocopa* 22.2⁰/₀₀ *M. rectirostris* about 32⁰/₀₀, *M. salinarum* 34⁰/₀₀, *M. wierzejski* 12.2⁰/₀₀, *Macrothrix hirsuticornis* about 30⁰/₀₀, *Alona rectangula* 12.6⁰/₀₀.

These are examples of further penetration of fresh-water species into saline waters in warm regions as well as colonization of these saline waters by species from temporary pools *(Moina)*.

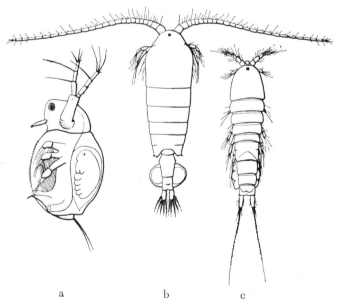

a   b   c

Fig. 28. Entomostraca from small bodies of water with fluctuating salinity. a, *Moina microphthalma;* b, *Diaptomus (Arctodiaptomus) salinus;* c, *Cletocamptus confluens*. After SARS.

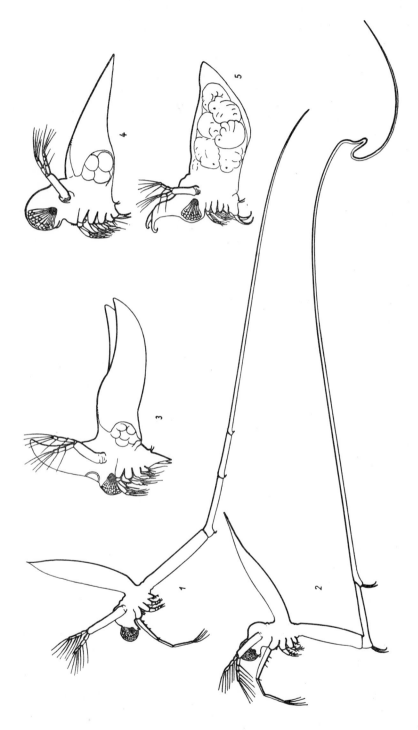

Fig. 29. Specific planktonic Cladocera from the brackish water of the Caspian Sea. 1, *Apagis longicaudata*; 2, *Cercopagis gracillima*; 3, *Evadne maximovitschi*; 4, *E. camptonyx*; 5, *E. hircus*; enlarged. After SARS 1927.

In spite of this moderate euryhalinity of fresh-water Cladocera several lines of immigration have penetrated into brackish water and the sea where distinct species and genera have developed. These are exclusively planktonic Cladocera and there are none of these in the marine benthos.The marine Polyphemidae are very plentiful, the genera *Podon* and *Evadne* have a world-wide distribution in the sea and in brackish water; in the pontocaspian area there are the quaint genera *Cercopagis,* and *Apagis* with numerous species (see fig. 29). The Polyphemidae have probably invaded the brackish water and the sea along two lines. A third one is represented by the Sididae; the species *Penilia avirostris* has a world-wide distribution in warm seas, chiefly in coastal areas, but by no means restricted to lagoons. I found it in the Red Sea, the Atlantic of South Brazil and the Pacific Ocean of Central America, in every case in plankton hauls near the shore. *Penilia* is a marine genus which in the Black Sea invades brackish water as well.

The continental salt waters also seem to have their own halophilous species, e. g. *Moina salinarum.*

The A n o s t r a c a have the peculiar feature of developing species and genera capable of tolerating high salinities (*Artemia salina,* the brine shrimp over 200⁰/₀₀!). But they only occur in brackish and salt pools, not in the sea, but are found in saline waters near the shore. It is strange that this undoubtedly ancient order should be restricted to small bodies of water which frequently dry out periodically. It has been suggested that avoiding fishes has driven these relatively large (mostly 1 cm) animals which swimm in a conspicuous manner, into water from which fishes are absent. In the continental salt waters — especially of arid regions — there occur a number of species from the genera *Parartemia, Branchinella, Branchinecta, Branchinectella* a few of which may even prefer a certain salinity (see LINDER, LÖFFLER). KERTESZ (1955) found 5 species in the soda lakes of Hungary.

The fresh-water C o p e p o d a behave differently in the individual environments: open water, bottom layer and subterranean region. In open water, that is in the pelagic zone and the upper region of vegetation, several species of Cyclopidae enter oligohaline and, in parts mesohaline water. Various authors (see HERBST 1951, REDEKE 1935c, REMANE 1934a, LEVANDER 1901b and others) call the following species euryhaline (mostly of the 1st degree): *Eucyclops serrulatus, Paracyclops fimbriatus, Cyclops viridis, C. bicuspidatus, C. bisetosus, Mesocyclops leuckarti, M. oithanoides, M. hyalinus,* in parts also *C. insignis, C. speratus;* these are barely 20% of the Central European Cyclopidae. But they have been reported from numerous brackish waters, even outside Europe. They mainly inhabit bays, pools and lakes near the sea and brackish pools, usually in small numbers. It is possible that reinforcement from fresh water maintains the numbers of some species. *C. bicuspidatus,* especially as var. *odessanus* and *C. bisetosus* are undoubtedly species which can maintain themselves in brackish water where they are widespread. *C. bicuspidatus* is even euryhaline of the 3rd degree (up to about 10⁰/₀₀) and has often been considered to be halophilous. The Cyclopidae in the inland saline waters of arid regions are more tolerant of salt than in the North. LÖFFLER (1956) mentions the following species from inland saline waters as tolerant of high salinity: *Paracyclops fimbriatus* 31⁰/₀₀, *Cyclops ladakanus* 12.8⁰/₀₀, *C. viridis* about 18⁰/₀₀, *C. bicuspidatus* about 32⁰/₀₀, *C. bisetosus* about 56⁰/₀₀, *Metacyclops mendocinus* 12.2⁰/₀₀, *M. minutus* 19.6⁰/₀₀, *Apocyclops dengizicus* 44.6⁰/₀₀, *Apo-*

cyclops panamanensis 74.8⁰/₀₀. The genus *Halicyclops* is probably of marine origin and hardly an immigrant from the fresh-water area.

The Diaptomidae behave very differently in sea and inland saline waters. In northern and Central Europe at least the fresh-water species of *Diaptomus* are largely intolerant of salt, in contrast to the Cyclopidae they are practically absent from records, even of oligohaline waters. REDEKE (1935 a) and S. JAECKEL (1962) report *Diaptomus gracilis* from oligohaline water.

Arctodiaptomus wierzejskii and *A. salinus* occur in brackish water in the inland saline of Central Europe, near Halle. *A. salinus* exibits halophilous tendencies. In Southeastern Europe, Central Asia and North Africa the species is common in brackish water and is also found in Lake Aral and the northern part of the Caspian Sea; in the South it is holeuryhaline. In addition SARS mentions other Diaptomidae (*D. incrassatus, D. bacillifer, D. asiaticus* and others) from the brackish waters in Central Asia. As for other groups, adaptation to brackish water has been specially successful for Diaptomidae in arid regions. From such regions LÖFFLER (1961) mentions a high upper salinity limit for the following species: *Boeckella propoensis* 61⁰/₀₀, *Metadiaptomus transvaalensis* about 16⁰/₀₀, *Arctodiaptomus (Diaptomus) bacillifer* 10⁰/₀₀, *A. salinus* 44.6⁰/₀₀, *A. spinosus* 25⁰/₀₀, *Diaptomus sicilis* 35⁰/₀₀. The family of Pseudodiaptomidae with *Calanipeda (Popella)* which inhabits the same areas as *D. salinus* and the Pseudodiaptomidae from tropical estuaries will be discussed with the brackish-water animals.

The bottom-living Harpacticidae of the fresh water are intolerant of salt and only a few of the fresh-water species have been recorded from oligohaline water: *Canthocamptus staphylinus, C. microstaphylinus, Atheyella crassa, Bryocamptus minutus, B. pygmaeus* (see NOODT 1956); but these are only scattered isolated records which hardly transgress the oligohalinikum. *C. staphylinus* occurs in the Kiel Canal up to 4⁰/₀₀ (SCHÜTZ 1960).

But in the subterranean coastal ground water fresh-water Harpacticidae occur as far as mesohaline regions. It came as a surprise to find *Parastenocaris*, a predominantly fresh-water subterranean genus of Copepoda in the marine coastal ground water: *P. vicesima* in Kiel Bay (KLIE 1934 b). Later studies have shown three further fresh-water species of *Parastenocaris* to be widespread in the brackish coastal ground water of Central and northern Europe (see NOODT 1956); in addition to *P. vicesima* they were *P. phyllura* and *P. fontinalis*. In America *P. panamericana* is largely fresh-water euryhaline (communication by NOODT).

Apart from the halophilous *Cyclops bicuspidatus* and *Arctodiaptomus salinus*, there are no brackish-water or marine Copepoda of fresh-water origin. However, the phylogeny of the Canthocamptidae and Pseudodiaptomidae has not been completely elucidated and further work may disclose species with a fresh-water ancestry. It may be that *Cletocamptus*, a genus widespread in saline pools, is a Harpacticid which has developed in salt pools from a fresh-water stock, as have the halobiont Anostraca found in such places. In continental saline waters there is more intensive formation of species. The genera *Nitocra* and *Nitocrella*, widespread in the marine region and especially in brackish water, partly invade fresh water and contain a few fresh-water species. This group the taxonomy of which is difficult will be mentioned later (cf. NOODT 1957).

The fresh-water Ostracoda are a little more tolerant of salt. In oligohaline and in part in miomesohaline water the following are not uncommon: fresh-water species of the genera *Heterocypris*, e. g. *H. incongruens, Cyclocypris* e. g. *C. laevis, Candona* e. g. *C. neglecta, C. candida, C. compressa, C. levanderi, Herpetocypris*, e. g. *H. chevreuxi*, species of *Eucrypis* and *Darwinula*. *Darwinula stevensoni* even forms large colonies in the mud of oligohaline and slightly mesohaline lakes and *Candona neglecta* is the only fresh-water crustacean of the Baltic to invade greater depths. But the main habitat of the fresh-water euryhaline Ostracoda is the shore with its brackish pools, ditches and shallow bays. In that region the fresh-water immigrants have speciated e. g. *Heterocypris salina, Candona angulata* in the South species of *Eucypris* e. g. *Eucypris inflata.*

The continental saline waters of arid regions contain larger number of species of Ostracoda, especially Cyprididae (see LÖFFLER 1961, table); particularly from the genera *Iliocypris, Eucypris, Heterocypris* and *Cypridopsis*. Most species are restricted to the Oligohalinikum and Miomesohalinikum. Higher maximal values are attained by: *Candona neglecta* 15⁰/₀₀ S, *Eucypris inflata* 8—110⁰/₀₀, *Heterocypris barbara* about 88⁰/₀₀, *H. incongruens* about 12.5⁰/₀₀, *H. salina* 20 (25?)⁰/₀₀, *Cypridopsis inaeguivalva* about 16⁰/₀₀, *Potamocypris arcuata* about 13.5⁰/₀₀, *P. steneri* 1—12⁰/₀₀, *Darwinula stephensoni* 15⁰/₀₀.

The higher Crustacea of the fresh water most of which have close relations in the sea, are predominantly fresh-water stenohaline and barely enter the Oligo-halinikum or else reach their limit there. This applies e. g. for the crayfish *(Astacus)*, the fresh-water species of *Gammarus* etc. There are also a few euryhaline species. Thus the water louse *Asellus aquaticus* is found in the Mesohalinikum of strand waters. The Pontic Amphipod *Corophium curvispinum* has spread in North Germany within the last decades where it has not yet appeared in brackish water, but it does occur in brackish water in the South. The same holds true for the fresh-water prawn *Atyaephyra desmaresti* which has immigrated from the Southwest (see THIENEMANN 1950) and has been found in the strand waters of the Mediterranean up to 18⁰/₀₀ S. It is clear that the migrating Malacostraca are euryhaline such as the mitten crab *Eriocheir sinensis* and the tropical Palaemonida e. g. *Cryphiops* (HARTMANN 1957); all of which alternate between fresh water and sea. It is of greater interest that some species of the rich subterranean fauna are euryhaline and occur in the coastal ground water e. g. *Pseudoniphargus africanus* (BALAZUC and ANGELIER 1951). The same authors report that in the shore ground water of the etangs in the south of France species of *Niphargus* occur regularly at 3—22.7⁰/₀₀ S. GERLACH and SIEWING even found a Bathynellida in the brackish water of the mouth of the Amazon: *Thermo-bathynella amyxi*, a slightly euryhaline fresh-water species from this interesting group of which many species have a worldwide distribution in the interstitial ground water (NOODT 1964). The crustacean *Potamobius pachypus* is a decapod species of the brackish water with a fresh-water origin. It inhabits the Caspian Sea down to a depth of 60 m, while *P. leptodactylus* is confined to the parts of the northern Caspian where the water has a reduced salinity.

The Arachnida which are so adaptable with regard to climate and ecology have a marked salinity barrier in their fresh-water representatives. The water spider *(Argyroneta aquatica)* which is only facultatively aquatic occurs in the *Enteromorpha* layers of brackish strand waters. The majority of the numerous species of water mites (Hydrachnella) are stenohaline-fresh-water. Isolated records compiled by VIETS (1925) have been reported from brackish water. This list includes 44 species and has become even larger to-day, but most of them show the character of "chance penetration" and are designated as haloxen by VIETS. This presumably applies to records reported by BEADLE from saline waters in North Africa (*Eylais megalostoma, Arrhenurus pervius* at about 10⁰/₀₀, *Eupatra semiperforata* at about 25⁰/₀₀). Only a few species have been found in brackish water more frequently e. g. *Diplodontus despiciens, Hydrophantes dispar, H. ruber, Piona uncata* so that they are presumably euryhaline. No species formation has taken place in brackish water. It is all the more surprising that in full-strength sea water up to over 40⁰/₀₀ there are special genera of Hydrachnella *(Pontarachna, Litarachna)* which are common. In the North a wide gap separates them from the fresh-water forms, but in the Black Sea they also invade brackish water.

This behaviour also contrasts with that of the terrestrial mites which not only penetrate into the phytal zone of the tidal area, but also secondarily adopt a nearly aquatic way of life in the coastal ground water, e. g. the Rhodacaridae. Evidently the marine mites (Halacaridae) have invaded the sea from the land; there many species (about 250 are known) have colonized a large number of habitats in some depth — the only case of a successful colonization of truly marine regions outside the coastal zones which has been achieved by invertebrate land animals. In the sea these Halacaridae behave like true marine animals, displaying a reduction of species in brackish water. It is all the more astonishing that a twin family the Limnohalacaridae which immigrated into fresh water at an early period behave like true fresh-water animals in their intolerance of salt. Only in the region of the Black Sea and the Caspian do the two families meet, *Caspihalacarus hyrcanus* from the family of Limnohalacaridae and several marine species of the Halacaridae sens-strict. Might the Limnohalacaridae have entered fresh water from the common terrestrial ancestor, direct from the land?

Insects. Only aquatic insects can be considered here, not the many inhabitants of the shore of which there are large numbers of species on the sea coast. Even at a glance three groups can be recognized which vary considerably in their behaviour towards salt water.

The first group comprises markedly fresh-water stenohaline species. The Ephemerida and Plecoptera belong here. They are completly absent from the lists of insects from mesohaline water; even LINDBERG's thorough studies of the insects in the brackish waters of Finland showed an absence of these two groups. JOHANSEN (1918 b) mentions isolated records of Ephemerida larvae from the oligohaline water of the Randersfjord: *Ecdyurus volitans* at S surface 0.5, bottom 3.2⁰/₀₀; *Caenis* sp. and *Baetis* sp. at S surface 1.5, bottom 1.6⁰/₀₀. But these are isolated records at the boundary[8]. The Neuroptera show similar behaviour; only the larva of *Sialis lutaria* has been observed in brackish water up to about 3⁰/₀₀ (JOHANSEN). The second

---

[8] A record of Ephemeridae larvae at higher salinity comes from warm regions. SCOTT, HARRISON and MACNAE (1952) report a larva of *Cloëon* at 19.7⁰/₀₀ from a river mouth (Klein-River) in South Africa.

group includes orders which in addition to numerous stenohaline fresh-water
species have a fair number of euryhaline species of 1st, 2nd and in parts 3rd degree;
they are widespread in brackish water up to about 10—15⁰/₀₀. But no distinct species
are formed in brackish water. The Odonata, Lepidoptera and, in particular, the
Trichoptera belong here. Because of the necessity of hatching from a solid substratum
many are restricted to the shallow areas of the sea and occur particularly in bays
and lagoons. Only in exceptional cases do they transgress the salinity limit of about
$15^0/_{00}$.

Among the Odonata *Ichnura elegans* is specially euryhaline (3rd degree). Larvae
of this species have been found in water with over $12^0/_{00}$ several times. More locally
Orthetrum cancellatum penetrates into this region (K. LARSEN 1936, Dybsoefjord,
$13^0/_{00}$). JOHANSEN also records *Cordulia aenea* from the Randersfjord.

Odonata larvae of the 2nd degree of euryhalinity are more common. From the
range 3—$6^0/_{00}$ LINDBERG mentions: *Enallagma cyathigerum, Agrion armatum, A. pulchellum, A. hastulatum, Erythromma naias, Brachytron hafniense, Aeschna coerulea*
(9.4), *A. juncea* (10.9), *A. grandis, A. cyanea, Libellula quadrimaculata, Sympetrum
striolatum, S. vulgatum, S. flaveolum, S. danae, Leucorrhinia rubicunda*. These are
about half the species from the fresh water of the corresponding region.

Among Trichoptera the larvae of the genus *Oecetis* (*O. ochracea*, K. LARSEN,
SEIFERT), *Triaenodes Raiteri* attain euryhalinity of the 3rd degree. For the less saline
marine areas of Finland SILFVENIUS mentions 33 species. In water over $3^0/_{00}$. LIND-
BERG found *Oecetis furva, Molanna angustata, Phryganea striata, Grammotaulius
atomarius, Limnophilus flavicornis, L. decipiens, L. lunatus, L. ignavus, Colpotaulius
incisus*. From the Danish Wick near Greifswald at 5—$6^0/_{00}$ SEIFERT (1938) reports
Phryganea grandis, Agraylea multipunctata, Oecetis ochracea and *Triaenodes bicolor*.
It is interesting that the Trichoptera have also transferred to marine algae and occur
in large numbers, e. g. on *Fucus* (8 specimens on 1 kg *Fucus* in calm areas, according
to SEGERSTRÅLE.

Only a few species of Lepidoptera have aquatic larvae so that the number of
euryhaline fresh-water caterpillars is small. The well known moth *Acentropus niveus*
is widespread in brackish water, with both types of females (winged and wingless
ones, the latter remain under water). It lives on marine Potamogetonaceae and on
Zostera marina, up to about $15^0/_{00}$ S. Locally the caterpillar of *Paraponyx stratiotata*
invades brackish water.

The aquatic Hemiptera form a transition to the next group. Many fresh-water
species of the genus *Sigara* are euryhaline of the 2nd degree (*S. striata, S. distincta,
S. semistriata, S. praeusta* and others). *S. hieroglyphica (lateralis)* is even euryhaline
of the 3rd degree. Other genera are less common in brackish water *(Corixa affinis,
Cymatia bonsdorffi, C. coleoptrata, Micronecta, Notonecta, Aphelocheirus)*. But it
is significant that species of *Sigara* penetrate into hyperhaline waters (70⁰/₀₀) and
in our region *S. lugubris-stagnalis* has differentiated into a true brackish-water animal.
Physiologically, too, this species occupies a special position (CLAUS).

The third group of insects inhabiting salt water is represented by orders which
1) transgress the 15⁰/₀₀ salinity limit in many species and often invade regions of
50—200⁰/₀₀ and 2) have produced specific species in salt water. The Coleoptera and
Diptera belong to this group.

Among Coleoptera the Haliplidae, Dytiscidae and Hydrophilidae are widespread and common in shallow coastal regions and strand waters. Individual interesting forms of the Chrysomelidae, Curculionidae and Staphylinidae enter the marine region.

LINDBERG (1948) mentions no fewer than 100 species of aquatic beetles from the brackish waters of Finland. Of these no less than 84 species enter mesohaline water; 28 species transgress the $8^0/_{00}$ limit and penetrate partly to over $20^0/_{00}$ in inland saline waters.

Every group of aquatic beetles has produced distinct species in salt water, *Haliplus apicalis* in the Haliplidae, *Coelambus flaviventris* in the Dytiscidae; the halophilous and halobiont Hydrophilidae have specially large numbers *(Berosus spinosus, Enochrus bicolor)* and in particular species of *Ochtebius* which occur everywhere in different places in inland saline waters, salines, rock pools, shore pools up to $270^0/_{00}$ S. (HASE for *Ochtebius quadricollis*). Some species have a tendency to holeuryhalinity.

The Chrysomelide *Haemonia mutica* is interesting as it lives clinging to plants in the sea or brackish water even as imago. It settles on Potamogetonaceae and *Zostera* and lives in 3 to about $15^0/_{00}$ S.

Brief mention will be made of those genera which have invaded the eulittoral zone of brackish and marine waters from the land; they survive the period of high tide under stones, in burrows or between grains of sand. They are Carabidae (*Aepus marinus, Aepopsis Robini, Bembidion laterale,* species of *Dyschirius*). The larva and imago of the weevil *Mesembriorrhinus eatoni* live in the Kerguelan in the *Enteromorpha* zone of the shore.

The beetles have sent several migration routes from fresh water into brackish water and sea; in so far as they have come from fresh water — and not from the land — they are confined to quiet zones of the shore, shore pools and inland saline waters; only *Haemonia* is a true inhabitant of the phytal zone, but restricted to shallow waters.

The Diptera have shown the most far-reaching adaptation to saline habitats. Among them is a specially large number of holeuryhaline species (see p. 58). In their distribution in salt water and the sea they are not confined to the shallow calm coastal water and salt pools; they penetrate into the Wattenmeer *(Trichocladius psammophilus, Scatella subguttata, Hydrophorus praecox)* and the rocky zone of breakers of the sea *(Clunio)*, they also colonize the sand and mud regions of the sea down to about 15—50 m depth. Together with *Nereis*, Tanaidae and others, their shells have a share in producing mudstone on the rocks of the sea shore; they live in *Zostera*, on *Fucus* and Red Algae. They occur in the Baltic, in the highly saline Red Sea, on the Atlantic coast of Brazil, in the Antarctic and on oceanic islands. They also inhabit inland saline waters and salines up to high salt concentrations. In a salt lake in Asia SUWOROW found larvae of Chironomidae up to $285^0/_{00}$. *Ephydra* larvae are common in highly concentrated salines at $250^0/_{00}$. In the Baltic *Chironomus* larvae are not uncommon at 50 m depth (HESSLE 1924).

These Diptera larvae, especially midges, had important predispositions for this conquest of the marine environment, far-reaching in an ecological sense. Their pupae are able to hatch directly on to the free water surface while most other aquatic insects — excluding Trichoptera — require the shore or at least a solid starting point for the hatching of the imagines. The Diptera larvae are able — more than any other insects — to emancipate themselves from obtaining air at the free surface; their tracheal system is reduced and they have true aquatic respiration.

According to REMMERT (1955) the following Diptera larvae are fresh-water euryhaline of the 2nd or 3rd degree: *Culex pipiens* 0—15⁰/₀₀, *Sphaeromias fasciatus, Chironomus plumosus, Ch. thummi, Ch. annularius* 0—8⁰/₀₀, *Limnochironomus nervosus* 0—8⁰/₀₀, *Polypedilum nubeculosum* 0—8⁰/₀₀, *Cryptochironomus* sp., *Eucricotopus sylvestris* 0—10⁰/₀₀, *Odontomya viridula* 0—28⁰/₀₀. Many more euryhaline species can be found in THIENEMANN's (1954) comprehensive survey.

The high number of halobiont Diptera which have developed species or even genera in salt water is more important. In his major work "Chironomus" (1954, p. 576—618) THIENEMANN gives a comprehensive survey of the Nematocera of saline areas. Among the Diptera larvae he describes as primary marine animals those groups which have developed their own taxonomic and ecological forms in the sea and have undoubtedly had a long development in the sea though they originally came from fresh water. For the exclusively marine group of Clunionariae a total of 42 species is listed (*Clunio* with 14 species, *Eretmoptera* with 2, *Tethymyia* with 1, *Belgica* with 1, *Thalassomyia* with 7, *Telmatogeton* with 12, *Paraclunio* with 2, *Psammatomyia* with 1 and *Halicystus* with 1). It is an interesting fact that some of these marine genera have evidently returned into fresh water, into fast flowing mountain streams (5 species in Hawaii). Among the Tanytarsariae the group "*Halotanytarsus*" belongs to the true marine Diptera (6 species). Here, too, females with reduced wings develop and remain purely marine *(Pontomyia)*.

There follows a long list of "secondary" marine Diptera. These are individual halobiont species from genera in which the remaining species inhabit fresh water. The number of halobiont and halophilous species of Diptera of this kind is very large as shown in THIENEMANN's compilation. They belong to the Ceratopogonidae (*Dasyhelia* and *Culicoides* species), Orthocladiinae (above all species of *Trichocladius, Eucricotopus* and *Smittia*), Chironomariae *(Chironomus salinarius, Ch. halophilus)*.

Because of their mobility fishes can span large areas with ease and this makes it even more difficult to demarcate their occurrence than for other groups. Some general points may be made: The majority of fresh-water fishes also inhabit slightly brackish waters up to about 5⁰/₀₀; parts of brackish Haffe, the eastern and northern parts of the Baltic contribute quite a considerable yield of "fresh-water fish". But the spawning areas are far more restricted as many of these fishes migrate into fresh water for oviposition. Nearly all manuals of fisheries, e. g. EHRENBAUM (1936), BERG (1933) contain lists of fresh-water fishes found in brackish waters.

A regular occurrence of fresh-water fishes above 8⁰/₀₀ is rare. Apart from migratory fishes the most euryhaline fish of northern Europe is the perch *Perca fluviatilis* which is often found in water of 15⁰/₀₀. The pike-perch *Lucioperca sandra* is quite tolerant of salt, while the third percide of northern Europe, the pope *Acerina cernua* is far more sensitive to salt. It was only after the damming-up of the Zuiderzee during the lowering of salinity that pope developed in large numbers there. That adult Salmonidae are relatively tolerant of salt is evident from the fact that most species have populations which spend a special period of growth in the sea e. g. the sea trout *Salmo fario* f. *trutta* or many Coregonae.

In the warmer zones many teleosts in particular (Cyprinodontidae sens. lat. and Cichlidae) penetrate into brackish areas, even into concentrated salt water (see p. 58). With increased salinity of Lake Möris (Lac Quarun) in Upper Egypt — its present salinity is about $25^0/_{00}$ — two species of *Tilapia (T. nilotica* and *T. zillii)* remained; they still reproduce in the lake now (NAGUIB).

To my knowledge only the fresh-water fishes of the Pontocaspian area have produced halobiont species in the sea and brackish water. Here *Lucioperca marina* lives in the Black Sea and Caspian Sea; in the latter only in brackish water according to BERG; the subspecies *maeotica* of *Percarina demidoffi* inhabits the Sea of Azov, but not its rivers (BERG). The Cyprinidae have a subspecies *caspius* of *Barbus brachycephalus* in the Caspian Sea and Lake Aral, a subsp. *taeniatus* of *Aspius aspius* in the southern part of the Caspian and Lake Aral, of *Rutilus rutilus* a subsp. *heckeli* in the brackish water of the Black Sea and Sea of Azov and a subsp. *caspius* in the Caspian, a *Cobitis caspia* in the Caspian etc. These Cyprinidae still ascend the rivers for spawning so that differentiation into halophilous races and species is still in its beginning. Among the Cyprinodontidae *Cyprinodon (Aphanius) fasciatus* has halophilous tendencies. Altogether fresh-water fishes attain higher degrees of euryhalinity only to a moderate extent and there is little tendency to form immigration lines with distinct species in salt water.

Amphibia are known to be very sensitive to salt, the Urodela and tadpoles of Anura are usually absent from the oligohaline region; only *Bufo calamita* and *B. viridis* are somewhat more resistant to salt and in continental saline waters of arid regions *Rana ridibunda* has also been listed as a brackish-water inhabitant.

Plants. Fundamentally, plants display the same behaviour as animals: high sensitivity of fresh-water species to increased salinity; here, too, the degree of reaction varies from group to group.

The Cyanophyceae contain many strongly euryhaline species; altogether this group shows the highest tolerance of salt content and salinity fluctuations. The other extreme is provided by the Desmidiaceae which are fresh-water stenohaline to a high degree and are indicator forms of fresh water. Only a few species have been recorded from oligohaline water. LIEBETANZ (1925) records only *Cosmarium constrictum, C. laeve* and *Closterium acerosum* var. *elongatum* from inland brackish waters. According to that author the latter species is resistant to NaCl solutions up to 2%.

The special position of the Desmidiaceae can be seen from the following table. Based on REDEKE's survey "Synopsis van het Nederlandsche Zoet-en Brackwater-

	Total number of fresh-water species	of these in the oligohaline	of these in the mesohaline
Chlorophycea	97	41	19
	100%	42%	20.0%
Desmidiacea	278	3	1
	100%	1%	0.3%
Chrysomonadina	37	12	2
	100%	32%	5.4%
Euglena	66	37	22
	100%	56%	33.0%

plankton 1935" it lists the number of species confined to fresh water and of fresh-water species from different groups of lower plants which also inhabit brackish water (oligo- and mesohalinikum).

The Euglenae and Diatomaceae are intermediate in their behaviour. Diatoms have large numbers of species everywhere and are important forms indicating sa-linity zones. According to personal communication by Dr. BROCKMANN the fol-lowing fresh-water diatoms penetrate up to about $3^o/_{oo}$: *Melosira arenaria, Thalas-siosira fluviatilis, Cyclotella meneghiniana, Caloneis amphisbaena, Diploneis ovalis, Anomoeoneis sphaerophora, Navicula cryptocephala, N. oblonga, N. pusilla, Epithemia*

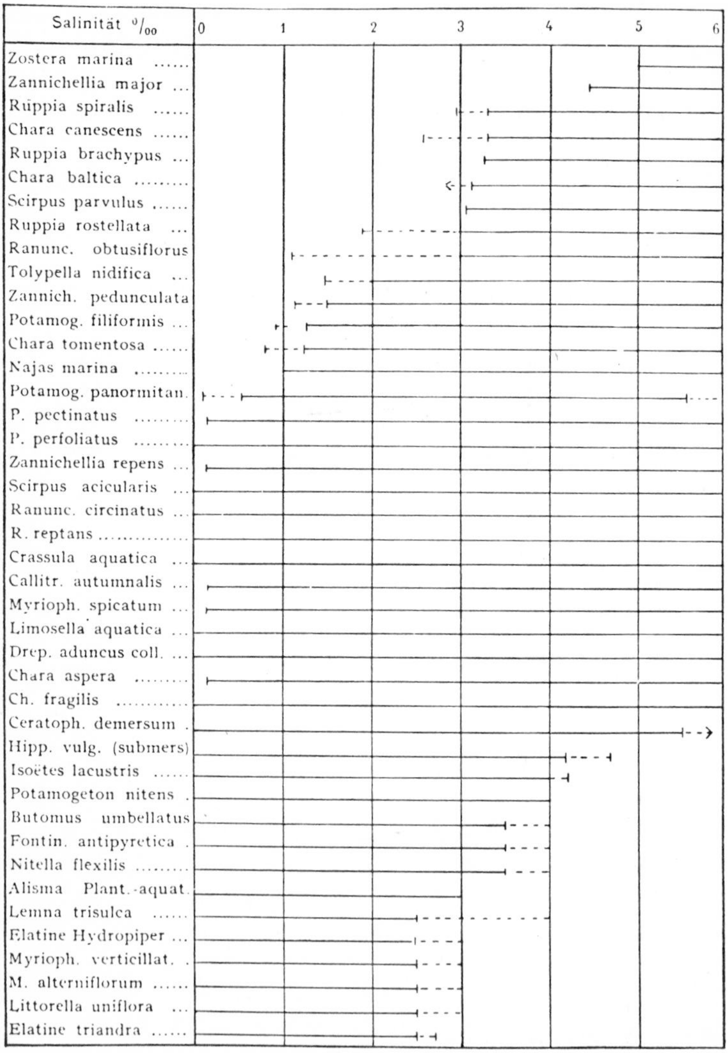

Fig. 30. Areas of occurrence of higher plants in the northern Baltic (coast of Finland.) Salinity region $0—6^o/_{oo}$. After H. LUTHER 1951.

turgida, E. sorex, Gomphonema olivaceum. Fresh-water euryhaline of the 2nd degree, reaching to about 8⁰/₀₀ are: *Synedra pulchella, Diatoma elongatum, Cocconeis pediculus, Rhoicosphenia curvata, Mastogloia smithi, Gyrosigma acuminatum, Fragilaria inflata, Navicula scutelloides.* Euryhaline of the 3rd degree up to about 15—20⁰/₀₀ is *Cylindrotheca gracilis.*

The fresh-water Green Algae are also fairly tolerant. This applies to the planktonic and the benthic forms. In the plankton species of *Pediastrum* occur up to over 5⁰/₀₀ and, together with Rotifera and Cyanophyceae they form a characteristic plankton. The benthic forms often penetrate far into the Baltic (over 5⁰/₀₀) and frequently

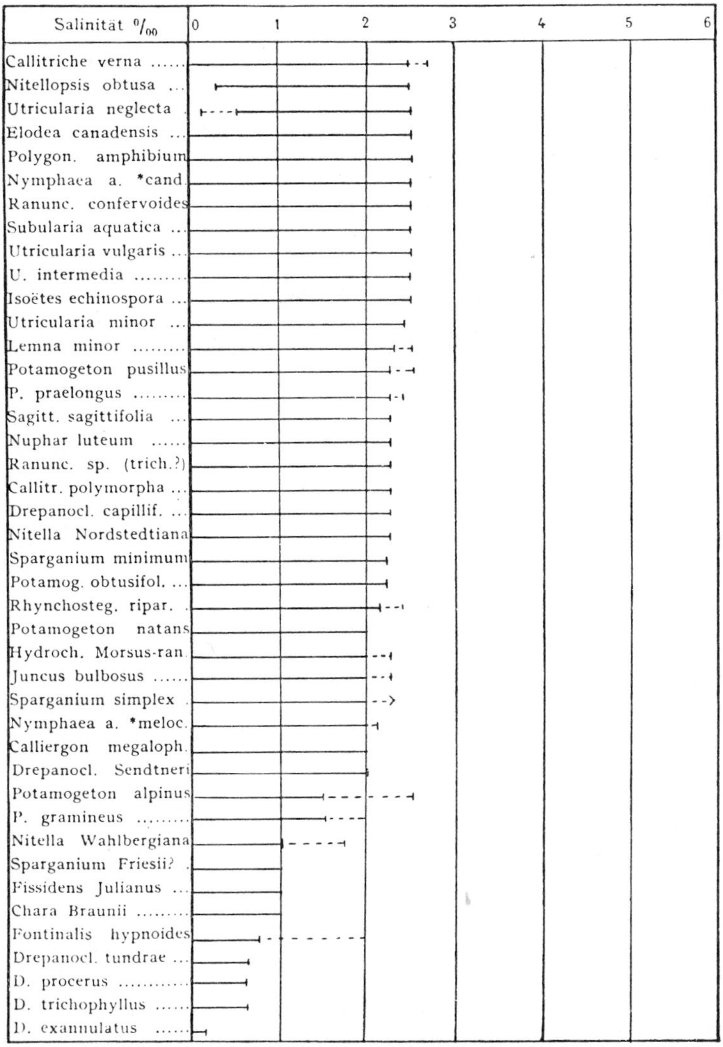

Fig. 31. Areas of occurrence of higher plants in the northern Baltic. Continued. After H. Luther 1951.

form large stands. This is especially the case for species of *Cladophora (Cl. glomerata* to over 15⁰/₀₀, *Cl. aegagropila), Ulothrix zonata, Aphanochaeta repens, Bulbochaete rhadinospora* and forms of *Vaucheria* the species of which cannot always be identified in salt water as there is no sexual reproduction. The same applies to forms of *Spirogyra* and *Zygnema*; the cushions of their filaments are found even in mesohaline brackish water. As in fresh-water lakes there are meadows of Characeae in brackish-water lakes and seas; in part they consist of the same species *(Chara tomentosa, Ch. aspera, Ch. fragilis, Nitella flexilis)*; but there are also specific brackish-water species *(Tolypella, Chara crinita, horrida)*.

The behaviour of the higher plants is illustrated in figs. 30 and 31 in which the salinity ranges of Characeae, aquatic mosses and flowering plants in the Baltic, that is in the Finnish waters, are depicted. In conformity with the maximal salinity of the region, the range does not go beyond 6⁰/₀₀ S. The true fresh-water plants such as *Zostera, Ruppia, Zanichellia* in part and *Tolypella* among the Characeae (excluding the genera which wholly or predominantly colonize saline waters) barely transgress the 6⁰/₀₀ limit.

Most groups of plants which have invaded salt water from fresh water have shown little further development in the new environment.

True, the Green Algae and Characeae have developed a number of special forms in brackish water which may be considered as specific brackish-water races or species *(Vaucheria, Cladophora* etc.*)*; among Characeae *Tolypella* may be a special genus originated in the marine region; but in general these forms are but little different from their fresh-water relations. Strangely enough the flowering plants occupy a special position. Not only have the Potamogetonaceae and Hydrocharitaceae reached the fully marine region, they form extensive stands in it and have developed specific genera e. g. *Zostera, Posidonia, Diplanthera, Cymodocea* and others which behave like true marine organisms.

Summary

All groups of fresh-water organisms, animals as well as plants, are highly sensitive to salt water. The reduction in numbers of species is so rapid that already in the oligohalinikum only a fraction of the species is able to survive; in the mesohalinikum fresh-water species are rare and only a few species transgress the 8⁰/₀₀ boundary.

The tolerance of increasing salinity varies from group to group. Intolerance is shown by e. g. the Thecamoebae, the fresh-water Ciliata, the fresh-water Rhabdocoela among the Turbellaria, the fresh-water bivalves, among aquatic insects the Ephemerida and Plecoptera in particular, somewhat less the Odonata. Among plants the Desmidiaceae are especially fresh-water stenohaline.

A greater tolerance of salinity changes is displayed by the Cyanophyceae, among animals chiefly the Oligochaete families of Enchytraeidae, Tubificidae, and the aquatic Diptera larvae. These are mainly groups which have evolved from semi-terrestrial species or are able to live in most soil.

The number of immigration lines varies considerably from group to group. The Oligochaeta, Rotifera, Diptera larvae, Chaetonotidae send many lines from fresh water into the sea; only very few the fresh-water molluscs, Crustacea, Arachnoidea.

Most lines of immigration only lead to the formation of new races and species in brackish water which in their organization still resemble their fresh-water relations. A richer phylogenetic development in the marine region is displayed by the Hirudineae, Cladocera, Oligochaetae, Diptera larvae and, in particular the Halacaridae (marine mites) which have probably entered salt water from the moist soil. Among the families of flowering plants the Potamogetonacae and Hydrocharitaceae have undergone special development in the sea.

Ecologically the fresh-water immigrants can be divided into a thalassic and a littoral-continental group. The latter is confined to shallow coastal stretches, brackish pools, near the margin and inland salt pans. They comprise the majority of the fresh-water immigrants. The thalassic group penetrates into the sea itself; many Cladocera (Polyphemidae) and Rotifera *(Synchaeta, Trichocerca)* into the plankton; the Halacaridae into all regions of the benthos and phytal zone and the Hirudineae *(Ichthyobdella)* are ectoparasites.

Regions of large salinity fluctuations present no obstacle to the invasion of euryhaline fresh-water animals; on the contrary, they are densely populated. After crossing the brackish-water barrie, fresh-water immigrants frequently reach regions of extremely high salt content.

D. Penetration of marine animals and plants into brackish water

As shown by the curve of diminishing number of species (p. 19) marine organisms invade brackish water with a relatively slow and uniform reduction of species and a few species almost reach the fresh-water boundary. For the sake of clarity this wide range has been subdivided into several stages (STAMMER 1935, REMANE 1941):

1) Stenohaline marine organisms or halobionts. They live in the range of full-strength sea water ($35—40^0/_{00}$) up to nearly $30^0/_{00}$. They occupy the euhaline region and have been designated "Euhalobien" by KOLBE. I wish to alter my previous demarcation of this group at $25^0/_{00}$ S and put the lower limit for stenohaline marine animals at $30^0/_{00}$ S, bringing it into line with the classification of salinity grades (p. 5).

2) Euryhaline marine organisms or halobionts of the 1 st degree (I). They range from the sea into salinity ranges between 30 and 18 ($15^0/_{00}$).

3) Euryhaline marine organisms of the 2nd degree (II) = pleiomesohaline marine animals (STAMMER). They occur from the sea into the pleiohaline mesohalinikum, that is into the 15 (18) to $8^0/_{00}$ S range.

4) Euryhaline marine organisms of the 3 rd degree (III) = meimesohaline marine animals (STAMMER). They penetrate from the sea into the $8—3^0/_{00}$ S range.

5) The interesting marine organisms which can survive in the oligohalinikum that is beyond $3^0/_{00}$ S may be called oligohaline marine organisms or euryhaline marine animals of the 4th degree.

The number of stenohaline marine animals is vast. It is important to note that not only many species, but whole groups of animals are stenohaline. These are, among Protozoa, the class of Radiolaria with several thousand of species; the class of Hexactinellida (Triaxonia) among the Porifera; among Cnidaria the Siphonophora in the pelagic region, in the benthos the stony corals (Madreporaria) and most of the Alcyonaria, Gorgonaria, i. e. most of the Octocorallia, as well as the Ptero-branchia, Pogonophora; among Crustacea the Hoplocarida (*Squilla*) among the Chelicerata the Xiphosura *(Limulus);* among the molluscs for example the Scapho-poda and Solenogastra. Among the Echinodermata the Crinoidea are restricted to fully-marine regions. Among the Chordata the Salps (Thaliacea) are stenohaline marine. Much larger than the number of classes and orders mentioned above is the number of families which are stenohaline marine animals such as among Echino-dermata, teleosts etc.

Plants differ in their behaviour form the animals. There is no class of algae re-stricted to full-strength sea water and any series that might be termed stenohaline come from groups poor in species.

There are large numbers of euryhaline marine animals of the 1st degree. The Belt Sea (Kiel Bay) has about 200 species of Copepoda, over 300 species of Nematoda, about 100 species of Polychaeta which mainly belong to this group. But even in this area (up to about $15^0/_{00}$ S) some fairly large groups reach their boundary; the Cal-cispongia among the Porifera, almost all the Actinaria among the Cnidaria, as well as Lucernida and especially the Pennatularia; the Leptostraca *(Nebalia)* among the Crustacea, the Cyclostomata among the Bryozoa, the Echinoidea, Ophiuroidea, almost all the Asteroidea (only *Asterias rubens* somewhat transgresses this barrier in our region) and Holothuria, the Enteropneusta; among molluscs the Solenogastra, Cephalopoda and Placophora (Chitonidae), among Chordata the sea squirts (As-cidiae) and Acrania *(Branchiostoma)*.

Euryhaline marine animals of the 2nd degree penetrate into typical brackish water; their ecological limit is between 15 and $8^0/_{00}$. Compared with the previous group the number of their species has been considerably reduced. Of the molluscs of the Baltic the following belong here: *Littorina littorea* $9.5^0/_{00}$, *Lacuna divaricata* $8—9^0/_{00}$, *L. pallidula* $13^0/_{00}$, *Zippora membranacea baltica* $8^0/_{00}$, *Rissoa inconspicua* $8—10^0/_{00}$, *Brachystomia rissoides* $10^0/_{00}$, *Palio nothus* $14^0/_{00}$, *Scrobicularia plana* $10^0/_{00}$, *Mya truncata* $10—11^0/_{00}$, *Syndosmya alba* $12^0/_{00}$, *Cardium fasciatum* $14^0/_{00}$, *Aloides (Corbula) gibba* $14^0/_{00}$. (Figures after S. JAECKEL jun., 1952).

An Actinian *Halcampa duodecimcirrata* somewhat transgresses the $8^0/_{00}$ S limit; in some brackish areas *Actinia equina* also occurs in this range. Several genera of Polychaeta inhabit this region and may form stands. Compared with the previous region the number of species has been enormously reduced. The genera *Nephtys*, *Aricidea suecica*, species of *Rhodine*, *Harmothoe imbricata*, *Pholoe minuta* and others have their boundary in this area. Among the microfauna there are still very many marine species of this type, e. g. among the Copepoda, Ostracoda.

The euryhaline marine animals of the 3rd degree are more interesting since they are marine outposts in regions of low salinity (see p. 98 on brackish-water animals).

Among Protozoa the number of Foraminifera is still fairly high (see HOFKER 1922, RHUMBLER 1935, ROTTGARDT 1952). According to LUTZE's (1965) thorough study of the Baltic Foraminifera the following species transgress the $8^0/_{00}$ boundary: *Jadammina polystoma, Trochammina inflata, Miliammina fusca, Cribrononion astelundi* (often called *Nonion depressulum*). *Cr.* cf. *alvarezianum, Cr. excavatum excavatum*

(= *Polystomella striatopunctata*). Some of these species reach about 2⁰/₀₀ S. In warmer regions the number of these euryhaline Foraminifera appears to be even greater. ROTTGARDT (1952, p. 215) stresses: „Kalkschalige Foraminiferen können also unter dem Einfluß wärmerer Temperaturen eine größere Abnahme des Salzgehaltes, größere Salzgehaltsschwankungen ertragen und bedeutend weiter in fast limnisches Gebiet vordringen" [under the influence of higher temperatures Foraminifera with a calcareous shell are able to tolerate a greater lowering of salinity, larger salinity fluctuations and penetrate much further into an almost fresh-water region]. (Fig. 32).

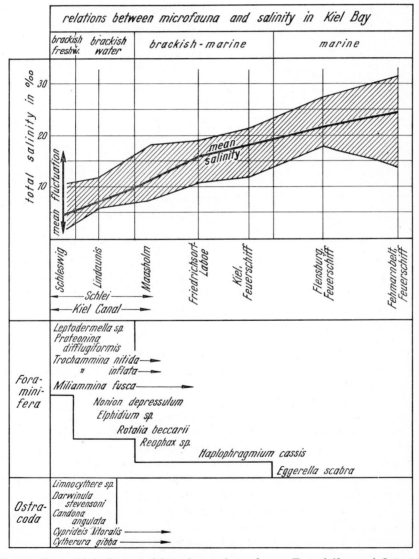

Fig. 32. Characteristic forms and boundary regions of some Foraminifera and Ostracoda in the area of Kiel Bay. After ROTTGARDT 1952.

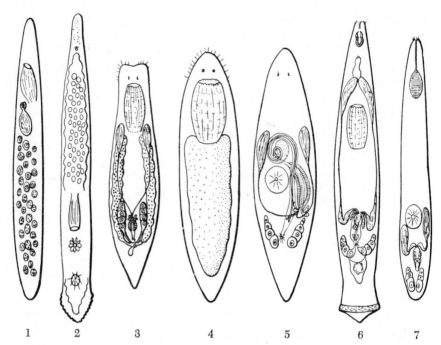

Fig. 33. Typical, strongly euryhaline marine Turbellaria. 1, *Archimonotresis limophila;* 2, *Monocelis lineata;* 3, *Provortex balticus;* 4, *Pseudograffilla arenicola;* 5, *Promesostoma marmoratum;* 6, *Baltoplana magna;* 7, *Placorhynchus octaculeatus.* Enlarged.

The number of euryhaline Ciliata of the benthos is very large, but more detailed studies are still lacking. MÜNCH and PETZOLD (1956) found a number of extremely euryhaline species in the low-salinity coastal ground water, e. g. *Uronychia transfuga* still at 1.6⁰/₀₀, *Prorodon moebiusi* up to 19⁰/₀₀ S.

In the cultures of Ax and Ax the following marine species were found at 3⁰/₀₀ S: *Placus socialis, Trachelocerca phoenicopterus, T. fusca, Cristigera setosa, Pleuronema coronatum, Dysteria neesi, Blepharisma clarissimum, Euplotes balteatus, Prorodon morgani.* The pelagic Tinitinnae a typically marine groups of Ciliata with only a few outposts in fresh water *(Tintinnidium fluviatile, Tintinnopsis lacustris)* are still of significance in the meiomesohaline plankton and, in part, in the oligohaline one e. g. *Tintinnopsis tubulosa, T. beroidea, T. campanula, Helicostomella subulata.*

Some Porifera of marine character have been recorded for Lake Chilka in India which has a reduced salinity at times *(Laxosuberites)* (ANNANDALE); but it is difficultat to classify them. In northern Europe marine sponges no longer occur in the meio-mesohalinikum. But here the two polyps *Clava multicornis* and especially *Laomedea loveni* invade these regions. The common jellyfish *Aurelia aurita* still occurs here with polyp and medusa and the Ctenophore *Pleurobrachia pileus* is euryhaline of the 2nd degree.

There is a rich fauna of marine Turbellaria in low-salinity areas. According to a communication by P. Ax there are in the Baltic region alone over 50 marine Tur-bellaria of the 2nd and 3rd degree of euryhalinity. From about 12⁰/₀₀ onwards only the order of Polycladida is absent; the Acoela are represented with about 9 species. The bulk of marine euryhaline species is provided by the Seriata (*Coelogynopora biarmata, Bothriomolus balticus, Monocelis lineata* and others) and the Rhabdocoela (*Provortex balticus, P. karlingi, Baicalellia brevitubus, Pseudograffila arenicola, Proxenetes flabellifer, Promesostoma marmoratum, Pr. baltica*); the Kalyptorhynchia so sparsely represented in fresh water are abundant as far as the coast of Finland. In the Meiomesohalinikum and in part in the oligohaline region the following occur: *Polycystis crocea, Acrorhynchus robustus, Cicerina brevicirrus, Paracicerina maristoi, Uncinorhynchus flavidus, Neognathorhynchus lobatus, Baltoplana magna, Thylacorhynchus conglobatus, Pseudomonocelis cetinae, Paramonotus schulzei* and others. Studies by A. LUTHER, KARLING and AX have shown that an unexpectedly rich fauna of marine Turbellaria inhabits the brackish waters of Finland. In the southern brackish waters (France, Black Sea), too, the Turbellarian fauna is uncommonly rich.

The Rotifera also invade slightly brackish waters with a few euryhaline marine animals of the 3rd degree. This applies particularly to pelagic species, e. g. *Synchaeta baltica, S. triophthalma, Trichocerca marina.* According to a communication by S. GERLACH the following euryhaline marine Nematoda still inhabit the oligohaline region: *Anoplostoma viviparum, Visciosa viscosa, Paracanthonchus caecus, Hypodontolaimus balticus, Ascolaimus elongatus, Tripyloides marinus, Eleutherolaimus stenosoma* and *Theristus setosus* — unless the latter is holeuryhaline. The Gastrotricha Macrodasyoidea enter this zone with *Turbanella cornuta T. hyalina, Dactylopodalia baltica.* The Polychaeta have relatively few strongly euryhaline marine animals: *Neanthes (Nereis) diversicolor, Harmothoe sarsi, Polydora ciliata* attain the boundary meso-

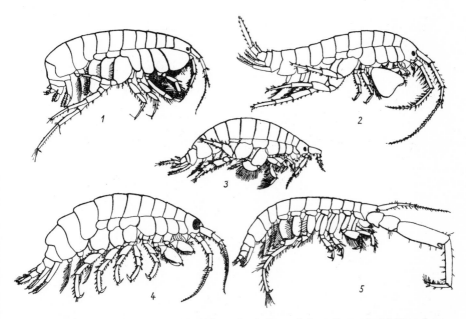

Fig. 34. Some Amphipoda from brackish water. 1, *Leptocheirus pilosus;* 2, *Melita palmata;* 3, *Bathyporeia pilosa;* 4, *Calliopius laeviusculus;* 5, *Corophium volutator.* Enlarged. From SARS.

halinikum oligohalinikum; *Fabricia sabella* (5—6⁰/₀₀) and *Pygospio elegans, Terebellides stroemii, Stygocapitella subterranea, Diurodrilus subterraneus* extend into the miomesohalinikum.

There is also a scarcity of such forms in the Nemertina (*Lineus ruber*).

Many groups of marine Crustacea provide a number of species in the regions below 8⁰/₀₀ S. In the North Sea and Baltic the Decapoda are poorly represented: *Crangon crangon, Leander adspersus.* In warmer countries the strongly euryhaline marine Decapoda are more numerous. There crabs like *Callinectes sapidus* even invade lagoons of low salinity and, above all Peracarida penetrate oligohaline waters where they provide a considerable catch. Strongly euryhaline Peracarida are far more nu-

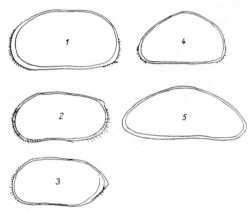

Fig. 35. Outline of the left shell of strongly euryhaline marine Ostracoda. 1, *Cyprideis litoralis* (conditionally euryhaline); 2, *Leptocythere castanea;* 3, *Cytherura gibba;* 4, *Xestoleberis aurantia;* 5, *Cytherois fischeri.* Enlarged. From SARS.

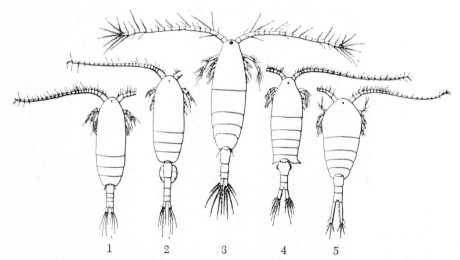

Fig. 36. Common euryhaline planktonic Copepoda (Calanoidea) from brackish water. 1, *Paracalanus parvus;* 2, *Pseudocalanus elongatus;* 3, *Acartia longiremis;* 4, *Centropages hamatus;* 5, *Temora longicornis.* Enlarged. From SARS.

merous, but cannot always be distinguished from true brackish-water species. The following nearly reach the fresh-water region (1—3⁰/₀₀): the Amphipoda *Corophium volutator, Gammarus locusta, G. oceanicus* (SEGERSTRÅLE 1959), the Isopoda *Jaera ischiosetosa* and *J. praehirsuta* (cf. HAAHTELA 1965), the Mysidae *Neomysis integes (= vulgaris)*; up to about 5—6⁰/₀₀ *Idotea baltica, I. granulosa, Jaera albifrons syei, Calliopius laeviusculus, Praunus flexuosus, Pontoporeia femorata.*

The following are strongly euryhaline: the Cirripede *Balanus improvisus,* the Ostracoda (fig. 35) *Heterocyprideis sorbyana, Leptocythere pellucida, L. lacertosa!, L. baltica, L. castanea, Cytherura nigrescens!, L. impressa, Hirschmannia viridis, Xestoleberis aurantia?, Cytherois fischeri?, Paradoxostoma variabile* (after ELOFSON 1941). The species with ! attain 5—7⁰/₀₀, the others 2—3⁰/₀₀ S. NOODT (1953) records *Loxoconcha pusilla* and *Cytherois arenicola* from the Finnish coast (5—7⁰/₀₀). A large number of marine Copepoda is euryhaline of the 3rd degree. The following are planktonic: the Cyclopid *Oithona similis,* the Calanoidea *Centropages hamatus, Acartia bifilosa, A. longiremis, Temora longicornis, Pseudocalanus elongatus,* to a lesser extent *Paracalanus parvus* (fig. 36); in the phytal and benthic zones: *Ectinosoma curticorne, Pseudobradya minor, Arenosetella germanica, Stenhelia palustris, Amphiascoides debilis, Nitocra typica, N. fallaciosa, Paramesochra holsatica, Remanea arenicola, Paraleptastacus spinicauda, Arenopontia subterranea, Huntemannia jadensis, Paronychocamptus nanus, Platychelipus littoralis* (see figs. 37, 38 and p. 132). Among the Cyclopidae a few species of *Cyclopina* enter this region (NOODT 1955).

The majority of marine pelagic Cladocera still occur in the meiomesohaline and, in part, oligohaline regions such as *Podon intermedius* up to 3.5⁰/₀₀, *P. leuckarti* up to 6.1⁰/₀₀, *Evadne nordmanni* up to 1.3⁰/₀₀, *E. spinifera* up to 8.5⁰/₀₀, *P. polyphermoides* has even been observed in fresh water locally (see RAMNER, T. N. O.).

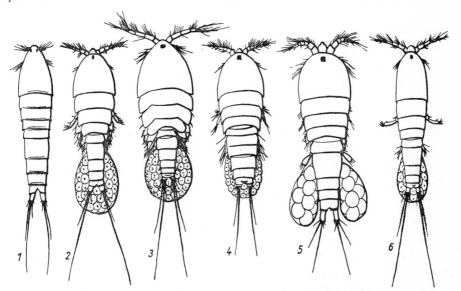

Fig. 37. Common, euryhaline Copepoda-Harpacticoidea from brackish water. 1, *Ectinosoma curticorne;* 2, *Tigriopus fulvus* (characteristic form of saline rock pools); 3, *Idyaea furcata;* 4, *Dactylopusia vulgaris;* 5, *Stenhelia palustris;* 6, *Harpacticus chelifer.* Enlarged. From SARS.

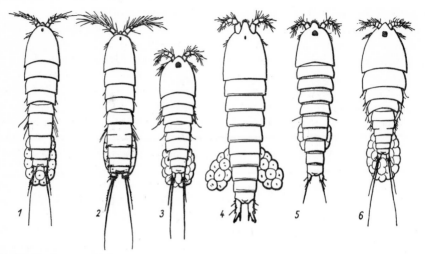

Fig. 38. Further euryhaline Copepoda-Harpacticoidea from brackish water. 1, *Mesochra lilljeborgi;* 2, *Nitocra typica;* 3, *Laophonte nana;* 4, *H. ntemannia jadensis;* 5, *Nannopus palustris;* 6, *Tachidius brevicornis.* Enlarged. From SARS.

A fairly large number of Halacaridae are found below 8⁰/₀₀: *Halacarellus basteri, H. balticus, H. floridearum, H. inermis, H. anomalus, Copidognathus oculatus, C. brevirostris, Rhombognathopsis seahami.*

The marine molluscs show a considerable decline in numbers in the Meiomesohalinikum. The following reach their boundary in the 8—6⁰/₀₀ S area: the bivalves *Astarte borealis, Macoma calcarea, Cardium exiguum,* the Prosobranchia *Zippora membranarea* f. *baltica, Littorina saxatilis* f. *tenebrosa, Hydrobia ulvae* and the Opisthobranch *Stiliger mariae;* the following penetrate into the 3—5⁰/₀₀ area: the bivalves *Mytilus edulis, Macoma baltica, Mya arenaria* and *Cardium edula* and the Opisthobranchia *Embletonia pallida, Pontolimax capitatus* and *Alderia modesta.*

Up to about 5⁰/₀₀ S marine fishes provide the largest number of species with the exception of lagoons and shore pools where the euryhaline fresh-water species frequently predominate. Specially euryhaline groups are: the Syngnathidae with *Syngnathus typhle, Nerophis ophidion,* the Pleuronectidae with *Rhombus maximus, Pleuronectes platessa, Pl. flesus, Pl. limanda,* species of *Ammodytes, Gadus morrhua, Belone, Spinachia, Zoarces viviparus, Centronotus gunellus, Cyclopterus lumpus, Liparis vulgaris, Gobius niger, G. minutus, G. flavescens, Cottus scorpius, C. bubalus,* the herring etc. These species, strongly euryhaline in the Baltic, also occur in southern brackish waters or are represented there by related species (Syngnathidae, Gobiidae, Clupeidae). In addition there are euryhaline species from other families, e. g. Mugilidae, Tetrodontidae and in warm regions some Selachia, especially rays of the family Trygonidae; they are common in brackish water and even invade fresh water.

Low-salinity lagoons on the coast of Lousiana have a salinity of 0.2—2.7⁰/₀₀ on the whole, while there is a local and temporary rise to 4⁰/₀₀ (Little Bay); GUNTER and SHELL (1958) found a strongly marine fauna of fishes there. There were 15 marine species as compared with 10 fresh-water ones. In the marine species the number of individuals was considerably higher than in the fresh-water ones. The marine euryhaline species were *Elops saurus, Anchoa mitchelli, Brevortia patronus, Galeichthys felis, Bagre marine, Mugil cephalus, Membros martinica, Pogonias cromis, Micropogon undulatus, Cymoscion arenarius, Leiostomus xanthurus, Trinectes maculatus, Menidia beryllina, Polydactylus octonemus, Citharichthys spilopterus.*

It is interesting that the majority of euryhaline fishes do not spawn in higher saline regions, but that reproduction takes place even at low salinity.

While the fresh-water euryhaline fishes migrate into low-saline areas for spawing, the marine fishes of the brackish water reproduce in that region. In some cases, e. g. the herring, the eggs and juvenile stages have a particularly wide salinity range. Fertilisation, development and hatching take place in the range of 5.9—52.5 ^{0}/$_{00}$ S.

Plants. In contrast to the animals, marine plants have no large stenohaline marine groups; at least a few species of all the bigger units enter the meso- or oligohaline regions.

The marine Peridinaea and Diatomaceae (e. g. *Cocconeis scutellum, Coscinodiscus excentricus*) have a good number of strongly euryhaline species. The typical algae are also well represented. They reach the meiomesohaline and, in part, oligohaline regions. According to LEVRING and WAERN the position in the Baltic is as follows:

Red Algae

    a) as far as the Arkona-Bornholm Sea (about 6—8 ^{0}/$_{00}$): *Delesseria sanguinea, Polysiphonia urceolata, Plumaria elegans, Chantransia hallandica, Ch. humilis, Ch. efflorescens, Chondrus crispus, Rhodymenia palmata;*

    b) into the southern Gotland Sea, including Gotland: *Phycodrys sinuosa, Harveyella mirabilis, Polyides rotundus;*

    c) into the northen Gotland Sea, the Alands Sea: *Ahnfeldtia plicata, Audouinella efflorescens, Callithamnion roseum, Ceramium arborescens, C. rubrum, Rhodomela subfusca;*

    d) into the Botten Sea or the central parts of the Gulf of Finland: *Bangia fuscopurpurea, Rhodochorton Rothii, Hildenbrandia prototypus, Furcellaria fastigiata, Phyllophora Brodiaei, Polysiphonia violacea, P. nigrescens.*

It can be seen from this list that a relatively high number of Red Algae enters the brackish-water areas of 4—6 ^{0}/$_{00}$ S.

Brown Algae

    a) into the Arkona-Bornholm Sea: the Seaweeds *Laminaria saccharina, Desmarestia viridis;* also *Streblonema aequale, Phaeostroma pustulosum, Ectocarpus penicillatus, E. tomentosus, Hecatonema terminale, H. reptans, Myrionema strangulans, Ralfsia verrucosa;*

    b) into the southern Gotland Sea including Gotland: *Fucus serratus, Myrionema balticum, M. globosum, Sphacelaria cirrosa, Leptonema fasciculatum, Stilophora rhizodes;*

    c) into the northern Gotland Sea, Alands Sea, south coast of Finland and the Gulf of Riga: *Halopteris scoparia, Waerniella lucifuga, Microsyphar polysiphoniae, Streblonema oligosporum, Petroderma maculiforme, Lithoderma Rosenvingii, Ascocyclus magni, Elachista fucicola, Myriotrichia repens;*

d) into the Botten Sea and the inner parts of the Gulf of Finland: *Fucus vesicu-losus* (up to about 2º/₀₀), *Chorda filum* (up to about 2º/₀₀), *Ch. tomentosa, Sphacelaria radicans, Sph. plumigera, Sph. arctica, Pylaiella rupincola, Ecto-carpus confervoides* (up to 2º/₀₀), *Stictyosiphon tortilis, Dictyosiphon foenicu-laceus* (up to about 2º/₀₀), *D. chordaria* (up to about 3º/₀₀).

The Green Algae are quite tolerant of salt. Of the larger forms *Ulva lactuca* reaches into the Arkona Sea, *Chaetomorpha linum, Monostroma Grevillei* into the southern Gotland Sea. Species of *Enteromorpha* and numerous forms growing in the spray-zone (Geolittoral), or else on or in other plants, reach into the meiomeso-halinikum or oligohalinikum.

The marine flowering plants are, in part, only weakly euryhaline. It is interesting that *Zostera nana* which in the shore region reaches much higher into the eulittoral than *Zostera marina*, attains its limit much sooner than *Zostera marina*; the latter occurs up to about 3º/₀₀ S.

To sum up, the marine organisms contain a number of stenohaline orders and classes. This applies mainly to the animal kingdom, hardly to the plants. In addition there are many euryhaline groups and species which transgress the middle region of 17—18º/₀₀ S over a wide front, with a fairly uniform reduction in the number of species. In many habitats the marine species are still dominant at 5—6º/₀₀ S. That holds true especially for the fauna of the sand and the coastal ground water, but for the rocky regions as well: numerous marine organisms still attain the bound-ary of the oligohalinikum at 3º/₀₀, a number penetrate even further, e. g. the sea-weeds *Fucus vesiculosus, Chorda filum,* the Myside *Neomysis integer,* the Isopoda *Jaera ischiosetosa* and *J. praehirsuta,* the Amphipod *Gammarus oceanicus.*

On the whole the smaller forms tend to be more euryhaline than the large ones. This is striking, for example, in the Turbellaria; their large representatives (Poly-clada) are far less euryhaline than the small Acoela, Rhabdocoela etc. Similarly, the large Crustacea (Decapoda, Amphipoda, Isopoda, Cirripedia) show a far more pronounced reduction in areas of lower salt content than the Copepoda and Ostra-coda. This is the opposite to what is found in fresh-water organisms in which ap-parently larger species of a group (molluscs and others) are more euryhaline than smaller ones.

Several groups of marine organisms achieved a transition into fresh water in the recent geological past (that is Tertiary-Pleistocene). The new immigrants can be distinguished from the old ones (Rotifera, Phyllopoda, Ostariophysi) by the small number of their species found in fresh water and the close relationship of these fresh-water species with marine forms. Most of them still belong to the same genus as their nearest marine relations. In part, these new immigrants enable us to detect the ecological routes along which the transfer of marine organisms into fresh water took place. Rivers come to mind as migration routes in the first place; but these allow immigration only to those organisms which can make headway against the current. Presumably a number of fishes, higher Crustacea (Malacostraca, especially Decapoda) and Gastropoda have taken this route. At first it frequently results in migratory forms returning to the sea or brackish water for spawning. Even more commonly than by fishes such behaviour is displayed by the decapod Crustacea, e. g. the mitten crab *Eriocheir,* as well as by American river Palaemonidae, e. g.

Cryphiops and apparently *Macrobrachium*. According to HARTMANN (1957b) the South American *Cryphiops caementarius* may be able to spawn in fresh water.

To a large extent transition into fresh water may be accomplished by a gradual freshening of marine waters, provided the decrease in salt content takes place very gradually. Two such cases are known in detail: the Baltic area with its "relict forms" in fresh water (see p. 141) and the Pontocaspian region in which a particularly large number of successful transitions have been achieved (see p. 148). In such parts of relict seas transfer becomes possible for all groups, not only the large and actively mobile ones. Adaptation to life in rivers, to fresh-water lakes or small bodies of water may take place. Cut-off lagoons which have become fresh-water lakes offer another possibility of transfer from the sea into fresh water. This is borne out by records of marine Ostracoda (HARTMANN 1959) and Nematoda (GERLACH) in Lake Nicaragua, of marine Copepoda in Lake Iznik (NOODT 1954) and others.

Isolated lakes and saline pools along the shore are a third type of ecological migration route. Even in their saline phase colonization presupposes a tolerance of large temperature fluctuations, of occasional O_2-deficiency and strong insolation. In contrast to the migratory route in the opposite direction from fresh into salt water, clear examples of organisms which have used this migration route from the sea into fresh water are not yet available; there is some probability that this path has been taken by the Polychaete *Manayunkia*, some Copepoda and Ostracoda (Cytheridae), possibly also by *Hydra* and Hydrobiae.

There is ample evidence for the migration routes leading from the marine sand into subterranean fresh-water areas: One of these, for example, leads from inhabitants of the marine interstitial system (interstitial fauna) to forms living in the interstices of river sands. P. AX (1957) has been able to show that the "fresh" sands of the banks of the river Elbe contain numerous marine elements far beyond the brackish-water region; the fresh-water Otoplanidae have undoubtedly taken the route marine sand — river sand. Not only the sands of the banks, but also the shifting sands of the rivers are colonized by species of marine origin. RIEMANN (1965) found several Turbellaria Proseriata in the fresh-water region of the mouth of the Elbe: *Bothriomolus balticus, Paramonotus hamatus,* distinct species of *Pseudosyrtis* and a few Nematoda of marine character *(Theristus heteroscanicus)*.

Another route leads direct from the marine region into the interstitial coastal ground water (see p. 175) which is often in contact with subterranean fresh water. Polychaeta of the family Nerillidae (*Troglochaetus* in ground water) and others have probably taken this route. Many small Crustacea, e. g. the Isopoda *Microcerberus, Microcharon*, the Amphipod *Bogidiella* whose occurrence in subterranean fresh water had been inexplicable for a long time, have now proved to be fresh-water outposts of the subterranean fauna of the sea shore, where close relatives occur in large numbers and are widely distributed (see p. 174). KARAMAN's hypothesis that these marine elements in subterranean fresh water owe their origin to a direct immigration from the sea into the ground water has proved to be correct. Presumably Copepoda, Ostracoda (Kliellinae!) and in part, the groups of *Niphargus*, have immigrated along this route — marine sand — marine coastal ground water — fresh-water ground water.

The problem as to how far larger forms have penetrated from surface marine areas into subterranean fresh water by means of connecting fissure waters in Karst areas (e. g. the tubicolous Polychaete *Marifugia cavatica* see THIENEMANN 1950) will not be discussed here.

Immigration of marine organisms into fresh water has been achieved by a variety of ecological routes.

Resistance to concentration and dilution are generally combined. So far the degrees of steno- or euryhalinity have only been determined from sea water of $35^0/_{00}$ into brackish water. It is remarkable that species which are euryhaline in that sense are at the same time highly resistant in the opposite direction, that is in concentrations beyond $35^0/_{00}$ S, e. g. *Cardium edule*.

In the etangs of the south of France P. Ax (1956) found several euryhaline Turbellaria which occurred as far as 12—$16^0/_{00}$ S, as well as in $49^0/_{00}$ S, e. g. *Monocelis lineata, Promesostoma gallicum, Pseudograffilla arenicola, Convoluta schultzii*.

MARS (1950) has recorded the euryhalinity of some molluscs from the coast of the south of France according to their occurrence in etangs of varying salinity, by giving a mean value between the extreme records as well as deviations from this mean, both upwards and downwards. Thus he found for *Cardium edule* 32 ± 28 (that is ranging from 4—$60^0/_{00}$), for *Brachydontes marioni* 24 ± 18, *Rissoa labiosa* 23 ± 13, *Mytilus galloprovincialis* 25 ± 13, *Loripes laeteus* 28 ± 15, *Gibbula adamsoni* 27 ± 12, *Ocinebra erinaceus* 30 ± 9, *Murex trunculus* 32 ± 6, *Rissoa ventricosa* 34 ± 4, *Muricopsis blainoillei* 37 ± 2.

In his studies of the Laguna Madre HEDGPETH (1947) reports a number of species which combine resistance to concentration and to dilution. For the medusae salinity ranges were *Aurelia aurita* 15—60, *Nemopsis bachei* 22—78 (in Europe both species attain much lower values, about 7—8), *Stomolophus meleagris* 10—60, *Gonionemus murbachi* 20—60, the Crustacea *Penaeus aztecus* 5—75, *P. duorarum* 5—70, *Callinectus sapidus* 3—60, the fishes *Elops saurus* 3—75, *Anchoa hepsetus* 3—80%, etc.

This combined resistance to concentration and dilution in marine animals is hardly universal, but it is remarkably common.

E. Species confined to brackish water (Hyphalmyrobies)

The existence of animals and plants which thrive in brackish water, but are not found in the sea or fresh water has long been known. These species have a boundary for the two environments of sea and fresh water, they are genuine brackish-water organisms (VÄLIKANGAS) = specific B.

From the above remarks it is easy to define specific brackish-water organisms. But there are often great practical difficulties in establishing the fact whether a species actually belongs to this group. Quite a few of those usually designated as brackish-water species occasionally inhabit fresh water as well such as *Cordylophora caspia*.

In order to include such cases in the definition, the narrower concept (absolute boundary of occurrence towards euhaline sea and fresh water) will have to be supplemented by a wider definition which would read: Brackish-water species are those which abound in brackish water and occur only occasionally in the sea or fresh water (see REMANE 1934).

But the practical application of this wider definition also presents difficulties beyond those discussed on p. 52.

a) Halophye species

A number of typical fresh-water organisms may occur in abundance in weak brackish water. They seem to thrive particularly well in this region for their density per area or unit volume reaches specially high values. VÄLIKANGAS (1926, p. 200) in particular, drew attention to this phenomenon in his studies of the plankton in Finnish waters: „Die große Mehrzahl der als Süßwasserformen angeführten Organismen zeigt trotz ihrer Süßwassernatur eine weit stärkere Entwicklung in

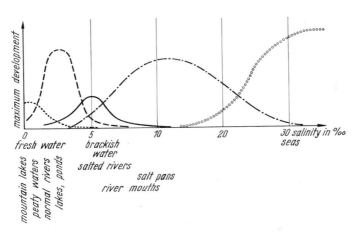

Fig. 39. Types of diatoms differing in their behaviour to salt content. From left to right: fresh-water species, halophye species, oligohaline brackish-water species, mesohaline brackish-water species. "Marine" species. After KOLBE 1932.

salzreicherem Wasser. So lagen die Grenzen der Hauptentwicklung bei *Oscillatoria Agardhii* zwischen etwa 3 und 5,35⁰/₀₀, bei *Richteriella* zwischen etwa 3,1 und 3,5⁰/₀₀, bei *Euglena* zwischen 4 und 5,25⁰/₀₀, bei *Cyclotella laevissima* zwischen (2,8) 4,2 und 5,2, bei *Nitzschia acicularis* etwa zwischen 3,0 und 4,8. Ähnlich verhielten sich die in der Tabelle nicht aufgenommenen *Anabaena spiroides, Scenedesmus obliquus* und *S. quadricauda, Cryptomonas* u. a. Bei den Zooplanktonen waren die Verhältnisse analog, so die entsprechenden Zahlen bei *Tintinnidium fluviatile* etwa 3,45—4,95, bei *Floscularia* etwa 3,65—4,8, bei *Asplanchna Brightwelli* 4,0—4,2, bei *Triarthra longiseta* 4,0—5,4 (absol. Maximum um etwa 4,0—4,2⁰/₀₀) und *T. brachiata* um 3,4⁰/₀₀, bei *Brachionus pala* 3,2—4,2 (bzw. 5,25) und *B. angularis* 3,3—4,1, bei *Anurea aculeata* (Süßwasserform) 3,5—5,45 und *A. cochlearis* (f. *minor*) etwa 2,3—4,0⁰/₀₀, um nur die wichtigsten Formen zu nennen.“ [In spite of their fresh-water character most of the organisms listed as fresh-water forms show far more vigorous development in saline water. The boundaries for their main

development were: for *Oscillatoria Agardhii* between about 3 and 5.35⁰/₀₀, for
Richteriella between about 3.1 and 3.5⁰/₀₀, for *Euglena* between 4 and 5.25⁰/₀₀,
for *Cyclotella laevissima* between (2.8) 4.2 and 5.2, for *Nitzschia acicularis* about
between 3.0 and 4.8. Similar behaviour was exhibited by *Anabaena spiroides, Scenedesmus obliquus* and *S. quadricauda, Cryptomonas* etc. which have not been included
in the table. Conditions for the zooplankton were analogous, the corresponding
figures being for *Tintinnidium fluviatile* about 3.45—5.95, for *Floscularia* about
3.65—4.8, for *Asplanchna Brightwelli* 4.0—4.2, for *Thriarthra longiseta* 4.0—5.4,
(absolute maximum about 4.0—4.2⁰/₀₀), and *T. brachiata* around 3.65—4.8, for
Brachionus pala 3.2—4.2 (5.25 resp.), and *B. angularis* 3.3—4.1, for *Anurea aculeata*
(fresh-water form) 3.5—5.45 and *A. cochlearis* (f. *minor*) about 2.3—4.0⁰/₀₀, to mention only the most important forms]. Kolbe (1932, p. 267) stresses the same phenomenon in his Ecology of Diatoms ,,Ein in biologischer Beziehung eigenartiges
Verhalten zeigen einige euryhaline Formen des Süßwassers. Sie werden durch geringe Salzmengen in ihrer Entwicklung stimuliert und erreichen an entsprechenden
Standorten unter Umständen enorme Individuenzahlen. Dabei kann man sie in
ökologischer Hinsicht auf keinen Fall etwa zu den mesohaloben Formen rechnen,
denn sie haben ihr Hauptverbreitungsgebiet im Süßwasser und gehören zu den
ständigen Bewohnern von Seen und anderen süßen Gewässern des Binnenlandes.
Zu solchen Formen rechne ich z. B. *Caloneis amphisbaena, Cyclotella Meneghiniana,
Diatoma elongatum, Epithemia sorex, Epithemia turgida, Gomphonema parvulum,
Navicula pusilla, Nitzschia inconspicum, Nitzschia microcephala* u. a.“ [Some euryhaline fresh-water forms display a peculiar biological behaviour. Small amounts
of salt stimulate their development and in appropriate localities they may sometimes attain enormous numbers of individuals. But from an ecological point of
view they can by no means be counted as mesohalobionts since their main area
of distribution is in fresh water and they are among the permanent inhabitants
of lakes and other continental fresh waters. I count among these forms e. g. *Caloneis
amphisbaena, Cyclotella Meneghiniana, Diatoma elongatum, Epithemia sorex, Epithemia turgida, Gomphonema parvulum, Navicula pusilla, Nitzschia inconspicum, Nitzschia
microcephala* and others].

In cultures with graded salinities Ax and Ax found that in several fresh-water
Ciliata 1⁰/₀₀ S had a stimulating effect on their rate of division. *(Metopus spiralis,
Frontonia leucas, Spirostomum ambiguum, Sp. intermedium, Paramecium.)* Kolbe
calls these fresh-water forms "halophilous". But as this term has been previously
used in a totally different sense I suggest the name "halophye" species. No detailed explanation of the behaviour of halophye species will be given (see fig. 39).

A similar phenomenon is found in marine organisms reaching their highest density
in salinities somewhat below that of sea water. This is particularly marked in the
region of river mouths and in lagoons with well circulating water (see p. 168). A similar situation obtains in shallow lagoons where an abundance of nutrients, partly
high temperature, and the reduced number of enemies allow some marine euryhaline
species to develop in large numbers in water of diminished salt content (the oyster
Crassostrea, the Cirriped *Balanus improvisus*, species of *Penaeus* etc., see p. 164).
In these marine "myophye" species there is a clear correlation with the greater supply
of nutrients in this region.

b) Regional brackish-water species

A number of species meet the requirements of true brackish-water forms only in part of their area of distribution, while they are truly fresh-water in other regions. This may be due to two causes. 1) In certain parts of the sea the particular environmental conditions confine the species to brackish water and prevent it from colonizing fresh water as it would normally do (Pseudohalobionts). 2) The species consists of several physiological races, one or a few these are restricted to brackish water in their distribut on.

1) Pseudohalobionts. LINDBERG (1948) uses this term for species „die (im nördlichen und mittleren Teil des Baltischen Meeres) in salzigem Wasser anzutreffen sind und hier in ihrer Lebensweise mit den Halobionten übereinstimmen. Schon im südlichen Teil des baltischen Gebietes wie auch in sonstigen Teilen ihrer Ausbreitungsgebiete sind es jedoch keine Halobionten, nicht einmal Halophilen, sondern typische Limnobionten" [which in the northern and central parts of the Baltic can be found in saline water where their way of life resembles that of the halobionts. But in the southern parts of the Baltic area and other southern regions of their distribution they are not halobionts, not even halophilous, but typical limnobionts]. LINDBERG includes the following insects: *Gerris thoracicus, Haliplus obliquus, H. confinis* race *pallens, H. immaculatus, H. apicalis, H. flavicollis, Noterus clavicornis, Coelambus parallelogrammus, Deronectes depressus* race *lutescens, Agabus nebulosus, A. conspersus, Enochrus melanocephalus* (*E. quadripunctatus* race *fuscipennis*). Among molluscs *Theodoxus fluviatilis* belongs here. A. LUTHER (1955) mentions the Turbellarian *Microdalyellia fusca,* the sponge *Ephydatia fluviatilis*; H. LUTHER (1951, p. 103) the plants *Potamogeton filiformis, Najas marina* and *Chara tomentosa,* all fresh-water organisms which avoid fresh water in Finland.

There may be different reasons for this pseudohalobiosis. It may be that temperature or a long period of freezing-over exclude the organisms from fresh water. It seems improbable, at least for insects, that the tolerance of fresh water should be physiologically reduced as a result of the temperature-salinity relation. A. LUTHER (1955, p. 108) assumes that such regional brackish-water species are prevented from colonizing the fresh water in Finland because of the more or less acid water, poor in electrolytes, from the Fennoscandian Precambrian rocks.

2) Brackish-water races of euryhaline species. The regional restricrion of some forms to brackish water is undoubtedly due to the existenc_ of distinct brackish-water races within euryhaline species. We know the importance of "origins" for many plants, for example in forestry; this proves that widespread species possess many "ecotypes" closely adapted to the special conditions of the locality; a euryoecious species may consist of a mosaic of stenoecious races or populations. The prawn *Palaemonetes varians* provides the best example. In western and northern Europe it is a true brackish-water animal, living in large numbers in small bodies of brackish water (brackish ditches, brackish lakes) on the sea shore. Its dependence on a certain salt content is proved by the fact that it disappears when the salinity of the water is lowered. On the island of Amrum it appeared in large numbers in brackish ditches after the dyking of a salt meadow; further freshening destroyed the stand completely. After the closing-in of the Zuiderzee it retreated to the brackish Northwest

corner (BEAUFORT 1953). In southern Europe, however, numerous populations thrive in pure fresh water (Lake Garda etc.); these differ from the brackish-water race by their larger eggs and a different type of development (f. *macrogenitor*). For a similar situation in *Gammarus duebeni, Gyratrix hermaphroditus* see p. 59.

By means of experiments on resistance PORA (1946) examined the survival in various salinities of the strongly euryhaline prawn *Leander squilla*. With a survival range from nearly 0 to about 40⁰/₀₀ this species is almost holeuryhaline in the Mediterranean and the Black Sea. A study of different populations from Naples, from the Bosphorus and, within the Black Sea, from Varna, Agigea, Sulina, Odessa and Lake Mangalia revealed that each of these populations had a limited range of tolerance which comprised only about ¼ of that of the species as a whole; the population from Naples for instance a range of 28—40⁰/₀₀, that from Varna from 18—30, the population from Odessa from 2—12⁰/₀₀ (fig. 40).

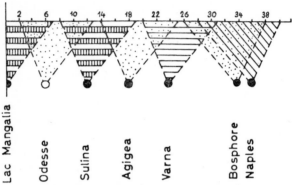

Fig. 40. Salinity tolerance of different populations of the Decapod *Leander squilla* in the regions of the Mediterranean and the Black Sea, as tested in resistance experiments. The figures in the upper margin denote salinity values, the names indicate the place of origin of the population. After PORA 1946.

No doubt such cases are not uncommon. As mentioned before, according to experiments by GRESENS the fresh-water polyp *Pelmatohydra oligactis* has its upper limit of permanent viability at 3⁰/₀₀ in the fresh-water population from the Ryckfluss near Greifswald; but the brackish-water population of the Greifswalder Bodden has an upper limit of 8⁰/₀₀ and a lower of 4⁰/₀₀. The areas of existence of the two populations were mutually exclusive! SEIFERT (1938, p. 230) reports that the brackish-water populations of the common jellyfish *Aurelia aurita* from the Baltic near Greifswald reproduce normally in the range of 4—16⁰/₀₀. At 16⁰/₀₀ „die Planulae von Aurelien aus dem Greifswalder Bodden entwickeln sich nicht mehr zu Scyphistomen" [Planulae of *Aurelia* from the Gr. B. no longer develop into Scyphistoma]. But *Aurelia aurita* occurs and develops quite normally not only in the North Sea, but also in the Red Sea at 40⁰/₀₀! In the brackish-water polyp *Cordylophora caspia*, too, the fresh-water populations appear to have a different range of reactivity, corresponding to the habitat, from those of the typical brackish-water populations (KINNE 1957). Where no genetical experiments are as yet available there always remains the possibility that these populations have individual adaptations acquired over a long time and these have not been reproduced in the resistance experiment. They may be dauermodifications. But the majority of cases are probably based on genetic differences. No doubt there exists a large number of similar cases. However, the genetic experiment is required for a clear-cut evidence of the existence of ecological races. This is available in a few cases, e. g. the Copepod *Tigriopus fulvus*; crossing experiments revealed a number of genetically very different races in the Mediterranean, while a uniform type colonizes the Atlantic coast from Spain to Norway (BOZIC 1956).

3) Regional restriction of halolenitobionts to brackish water. A wealth of species from the groups of Turbellaria, Copepoda, Ostracoda and Rotifera are confined to areas of calm water on the sea shore, that is shallow bays, pools etc. KUNZ (1938) designated them as littoral halolenitobionts.[9]

In humid areas such habitats are naturally brackish as the effects of precipitation reduce their salt content compared with that of the adjoining seas. The species living here have a boundary towards the sea and towards fresh water; thus, by definition, they are brackish-water organisms. But the boundary towards the sea may be determined by the varying conditions of the biotope and not by salt content. Only investigations in salt works or highly saline lagoons and salt pools of arid regions will decide whether an individual case is a true brackish-water species or a euryhaline halolenitobiont. P. Ax (1951, 1956) has carried out such comparative studies on Turbellaria from the coasts of the North Sea and Baltic on one hand, and the French Mediterranean coast on the other hand. He found that in the North Sea and the Baltic a number of species disappeared at a salinity boundary of about $15^0/_{00}$, e. g. *Tvaerminnea karlingi, Vejdovskya pellucida, V. ignava, Macrostomum spirale, Pseudomonocelis cetinae*. These are true brackish-water species. But others occurred in the Mediterranean in etangs and salt works at $30—49^0/_{00}$ S, e. g. *Monocelis lineata, Pseudograffilla arenicola, Promonotus schultzei;* they are euryhaline halolenitobionts. The same dichotomy is found in Copepoda, Diptera larvae (see REMMERT 1955), Ostracoda (*Cyprideis litoralis* p. 58).

4) Eulittoral brackish-water species. The littoral brackish-water organisms bear some resemblance to the previous group. They are also linked with it by transitional forms (*Cyprideis litoralis*). These are species which in the sea itself have their sole or main development in the tidal area, but in brackish water they advance much further ecologically and penetrate to greater depths (*Mytilus edulis, Nereis diversicolor* etc.). The main representatives of whole communities, e. g. the *Macoma-baltica* association and the *Halammohydra* association are eulittoral brackish-water species. If one adopts the wider definition of brackish-water organisms, they might be counted among them. In the sea they colonize a narrowly delimited area which frequently has a somewhat reduced salinity compared with the open sea, while they are almost universally distributed in brackish water, mostly with large numbers of individuals. The classification of these species is quite difficult.

---

[9] The term halotelmatobionts coined by SCHÄFER (1936) is narrower. It only includes the inhabitants of marine shore pools, while those of Vaucheria cushions, of collections of detritus of organic origin on the bottom of shallow lakes and the coastal ground water also belong to the halolenitobionts.

The group is certainly not biologically uniform. Some are undoubtedly true brackish-water organisms (*Manayunkia aestuarina* and others); many have probably been excluded by competition from marine regions rich in species and can only survive in the extreme environment of the tidal region where competition is slight. (Compare the analogous situation of many shore plants). Unless a different kind of behaviour has been directly demonstrated these eulittoral brackish-water animals have here been included with the euryhaline marine animals; for most of them this is certainly correct. Some, e. g. *Nereis diversicolor* are often considered to be true brackish-water animals. But CH. HEMPEL (personal communication) found sexually mature animals of this species in 3—4⁰/₀₀ salinity. However, in the population of this species in the northern part of the Baltic the larvae only develop at a salinity above average, the optimum being about 13⁰/₀₀ (SMITH 1964).

Several true brackish-water species occur among the Cnidaria, but only Hydrozoa; so far as we know they are lacking in Scyphozoa and Anthozoa. Three brackish-water polyps occur and are widespread among the Hydrozoa which do not form medusae.

1) *Protohydra leuckarti*. This small polyp (up to 2 mm long) without tentacles, is able to crawl about between sand grains and algae by means of peristaltic movements; it has recently been found in brackish waters everywhere. It occurs in shallow bays, lagoons and the sandy regions of the sea shore to a depth of several metres (Kiel Bay). Its range evidently extends from about 4—5⁰/₀₀ to over 20⁰/₀₀ (found in a shore pool of the North Sea on the Kniepsand near Amrum). Eggs form direct on the polyp, but sexually mature animals have rarely been seen (KOLLER 1927, WESTBLAD 1930); it mostly reproduces by transverse fission, more rarely by budding (E. SCHULZ 1950).

2) *Cordylophora caspia* forms richly branched colonies everywhere in brackish water, on posts and on plants (also *Potamogeton*); locally it occurs in fresh water, as low, turf-like colonies in running water. Apart from these fresh-water colonies *Cordylophora* also thrives in the salinity range 1—15⁰/₀₀. In its physiology *Cordylophora* is a brackish-water organism, although the optimal salinity lies outside the area of its natural range of existence, as KINNE (1956) has shown. According to this author the optimum is at 16.7⁰/₀₀ when growth, reproduction of polyps are at their best, length of hydranths and number of tentacles greatest, movement and enzyme activity at their best.

However, under experimental conditions *Cordylophora* can survive for months in fresh water and sea water (35⁰/₀₀) within a temperature range of 8—20° C (KINNE 1956).

3) *Perigonimus megas* KINNE forms colonies which externally resemble those of *Cordylophora* and it has sometimes been mistaken for it. It was discovered in the Kiel Canal where it also lives in the 1—10⁰/₀₀ range. It also occurs in the Zuiderzee and the Baltic. No doubt the hydroid polyp from the brackish water in India (Lake Chilka) described by ANNANDALE as *Bimeria fluminalis* is closely related to it (see fig. 41).

Among the Hydrozoa with polyp and medusa the family of Moerisidae is of special interest, since the whole family is almost entirely confined to brackish water, a rare phenomenon. According to the survey by KRAMP (1961) they belong to the

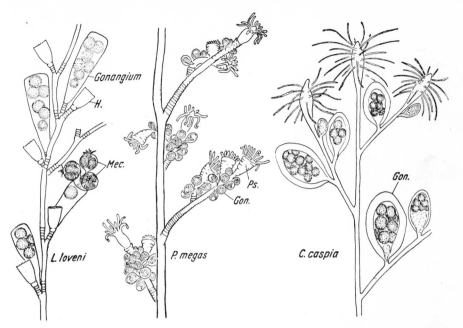

Fig. 41. Colonies of hydroids occurring in brackish water. *Laomedea loveni,* a strongly euryhaline marine species (H. = hydrotheca; Mec. = medusoid-meconidia); *Perigonimus megas;* evidently a specific brackish-water species (Gon. = gonophores, Ps. = pseudotheca); *Cordylophora caspia.* Enlarged. After KINNE 1956.

order of Limnomedusae whose habitat extends from the fully-marine range to fresh water. The Moerisidae live predominantly in lagoons, some almost in fresh water (*Halomises*), the majority in brackish-water lakes (Lac Quarun, Sea of Azov), lagoons or river mouths. This applies to the genera *Moerisia, Odessia, Ostroumovia.* Several species live in the Pontocaspian area (*M. pallasi,* Caspian Sea; *Odessia maeotica* Black Sea, but also Naples, South of France, Azores; *Ostroumovia inkermanica* Black Sea, but also India). VALKANOV (1938) produced a good monograph of the Moerisidae. The family has world-wide distribution from America (Antilles: *Halmomises*) to Japan (*Moerisia,* UCHIDA 1951). But it prefers warmer seas and is absent from the coasts of northern Europe.

KRAMP (1961) places the true fresh-water medusae into other families; the well known *Craspedacusta* with the polyp *Microhydra* into the Olindiadidae which contain many purely marine species as well as inhabitants of lagoons (*Gonionemus*). The fresh-water medusa *Limnocnida* forms a family of its own, occurring in Africa with 2 species (Tanganjika, Lake Victoria, Congo); it also occurs in India.

The polyps of the Moerisidae are usually solitary, partly basal with a short pericarp covering. Reproduction by means of lateral buds is widespread; they detach themselves before the tentacles have formed.

Two special medusae, *Eucheilota flevensis* and *Nemopsis bachei* live in the brackish water of the Zuiderzee. The latter has certainly been introduced from the Atlantic coast of America; it is now found in the lower Elbe as well (KÜHL). It is euryhaline.

Turbellaria are rich in brackish-water species as has been specially demonstrated by the ecological studies of P. Ax (1951—1957). Ax records no fewer than 25 species from the European brackish waters which are undoubtedly specific brackish-water species, in addition to 20 species so far found in brackish water only; presumably these are in part distinct brackish-water species. It is of ecological importance that most of them are bound to specific substrata; only a few are eurytopic, e. g. *Promesostoma bilineatum, Enterostomula catinosa, Proxenetes plebeius*. Most of the species live in sandy soils, some in special biotopes, in coastal groundwater e. g. *Coelogynopora schulzii, Pseudosyrtis subterranea, Macrostomum curvituba, Prognathorhynchus canaliculatus*. Only 3 species are typical of the zone of vegetation or the mussel banks: *Enterostomula graffi* and the two Triclads *Pentacoelum fucoideum* and *Sabussowia punctata*. Within the Turbellaria no distinct brackish-water species have been found so far in the orders of Polycladida and Acoela. *Stylochus flevensis* of the Zuiderzee is probably, like several "endemisms" of this area, an introduced species whose ecological evaluation is still uncertain. Among the Acoela one species (*Mecynostomum auritum* M. SCHULTZE) is widespread and common in brackish pools, but likely to be euryhaline. The genus *Macrostomum* contains true brackish-water species, e. g. *M. curvituba, M. hamatum* BL., *M. spirale, M. tenuicauda, M. timavi, M. longistyliferum, Dolichomacrostomum uniporum* BH.; among the Prolecithopora Ax mentions *Enterostomula graffi* and *E. catinosa*, among the Seriata *Pseudomonocelis cetinae, Paramonotus hamatus, Archiloa rivularis, Archiloa westbladi, Minona trigonopora, Coelogynopora schulzii, Pseudosyrtis subterranea* and the previously mentioned Triclada *Pentacoelum fucoideum* and *Sabussowia punctata*. From the coast of Brazil MARCUS (1950) reports that species of *Promonotus* (*P. villacae* and *P. erinaceus*) settle preferentially in brackish water; in Lake Aral and the Caspian resp. *P. orientalis* and *P. hyrcanus* are found (BEKLEMISHEV 1927). Among the R h a b d o c o e l a the marine family of Provorticidae provides several specific brackish-water species, e. g. *Vejdovskya pellucida, V. ignava;* the following are known only from mesohaline brackish water *Provortex pallidus* (Finland, Kiel Bay: Bottsand), *Canetellia beuachampi, Vejdovskya helictos* (Mediterranean), *Haplovejdovskya subterranea* (Kiel Bay, Finland), *Baicalellia subsalina* (Kiel Canal), *B. posiete* (East Asia). The Dalyellidae contain the following species: *Halammovortex nigrifrons* (Finland, Frische Nehrung), *Axiola remanei* (Kiel Canal, Frisches Haff), the Typhloplanoidea: *Beklemischeviellea contorta* (Baltic, Belt Sea, Lake Aral), *Tvaerminnea karlingi* (Baltic, Kiel Bay, Mediterranean), *Coronhelnis multispinosus* (Finland, Kiel Bay, Elbe), *C. lutheri* (Kiel Bay), *Promesostoma bilineatum* (Black Sea, Kiel Bay), *P. plebeius* (Black Sea, mesohaline lagoons of the Mediterranean), *Thalassoplanella collaris* (Finland, Kiel Bay), *Castrada subsalsa* (Finland), *Opistomum immigrans* (etangs south of France) and probably *Haloplanella curvistyla* (Finland) and *H. obtusituba* (Finland). The Kaloptorhynchia, rich in species, have fewer representatives. A number of species have so far been found in brackish water only, but in most cases only in one sea or one marine area, so that according to Ax it has not yet been possible to make an ecological evaluation. From Finland there are: *Gnathorhynchus krogeri, Prognathorhynchus campylostylus, Odontorhynchus lonchiferus, Thylacorhynchus pyriferus,* all discovered by KARLING (1931, 1947, 1952); from Lake Aral: *Phonorhynchoides flagellatus* and *Koinocystis relicta* (BEKLEMISHEV 1922);

from the Black Sea: *Schizorhynchus tataricus* and *Rogneda tripalmata* (BEKLEMISHEV 1922). In the brackish water of Finland and Kiel Bay or Kiel Canal the following have been found: *Zonorhynchus tvaerminnensis, Prognathorhynchus canaliculatus, Clyporhynchus monolentis*. These species are probably distinct brackish-water species (see fig. 42).

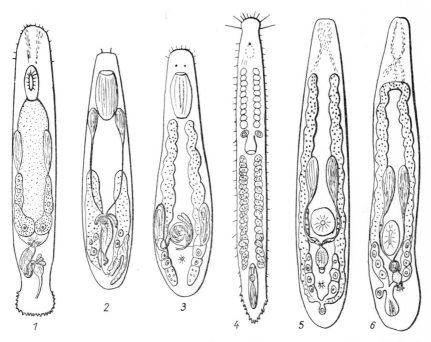

Fig. 42. Some specific brackish-water Turbellaria. 1, *Macrostomum curvituba;* 2, *Vejdovskya ignava;* 3, *Canetellia beauchampi;* 4, *Pseudosyrtis subterranea;* 5, *Coronhelmis lutheri;* 6, *Tvaerminnea karlingi*. Enlarged drawing P. Ax.

Rotifera. The rotifers have a large number of brackish-water inhabitants, among them a number of brackish-water species. The planktonic organisms will be considered first: The common and widespread species of *Brachionus plicatilis* and *Hexarthra (Pedalia) fennica* have to be placed with the holeuryhaline animals. But there are a number of specific species left, e. g. *Keratella eichwaldi* which in my opinion should be separated from *K. cruciformis* as a species of its own. Undoubtedly a number of *Synchaeta* species belong here, most likely *S. monopus,* common in the Baltic, *S. fennica* and *S. bicornis,* known more locally, *S. littoralis,* possibly *S. gyrina* and others. *Rhinoglena fertöensis* is only known from weakly brackish inland waters in Europe (Lake Fertö, Süße See near Halle).

There are numerous species of the phytal zone and the bottom, but their ecological requirements are not so well known. But the following deserve to be mentioned here: *Proales similis, Euchlanis plicata, Encentrum rousseleti, Aspelta baltica, Erignatha sagitta* and others. It is strange that the Bdelloidea, rich in species which have also penetrated into the sea, do not appear to have produced any brackish-water species.

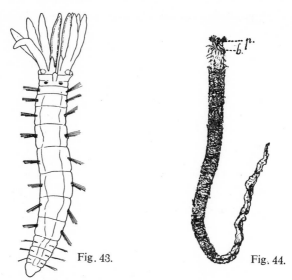

Fig. 43.   Fig. 44.

Fig. 43. The brackish-water Polychaete *Manayunkia aestuarina* (Fam. Sabellidae). Dorsal side. Length 1.5—2 mm. Related species occur in fresh water. (Lake Baikal, North American Lakes). Drawing BANSE 1956.

Fig. 44. The Pontocaspian brackish-water Polychaete *Hypania invalida* (Fam. Amphare-tidae). It penetrates far into the rivers, in the Danube above the Iron Gate. About 2.75×. After MOTAS & BAČESCU 1938.

Gastrotricha. Among the Macrodasyoidea *Turbanella lutheri* (Baltic region) is a species found only in brackish water; other species so far known only from Kiel Bay cannot yet be assessed regarding their salinity requirements. This also applies to the Chaetonotoidea of the brackish water.

Nematoda. According to a communication by S. GERLACH the following are brackish-water species inhabiting the meso- and oligohaline regions: *Enoplolaimus derjugini, Adoncholaimus thalassophygas, Oncholaimus conicauda, Paracyatholaimus intermedius, Microlaimus globiceps, Allgeniella guidoschneideri, Dichromadora geophila, Prochromadora orleyi, Theristus flevensis, Th. oxyuroides.* The following species penetrate farther into the marine area without reaching the euryhaline region: *Oncholaimus oxyuris, Paracyatholaimus proximus, Allgeniella tenuis, Axonolaimus spinosus.* A number of nematods of terrestrial origin inhabit the mostly brackish, moist sands of the sea shore in the supralittoral zone (Cyanophyceae-sand, cf. GERLACH 1954). Presumably some, possibly many, brackish-water species are marine euryhaline, corresponding to the Oligochaeta from similar areas. Among these are *Dorylaimus balticus, Tripyla cornuta, Mononchus spectabilis, M. rotundicaudatus, M. schulzi, Cephalobus strandi- cornutus, Rhabditis marina, Rh. ocypodis, Odontopharynx longicauda.*

Annelida (figs 43—45). Within the Annelida the Polychaeta and the Oligochaeta show rather different behaviour. The Polychaeta have but a few distinct brackish-water species: *Manayunkia aestuarina, Polydora redekei, Alcmaria romijni* (= *Microsamytha ryckiana*), *Streblospio strubsoli* and the tube-building species *Merceriella*

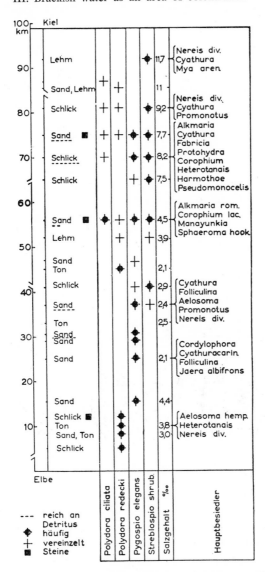

Fig. 45. Distribution of some Polychaetae in the Kiel Canal, from Kiel (100 km) to the Elbe, with data of salinity and records of the most important accompanying species. After CH. HEMPEL 1957.

enigmatica chiefly found in the South (Mediterranean, west coast of Europe). They mainly inhabit shallow pools and lakes on the shore; the strangely sluggish species *Stygocapitella subterranea* from the coastal ground water might be listed here. As in other groups there are some "relicts" among the Polychaeta of the Pontocaspian area; they live in slightly brackish water, but enter the fresh water of river mouths as well; above all the three Ampharetidae: *Parhypania brevispinis* (Caspian Sea), *Hypania invalida* (Caspian Sea, Limane and lower reaches of Dniester, Danube,

Bug, Dnieper up to 250 km upstream), *Hypaniola Kowalewskii* (Caspian Sea, parts of Sea of Azov with lowered salinity, Dniester, delta of the Danube); also *Manayunkia caspica* (see C. MOTAS and M. BAČESCU 1938).

The number of distinct brackish-water animals among the Oligochaeta is relatively higher: *Amphichaeta sannio, Paranais botniensis Limnodrilus heterochaetus, Tubifex costatus, T. nerthus* which penetrates far into the Danube, *Lumbricillus pagenstecheri* etc.

Nemertina. One species of Nemertina is a widespread and common brackish-water animal: *Prostoma abscurum.* In the Belt Sea and the Baltic it even enters shallow pools and lakes on the shore in which temperature and oxygen content show large fluctuations; it invades the northernmost parts of the Baltic (Botten Bay about $2^0/_{00}$ S) where it also occurs at greater depths (LASSIG 1964).

Molluscs. In the Pontocaspian area a considerable number of molluscs may be described as distinct brackish-water animals (p. 152), but only a few species from the rest of Europe may be listed here. Among bivalves only the immigrant *Congeria*

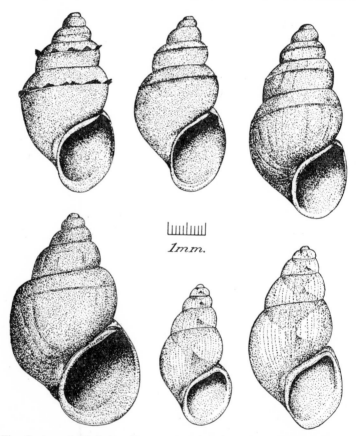

1mm.

Fig. 46. The Gastropod *Hydrobia (Potamopyrgus) jenkinsi* from different Danish waters. Upper row: Estuary of the Vond Aa, marine region of the Ringköpingfjord, estuary Kyllebakken. Bottom row: Estuary Egaa, fresh-water marl pit Kolstrup, fresh-water river Binderup Aa. After BONDESEN & KAISER 1949.

cochleata belongs here; in Europe it is restricted to brackish water, but in its African home there are isolated records from fresh water (see THIENEMANN 1950). On warmer coasts further species of *Congeria* and *Rangia* (Mactridae) are brackish-water species which partly invade fresh water (see THIENEMANN 1950). Among the Gastropoda *Hydrobia ventrosa* (= *H. baltica*) is widespread, a typical brackish-water species; an isolated record from the richly saline tidal area of the Canary Island Lanzarote (see BONDESEN and KAISER 1949) is reminiscent of similar occurrence of brackish-water animals in the uppermost tidal area of the sea. In the inland regions of Europe the immigrant *Hydrobia jenkinsi* (= ? *Potamopyrgus crystallinus*) prefers brackish water and invades fresh water more locally, while the British (? tetraploid) form is widespread in fresh water (fig. 46).

Among the Opisthobranchiata there is or was an endemic form in the Zuiderzee: *Corambe batava* which may belong here; in the Belt Sea and the Baltic *Alderia modesta* also behaves like a true brackish-water animal.

S. JAECKEL (1965) gives a survey of the formation of brackish-water forms in molluscs. About 30 genera of the Neritidae live in brackish and fresh water, mainly in the Indo-Pacific area as well as numerous Stenothyridae in that region, furthermore species of Iravadiidae, Fairbankiinae, Hemistomiinae.

Crustacea. There are large numbers of brackish-water species among the Crustacea. In the Copepoda, especial the Harpacticidae (see figs. 37, 38) we know many species whose main area of existence is in brackish water: in the North Sea

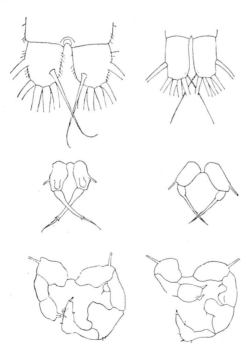

Fig. 47. Characteristics of the brackish-water Calanides *Acartia tonsa* (left series), compared with the euryhaline marine *Acartia bifilosa* (right series). From top downwards: Furca, 5th leg of female, 5th leg of male. Enlarged about 225×. After REDEKE 1935.

and Baltic regions, for example, species of *Schizopera* (*Sch. clandestina* and probably *Sch. ornata* and *Sch. inornata*), *Sigmatidium minor, Nitocra reducta, Mesochra aestuarii;* a tendency to holeuryhalinity is displayed by *M. rapiens, Itunella muelleri, ? Parepactophanes minuta, ? Paramesochra acutata, Horsiella trisetosa* and *H. brevicornis* (? holeuryhaline), *D'Arcythomsonia fairliensis.* LANG (1948) mentions further species from other regions. Among the pelagic Calanoidea *Acartia tonsa* has recently turned up everywhere in the brackish waters of European coasts, in parts in large numbers (fig. 47). Here the species behaves like true brackish-water organisms, but HEDGPETH (1953, p. 178) also records it from the hyperhaline lagoons of the Gulf of Mexico (as well as the Polychaete *Polydora ligni*). *Limnocalanus grimaldii* is brackish-fresh water in northern Europe, North America and in the Caspian Sea.

Table 6.

The distribution of the Ostracoda in the Schlei. After REMANE 1937 b from ROTTGARDT.

Salinity kleine Breite 2.8—5.6 Maasholm 7.—18. März	Kl. Breite	Gr. Breite	Missunde	Kieholm	Lindaunis	Sieseby	Arnis	Kappeln	Rabelsund	Maasholm
Xestoleberis aurantia BAIRD ……………			○	○	○	○	○	○	○	○
Cyprideis litoralis BRADY ………………	○	○	○	○	○	○	○	○	○	○
Leptocythere castanea G. O. SARS ………				+		+				
Cytherura gibba O. F. MÜLLER …………			/			/				
Loxoconcha gauthieri KLIE ……………	○	○	○	○	○					
Cytheromorpha fuscata BRADY …………	○	○	○	○						
Heterocypris salinus BRADY ……………					○					
Candona angulata G. W. MÜLLER ………	○	○								
Cyclocypris laevis G. F. MÜLLER ………	/	/								
Darwinula stevensoni BRADY & ROBERTSON ..	/	?								

Key to symbols: ○ common, / not uncommon, + rare, ? occasional.

ELOFSON (1941) mentions 7 Ostracoda as true brackish-water forms: *Cyprideis litoralis, ? Paracyprideis fennica, Leptocythere ilyophila, Cytheromorpha fuscata, Cytherura gibba, Loxoconcha elliptica* (= *L. gauthieri*), *L. baltica.* But *Cypridies litoralis* proves to be euryhaline (cf. p. 58). *Loxoconcha emelwardensis* and a few species which have entered brackish shallow pools and lakes on the shore from fresh water should be added: e. g. *Cypridopsis aculeata, Candona angulata,* species of *Eucypris* (fig. 48).

In the Cirripedia brackish-water species are unknown; their assessment in the Phyllopoda is partly uncertain. *Moina microphthalma* (fig. 28) might be counted here. For the Anostraca (p. 76) reported from continental brackish waters the limiting salinity has not yet been exactly determined.

The groups of Malacostraca are rich in distinct brackish-water species. In North and Central Europe the Isopoda belong here: *Mesidotea entomon,* the large Isopod of the Baltic which also occurs as relict in some Scandinavian and Finnish freshwater lakes (see p. 141), the green *Idotea viridis* in the phytal zone of the shore region; from the family of Sphaeromidae which is widespread in the extreme biotopes of the sea shore there are the species *Sphaeroma rugicaudum* and *Sph. hookeri*

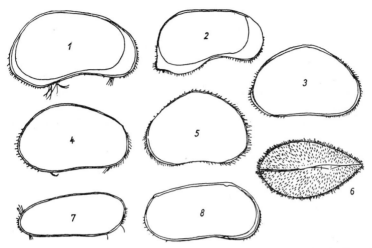

Fig. 48. Brackish-water Ostracoda and euryhaline fresh-water Ostracoda. Outline of shell: 1—5, 8, right side; 7, left side; 6, viewed from above both valves. 1, *Candona neglecta;* 2, *Candona angulata;* 3, *Heterocypris salinus;* 4, *Heterocypris incongruens;* 5, 6, *Cypridopsis aculeata;* 7, *Darwinula stevensoni;* 8, *Cytheromorpha furcata.* Enlarged. From SARS & GAUTHIER.

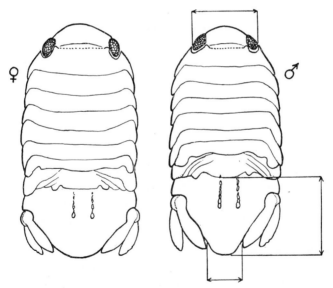

Fig. 49. The Isopod *Sphaeroma hookeri.* After KINNE.

(fig. 49); *Cyathura carinata* is burrowing into the ground. The Amphipoda are re-presented here with species of *Corophium,* e. g. *Corophium lacustre* (mesohaline), *C. multisetosum* (ditto), *C. insidiosum* (polyhaline), species of *Gammarus, G. zaddachi, G. salinus* (and in transition to holeuryhalinity *G. duebeni*), the relict form *Ponto-poreia affinis, Leptocheirus pilosus, Bathyporeia pilosa* (polyhaline) (fig. 34).

The Mysidae have many typical relict forms in the weakly brackish region, especially in the Pontocaspian area; they often invade fresh water; the same holds for the Cumaceae (see p. 149).

Brackish-water Decapoda are scarce with us. Only the prawn *Palaemonetes varians* could be mentioned which inhabits chiefly lagaoons and ditches on the shore in Europe, while in Italy it is found as relict form in fresh water. The immigrant Brachyura *Rithropanopaeus harrisi* is also a brackish-water animal. It was first found in the Zuiderzee and described as *Pilumnopeus (Heteropanope) tridentatus,* but has since been found in the Kiel Canal, in the Schlei and in the Black Sea (fig. 50). There seem to be a much larger number of brackish-water Decapoda in tropical waters. A species of *Potamobius* is indigenous in the Caspian Sea.

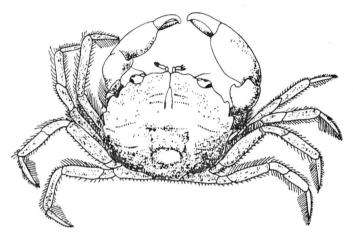

Fig. 50. The crab *Rithropanopaeus harrisi* (= *Pilumnopeus tridentatus*) which has been introduced into European brackish waters. Enlarged about 2×. From Thienemann 1950.

The majority of insects living in the sea or in salt pools are undoubtedly euryhaline fresh-water or euryhaline marine species. But there appear to be a number of true brackish-water species. Remmert (1955) records brackish water as the range of existence for the following Diptera larvae: *Culicoides algarum* 8—20⁰/₀₀, *Bezzia calceata* 8—15⁰/₀₀, *Chironomus halophilus* 8—15⁰/₀₀, *Cladotanytarsus mancus* 3—10⁰/₀₀, *Eucricotopus atritarsis* 5—18⁰/₀₀, *Procladius breviatus* 3—18⁰/₀₀. The beetle *Haemonia* also behaves like a brackish-water animal.

On the whole brackish-water species have not been produced by the large groups of marine animals which hardly enter fresh water or do so only to a slight extent. This applies to Echinodermata, Tunicata, Brachiopoda etc. But three species of Bryozoa are widespread: *Membranipora (Electra) crustulenta* which can even build Bryozoan reefs in brackish water (fig. 51); *Conopeum seurati* and *Victorella pavida.* In our latitudes *M. crustulenta* exhibits mass development in brackish water, but as Borg (1931) has shown it is not entirely confined to this region. In northern waters and in the Gullmarfjord it has been found in the sea; from the South there is a record from fresh water (two small rivers in Tunisia). Jebram (1968) cultivated

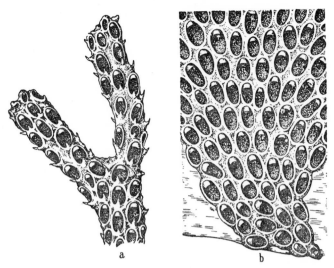

Fig. 51. The brackish-water Bryozoon *Membranipora crustulenta* (var. *baltica*); a, growing on the alga *Furcellaria;* b, on *Zostera*. Enlarged 12×. After SEIFERT 1938.

Membranipora crustulenta and *Conopeum seurati* at 35⁰/₀₀ S. Cultures with lawer salinity (7⁰/₀₀) degenerated.

There are no true brackish-water species among the fishes of the North Sea-Baltic area. Some approximation to such species is found in *Gobius (Pomatoschistus) microps* and *Platichthys flesus*. Among the relicts of the Pontocaspian region there are a number of fishes which are essentially brackish-water inhabitants, but also enter the lower reaches of rivers. To these belong *Mesogobius batrachocephalus, Gobius melanostomus, G. syrman, Knipowitschia longicaudata, Lucioperca marina, Percarina dermidoffi, Cobitis caspia*, etc. In tropical brackish waters, too, the number of fishes preferring this habitat is likely to be large.

Thanks to intensive studies the Diatomaceae among the unicellular plants, can be clearly classified according to the salinity of their environment. There is a large number of specific brackish-water diatoms. According to a communication by CHR. BROCKMANN the following are euryhaline brackish-water species found at about 2—20⁰/₀₀ S: *Melosira binderiana, M. nummuloides, Coscinodiscus commutatus, Synedra pulchella, Gyrosigma balticum, Diploneis didyma, Amphiprora alata, Hantzschia virgata, Nitzschia scalaris, Surirella striatula, Campylodiscus clypeus, C. echeneis*. For the upper brackish water between 8—20⁰/₀₀ the following are specific: *Synedra gailloni, Achnanthes brevipes, Diploneis smithi, Navicula peregrina, N. digitoradiata, Nitzschia circumsuta, N. punctata*. The following occur particularly in the lower brackish water between 2—10⁰/₀₀ (= about meiomesohalikum): *Cyclotella striata, Coscinodiscus normani, C. lacustris, Synedra affinis, Achnanthes delicatula, Mastogloia brauni, M. elliptica, Caloneis formosa, Diploneis interrupta, Anomoeoneis sphaerophora, Navicula crucigera, N. mutica, N. salinarum, N. rhynchocephala, N. viridula, N. humerosa, Amphora commutata, Rhopalodia gibberula, Rh. musculus, Bacillaria peradoxa. Coscinodiscus rothi* has often been reported from this range.

The percentage of brackish-water species among the Euglenoidea also seems to be relatively high. But only few species may be listed among the multicellular algae: WAERN (1952, p. 9) records in the Baltic the Red Alga *Ceramium tenuicorne* (closely resembling *C. radiculosum* from river mouths in the Adriatic), the Brown Algae *Ectocarpus confervoides "fluviatilis", Porterinema fluviatile* and among the Green Algae the Baltic endemic *Monostroma balticum*. Among the Characeae LEVRING (1940, p. 155) mentions *Chara crinita, Ch. horrida,Ch. baltica* (considered by WAHL-STEDT to be a non-encrusted form of a fresh-water species *Ch. intermedia*) and *Tolypella*.

I dare not decide as to how far higher plants, such as certain species or forms of *Ruppia, Zannichellia, Ranunculus* and the eel-grass *Zostera* may described as brackish-water species.

c) Survey of the behaviour of specific brackish-water organisms

Most groups of animals and plants have evolved distinct brackish-water species, but their relative numbers vary from group to group.

The lowest numbers are found in the marine groups which penetrate only into brackish water, but do not enter fresh water or do so only slightly. Thus distinct brackish-water species are absent from Echinodermata, Brachiopoda, Tunicata, Cephalopoda, Cirripedia, Porifera. Small numbers are found in Foraminifera, molluscs, Bryozoa, among plants in the Red, Brown and Green Algae; their numbers are greater in Copepoda, Ostracoda, Malacostraca.

An examination of the range of distribution in brackish-water organisms shows clearly that their optimum lies in the mesohalinikum. The salinity range from about $3^0/_{00}$ to $18^0/_{00}$ is the habitat of over 90% of the brackish-water organisms. Only rarely do species transgress this range into higher salinity grades. Conversely, quite a few species enter fresh water with some of their populations or wander into the lower reaches of rivers. That is true for Cnidaria *(Cordylophora caspia!)*, Turbellaria *(Macrostomum curvituba, Vejdovskya ignava, Coronhelmis multispinosus, Pseudomonocelis cetinae* (cf. Ax 1957), Nematoda *(Paracyatholaimus intermedius, Theristus flevensis)*, Ostracoda *(Cytheromorpha fuscata)*, Copepoda (species of *Nitocra)*, Isopoda *(Mesidotea entomon)* etc. This is specially true of the Pontocaspian relicts.

Thus the distribution of brackish-water species displays the same asymmetry as the fewest number of species: The greatest number of species does not occur at a salt concentration half that of sea water, but at a concentration near that of fresh water.

DAHL (1948, 1956) mentions a number of species which are typical of the 16.5 to $30^0/_{00}$ range, that is brackish-water animals of that highly saline zone. DAHL lists the Amphipoda *Corophium insidiosum, Bathyporeia pilosa, Melita palmata,* the Copepoda *Laophonte baltica, Zaus spinatus, Harpacticus chelifer,* furthermore *Nereis diversicolor* and *Balanus improvisus*. DAHL also points to the oyster banks of *Crassostrea virginica*. The oviparous oysters of the genus *Crassostrea* penetrate much further into water of low salinity $(12—20^0/_{00})$ than the species of the genus *Ostrea;* they also settle more easily on soft substrata. Their maximum occurrence is in bays, lagoons and estuaries with a reduced salt content (KORRINGA 1956, HEDGPETH 1953).

Though it is quite feasible for specific species to exist in the polyhaline region DAHL's list must be regarded as too optimistic. The Copepoda *Laophonte baltica, Harpacticus chelifer* and *Zaus spinatus* have, for instance, been found on several occasions in the North Sea near Heligoland (KLIE 1949, 1950). They reach their greatest abundance in coastal waters, but the first two live in the stand of Red Algae of the Western Mole, that is in the fully marine area. Therefore the species are described as euryhaline marine (see NOODT 1956). The oyster *Crassostrea* displays similar behaviour. It occurs everywhere on rocky sea shores — that appears to be its original biotope; in the lagoons it only forms banks (reefs) under the conditions prevailing there: abundance of nutrients, currents, shallow waters with horizontal floors etc. KORRINGA (1956) stresses: "The low salinities as such do not bring any advantages, but it is the biological richness of estuarine waters and the escapement (sic) from many parasites and predators which render many estuarine bodies of water so excellently suited to the oviparous oysters." Its behaviour resembles that of *Mytilus* and it is likewise rather euryhaline marine. *Nereis diversicolor* and *Balanus improvisus* range over the whole of the mesohalinikum into the oligohalinikum; they are strongly euryhaline marine animals or perhaps brackish-water species. There remain the Amphipoda *Melita palmata, Bathyporeia pilosa* and *Corophium insidiosum.* The first two extend into the middle of the mesohalinikum (about $6^0/_{00}$ S), the last two are twin species of marine representatives *(Corophium bonelli, Bathyporeia robertsoni)* which are difficult to separate. A special species of the polyhaline region might be *Corophium indisiosum,* the other two Amphipoda may be specific for the polyhalinikum and pleiomesohalinikum.

Whatever the ecological evaluation of these species may be, there is no doubt that incomparably fewer species are found in the polyhalinikum as compared with the mesohalinikum or even the meiomesohalinikum. In any case there is a marked asymmetry.

The region of the fewest species is one where evolution of distinct brackish-water species proceeds most vigorously.

d) Origin of brackish-water species

The above-mentioned species are usually so closely related to fresh-water or marine species that it is possible to trace their origin. They either come from fresh water and are "limnogenous" or else from the sea "thalassogenous". Examples of organisms of fresh-water origin are many brackish-water Oligochaeta, especially the Naididae and Tubificidae, many Rotifera, Diptera larvae, some Ostracoda, e. g. *Cypridopsis aculeata, Candona angulata* and others; of marine origin are brackish-water Polychaeta, Bryozoa, Malacostraca excluding *Potamobius,* most Copepoda, Turbellaria, Nematoda etc. A comparison of the numbers of brackish-water species of fresh-water and of marine origin once again reveals the "asymmetry". Most of the species are of marine origin, even in the meiomesohalinikum. I have estimated their number to be 3/4 to 4/5 of the total number of species (REMANE 1941, p. 28). This preponderance of species of marine origin even applies to the Pontocaspian brackish-water fauna with its Mysidae, Cumaceae, Polychaeta etc. Even in slightly brackish water there are far more species of marine than of fresh-water origin, e. g. Cladocera

Cercopagis, Apagis etc. Now a peculiar situation arises. The typical brackish-water species are mainly of marine origin, the inhabitants of highly saline waters of the hyperhaline region are chiefly of fresh-water origin, as BEADLE has emphasized. This contrast however, is not based on salt content, but on two other factors, one regarding the ecology of distribution, the other a physiological one. For the ecology of distribution it is important that all hyperhaline waters are isolated single bodies of water, either small bodies of water (pools, springs) or the extreme end regions of lagoons in dry areas (Laguna Madre, Siwash). Naturally organisms capable of some means of aerial dispersal are predisposed to colonize such isolated localities. That is true for a large number of fresh-water species with their resistant stages, but very few marine organisms for whose larval dispersal continuous, large sheets of water are a prerequisite.

It is physiologically significant that fresh-water organisms having once transgressed the critical brackish-water zone, acquire an extreme lack of sensitivity to varying and high salinities whereas marine organisms rarely attain such a high degree of independence of the surrounding salt concentration. It would be worth examining in detail what kind of reactions account for this difference.

IV. The biotopes in brackish water

Fresh water and sea not only differ in their salt content and the different species of plants and animals inhabiting them, but also in the structure of their habitats; this, in its turn, is caused by the differing types of life forms. This may be illustrated by an example: in the sea there are large numbers of boring organisms which penetrate and decompose shells and rocks. Among these borers are bivalves (Pholadidae, Lithodomidae, *Saxicava*), Polychaeta *(Polydora ciliata, Dodecaceria concharum),* Bryozoa *(Terebripora, Spathipora),* sponges *(Clione),* sea urchins *(Strongylocentrotus),* nor must the boring endolithic Cyanophyceae be forgotten. The fact that organisms boring in rock or chalk are confined to the sea results, geologically, in a different type of erosion and biologically in a different kind of colonization of the rocky bottom. Only a few boring organisms occur in brackish water, *Polydora ciliata* penetrating farthest. It may be worth giving a short survey of habitats in brackish water, bearing in mind the specific properties of sea and fresh water.

A. Phytal zone (plant and animal epibiosis) „Aufwuchs"

1. Conditions of vegetation

There is a vast difference between the macroflora of the sea and of fresh water due to differences in the substratum. Its base in the sea is mostly solid rock. Accordingly, rocky regions are densely populated; this density of colonization immediately decreases on gravelly bottoms; on soft floors primary colonization only takes place where single projecting stones or bivalve shells offer a solid surface for attachment (but cf. p. 119). In fresh water stones, rocks or solid substrates are colonized — at the most — by filamentous algae or an algal covering, a very small amount of vegetation compared with the masses of kelp in the sea. In fresh water any stands of plants bound to a substrate develop mainly on soft floors and fine sand. This is due to the fact that the plants forming the vegetation in the sea are chiefly haptophytes while in fresh water they are rhizophytes that is plants "die sich mit Wurzeln Grund- achsen, Rhizoiden oder anderen von den Wassersprossen abweichenden Organen in ± feinkörnigem Boden befestigen" (H. LUTHER 1949, p. 7) [which fasten themselves in ± fine grained soil by means of roots, rhizomes, rhizoids or other organs differing form the aquatic shoots]. It is interesting that those rhizophytes which settle on soft or fine-sand floors in the sea as well, are higher plants or Characeae which have entered the sea from fresh water. The marine seagrass swards consist of flowering plants, that is Potamogetonaceae *(Zostera, Posidonia, Ruppia)* and Hydrocharidaceae *(Halophila, Thalassia)*. It is also significant that in the sea these flowering plants form dense close stands which we call seagrass swards. In fresh water the ribbon-type of leaves which leads to the formation of swards is far less common and on the average the stands of rhizophytes are much looser, showing less covering. The dense stands of "seagrasses" helps to collect sediment. In this way a layer of sapropel may develop at the bottom of the swards, even on sandy soils, in which inhabitants of mud bottoms deficient in oxygen may occur *(Capitella, Halicryptus)*.

Brackish water contains both the marine and fresh-water type of benthic plants at the same time. The marine haptophytes extend to about $15—18^0/_{00}\,S$, with their large kelps *(Laminaria)*, though sublittorally. Kelps of medium size such as *Fucus* penetrate as far as the boundary region of mesohaline and oligohaline and on the rocky coasts of Finland they still develop large stands of marine character. The membraneous algae, e. g. *Enteromorpha* (locally in fresh water), *Monostroma* and others extend just as far. The rhizophytes of the brackish water, on the other hand, are rich in species and distributed over a wide area as compared with the sea. They are rich in species since, in addition to marine flowering plants such as *Zostera marina, Ruppia maritima* there are fresh-water species which form stands far into the mesohalinikum *(Potamogeton pectinatus, P. perfoliatus, Myriophyllum)* and Characeae form swards just as they do in fresh water.

These special features are mainly found in cold and temperate seas while the contrasts are less marked in warmer regions. Here large kelps such as *Laminaria* are missing, *Fucus* is absent, and their place is taken by medium-sized, squarrose kelps such as *Cystoseira, Sargassum, Laurencia, Turbinaria* which usually form but loose stands. But there is a larger proportion of marine algae which anchor themselves in the soft floor by means of a network of rhizoids, they are thus primary rhizophytes of the sea. In the first instance species of *Caulerpa* must be mentioned here whose rhizoids sprout "roots" and runners even in areas of fine sand. To my knowledge these rigid algae only form loose stands. Other rhizophytes among the algae mainly colonize the seagrass swards as associates, e. g. species of *Penicillus, Udotea, Halimeda*. Lastly in the warmer seas the seagrass swards are represented not only by the *Zostera*- type with ribbon-shaped leaves, but also by species with oval leaves such as *Halophila*.

The primary fresh-water type and the marine type of bottom vegetation inter- mingle in a complicated fashion and interesting mixtures of associations arise. Thus in stands of seagrass, partly even in stands of *Potamogeton* there occur marine algae, membraneous forms *(Ulva, Monostroma)* as well as tufted ones *(Polysiphonia,*

Ceramium), filamentous algae *(Chaetomorpha)*, even kelps *(Chorda, Fucus)*. Sometimes these algae are growing on the seagrasses, often they lie loosely between them.

The most important feature of the brackish-water vegetation are the carpets of plants which in felt-like interweaving, lie on the floor or else penetrate the surface layers of the bottom. Corresponding forms of vegetation are not entirely lacking in the sea or fresh water, but they reach the peak of their development in brackish-water lakes or pools. Plants from various groups may take part in the formation of these carpets.

I begin with a description of such a carpet by R. W. KOLBE (1927, p. 9) from the brackish Mellensee (Sperenberg area): ,,Von größerer Bedeutung... ist die eigenartige Beherrschung des ganzen nördlichen Teils durch *Vaucheria dichotoma* AG. Die derbwandige Siphonee bildet auf quadratkilometergroßen Flächen des flacheren nördlichen Teils dichte Bestände, die den Grund mit einem zusammenhängenden, für Gasblasen undurchlässigen Teppich bekleiden. Die Behinderung des Gasaustauschs zwischen Grundschlamm und freiem Wasser bewirkt eine Störung der normalen, im ersten vor sich gehenden Fäulnisprozesse. Die sichtbare Folge davon ist — neben der üblichen Bildung von CH_4 — eine starke H_2S-Entwicklung, die stellenweise große Gasansammlungen hervorruft, welche den darüberliegenden *Vaucheria*-Teppich ,,hutpilzartig" in die Höhe und bis an den Wasserspiegel treibt. Die etwa 20—50 cm in Durchmesser haltenden, von oben gesehen kugeligen, im Sommer in großen Mengen anzutreffenden Massen werden von den Einwohnern bezeichnenderweise ,,Totenköpfe" genannt; der Eindruck wird dadurch erhöht, daß die Kuppe dieser *Vaucheria*-,,Blasen" — wohl durch die Einwirkung des H_2S vergiftet — aus abgestorbenen Fäden besteht und weißlich erscheint.

Eine weitere Folge der H_2S-Atmosphäre, die vom *Vaucheria*-Teppich ausgeht, ist die große Organismenarmut im nördlichen Teil des Sees (die Fischer klagen über das Zunehmen des Fischsterbens) und der Reichtum an Schwefelbakterien. Es genügt, den Teppich anzuheben (Stücke herauszureißen ist infolge seiner gummiartigen Zähigkeit kaum möglich), um weißliche, weinrote oder violett-rosa Wolken von *Chromatium, Beggiatoa, Thiothrix* und *Lamprocystis* aus tieferen Teilen des *Vaucheria*-Rasens aufzuwirbeln. Die *Vaucheria* scheint sich durch intensives Spitzenwachstum vor der Einwirkung des H_2S zu schützen, während die älteren Teile der Fäden ständig absterben und durch die halbverwesten Reste den Teppich weiter zu verkleben helfen.

Merkwürdigerweise beherbergt die Region der Fadenspitzen eine ziemlich reiche mikroskopische Lebewelt: eine sehr große Zahl epiphytischer Diatomeen siedelt sich hier an, und festsitzende Infusorien überziehen oft die Fadenenden mit einem dichten, weißen Flaum. Auch größere Watten von Grünalgen (beobachtet wurden *Spirogyra* und *Oedogonium*) sind in dieser Region nicht selten. Es scheint demnach, daß das Gefälle das Partialdrucks von H_2S sehr steil, und die Region der Fadenspitzen bereits von diesem Gas frei ist; auch merkt man selbst an windstillen Tagen nichts von H_2S-Geruch, solange man nicht den *Vaucheria*-Rasen anhebt. Das genauere Studium dieser eigenartigen Verhältnisse erscheint mir sehr lohnend." — [Of greater significance... is the strange dominance in the whole of the northern area of *Vaucheria dichotoma* AG. Over areas of square kilometres in the shallower northern parts the tough-walled Siphonea forms dense stands; these cover the ground with a coherent carpet impenetrable to gas bubbles. The exchange of gases between the bottom ooze and the open water being impeded, this results in a disturbance of the normal processes of decay in the former. Apart from the usual formation CH_4 — there is another visible consequence, a vigorous production of H_2S; in places this results in large accumulations of gas, pushing the *Vaucheria* carpet above it upwards like a "mushroom", driving it as far as the water surface. These masses about 20—50 cm in diameter, appear spherical when seen from above; it is significant that the inhabitants call them "death heads"; this impression is enhanced by the fact that the dome of these *Vaucheria*-"bubbles" — presumably poisoned by H_2S — consists of dead filaments and that it appears whitish.

A further consequence of the H_2S atmosphere emanating from the *Vaucheria* carpet is the great scarcity of organisms in the northern part of the lake (the fishermen complain about increasing number of deaths among the fish), and an abundance of sulphur bacteria. It is sufficient to raise the carpet (it is hardly possible to tear off pieces because of its rubbery toughness), in order to stir up whitish, wine-red or

Fig. 52. Algal carpet of *Vaucheria* and Cyanophyceae from the brackish-water Watt in the Harringvliet (Delta of Rhine). June 1952. Part of the carpet has been torn and rolled back by the tide. (Photograph Dr. D. König, Kiel.)

violet-pink clouds of *Chromatium, Beggiatoa, Thiotrix* and *Lamprocystis* from the deeper parts of the *Vaucheria* turf. *Vaucheria* appears to protect itself from the effects of H_2S by an intensive apical growth, while the older parts of the filaments continuously die off and, by their half-decayed remains, help to glue this carpet even more closely.

Strangely enough, the region of the tips of filaments harbours a fairly rich collection of microscopic life; a very large number of epiphytic diatoms settle here and sessile Infusoria frequently cover the ends of the filaments with a dense, white fluff. Also larger woads of Green Algae are not uncommon in this region (*Spirogyra* and *Oedogonium* have been recorded). It thus appears that the gradient of partial pressure of H_2S is very steep and that the region of the tips of the filaments may already be free of this gas. Even on calm days no smell of H_2S can be detected so long as the *Vaucheria* turf has not been raised. A more detailed study of these peculiar phenomena should be very rewarding].

Fig. 53. Top view of the algal carpet in the brackish water of the mouth of the Eider, West coast of Schleswig-Holstein. The light points are Nudibranchs sitting on the algae *(Alderia modesta)*. Photograph Dr. D. KÖNIG, Kiel.

Such tangled carpets of *Vaucheria* (often *V. Thuroti*) with Cyanophyceae and a substratum rich in H_2S, with purple bacteria, are very widespread, especially in calm bays with a mud bottom, often also on the edges of brackish-water ditches. Figs. 52 and 53 are photographs of the brackish water from the mouths of the Rhine and Eider, the carpets have been partly torn by the flood and rolled over. They are also found in the Wattenmeer; on them live the amphibious Nudibranchiata *Alderia modesta* and *Limapontia capitata* among them Nematoda, Copepoda and Foraminifera. The species composition of the plants forming the carpet varies; with the predominance of certain groups they range from the typical *Vaucheria* carpets to intermediate forms of Diatomacea-Cyanophyceae mats which often cover the mud in the Wattenmeer (cf. WOHLENBERG 1939), to patches of brilliant red purple bacteria which are so characteristic of brackish-water bays floored with fine sand or ooze. This bacterial layer consists mainly of *Lamprocystis roseo-persicina*, *Thiopedia rosea* and species of *Chromatium*; it is mixed with Cyanophyceae and diatoms (J. GIETZEN 1931); see also p. 129.

The Farbstreifenwatt also represents a stratified zone of Cyanophyceae—Thiorhodaceae vegetation. Here the layers of vegetation lie under a cover of a stratum of fine sand, but they can also stretch over a wide area (see E. SCHULZ 1937, E. SCHULZ and H. MEYER 1939, C. HOFFMANN 1942, REMANE 1957).

It is well known that many marine algae, when forcibly detached from their substratum, are capable of continuing their growth, either floating or lying on the bottom. The best known examples are the drifting *Sargassum*-kelps of the Sargaso Sea. A similar phenomenon occurs to a much larger extent in the Wattenmeer and in bays of brackish seas and lakes. In contrast to the *Sargassum*-kelps these stands of algae and kelps are deposited over the bottom where they continue to grow vegetatively; sexual reproduction is frequently lost and in species carrying air bladders *(Ascophyllum, Fucus)* these are usually absent; the growth form often changes in other respects as well. *Fucus* in particular, takes part in these algal colonies, in the Baltic *F. vesiculous* forms the "loose-lying *Fucus* association (Svedelius)" (A. LEVRING 1940). But *Phyllophora Brodiaei, Furcellaria, Polysiphonia, Monostroma, Enteromorpha* and *Cladophora*-species, *Spirogyra, Zygnema, Pylaiella* also share these deposits of algae; they often insert themselves in *Zostera*- and *Potamogeton* meadows; where the mussel *Mytilus* comes in contact with such kelps, especially species of *Fucus,* it spins its byssus threads round the algal thalli and anchors them; thus a *Fucus-Mytili* formation arises in the Wattenmeer (NIENBURG 1925) and, in a somewhat altered form, in the sublittoral of the Belt Sea and the Baltic. Locally Green Algae *Cladophora aegagropila = Aegagrophila Martensis* f. *biformis* (see HAYREN 1950) form loose-lying colonies of round bodies up to several centimetres in diameter. In some parts of the Black Sea the Red Alga *Phyllophora* forms extensive carpets at depths from 27 to about 45 m. Red Algae of the genus *Polysiphonia* are dominant in the algal carpets covering, for instance, large stretches of the brackish Lac Quarun in Egypt (see M. NAGUIB).

In bays and ditches detached masses of *Enteromorpha* (especially *E. intestinalis*) may be pushed to the surface by bubbles of gas where they form a dense layer of vegetation; the water in the interstices warms up quickly. The layer of water underneath it is cut off from circulation and from the surface in the same way as this happens in fresh water by dense covers of *Lemna*. As a result considerable decay and putrefaction take place below the *Enteromorpha* layer in brackish water.

Brackish-water vegetation also differs from that of fresh water by the absence or regression of certain types of life forms. Stands of rhizophytes with floating leaves such as the water lilies *Nymphaea, Nuphar,* furthermore *Potamogeton natans, Polygonum amphibium* and several species of *Ranunculus* so characteristic of fresh water disappear even in the oligohaline region. According to H. LUTHER (1951) *Nuphar, Nymphaea, Potamogeton natans* stop at $2^0/_{00}$ S, only *Ranunculus obtusiflorus* transgresses $6^0/_{00}$

as a brackish-water species. The group of Akropleurostophytes shows a similar behaviour, that is plants which float near the water surface and produce floating leaves. With us this group is represented by species of *Lemna* (*L. polyrrhiza, L. minor, L. gibba*), *Hydrocharis, Limnanthemum, Salvinia* and others. Over pools and bays they may from a layer of vegetation completely covering the surface; but they are practically absent from the oligohalinikum *(Hydrocharis morsus ranae* up to $2^0/oo$). Of the Mesopleustophytes *Utricularia* and *Azolla* only reach the region of $2—3^0/oo$ S, *Lemna trisulca* $3—4^0/oo$, *Ceratophyllum demersum*, however, $6^0/oo$. These species are of no significance in brackish water, but the formation itself may be very effectively replaced by the floating masses of *Enteromorpha*. The extensive stands of Benthopleustophytes (plants lying locsely on the bottom) which occur in brackish water have already been mentioned on p. 123. They are mainly derived from marine plants, only the loose-lying balls of *Cladophora aegagropila* come from a fresh-water species.

It may be worthwhile considering the behaviour of those aquatic plants whose leaves or shoots rise above the water (Helophytes after H. LUTHER 1951). The floating Helophytes represented in the tropics by *Eichhornia, Ceratopteris, Pistia* and others appear to be equally intolerant of salt as our plants with floating leaves. But the Helophytes of the banks which root in the soil near the shore line behave differently. A reed stand (Phragmitetum) extends far into brackish water and may even form stands in the calm bays of the Wattenmeer in the North Sea. However, the number of species in this plant community decreases rapidly on transition into the oligohalinikum. At $3—4^0/oo$ S *Alisma, Sparganium, Typha, Butomus* etc. have disappeared. The species producing the formation *Phragmites communis* still forms extensive stands of reeds in brackish water and locally it still occurs in the Wattenmeer of the North Sea where it is often washed by water of $30^0/oo$ salinity. But in considering the distribution of these Helophytes in salt water it must be borne in mind that the tolerance of the underground roots or rhizomes may be very different from that of the shoots or leaves above ground. Reeds, among others, are found in more saline waters only where the salt content of the water at the bottom is slight so that the rhizomes live in low saline water (limit about $6—8^0/oo$?). Only the parts above ground tolerate more saline water (about $30^0/oo$).

There is, however, a distinct brackish-water reed-like formation, the Scirpetum maritimi (WI. CHR.) Tx. in which the reed recedes and *Scirpus maritimus*, locally *Sc. tabernaemontani, Sc. triqueter* and *Sc. americanus* are dominant. This *Scirpus* reed formation is more common in highly saline brackish waters than the Phragmitetum; but it does not offer the same area of attachment for sessile marine animals as does the reed. In many seas species of the genus *Spartina* (Gramineae) form reedlike stands in the tidal areas of sheltered shores and in lagoons, e. g. *Sp. brasiliensis* in lagoons of Brazil. An apparently newly produced species, *S. townsendi,* as an alloploid hybrid — has also spread out in the North Sea or has been planted for land reclamation; by spreading out in the meadows of *Puccinella maritima* it has become a weed (cf. KÖNIG 1948).

A specific plant formation of marine and brackish areas of sheltered water are the glasswort meadows (Salicornietum herbaceae) which are formed by the annual species of glasswort *Salicornia herbacea;* they penetrate so far into the water that they have to be considered here.[10] They depend on the existence of salt water and so are already less common in the Belt Sea and absent from weaker brackish water. Since glasswort stabilizes mud it is of great importance for land reclamation and is being sown (A. WOHLENBERG 1939). As a result the Salicornietum has spread extensively in the German Watt area.

The salt marshes adjoining on the landward side are chiefly covered by a closed cover of Gramineae, but also by flowering plants *(Aster tripolium, Plantago maritima* etc.)*. In these salt marshes a specific brackish-water fauna may inhabit the upper layers of soil. The tropical mangrove forests are most markedly developed in estuaries and lagoons with brackish water. The stilt roots of the *Rhizophora*-banks are the substratum for a special flora of Red Algae *(Bostrychietum),* but also for bivalves (oysters) and Balanidae. One might look to such areas for the origin of the wood-boring bivalves (Teredinidae).

2. Colonization of the phytal zone

The phytal zone of the sea has been colonized by a large number of sedentary animals (Porifera, Hydrozoa, Polychaeta, e. g. *Spirorbis, Pomatoceros* etc., Tunicata) which frequently settle on the algal thalli over wide areas. In addition there are numerous sedentary Ciliata of the Peritricha and Heterotricha *(Folliculina)* and sedentary Foraminifera *(Crithionina, Ophthalmina* etc.). The phytal zone of the fresh water may locally be richly colonized by the sessile animals of the microfauna (Peritricha, less commonly Heterotricha, Rotifera), but the sessile macrofauna is negligible. Porifera and Bryozoa settle on the stems of reeds, and leaves of water lilies may also be colonized by Bryozoa, but the number of species and the surface area of colonization of the fresh-water phytal zone are only a small fraction of those found in the marine phytal. In this respect brackish water occupies an intermediate position; the number of species of sessile animals and plants has decreased, but the intensity of colonization per surface area may still be high, especially by the polyps *Cordylophora caspia* and *Laomedea loveni* as well as the Bryozoon *Membranipora crustulenta.* In the polyhaline region of the Belt Sea the thalli of *Fucus* may have a dense covering of *Spirorbis,* less commonly of Tunicata.

As has been described for plants, the animals of the phytal zone form varied mixtures of associations. Fresh-water animals such as *Ephydatia, Hydra,* Trichoptera-larvae, *Limnaea, Physa* live on sea-weeds, e. g. *Fucus* and *Furcellaria,* marine or thalassogenous animals, e. g. *Cordylophora* on fresh-water plants such as *Potamogeton perfoliatus.*

It would be rewarding to investigate whether some regularity of occurrence may still be traced in this mixture of marine, fresh-water and brackish-water species.

3. Reef formation

Formation of reefs by sessile animals is typical of the sea. Apart from corals and Hydrozoa reef builders are found among Polychaeta, with their lime or sand

---

[10] The perennial shrub-like species of glasswort (S. *fruticosa, S. macrostachya* and others) which cover the salt floors in nearly worldwide distribution from the Mediterranean, only appear higher up outside the water: they roughly correspond to the *Puccinella maritima* meadows in northern areas.

tubes, especially the Sabellariidae (honeycomb worms). They can still be found on the North Sea coast; here, too, the small Polychaete *Fabricia sabella* produces locally thick, reef-like masses of sand. Though some of these Polychaeta enter the brackish water of the Baltic — esp. *Fabricia sabella* — no mass colonization takes place. *Fabricia* lives mostly as solitary organism, occasionally in "stands" of a few animals.

Fig. 54. Formation of reefs by the brackish-water Bryozoon *Membranipora crustulenta.* Waldhusentief on the Island of Pellworm (West coast of Schleswig-Holstein). The water level is lowered, so that clumps of Bryozoa partly emerge from the water. In the centre there are several *Haematopus* (oyster catchers) which give an indication of scale. Photograph Dr. D. König, Kiel.

The massive stands of tube-building Polychaeta are more numerous in the brackish water of warmer regions, e. g. the Serpulid *Mercierella enigmatica* which builds calcareous tubes (Mediterranean and the west-European Atlantic coast, single record in Kiel Bay). But a new phenomenon, Bryozoan reefs, may occur in brackish water; the brackish-water Bryozoon of world-wide distribution, *Membranipora crustulenta*, forms dense masses in calm bays; they have been described from the Zuiderzee (see REDEKE 1922) and from the German North Sea Island Pellworm (KÖNIG 1956) (figs. 54, 55). KÖNIG found the "reefs" to be about 3 m wide and 40 to 85 cm high; *Entermorpha intestinalis,* for example, and brackish-water diatoms were growing on them. *Hydrobia stagnalis, Gammarus zaddachi* as well as beetles *(Haliplus apicalis, Helophorus minutus, Cymbiodytes marginella, Anacaena limbata, Ochthebius impressicollis)* inhabited them.

Fig. 55. *Membranipora crustulenta*. Reef in the Waldhusentief on Pellworm (Photograph Dr. D. König, 25. 6. 53). Scale indicated by a chick of *Charadrius alexandrinus* (Kentish plover) which is crouching among the clumps. This species is a typical inhabitant of the brackish-water shore region.

B. Sandy floors

The formation and preservation of pure sandy bottoms depends on moderate, changing movement of water; therefore they are far more widespread on sea shores, including those of brackish-water seas, than in fresh water where they occur only on wave-beaten shores of large lakes and river banks.

Several very characteristic coenoses colonize the sandy areas of the sea; any differences are due to grain size, wave movement and connected with these, the content of sediments in the sand. Following these animal communities into the brackish-water region one is struck by the severe depletion and change of species. These phenomena are far more pronounced on sandy bottoms than in any other formations. The rich macrofauna of bivalves, echinoderms and crustaceans living in the coarse to gravelly sands in the North Sea and in the Canal *(Spatangus pur-pureus—Venus fasciata* community and also the *Echinocardium cordatum—Venus gallina* association which colonizes the finer sands, see REMANE 1941) have been reduced to small relicts even at 18⁰/₀₀ S. Their last outposts are in the Belt Sea, e. g. the Myside *Gastrosaccus spinifer,* the sea urchin *Echinocyamus pusillus,* the Amphipoda *Bathyporeia pelagica* and *Haustorius arenarius,* the Cumaceae *Pseudo-cuma longicornis* etc. The Amphipod *Phoxocephalus holboelli* (Öland, see FORSMAN 1956) penetrates to about 7—8⁰/₀₀, the Polychaeta *Pygospio elegans, Scoloplos armiger* penetrate even further as does the Isopod *Eurydice pulchra* in pure fine sand. In the Black Sea *Gastrosaccus sanctus, Branchiostoma,* species of *Glycera* etc. are found further in low saline water than in the North. The absence of the typical macrofauna of the purely marine regions is only partly made good 1) by the macrofauna of the *Macoma*-community colonizing these sandy areas: *Macoma baltica, Mya arenaria, Cardium edule* and, in the fine sands, the shrimp *Crangon crangon.* 2) by distinct brackish-water species. Under this heading only the rather euryhaline Amphipod *Bathyporeia pilosa* may be listed which in this region inhabits the strand areas of fine sand and moving water as do its close relations in the sea; it occurs to about 6⁰/₀₀ S. No fresh-water froms appear in the sandy zone of the mesohalinikum and, to my knowledge they have hardly been reported from the oligohalinikum. Thus the sandy areas of brackish water are very poor in species of the macrofauna. As many species display an increasing eurytopy in brackish water, a number of them penetrate into sands rich in detritus from their ecological centres of soft substrata or from the boundary area soft bottom—sand; these are e. g. the Isopod *Cyathura carinata, Nereis diversicolor* and elements of the Baltic *Pontoporeia—Mesidothea* association (see p. 131).

There are both differences and similarities between the microfauna and the macrofauna. The differences lie in a much greater abundance of species in brackish water, especially in marine Turbellaria, Nematoda and Copepoda. Undoubtedly there is less reduction in the number of species than in the macrofauna. As far as can be judged the marine element is particularly pure and rich in species, penetrating as far as the oligohalinikum. At any rate investigations of the fauna of sands on the Finnish coast by KARLING, PURASJOKI, W. NOODT (1953), AX (1954) have shown that even there the microfauna is of far more marine character than the fauna of the phytal zone and that of the littoral soft floors. The gravelly marginal strip which forms in places with regularly occurring breakers, the Otoplanen zone (see REMANE) is typically marine as far as the lower mesohalinikum, with its characteristic Oto-planidae and species of *Coelogynopora;* in brackish water, however, the Otoplanide *Bothriomolus balticus* becomes dominant while it is only local in the North Sea. In fresh water the same habitat is occupied by species of *Macrostomum* and possibly *Dochmiotrema limicola.* As species of *Macrostomum* occur more frequently in the Otoplanen zone of the Finnish coast than in more saline waters this might be conside-red as the beginning of an influence of fresh-water elements in this zone.

The proportion of distinct brackish-water species in the microfauna may be much higher than in the macrofauna. A large number of species of Turbellaria, Nematoda, but also Ostracoda, e. g. *Cytherideis baltica* and *Microcytherura affinis* in the southern Baltic (Kolberger Kessel, Bay of Danzig, see KLIE 1938) are known only from the sands of the Belt Sea and the Baltic. The areas of the Belt Sea and the Baltic have been very thoroughly studied regarding the microfauna of the marine sands, most species have been newly discovered within the last 30 years so that an accurate survey of their distribution in other areas has not yet been possible. But on several occasions Turbellaria have been found in brackish-water regions and Ax (1956) describes about 20 species as brackish-water ones inhabiting sand. Undoubtedly a large number of sand dwellers from the brackish-water species of the microfauna occur in coastal ground water (see p. 175); this harbours many distinct species and forms part of the sandy regions.

As happens in the macrofauna, numerous microorganisms which inhabit the eulittoral in the sea also occur in brackish water. As mentioned before the *Halammohydra* coenosis, typical of medium and coarse sands of the sublittoral in the Belt Sea, inhabits the eulittoral in the North Sea with most of its species; it is found on the undercut slope in running water. In contrast to the macrofauna there are no immigrants from areas of soft substrata in the microfauna of pure sandy regions in brackish water. This can easily be explained since the microfauna of the sand lives in the interstitial system of the sand and this can only be occupied in places where it has not been filled up by large amounts of sediments.

C. Unconsolidated sea floor

The areas of soft floors with Gyttja, Mudd etc. occur in the sea in two situations which are usually wide apart. One is in the calm depths and hollows, the other in lagoons and the shallow zones of the shore which are protected from breakers by bars or islands. The two areas differ in their environmental conditions. The soft floors of the deep share the relatively constant environmental factors of temperature, salt content and illumination with their surroundings; the littoral soft floors are exposed to large fluctuations in these factors. The light intensity and high humidity here, even when uncovered, allow an abundant growth of lower assimilating plants (Diatomaceae, Cyanophyceae) which are absent from the soft substrata of the deep. In addition the arrival of plant parts which have been ground up on the shore or in the breaking waves produces a local accumulation of vegetable remains in or on the soft bottom. As a result of higher temperatures and the high incidence of material of organic origin extensive processes of decomposition and, locally, oxygen depletion take place. On the whole the littoral soft floors are poorer in salts because of their vicinity to the shore and the effects of precipitation. But that does not apply universally. In the Red Sea on the Island Abu Mingar I found an Avicennia—mangrove association with true littoral soft floor, but a salinity of $44—45^0/_{00}$.

There are great biological differences between these two zones of soft substrata. Typical of the deep zone is an abundance of Echinodermata, Isopoda, Cumaceae, while in the littoral areas of soft floors the Polychaeta are dominant as well as a few Gastropoda which are able to crawl on the surface of the soft floor *(Littorina litorea, Nerita,* species of *Hydrobia, Haminea* etc.). The microfauna, especially Copepoda and Nematoda, has special local mass development and forms of fresh-water origin, in particular Oligochaeta of the genera *Peloscolex, Clitellio* and others which often produce densely populated colonies.

In brackish water the contrast between the deep and the littoral zones is mitigated. Owing to the smaller size of the brackish waters and the resultingly reduced intensity

130   Ecology of brackish water

of the breakers the upper limit of the deep soft floors is higher. It is important that in the more or less stagnant deep zones oxygen depletion is an additional factor which inhibits life; this affects the areas of soft substrata and may lead to the disappearance of the fauna in brackish-water floors.

Colonization of the sublittoral soft floors from marine regions into brackish water presents a complex picture. The reduction in number of species is particularly marked in Echinodermata, but also noticeable in Amphipoda and bivalves. With the brittle star *Ophiura albida* the echinoderms have their outpost in the deep zones of the southern Baltic Basin. In the Black Sea the corresponding region has higher salinities and *Amphiura, Cucumaria orientalis* and *Synapta hispida* still occur there (cf. Caspers 1951, p. 126). In addition there are a number of other elements. Among them, in the first place, there are forms which inhabit the eulittoral or et least live near the surface in the sea. That applies particularly to the banks of bivalves. In the Baltic *Mytilus edulis* forms stands in the sublittoral area of soft substrata (down to over 100 m), in the Black Sea *Mytilus galloprovincialis* (to 183 m). Another bank-forming bivalve, originally belonging to the rocky region, can be found in the area of soft floors in brackish water. These are species of the genus *Modiolus* which form stands below the *Mytilus*-zone in rocky regions (*Volsella*—association of Gislen 1930). In the Baltic area it still reaches the soft floors of Kiel Bay, in the Black Sea it is typical of *Modiolus-phaseolinus* ooze (biocoenosis of the *Phaseolinus*—Schlick [mud] of Sernow 1913). In the Black Sea it extends from a depth of about 70 m to about 180 m, near the boundary of the azoic zone (figs. 56, 57). Caspers (1951) has made a list of species of this region, partly following Sernow and Borcea. In addition there

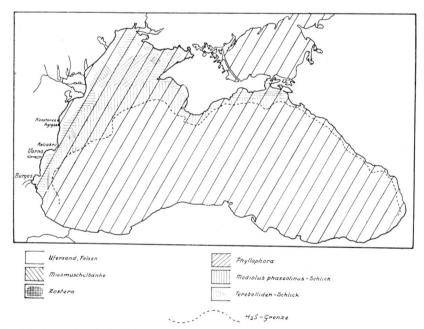

Fig. 56. Areas of colonisation on the bottom layer of the Black Sea. The region within the inserted H₂S boundary is azoic. After Sernow from Caspers 1951.

is the Pontocaspian *Mytilaster lineatus* which is dominant in the Caspian Sea and the Sea of Azov. As young bivalves settle on the shells of animals already attached or else on dead shells, they may form enormous masses where there is an ample supply of nutrients. The formations of bivalves just described belong primarily to the rocky floor, but as they are members of the epifauna of the littoral soft floors (partly *Macoma* and *Abra* community) they represent a transition to those forms which, from the littoral areas, invade the sublittoral ones in brackish water.

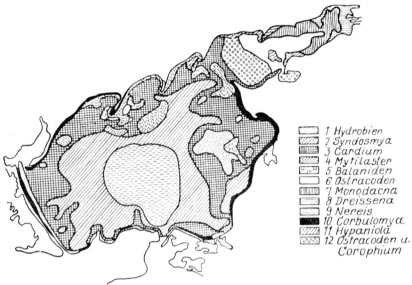

1 *Hydrobien*
2 *Syndosmya*
3 *Cardium*
4 *Mytilaster*
5 *Balaniden*
6 *Ostracoden*
7 *Monodacna*
8 *Dreissena*
9 *Nereis*
10 *Corbulomya*
11 *Hypaniola*
12 *Ostracoden u. Corophium*

Fig. 57. Distribution of the characteristic species in the Sea of Azov (area of shallow brackish water). From SHADIN 1952.

As in sandy regions elements of the Infauna of the littoral *Macoma* and *Abra* community penetrate the deeper soft substrata. Because of its salinity requirements *Abra (Syndosmya) alba* is found predominantly in the polyhalinikum or the most highly saline mesohalinikum, that is in the Belt Sea with *Cardium fasciatum* and frequently with a rich development of the Polychaete *Terebellides stroemii*. SERNOW (1913) mentions a Terebellidenschlick [mud] in the Black Sea; a similar formation occurs in the Belt Sea. In the Baltic area there are especially in the soft substrata a number of species which are either absent from the North Sea or else occur only locally; but they are widespread in the northern-boreal or arctic regions. The famous "glacial relicts" of the Baltic belong here. The arctic-baltic animals sensu stricto include *Mesidothea entomon, Pontoporeia affinis, Paracyprideis fennica* and, sensu lato, *Macoma calcarea*, species of *Astarta* etc. As a result of these various additional components the deeper zone of soft show not only a general impoverishment, but a peculiar change of the faunal elements until in less saline regions the arctic animal community of *Mesidotea—Pontoporeia* association (HESSLE 1924) assumes dominance, a mixture of species of varying origins in which also fresh-water elements appear (Chironomidae larvae, *Candona neglecta*) in deeper soft floors.

The macrofauna of the littoral soft substrata shows a rapid reduction from the sea into brackish water; but *Nereis diversicolor* and *Corophium volutator* which are particularly resistant to oxygen deficiency persist, as well as the euryoecious bivalves of the *Macoma* community: *Macoma baltica, Mya arenaria*, in part *Cardium edule*. In addition there is an irregular occurrence of brackish-water forms: the small Gastropod *Hydrobia ventrosa*, the bivalve *Cardium lamarcki*, the Isopoda *Cyathura carinata* and *Heterotanais oerstedi* and in places with stands of vegetation *Idotea viridis*, *Sphaeroma* and others. The animals of the meso- and microfauna are more typical, the Polychaeta *Manayunkia aestuarina, Streblospio shrubsoli, Alcmaria romijni*, Ostracoda (mass development of *Cyprideis litoralis, Loxoconcha elliptica, Cytherura gibba* and Copepoda. Therefore the zone is called the *Cyprideis litoralis-Manayunkia aestuarina*-coenosis (REMANE 1941).

D. Plankton

The productivity of the brackish-water areas may vary considerably. On p. 170 it will be shown that in brackish-water regions near the coast the plankton may be very abundant. On the other hand the central masses of the brackish-water seas are very poor in plankton (see DIETRICH and KALLE 1957). That applies both to the Baltic and the Black Sea. As vertical circulation is inhibited this leads to an accumulation of nutrients in the deep region, poor in life, and to a rapid consumption of the materials in the upper stratum poor in salts. The amount of plankton only increases in the zone in which nutrients rise as a result of the slow horizontal circulation, that is, for instance, in the Bottensee and Bottenwiek. There are not only quantitative, but also qualitative differences between the plankton of the central masses of brackish-water seas and that from the vicinity of the coast. In the Baltic the central brackish-water plankton — as I call it — is distinguished from that of the littoral brackish water by the following differential species: the Medusae *Sarsia tubulosa* and *Halitholus cirratus*, the Ctenophore *Pleurobrachia pileus*, *Sagitta*, the Appendicularia *Oikopleura dioica*, the Copepoda *Pseudocalanus elongatus, Paracalanus parvus, Centropages hamatus, Temora longicornis, Acartia clausi* and *Oithona similis*, the Cladocera *Evadne spinifera, Podon Leuckarti, P. intermedius*, the Rotifera *Synchaeta monopus* and perhaps *S. baltica*. The littoral plankton on the other hand, also has its typical differential species which attain their maximum development here, e. g. the Copepoda *Eurytemora velox, E. hirundo, E. hirundoides* and *Acartia tonsa*, the Rotifera *Pedalia fennica, Brachionus plicatilis* as well as most of the euryhaline fresh-water planktonic organisms which enter brackish water, e. g. *Brachionus angularis, B. calyciflorus, B. capsuliflorus, Filinia maior, F. longiseta*. A number of species are, of course, common to both regions. In enclosed bays and brackish lakes the central plankton appears with an inflow of salt water, while the coastal plankton represents the indigenous one. RENTZ (1940) refers to the central plankton of the Boddenwaters as the main plankton of the outside waters, to the coastal forms as the plankton of the inside water, and to common species as the basic plankton.

Various factors may account for the differences between coastal and central plankton. Euryhaline fresh-water planktonic organisms which only colonize the oligohaline region will obviously be restricted with it to the coastal region since the appropriate salt content will only be found there. Other forms which are tolerant of salt are probably restricted to the coastal waters in which microscopic phytoplankton and detritus are often very abundant. This may, above all, be true for ciliary feeders e. g. many Rotifera. Others again may be restricted to the coastal areas because of competition in the sea (cf. REMANE 1950). There is considerable agreement between the plankton of the Baltic and that of the Black Sea. The following general plan holds for the different salinity zones: in the oligohaline region there is a large amount of fresh-water plankton, with a number of brackish-water forms. In the sphere of the coastal plankton the mesohaline region also contains a fairly high number of fresh-water plankton. SEGERSTRÅLE (1939) lists in a plankton profile on the South coast of Finland Pellinge-Lill-Pernaviken, in a salinity range of 3.9 to 4.6$^0/_{00}$ 9 marine species, 11 brackish-water species and 18 fresh-water ones. Thus in the plankton, especially of the coastal regions, the fresh-water influence is more far reaching than in the communities of the bottom.

V. Types of brackish water

The wealth of special types of waters is as great in brackish water as it is in the fresh-water region. Nearly all types of fresh water have corresponding types in brackish water or water of varying salinity, only running water is rare and confined to the outflow of salt springs; on the other hand brackish-water seas such as the Baltic, the Caspian Sea etc. attain far greater dimensions than the largest fresh-water lakes. Recently a classification of brackish waters into types has been attempted (EMERY). The following classification appears to me the most suitable one:

A. Brackish seas. Large minor seas connected with the ocean by narrow straits (Baltic, Black Sea) or those which were isolated from the ocean in geological time (Caspian Sea, Lake Aral).

B. Estuaries. River mouths, mostly widened into a funnel shape, on the surface an outflow of fresh water, in the depth the saline countercurrent. Both mingle to form brackish bodies of water with oblique fronts. The deep layer with a good supply of oxygen, the bottom layer mostly of fine sands, which are transformed into "dunes" and moved about, especially in the tidal region. There are two variants: 1) Tidal estuaries with strong regular movement of the water masses at low and high tide. 2) Tideless estuaries in minor seas with slight tides (Baltic, Mediterranean) or without tides, but irregular water movements caused by wind damming and drifting off. It is obvious that supply of fresh water varies according to the climate.

C. Fjorde (= Förden). Mountain valleys or melt-water runnels which have formed a connection with the sea. The formation of these elongated bays is based on geological factors of past times: Their outlines and the bottom relief are irregular, deep troughs are often found in them and highly saline water will collect in these; it may poor in O_2 content. In the troughs or hollows mostly filled with mud, the supply of fresh water is relatively slight, but varied, e. g. in the mountain areas by the melting of the snow. Here, too, there are tidal fjords and those with slight tides.

D. Lagoons (Etangs, Strandseen, Haffs). Bays of the sea cut off by the migration of sand on the sea shores. Their maximal extent is usually parallel to the coast, the varying cutting-off from the sea is brought about by bars (Nehrungen). Usually little depth, very rich in nutrients, with large fluctuations in salinity, but mostly little stratification.

E. Shore pools. Small bodies of water which are reached by sea water at times, at least by spray water or by water seeping through the soil. Very rich and varied in species on rocky shores, but also in salt marshes, in sandy areas mostly periodic (flood). Great differences in substrata, change of water, bottom.

F. Salt marshes. The salt marshes with a dense cover of vegetation are already "land", but are flooded by sea water either by the tides or at a high water level. In the soil of the vegetation cover a zone of brackish water develops; this contains Cyanophyceae and a rich fauna.

G. Interstitial brackish water of the coast (coastal ground water). On sandy shores sea water seeps into the soil, in tidal areas there is a change of infiltration and outflow of the ground water. In brackish-water seas it is brackish and in places where the fresh soil water gets contact with this zone there also develops a subterranean brackish-water zone.

A. Brackish-water seas

According to the way in which brackish water is produced only minor seas with a restricted inflow of oceanic sea water may become brackish-water seas. But excessive amount of precipitation or the inflow of large rivers may reduce the salinity at least of the surface water in some marginal areas of the oceans to about $25^0/_{00}$ S. That holds true for the Pacific in the coastal region Panama-Columbia (Gulf of Panama) for the Gulf of Guinea on the West coast of Africa (Bay of Biafra) and the central parts of the Bay of Bengal (see DIETRICH and KALLE 1957). For an understanding of the behaviour of the organisms in the salinity range of 35 to $25^0/_{00}$, a biological investigation of these regions would be of great importance.

The typical brackish-water seas, however, are all true minor seas with narrow openings to the ocean or else basins completely isolated from the sea (Caspian Sea, Lake Aral). Common to all brackish-water seas is a reduction of the difference in level between low and high tide which attains a few centimetres at the most. In the Belt Sea, in the approaches of the Baltic, the tidal range is only 10—20 cm, at the opening of the Baltic proper only about half that amount. This is due to the character of minor seas displayed by these waters; but in the Baltic and the Black Sea this is further reinforced by the fact that separation from the ocean is increased by marine basins lying between them and the ocean, that is the Kattegat and the Mediterranean; these in themselves have reduced tides. Thus the typical tidal area of the coast of the open sea, with its regular alternation of low and high tide, is lacking in the brackish water. But because of their extension over wide areas these brackish-water seas have large irregular fluctuations of their water level, brought about by wind and damming-up of water. In the Belt Sea the difference between extreme high- and low-water levels measured in this century amounts to about 4.5 m! (cf. E. G. KANNENBERG 1956). In contrast to the high-water regions of river banks these marine extremes of water levels are more dependent on the weather and to a lesser extent on seasons or the climate; also the changes between high and low water may take place very rapidly, e. g. near Flensburg within 4 days (31st December 1904 to 3rd January

1905) from + 224 cm to —157 cm, near Travemünde within 3 days (6/7th to 9th January 1908) from —137 cm to +184 cm (KANNENBERG 1956).

The phenomena listed on p. 10 ff. as common for brackish water also apply to the brackish seas: Stratification of water and salt content, stagnation, oxygen deficiency, abundance of H_2S in deep water etc.

The brackish-water seas differ in their historical development and in their hydrography.

1. The Baltic

a) History

Geologically the Baltic is a very recent sea, having developed into one almost within historic times. Its area was still covered by the last Ice Age and the interglacial seas which existed in place of the Baltic were practically ruled out as habitats. The deciphering of the history of this Baltic is a scientific feat of Swedish (MUNTHE 1910, DE GEER 1910) and Finnish scientists (SAURAMO 1929). At the edge of the retreating ice a freshwater lake of vast dimensions was first formed — the Baltic Ice Lake — which existed from the beginning of the retreat of the ice until about 8000 B. C. (end of the Gothiglacial). During its later period glaciers were still covering the Gulf of Bothnia, so it filled the southern Baltic, but the sea level was far below the present one (over 80 m). The Belt Sea was land, southern Sweden had a land connection — over the Danish Islands — with Jutland and North Germany; similarly Bornholm had a land bridge with part of a penunsula jutting out in the South. But the shore line of the Ice Lake was above the present level in the Northeast of the Gulf of Riga and, above all, in the Gulf of Finland where it covered southern Finland including Lake Ladoga. Outflow occurred at first by a river course through the Öresund to the Kattegat which, at that period, was a sea in its eastern region; later it took place through central Sweden near Billingen. A transgression in this region led to a sudden heavy draining and a lowering of the water level by 26 m. The breakthrough resulted in a connection between the Ice Lake and the sea through central Sweden (region of Vänersee), and the infiltration of salt water. This created the saline Yoldia Sea, named after the arctic bivalve *Yoldia arctica* (now *Portlandia arctica*) which has been found in large numbers in clays near and north of Stockholm near the inflow of sea water and the margin of the glacier at that time. The diatom flora, too, reveals the period of the saline Yoldia Sea *(Campylodiscus clypeus)*. It is difficult to assess the degree of salinity of this sea. The bivalves themselves are smaller than the normal from. It is certain that this first brackish-water period of the Baltic was a short one, it only lasted about 500 (SAURAMO) to 700 years (MUNTHE). *Portlandia* itself probably only lived for 100 years near Stockholm (from MAGNUSSON-GRANLUND-LUNDQUIST): The uplift of Scandinavia which began shortly after the isostatic recovery and is, in part, continuing to the present day, soon broke the connection sea-Baltic Basin in central Sweden. For the second time the Baltic reverted to a fresh-water lake which has been called Ancylus Lake after the fresh-water Gastropod *Ancylus fluviatilis*, widespread at the time. The following diatoms are typical of that period: *Melosira arenaria, Cymatopleura elliptica, Gyrosigma attenuatum, Epithemia Hyndmannii* and others. The Ancylus Lake existed from 7800—6200 B. C. In magnificent water falls it discharged at first over central Sweden (Svea-Älf near

Digefen, Vänersee, Göta-Älf). But in the course of the general uplift of Scandinavia this region was raised to such an extent that the outflow came to a halt and the level of Lake Ancylus rose. This resulted in the coastal area of the southern Baltic being submerged and a new river-like outflow being formed, from about the area of the Darsser Schwelle through Fehmarnsund and the Great Belt. This prehistoric river is called Dana-Älf. At present it cannot be determined whether Lake Ancylus contained pure fresh water down to its greatest depths or whether any brackish-water zones had remained there.

Through eustatic rise of the sea level and the breakthrough of the Atlantic in the Channel (Dover—Calais) the North Sea originated (Flandrian transgression, Tapes-time); favoured by higher temperatures numerous Lusitanian species immigrated into it through the Channel. This transgression advanced further into the Belt Sae and led to a transgression of salt water into Lake Ancylus.

The Baltic turned into the saline Litorina Sea. It is significant that the water level was higher than at present so that salt water was able to flow through the Belts and Sound to a greater extent. In fact the fauna clearly indicates a higher salinity. *Littorina litorea* the Gastropod after which it has been named now only occurs in isolated localities near Rügen and Bornholm; at that time it extended as far as Finland and Hudiksvall (62° north). *Littorina saxatilis tenebrosa,* at present having the same limits as the previous species, then reached the northern end of the Gulf of Bothnia; the small Gastropod *Zippora membranacea* occurs as far as Greifswald today, but in that period it extended to the Aaland Islands and Uppsala. The peppery furrow shell *Scrobicularia plana* — more typical of the Litorina period deposits of the German Baltic than *Littorina,* now extends to the Bay of Mecklenburg (Wismar), previously to North Gotland. For sediments near Greifswald the following bivalves have been listed *Syndosmya alba, Saxicava artica, Corbula gibba* which are absent today, the Gastropoda *Gibbula cineraria* and *Littorina (Neritoides) obtusata* (see S. JAECKEL jun. 1949). An estimate of salinity based on differences in distribution and the larger dimensions of the molluscs of that period would yield a figure of 14—16⁰/₀₀ salinity for the central basin of the Baltic, that is twice the present value! The Belt Sea, too, was more saline at that time as shown by the occurrence of the oyster *(Ostrea edulis),* the Gastropod *Scala clathrus* and the shell-boring Polychaete *Dodecaceria concharum* (S. JAECKEL jun. 1949) in Kiel Bay. They are no longer found there today.

The salt phase provides important evidence that at least many of the boundaries of marine animals in the Baltic do not represent the fortuitous and transitory lines of progression of recent immigrants, but are border lines demarcated by the exigencies of the environment (salinity).

The climax of the Litorina Sea was about 4000 B. C. By a regression of the shore line and consequent reduction in the inflow of salt water through the Belts the Baltic became less saline and gradually approached its present-day condition. From about 2000 B. C. the Baltic is called Lymnaea Sea, so designated after the expansion of the euryhaline fresh-water snail *Limnaea ovata* had taken place; but it had also occurred in Lake Ancylus. The present era of the Baltic is called Mya Sea after the gaper *Mya arenaria.* Some authors (HESSLAND 1943) consider it a passively introduced immigrant from North America which only appeared after 1600 A. D., while others record its remains in the approaches to the Baltic as early as the Iron or even

Bronze Age (see S. JAECKEL jun. 1949). No hydrographic changes beyond the usual fluctuations have taken place in the Baltic in the Mya period.

This changing history of the Baltic provides an explanation of many of its biological peculiarities (see p. 141).

b) Hydrography

The special hydrographic features are based on the following data: The Baltic itself forms several deep troughs which are to some extent delimited from one another by rises, in the West the Arkona Basin (Rügen Basin) with a depth of 46—48 m, then the Bornholm Basin with a maximum depth of 105 m, the central basin with the Danziger Tief 113 m, the Gotlandtief 249 m and the Landortstief 427 m. In the North the Aaland Sea is bounded by islands, several rises (30—80 m) and deeps (219, 301 m). The northern part (Gulf of Bothnia) has two troughs, the southern Bottensee to a depth of 294 m and the northern Bottenwiek to a depth of 240 m. This structure of troughs and their position roughly in a row has a decisive influence over the course of the salt water coming in from the West which, according to its density, accumulates in the troughs. Consequently the deep water in the western troughs is much richer in salt content than in the eastern and northern ones, the deep water in the Arkona Basin may reach $24^o/_{oo}$, in the Bornholm Basin it often contains over $15^o/_{oo}$, in the Danziger Tief up to about $13,5^o/_{oo}$, in the Gotlandtief up to about $12.5^o/_{oo}$ and in the Landortstief $11^o/_{oo}$, in the depths of the Aaland Sea $7—8^o/_{oo}$, in those of the Bottensee $6—7^o/_{oo}$, in those of the Bottenwiek $4—5^o/_{oo}$. As shown in fig. 6 the salt content of the surface water is far more uniform in the western and central parts of the Baltic and only decreases to a marked degree in the interior of the eastern (Gulf of Riga and Gulf of Finland) and the northern bays (Gulf of Bothnia).

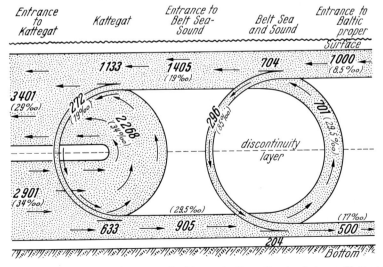

Fig. 58. Diagram of change of water and of intermingling of water between Kattegat and the Baltic, with salinity data. From STEEMANN-NIELSEN 1940.

The inflow of salt water occurs through two openings, one of them the Öresund. It is in direct communication with the highly saline Kattegat, but the depth of its swell is only 7 m and at its narrowest point it is only 4 km wide. Thus its influence is far less than that of the second entrance over the Darsser Schwelle between Gjedser and Mecklenburg, with a swell of 18 m and a width of 34 km. The salt water flows over the Darsser Schwelle after passing the great mixing zone of the Belt Sea, especially of Kiel Bay and the Bay of Lübeck. Here in the shallow Belt Sea which has a maximum depth of 30 m in this area, there is a mixing of inflow and outflow (fig. 58). Here a block of middle Belt Sea water has often formed which, according to weather conditions, is being pushed one way and another and partly transgresses the Schwelle to the East (WATTENBERG 1940). In particular there are superposition and interdigitation of different sheets of water which produce a changing and highly complex picture of stratification and distribution (cf. WYRTKI 1954). In shallow water the salinity curve may show rapid, almost vibrating deflections in the course of a few days and, within a few days, the salt content from the surface to the depth may snow far reaching differences, far more pronounced than in the Baltic proper or in the open sea.

The hydrography of the Baltic proper is in a large measure influenced by the existence of zonation into several successive basins and the relatively wide routes of supply of salt water. In contrast to the Belt Sea which is turbulent in this respect, the Baltic proper tends to show the following fundamental trends (see DIETRICH 1950).

In a horizontal direction salinity decreases quite gradually, on the surface from 8—9⁰/₀₀, in the Arkona Sea to 6⁰/₀₀, in the Aaland Sea and the Finnish coasts, in the Gulf of Riga it decreases to 4—5⁰/₀₀, in the Bottenwiek and the Gulf of Finland to nearly 0. Such a very gradual reduction in salinity over stretches of about 1500 km makes the Baltic region of cardinal importance for problems of brackish-water biology.

In contrast to this slight difference in salinities in a horizontal direction there is a strong vertical stratification which is far more pronounced than in the North Sea, also more marked than in the Black Sea, though less so than in the Belt Sea. The difference in salt content surface — sea floor at the height of the summer amounts to: (according to DIETRICH) in the North Sea itself under 0.25⁰/₀₀, mostly under 0.1⁰/₀₀, only in coastal areas, especially in the Skagerrak it is higher; in the Baltic it reaches over 8 in the Arkona and Bornholm Basins; in wide areas it attains over 4 and only decreases below 1 in the marginal areas. In the Belt Sea it rises above 15! The weakly saline water of the surface is mostly 40—80 m deep, in these depths there is a salinity discontinuity layer. This haline stratification shows a remarkable stability (DIETRICHS region B 2) and it is only regionally absent. (DIETRICHS region B 1). This makes it difficult for the bottom water to be renewed by convection. These conditions should lead to oxygen deficiency and accumulation of H₂S with biologically disastrous consequences, as is the case in the Black Sea. From time to time events tend to move in that direction[11]; but at irregular intervals they are interrupted by occasional strong

---

[11] The formation of an azoic zone, poor in oxygen, is most pronounced in the Gotland Basin (SHURIN and SEGERSTRÅLE 1956), from a depth of 80—150 m onwards; but in the Bornholm Basin there is also an occasional mass death of the bottom fauna. But these azoic periods are continually interrupted by normal ones. LUTZE 1965 found living Foraminifera in the greatest depths.

transgressions of salt water over the Darsser Schwelle; as a result, the bottom water
is renewed and the oxygen content rises steeply. This aperiodic renewal from the
depth is, of course, particularly frequent and intensive in the basins close to the influx,
that is the Arkona and Bornholm Basins, but the effects of these large transgressions
may reach as far as the Gotland Basin and they may influence the whole salinity

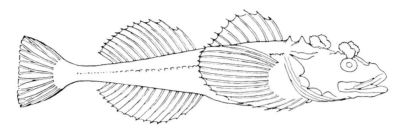

Fig. 59. *Cottus quadricornis,* an arctic-Baltic fish which inhabits low-salinity water in the
Baltic and fresh-water lakes in Scandinavia, Finland and Northern Russia. After Möbius &
HEINCKE 1883.

structure of the Baltic. KALLE (1943) has investigated such a big disruption in water
stratification in the Gotland depression 1933/34 and WYRTKI (1954) has studied the
big influx of salt into the Baltic in November/December 1951. In the last-mentioned
transgression of sea water about 220 cubic kilometres of highly saline water reached
the Baltic over the western ridges, with 4.4 milliards of tons of salt. Salinity in the
depth of the Bornholm trough from 70 m downwards rose to 21—21.5⁰/₀₀ S (normally

Table 7. Temperature of bottom water, as measured on the light ships Kiel, Flensburg
and Fehmarnbelt; monthly mean values (averages) based on continuous, daily observations
(records) 1937—1942. In degrees C; in () month.

	1937	1938	1939	1940	1941	1942
Light ship Kiel Highest Temperatures	13.77 (IX)	16.93 (VIII)	18.2 (IX)	12.55 (IX)	12.0 (VIII)	12.46 (IX)
Lowest	−0.23 (II)	2.42 (III)	2.2 (I)	0.5 (IV)	−0.11 (I)	0.6 (IV)
Light ship Flensburg Highest Temperatures	11.44 (XI)	12.48 (X)	11.67 (X)	11.09 (X)	12.62 (X)	11.39 (XI)
Lowest	−0.51 (II)	2.61 (XII)	1.98 (I)	−0.14 (III)	—	—
Light ship Fehmarnbelt Highest Temperatures	12.55 (X)	13.26 (X)	—	12.24 (IX)	12.14 (X)	12.62 (IX)
Lowest	0.10 (II)	3.62 (I)	—	—	0.20 (I)	1.50 (I)

The above values have been calculated by BANDL, Inst. f. Meereskde. d. Univ. Kiel;
made available by Inst. f. Meereskde.

14—18⁰/₀₀). The biological effects of such transgressions are very significant on account of the renewal of the water from the depth, increased salinity, supply and mobilization of nutrients. With increasing salinity animals extended their area northwards; in the deeper basins there occurs recolonization by bottom animals *(Macoma, Priapulus caudatus, Terebellides strömi, Scoloplos armiger, Diastylis rathkei* and others), there are increased returns from fisheries etc.

As an example of the complexities which may arise in water stratification I refer to the situation in the deep basins of the Baltic, Arkona—Bornholm—Gotland— Basins after Wüst and Brogmus (1955). Here we find a) a poor and brackish upper water; b) a very cold and brackish intermediate water with intermediate temperature minimum above the salinity gradient; c) an obliquely sloping cool deep water, about 20 m deep, within the larger layers of salinity transition and several temperature variations and d) somewhat warmer, highly saline water at the bottom (see fig. 5).

The annual fluctuation of salinity at the surface is about 0.5⁰/₀₀ in the large main area, and about 1.0⁰/₀₀ in the extreme marginal regions (Bottenwiek, Gulf of Finland) less than in the Belt Sea with 3—5 and the Kattegat with over 7⁰/₀₀, but it is greater than in the North Sea (excluding the marginal areas) (Dietrich).

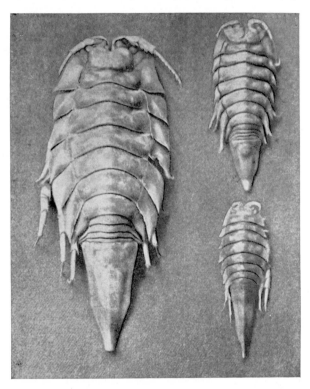

Fig. 60. Species of the arctic-Baltic-Caspian Isopod genus *Mesidothea*. Left: *M. sibirica*, the purely arctic relative of *M. entomon* (possibly ancestral form). Male from the Karia Sea at maximum size. Right upper *M. sibirica*. Female from the Karia Sea. Bottom right: *M. entomon* from Lake Vättern, Sweden (var. *vetterensis*). Male at maximum size. Natural size. After Ekman from Thienemann 1950.

Clearly, the temperature of the surface layer is determined by the northern latitude; in the northern regions of the Baltic it is only 12—14⁰C at the surface in August. It is significant that water from the depth is of a lower temperature than that from comparable situations in the North Sea or the open sea: from a depth of 30—40 m it usually remains below 5⁰C and EKMAN (1953) calls the temperature of this area of the Baltic "continually arctic", a fact which is of importance for the occurrence of the arctic fauna elements in the Baltic (see Table 7).

c) Biology

From the preceding chapters it has become evident that the Baltic does not merely harbour an impoverished fauna of the North Sea-Kattegat region. The peculiar features may be briefly summarized as follows: Communities of the coastal regions of the North Sea displace their area into deeper zones or extend it downwards. The first type of behaviour has been established for the *Halammohydra* coenosis which is rich in species. In the North Sea it lives predominantly in the undercut-slope areas of coarse sand of the tidal zone, but in the Belt Sea and Baltic in the sublittoral (REMANE 1955). The second type of behaviour is typical of the macrofauna of the *Macoma baltica* coenosis (in parts also the *Abra alba* coenosis). In the North Sea they are littoral with few records in the deeper parts, but in the Baltic they are found at over 100 m depth *(Macoma baltica, Hydrobia ulvae, Mytilus edulis,* similarly the Polychaete *Pygospio elegans, Scoloplos armiger, Harmothoe sarsi* (see p. 46) (figs. 61, 62).

A number of species mainly arctic in distribution occur in the Baltic as well, but not in the North Sea at the same latitude (or only at a few points). EKMAN, SEGERSTRÅLE have made a thorough investigation of this arctic-Baltic faunal element (see THIENEMANN's compilation 1950). It shows neither ecological nor historical uniformity. Two main groups can easily be distinguished.

I. The first group includes the large Isopod *Mesidothea entomon* (fig. 60), the Amphipod *Pontoporeia affinis,* the planktonic Copepod *Limnocalanus grimaldii-marcurus,* the Mysidae *Mysis relicta,* the fish *Myoxocephalus (Cottus) quadricornis* and, in a wider sense, *Coregonus albula.* Furthermore the seal *Phoca hispida.* The Amphipod *Gammaracanthus loricatus-lacustris* which belongs to this group from the point of animal geography, is absent from the Baltic itself, but widespread in fresh-water lakes in the Baltic region. In SEGERSTRÅLE's view the high temperatures during the Litorina-period have probably caused the extinction of this species in the Baltic itself.

The animals in this group are predominantly brackish-water species. The main area of their occurrence lies in the northern parts of the Baltic where they form the *Mesidothea entomon—Pontoporeia affinis* community. They occur only sporadically in the western Baltic.

During the Ice Age nearly all these species also inhabited fresh-water lakes. This applies not only for Europe and parts of Northern Asia; in North American lakes there were *Mysis relicta, Pontoporeia affinis, Limnocalanus, Myoxocephalus. Cottus thompsoni* takes the place of *quadricornis, Mesidothea, Gammaracanthus* are absent from North America. Compared with brackish-water populations those from fresh water are always smaller, in part exhibiting juvenile characters; in *Myxocephalus* the cephalic spines are reduced. Fresh-water populations are often considered to be separate species *(Limnocalanus macrurus, Gammaracanthus lacustris)*. According to available data this can not be justified. All species are found in the brackish water of the polar region where *Gammaracanthus loricatus* can tolerate a higher salinity (26—30⁰/oo). *Myoxocephalus* is designated marine thus it would be holeuryhaline. Some species also occur in the Caspian Sea (p. 154).

Some penetrating studies have been made to elucidate the problem of immigration of these species into the Baltic and into fresh water (EKMAN, HÖGBÖM 1917, SEGER-STRÅLE 1957, 1962). It is certain that invasion of the Baltic took place from the North. According to SEGERSTRÅLE's careful investigations the Siberian Ice Lake has played an important part in the expansion of the species; it formed a large basin east of the Urals. Through ice-dammed lakes in front of the receding ice the species reached the Onega dammed lake and from this they reached the Baltic. These species are supposed to have undergone a period in fresh water before reaching the brackish water of the Baltic.

II. The species of the second group of arctic—Baltic animals live mainly in the central and southern half of the Baltic, most of them occurring in the Belt Sea and Kattegat as well, and there are isolated records from the Norwegian coast. Nowhere do they invade fresh water and mostly avoid oligohaline water. To this group belong the bivalves *Astarte borealis* and *Macoma calcarea,* typical species of the shallow-water community (see SPÄRCK 1938), the Amphipod *Pontoporeia femorata,* the Ostracod *Paracyprideis fennica,* the Priapulida *Halicryptus spinulosus,* the medusa *Halitholus cirratus* with its polyp *Perigonimus cirratus* and the Myside *Mysis mixta* whose area extends further than that of the other species (Bottensee, North Sea). Among the Foraminifera LUTZE (1965) also found species of this group which he calls "arctic-boreal brackish-water Foraminifera". They are *Astramine sphaerica, Hippocrepina flexibilis, Cribrononion excavatum clavatum, Rheophax dentaliniformis regulans, Cr. incertum* subsp.?. The species *Cr. exc. clavatum* even invaded the depths of the Baltic basins (Landortstief!).

On the whole these species are far less sensitive to oxygen deficiency than those of the previous group. Their intolerance of oligohaline water makes it unlikely that they have survived in the Baltic from the Yoldia period through Ancylus Lake. As EKMAN (1953) points out they probably occurred in the region of the Kattegat before the Littorina period when it was a partly arctic bay and they have only penetrated into the Baltic with the Littorina submergence. This view is supported by their distribution in the very regions which only became Baltic at that time. The stand in the Skagerrak—Kattegat has been destroyed or narrowed down by subsequent changes in the region.

In contrast to the Pontocaspian area the Baltic is poor in endemic forms. This can readily be understood in view of its geological youth. LAKOWITZ (1929), however, recorded numerous endemic Baltic algae. A critical review (see HOFFMANN 1950, LEVRING 1940, WAERN 1952) reveals them to be forms of widely distributed algae. Only the large Green Alga *Monostrema baltica* has been retained as an endemic form. There is a larger number of animals which are only known nowadays from the Baltic and the Belt Sea. But they belong almost exclusively to the microfauna (Turbellaria, Rotifera, Gastrotricha, Nematoda). We may expect to find them in other brackish or marine regions. Yet the beginnings of a development of endemic forms can be detected. The Cladocera *Bosmina obtusirostris* forms a specific brackish-water form, var. *maritima;* it is highly probable that it originated in the Baltic itself after the Ancylus period. Some forms of *Keratella* may possibly be added. If this is a case of specific brackish-water variants arising from fresh-water animals, the Baltic may also have caused the development of separate fresh-water species from brackish-water animals during the postglacial period. The planktonic Copepod *Eurytemora lacustris* is widespread in the fresh-water lakes of the Baltic region; it is closely related to the brackish-water species *Eurytemora affinis* and has probably evolved from it in the Ancylus Lake or at any rate in a fresh-water period of the Baltic. A similar interpretation for the little Gastropod *Hydrobia (Amnicola)* seems less likely. The arctic-Baltic animals of the first group (p. 141) mentioned above, have also evolved several specific forms in their new fresh-water areas.

2. The Pontocaspian brackish seas

Three brackish seas, the Black Sea, the Caspian Sea and the Lake Aral are situated in the Pontocaspian region. They have a long period of history in common and, in complete contrast to the Baltic, they are ancient seas; but they differ in their hydrography and, in parts, in their colonization.

a) History

The Black Sea, the Caspian and the Sea of Aral are relics of an old marine area (Sea of Tethys) which extended through Europe and Asia (north of India) to the Pacific and remained remarkably constant in spite of frequent changes of shape in detail. In the Early Tertiary this Tethys was still a truly tropical sea with Nummulites, corals etc., in communication with the Western Pacific over Persia and Turkestan. In the middle Miocene a northern part separated from the Tethys and formed the Paratethys. This extended from France over southern Germany, the region of the Danube to Lake Aral; in the Rhone area and, in parts, probably over Anatolia, it was in communication with the Mediterranean. This Paratethys had already begun the transition to a brackish sea; the corals, Brachiopoda and Cephalopoda disappear. Euryhaline bivalves *(Cardium, Mactra, Tapes, Congeria)* and Gastropoda *(Rissoa, Trochus)* predominate in the sediments. The further fate of the sea is highly complex, owing to tectonic changes in this area and a fluctuating influx of fresh water (see ZENKEVITCH 1963). The following facts are important: There was no gradual reduction in salinity from marine to brackish, but the salinity underwent great changes. There are considerable fluctuations of water level in these areas. At its maximum the water level of the Caspian Sea was 80 m above the present one, thus reaching far northwards. At times it was lower so that only the southern part of the present-day sea contained water. At certain periods the Sea of Marmara belonged to the region of the Pontocaspian Sea. During the Pleistocene the Black Sea was twice in communication with the Mediterranean over the Bosphorus—Dardanelles. As a result there was a steep rise in salinity from about $7^0/_{00}$ at times to about $22^0/_{00}$ S, and an invasion of Mediterranean fauna took place. The Caspian and Black Seas also were connected several times over the Kumo—Manyah depression, even in most recent (postglacial) times.

A fauna has survived all these changes and has even differentiated into swarms of species. It prefers oligohaline water and invades fresh water in many cases, particularly in the large rivers. In contrast to the Baltic, an ancient and specific element of the fauna occurs here. Thus the following groups may be distinguished in this area: 1) the Pontocaspian brackish-water fauna (often called Caspian element or relict fauna), 2) the Mediterranean element, 3) an arctic fauna, 4) a fauna of fresh-water origin derived from the Pontocaspian fresh-water region (see p. 155).

These events in the Pontocaspian region are of special significance for the problem of brackish water as brackish-water seas have formed in the course of slow geological development, permitting an undisturbed, slow evolution of a rich, specific brackish-water fauna as well as making a specially successful transition from sea to fresh water possible; but there was less opportunity from fresh water to the sea.

b) Hydrography

For detailed data see ZENKEVITCH (1963). Hydrographically the Black Sea differs markedly from the Baltic. This is essentially due to two factors: 1) its much closer connection with highly saline seas; 2) the uniform depression in the Black Sea which in its central part forms an almost symmetrical basin of 245 m depth, five times that of the Baltic.

The Bosphorus represents the link with the Mediterranean through the Aegean and the Sea of Marmara. At its narrowest point it is only about 550 m wide, the shallowest point is 36 m. The influx and exchange of water are far less than in the Baltic with its wider connections. The Black Sea receives only small supplies of highly saline water. At its present inflow and outflow a renewal of its water could require 2500 years as the undercurrent through the Bosphorus — already depleted of oxygen — only provided 202 cubic kilometres a year. But its morphological structure as a uniform, deep basin results in the highly saline water collecting in the depth as

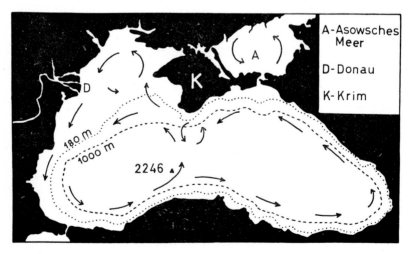

Fig. 63. The Black Sea with isobaths 180 m (about beginning of the azoic zone) and 1000 m. The arrows indicate the main currents. Simplified after CASPERS.

a relatively stable body. The consequences of this oxygen deficiency of the deep water and abundance of H_2S are known. Animal life ceases at depths of about 130—200 m. This azoic region comprises about 77% of the bottom area; thus animal life is restricted to the upper layer and the marginal zones. In this layer the main basin has a salinity of 18—21$^0/_{00}$ according to depth, and it is fairly stable; at greater depths it rises to 23$^0/_{00}$. Therefore the water of the Black Sea belongs to the polyhalinikum. Lower salinity is found in the zone of the shore itself up to 30 km, especially under the influence of river mouths and Limanen (lagoons) such as the minor seas. The annual supply of fresh water through the rivers amounts to 400 cubic kilometres. The shallow northern sections, that is the parts north of the Danube — Crimea and Sea of Azov — have a salinity which decreases rapidly towards the north (see fig. 64). Altogether the Black Sea consists of a mighty core of polyhaline and almost homo-haline water, surrounded by a belt of about 30 km wide on the shore, in which salinity is reduced and variable. There are special areas such as a relatively small region of increased salinity at the mouth of the Bosphorus, as well as larger areas in the North on either side of the Crimea where salinity decreases rapidly.

But the lower limit of the zooplankton is not found at a uniform depth everywhere. It is high, at 100—125 m, in the halistatic regions with little intermingling of water. According to the conditions of currents (cf. fig. 64) there are three regions in which the azoic limit is high (NIKITIN 1931). In the coastal regions the limit is usually found at a depth of 175, at 200 m only in 2.5% of the area. It is only in the region of influx of the Bosphorus that it is even deeper (about 225 m), but only in 0.33% of the area. Of the bottom animals the Polychaete *Melinna palmata* extends to a depth of 200 m (SERNOW 1913).

But while the deep layer, deficient in oxygen and rich in H_2S is azoic, it is not without living organisms. Here in the H_2S zone is the world of microorganisms. After a minimum at depths of 75—150 m their numbers per cubic cm are greater than in the surface layer. These microorganisms colonize the water down to the greatest depths (KRISS 1958).

Because of its geographical position temperature fluctuations are greater than in the Baltic. While winter temperatures of the surface water are but a little higher than those of the Baltic, the maximum temperatures of the marginal zones rise to 24—30^0 C (Sea of Azov).

The Sea of Azov is a large lagoon of the Black Sea. It is shallow, its maximum depth being 13.2 m. Its salinity is reduced, on an average to 11.2$^0/_{00}$. It decreases from the Kertsch Strait to the mouth of the Don from 7.5$^0/_{00}$ to 1$^0/_{00}$, it is only in the lagoon of Sivash that it exceeds 40$^0/_{00}$.

The Caspian (439,000 square kilometres) is the largest inland sea of the world. In spite of their common origin its morphology and hydrography differ considerably from those of the Black Sea. Its greatly reduced depth is of morphological importance; with about 1000 m this exceeds the depth of the Baltic, but is far below that of the Black Sea. Of hydrographic importance are the large amounts of fresh water coming in from the Volga and the Ural. As a result the shallower northern section has freshened and mainly oligohaline water. But even in the deeper central and southern parts salinity reaches only about 14$^0/_{00}$. Thus the Caspian Sea is mesohaline, in its northern part oligohaline (while the minor sea of the Karabugas Bay is hyperhaline with

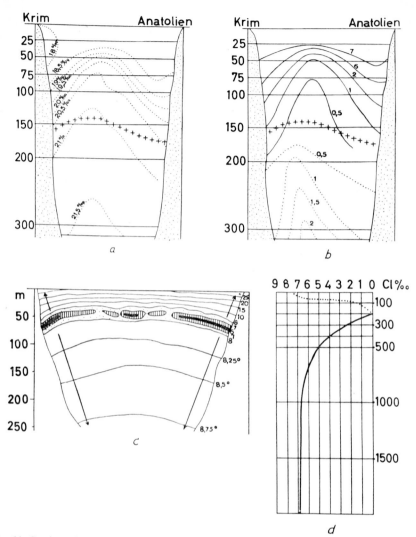

Fig. 64. Sections through the Black Sea between the Crimea and Anatolia (March 1924). a, salinity; b, O₂ and H₂S content; c, temperature; d, behaviour of O₂ (graph in dotted lines) and of H₂S (graph in solid lines) in the deep zones. The line +++ in a and b indicates the lower limit of aerobic life. From Pora.

17 0/00 S). Its chemistry is very different. In comparison with oceanic sea water there is a considerable increase in sulphates (7.1% in the open sea, 25% in the Caspian), in particular there is an increase in MgSO₄ content (1.66% in the Ocean, 2.36% in the Caspian Sea). The relative Ca-content is also higher (2.2% against 1.6% in the Ocean), but the NaCl content is reduced. The H₂S boundary which is also present is much lower than in the Black Sea, according to Knipowitsch it is at a depth of 600 m in the central Caspian and even deeper down in the southern part. The lower limit of the zooplankton is at 400—600 m, that of the benthic fauna at 400 m.

Lake Aral (64,500 square kilometres) is a shallow sea with a maximum depth of only 68 m. Its salinity also reaches $12^0/_{00}$ in the South, near the estuaries of the rivers the environment is almost fresh water. Deviations from the Ocean in respect of the salt composition are even greater, the proportion of SO_4 ions attains 19.76% of the total amount of salts, the Ca ions 7.6%. In the brackish water region Lake Aral is very poor in species of animals.

c) Faunal elements in the Pontocaspian brackish seas

As result of their different histories these seas also differ in their faunal elements from those of the Baltic. The existence of an old indigenous fauna is significant; it is represented by a relict of the Tertiary brackish-water fauna of those regions. It is the element of the Pontocaspian fauna. After communication with the Mediterranean through the Bosphorus an element of the Mediterranean fauna invaded the region to a large extent, particularly the Black Sea. The occurrence of nordic—Baltic animals in the Caspian and Lake Aral — already referred to on p. 143 —, is strange; They form the nordic—Baltic element of the fauna. The fresh-water organisms might be described as a fourth element of the fauna; they have evolved their own races and species in the brackish water of this area.

1) The Pontocaspian (Caspian) relict fauna. Numerous animal species are restricted to the region of these seas, being endemic for this area. The most interesting part of these endemic animals is represented by the Caspian relict fauna. It comprises all the animal groups which maintained themselves in the Tertiary seas (arising from the cut-off part of the Tethys) and have been preserved to this day through all the changes which these seas have undergone. These relict groups have not remained unchanged from Tertiary times; they have split into numerous species, thus producing new species in this area. A species may have differentiated as recently as the Pliocene or the Pleistocene; yet it belongs to the relict fauna provided it has evolved from a group of animals which had become isolated there.

It is not always easy to determine whether an endemic species or genus is a true relict or whether it derives from later immigration. Genera and species whose ancestors are found as fossils in the Tertiary seas are undoubtedly Caspian relicts. This holds above all for molluscs. Groups of closely related endemic species whose main development is in the Caspian may also be described as relicts. This applies to Mysideae, Amphipoda, Cumaceae etc. There is some uncertainty whether species and genera which are easily transported belong to the relict fauna; they may be carried through the air, transported by birds or by wind, and dispersed as resistant stages. Such animals could have penetrated into the region at any period and could have differentiated into their own species here. This applies particularly to the microfauna such as many Copepoda *(Ectinosoma, Schizopera* and others), to Ostracoda, many Turbellaria etc. There is some doubt whether species of *Loxoconcha* from the Caspian Sea are part of the relict fauna.

The Caspian relict fauna displays the following features:

a) In the Black Sea the species predominantly colonize the meiomesohaline and oligohaline regions. They are practically absent from the proper basin of the Black Sea. They concentrate in river mouths, Limanen (lagoons) and the Sea of Azov; in the Caspian Sea abundant numbers of species occur in the pleiomesohaline region as well. This concentration of relict species to brackish water of lower salinity has been accounted for by a stenohalinity of these brackish-water regions, as well as by their displacement through later Mediterranean immigrants. In one case the Pontocaspian element would have been pushed back through transgression of higher saline sea water; in the other by newly immigrant animals. It is possible that both factors have been in operation. The probability that salinity plays an important part is borne out by the fact that there has not been a simple pushing back of the Pontocaspian fauna, but there have been several advances and retreats connected with climatic changes in the most recent geological past. BIRSTEIN (1946) has in mind not so much the effects of salinity itself, but those due to differences in the ionic composition of sea water compared with the "Pontocaspian" water.

b) Though the Pontocaspian fauna is rich in species, it consists of a few groups of closely related species from a few families. The Mediterranean element belongs to many more families. Thus all the Pontocaspian Cumaceae belong to the family of Pseudocumidae, while the Mediterranean immigrants are derived from 5 families; among the Amphipoda only 2 families (Gammaridae and Corophiidae) share in the Pontocaspian relict fauna as against 16 families of marine immigrants. Among bivalves only descendants or the *Cardium* group (*Adacna, Monodacna* etc.) and *Dreissena* make any extensive contribution on the Pontocaspian relict fauna; among the teleosts Clupeidae and Gobidae etc. The swarms of closely related species and genera may will be explained by the Tertiary history of the region. The many bays and isolated basins of this brackish sea offered opportunities for a regional differentiation of vicarious species and when the marine regions were further displaced the species intermingled. The differentiation of these species appears to be continuing as shown by the formation of subspecies in the Sea of Azov (subsp. *maeotica*) and in the Caspian Sea.

c) A high percentage of the Pontocaspian relicts of marine origin have succeeded in invading fresh water, at least the lower reaches of rivers. While the nordic-Baltic relicts of the Baltic region which have penetrated into fresh water (see p. 142) are inhabitants of lakes, large numbers of Pontocaspian relicts live in rivers. In the Danube, even above the Iron Gates, mass colonies of Pontocaspian animals occur, e. g. the Polychaeta *Hypania invalida, Manayunkia caspia*, the Crustacea *Corophium maeoticum, C. robustum, C. curvispinum, Dikero-, Ponto-* and *Chaetogammarus, Jaera sarsi*, the mite *Caspihalacarus* etc., as has been established by BAČESCU (1948, 1966). Though their occurrence mainly coincides with the region of the former brackish sea a number of these immigrants have at a later date succeeded in extending their area in fresh water. I only mention *Dreissena polymorpha, Corophium curvispinum, Lithoglyphus naticoides* (see THIENEMANN 1950). Those emigrants as well as *Gammarus ischnus=tenellus* are far less tolerant of salt in the North Sea and Baltic regions than in their Pontocaspain home. Whether this is due to a change in tolerance to salinity brought about by other environmental factors or else to the fact that special fresh-water races of this species have emigrated has yet to be investigated.

The following are the most important Pontocaspian brackish-water animals: The Polychaeta are represented by a few Ampharetidae: *Parhypania invalida* (K. P.), *Hypaniola Kowalewskii* (K. P.). It is not quite certain whether *Manayunkia caspia*

(K) is a true relict. The genus is much more widely distributed than had been at first assumed, it even occurs in the mangrove swamps of southern Brazil (*M. brasiliensis* BANSE 1956). AX (1959) has studied the Turbellaria of this area and has made an accurate analysis of their zoogeographical relationships. About 16—17 species may be considered as relicts, among them there are 7 endemic genera (*Achoerus, Pseudoconvoluta, Annulovortex, Selimia, Thalassoplanina, Kirgisella, Phonorhynchoidea*). The relicts occur particularly in the Caspian Sea and Lake Aral.

The higher Crustaceae (Malacostraca) contain numerous Pontocaspian relicts among the Cumaceae, Mysideae and Amphipoda, but strangely enough not among the Decapoda and hardly any among the Isopoda. The Cumaceae are represented by about 20 Pontocaspian species. This, too, is an interesting fact since in other regions the Cumaceae avoid areas of low salinity and only one or two species penetrate into the Baltic proper. All Pontocaspian relicts of this groups belong to the family of Pseudocumidae; this is by means confined to this area, as has sometimes been stated, but is here particularly rich in species. The relicts belong to the genera *Schizorhynchus, Volgocuma, Pterocuma, Caspiocuma* and to the subgenus *Stenocuma* or *Pseudocuma*. All these groups are endemic in the area (see SARS 1903, 1927; DERJUGIN 1925 and BACESCU 1951).

The Mysidae also contain about 20 relict species. They all belong to the *Tribus Mysini*. Endemic genera are *Katamysis* (1 species), *Limnomysis* (1 species) and in the Caspian Sea *Caspiomysis* (16 out of 20 species), their subgenera *Pseudoparamysis* and *Metamysis* are entirely endemic. Of the 3 species of *Diamysis* one lives in the Black Sea as well as in the brackish water of the Mediterranean *(D. bahirensis)* — possibly the genus is Pontocaspian and this one species may have secondarily immigrated into the Mediterranean. *Hemimysis* (with *H. lamornae)* is widely distributed on European coasts; in addition there are two endemic forms in the region: *H. serrata* (Black Sea) and *H. anomala* (Black and Caspian Sea) whose origin has not yet been elucidated (see BACESCU 1954, p. 68).

Both in geographical distribution and in salinity tolerance the Mysidae cover a much wider range than the Pontocaspian Cumaceae. As in the latter, the maximum number of species and amount of individuals occur in the oligohaline and meiomeso-haline regions, but they extend much further, and with a larger number of species, into fresh water. A number of species must be considered fresh-water or fresh-water oligohaline, e. g. *Diamysis pengoi* (0—0.5⁰/₀₀ S), *Limnomysis benedeni* (0—5⁰/₀₀, in the Caspian Sea up to 10⁰/₀₀, *Metamysis ullskyi* = *M. strauchi, Paramysis kosswigi* (Anatolia), *P. intermedia, P. lacustris* and others. Only a few occur in the marine region of the Black Sea, e. g. *Paramysis pontica* at 6—100 m depth, *P. kroyeri, Paramysis agigensis* and *Diamysis bahirensis mecznikowi*. Among the Mysidae a number tolerate higher salinities in the Caspian than in the Black Sea, e. g. *Paramysis baeri, P. ullskii, Catamysis, Paramysis intermedia* etc. While the Caspian Sea has a fair number of indigenous species, there are but a few species of this group confined to the Black Sea.

Geographically the Pontocaspian Mysidae extend into the upper Volga, the middle reaches of the Danube and to the Anatolian waters. Their occurrence coincides almost entirely with the maximum extent of the Pontocaspian sea so that they can, to a large extent, be considered as relicts of these marine regions in the rivers. BA-

ČESCU (1940) has also pointed out that most species are hardly able to migrate against the current. Only a few, e. a. *Diamysis pengoi, Paramysis intermedia* are capable of doing so to a limited degree. Some such as *Limnomysis benedeni* might also be transported by plants.

The largest number of Pontocaspian forms is found among the Amphipoda, though only in the families of Gammaridae and Corophiidae. The former also have a number of specific genera, e. g. *Dikerogammarus, Gmelina, Gmelinopsis, Pontogammarus, Shablogammarus, Stenogammarus, Cardiophilus, Iphigenella, Niphargoides* and *Amathillina* (see CARAUSU, DOBREANU, MANOLACHE). According to CARAUSU 1943 the Amphipoda common to the Pontocaspian—Aralian fauna consist of 64 species and 9 subspecies, a total of 73 forms. Of these 32 forms are endemic for the Caspian Sea, that is restricted to this sea. In the Black Sea 38 forms (species and subspecies) are found, of these 25 occur in the Caspian as well. Of the remaining 13 forms 3 are endemic in the Sea of Azov.

„Die anderen 9 für die übrigen Gewässer des Schwarzen Meeres, in erster Linie Lagunen und Donaumündung, eine als Bachform, endemisch.

Fig. 65 (left). *Gammarus ischnus,* a representative of the Pontocaspian sub-genus *Chaetogammarus.* It occurs in the Caspian Sea, Sea of Azov, the marginal areas of the Black and Caspian Seas, in addition (? recently) in the Vistula where it is confined to fresh water. From THIENEMANN 1950.

Fig. 66 (right). The Pontocaspian Isopod *Jaera sarsi* which has penetrated far into rivers (Danube). From THIENEMANN 1950.

Für die rumänischen Gewässer konnte CARAUSU (1943) 32 Formen feststellen, und zwar für das Meer selbst, die Limane, Flußmündungen sowie die Flußläufe. Von diesen besiedelt die größte Zahl ihrer 21 Formen die Gewässer der Donaumündung, 17 Formen besiedeln die Küstenlimane, 10 Formen die Unterläufe der Flüsse, 7 Formen die Brackgewässer und nur 2 das Meer selbst" (KARAMAN 1953).

[Of the other 9 for the remaining waters of the Black Sea, in the first instance lagoons and mouth of the Danube endemic, one as a form from streams. — For the Roumanian waters CARAUSU (1943) established 32 forms, for the sea itself, the Limanen (lagoons), river mouths and river courses. Of these the largest numbers out of 21 forms colonize the waters of the Danube delta, 17 forms settle in the coastal lagoons, 10 forms in the lower reaches of the rivers, 7 forms in brackish waters and only 2 in the sea itself]. Here, too, the number of forms living in the marine region is much greater in the Caspian Sea. According to BIRSTEIN (1946) a fairly large number even avoid the slightly brackish areas of the Caspian, e. g. *Gammarus placidus, Dikerogammarus macrocephalus, D. oskari, Niphargoides weidemanni, N. grimmi, N. subnudus, Gmelinopsis aurita, Amathillina spinosa, A. maximiviczi, A. pusilla, Corophium spinolosum.* Some of these forms occur at a depth of 50—100 m in this mesohaline part of the Caspian Sea.

In contrast to the Mysidae the Amphipoda are capable of migrating upstream. In the Danube area of Yugoslavia KARAMAN (1953) found 11 Pontocaspian species, among them 4 species of *Corophium (C. robustum, C. maeoticum, C. curvispinum, C. chelicorne).* It is known that some of these species have reached waters of central and northern Europe in recent times, e. g. *Corophium curvispinum* and *Chaetogammarus tenellus (= ischnus)* (fig. 65).

While the Isopoda of west European regions have left some relict forms behind in subterranean waters, they display but few features worth notice in the Pontocaspian area. There are a few endemic species of widely distributed genera, e. g. *Jaera sarsi* (fig. 66), *Eurydice pontica, E. valkanovi.*

Some of them have no doubt been inhabiting the area for a long time such as *Jaera sarsi* which is widespread in the Black and Caspian Sea and now occurs in the Danube as far as Vienna, probably in part passively transported.

The Decapoda resemble the Isopoda in their behaviour.

There are numerous relict forms of molluscs in all the three seas. Among bivalves these concentrate round the Cardiidae related to *Cardium:* they have evolved the specific genera *Didacna* (7 species), *Monodacna* (3 species) and *Adacna* (4 species). They differ from *Cardium* by a gradual regression of the hinge and elongation of the siphons. The species of *Didacna* and *Adacna,* in particular, occur in the Caspian Sea, Lake Aral and Sea of Azov as far as the pleimesohalinikum (10—14⁰/₀₀ S). Species of *Monodacna* as well as *Adacna fragilis, A. relicta* penetrate the fresh water of the lagoons and river mouths.

The Dreissenidae are the second family of bivalves with their own development in the Pontocaspian area; its two genera *(Congeria* and *Dreissena)* prefer brackish and fresh water. Most of the 7 species of the genus *Dreissena* are mesohaline brackish-water animals of the Caspian Sea and Lake Aral; *D. bugensis* lives in the lagoon of the mouth of the river Bug and *D. polymorpha* is the famous wandering mussel which has penetrated far into European fresh waters from the Pontocaspian area.

Among Gastropoda (fig. 67) the family of Micromelaniidae has undergone abundant development and it contains many relict brackish-water species, e. g. the genera *Micromelania* (7 species), *Clessiniola* (3 species), *Caspia* (7 species), *Caspiella* (3 species). It is concentrated almost entirely in the Caspian Sea which contains about 19 species of this family; in Lake Aral also *Micromelania spica;* in the marginal areas of the Black Sea *M. lincta* and *Clessiniola variabilis,* both of which penetrate into fresh water of the lagoons and river mouths.

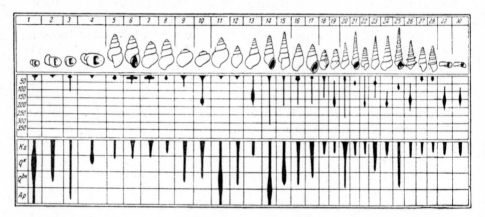

Fig. 67. The Pontocaspian Gastropoda. The figures and illustrations on the upper margin denote: 1. *Theodoxus pallasi;* 2. *Th. pallasi* var. *nalivkini;* 3. *Th. schultzii;* 4. *Th. schultzii* var. *juhovi;* 5. *Hydrobia ventrosa* var. *evanescens;* 6. *H. grimmi;* 7. *H. grimmi* var. *gemmata;* 8. *H. chrysopsis;* 9. *Lithoglyphus exiguus;* 10. *Zagrabica brusiniana;* 11. *Clessiniola variabilis;* 12. *Cl. martensi;* 13. *Cl. grimmi;* 14. *Caspia caspia;* 15. *C. inxata;* 16. *C. conus;* 17. *C. cincta;* 18. *C. gmelini;* 19. *C. pallasi;* 20. *Micromelania spica;* 21. *M. spica,* n. var.; 22. *M. nossovi;* 23. *M. turricula;* 24. *M. grimmi;* 25. *M. bakuana;* 26. *M. dimidiata;* 27. *M. sieversi;* 28. *M. pseudodimidiata;* 29. *Planorbis eichwaldi;* 30. *Pl. eichwaldi* var. *dybowskii.* The diagram in the middle shows the distribution of the species in depth, the lower one the regional distribution. KS = Caspian Sea, Qk = fossil quarternary deposits in the Caspian Sea. Qbk = fossil quarternary deposits in the Baku stratum, Ap = Peninsula Apcheron. From SHADIN, W. I.

Smaller centres of evolution are found in the area of the family Neritidae, of world-wide distribution; it has a strong tendency to invade brackish and fresh water from the sea. The fresh-water — brackish genus *Theodoxus* develops here the species *Th. transversalis* and *Th. danubialis* which extend from fresh to salt water; but also the separate brackish-water species *Th. pallasi* which occurs in the Caspian, Lake Aral and Sea of Azov, and *Th. schultzii* = *Ninnia schultzii* which is restricted to the Caspian Sea.

The Lithoglyphinea contain 3 species of *Lithoglyphus,* one of which, *L. caspius,* is a brackish-water animal of the Caspian Sea; another *L. pyramidatus* is found in southeast Europe, while the third *L. naticoides* extends from the brackish water of the Black Sea into the rivers and has recently colonized the rivers of central Europe.

The Hydrobineae, otherwise so widespread in brackish water, are here only represented by a small number of species. *Hydrobia acuta,* widespread in the Mediterranean, is found in the lagoons of the Black Sea, the widespread brackish-water species *H. ventrosa* also in the Caspian and Lake Aral. *H. grimmi* is an endemic form of the Caspian Sea, similarly the genus *Zagrabica* with one species, *brusiniana*.

The Melaniidae which are widespread in the fresh and brackish waters of warmer regions are represented by two species *(Fagotia acicularis* and *F. esperi)* in rivers and brackish lagoons of the Black Sea regions.

It is more difficult to assess the origin of fishes and to evaluate them as relicts since they actively cover long distances and often readily change between fresh and sea water. Undoubtedly a number of Gobiidae, represented by endemic genera and species in the Pontocaspian Sea, are relicts of the Tertiary fauna of this brackish sea. The Gobiidae are a marine family of world-wide distribution; the fishes are mostly rather small with little tendency to migrate; they often invade extreme habitats on the sea shore and in rivers. In the Baltic, however, only a few species penetrate into the meiomesohaline region, *Gobius niger, G. (Pomatoschistus) microps* and *G. (P). minutus*. But in the Pontocaspian region over 20 species occur, among them the endemic genera *Bubyo* (1 species), *Knipowitschia* (1 species), *Hyrcanogobius* (1 species), *Mesogobius* (2 species) *Proterorhinus* (1 species), *Caspiosoma* (1 species), *Benthophilus* (3 species) as well as several species of *Gobius*. Most of these forms are brackish-water species in the Caspian Sea, Sea of Azov and in river mouths, while several species show a tendency towards invading the fresh water of the rivers. The second place is occupied by the Clupeidae with the endemic genera *Caspialosa* (8 species) and *Clupeonella* (2 species). These Clupeidae penetrate more into the rivers. The Cyclostomata genus *Caspiomyzon* is also endemic for the Caspian Sea and there is a striking abundance of Acipenseridae, though only a few of them are confined exclusively to the region Black Sea, Caspian and Lake Aral *(Acipenser nudiventris, A. güldenstädti)*. The perches have here evolved the genus *Percaria* and the Cyprinidae originating in fresh water have produced separate brackish-water forms (see p. 83).

It is difficult to determine the zoogeographical origin of the microfauna and that of the lower invertebrates. The species usually have great opportunuties for dispersal and since the areas of maximum population density and greatest wealth of species need not be the areas of their origin, conclusions can only be drawn in very rare cases. Among the Porifera 4 endemic species are found in the Caspian Sea, the genera *Metschnikowia* with *M. intermida* and *M. tuberculata, Protoschmidtia flava* and *Amorphina caspia*. Occasionally they have been placed in a family of their own, the Metschnikowidae, and earlier with the Reneridae. On the other hand, the genera *Baicalospongia* from Lake Baikal and *Ochridaspongia* from Lake Ochrid have been considered to be related; both are fresh-water species. Among the Cladocera there is a striking number of species in the Polyphemidae (see p. 75). According to MORDUKHAI-BOLTOVSKOY (1965) there are 23 species in the area of the Caspian Sea, outside the region only 8 species. This family occurs both in the sea and in fresh water. Like all Cladocera, it was originally fresh water. The species belonging to the marine genera *Podon* and *Evadne* might be considered as marine Caspian relicts; the species connected with the fresh-water genera *Polyphemus* and *Bythotrephes* *(Cercopagis, Apagis* fig. 29) may be derived from fresh-water immigrants. The Oligochaete *Peloscolex svirenkoi* which is abundant in the lower inhabited bottom areas of the Black Sea (HRABE 1964) belongs to a genus which lives both in fresh water and in the brackish water of the seas. The record of a species of *Diamysis*

(D. americanus) and a species of *Hypania* from America (after BACESCU 1966) shows the difficulties inherent in zoogeographical analysis. It is remarkable that hardly any Pontocaspian relicts have been described among plants.

2) The arctic—Baltic immigrants. The arctic—Baltic animals belong to the most peculiar species of the fauna of the Caspian and Lake Aral. There are about 20 species whose nearest relatives occur only in the Baltic and Arctic or else in the Arctic alone *(Pseudalibrotus)*. Most of them still belong to the same species as their relations in the North, in part they are separate species, but closely related to the arctic—Baltic ones. The majority of these arctic species belong to the higher Crustacea, especially the Peracarida, e. g. the large Isopod *Mesidotea entomon* (subsp. *caspia*), the Amphipoda *Gammaracanthus loricatus* (subsp. *caspia*), *Pontoporeia affinis (*subsp. *microphthalmus)*, *Pseudalibrotus caspius*, *P. platycera*, 4 species of *Mysis* closely related to the northern *M. relicta*: *M. caspia*, *M. macrolepis*, *M. microphthalmus*, *M. amblyops*, the pelagic Copepod *Limnocalanus grimaldii*, the Salmonida *Stenodus leucichthys*, according to P. Ax (1959) the Turbellaria *Pentacoelum caspium*, *Beklemischeviella brebistyla*, *B. contorta* etc.

There has been much discussion about the route by which these immigrants from the Arctic or Baltic may have reached the Caspian—Aral region. SEGERSTRÅLE (1957, 1962) has made a close study of this problem, using extensive material from systematic animal geography and geology. He arrives at the following concept: Through the great Ice Age (Riss glaciation) a large brackish-water lake was formed in the west Siberian plain, east of the Ural; in it arctic animals evolved into the arctic—Baltic relict species. Over the area of the river Tobol this lake established connection with the Caspian—Aral area, thus making it possible for arctic animals to advance into this region. At the end of the last glacial period (Würm Ice Age) a later connection with the North was established. The Onega Ice Lake, then of large dimensions, reached into the vicinity of the large Caspian Sea which at that time extended as far as the upper Volga, near Kasan. In the Caspian Sea the arctic—Baltic animals occur mainly in deeper, cooler water (40—70 m depth). No purely freshwater populations have developed here; they are common in the North.

In the years 1954—56 the Baltic herring *(Clupeus harengus membras)* was transferred from the Baltic into Lake Aral by means of spawn. It flourished and became bigger than in its original home (ZENKEVITCH 1963).

3) The Mediterranean immigrants. Twice the Black Sea was in communication with the Mediterranean and as a result, the Mediterranean animals and plants reached the Black Sea, as well as the Caspian and Lake Aral, with the influx of salt water. The first connection was established in the middle Pleistocene, the second in the Ice Age. There are 1,500 Mediterranean species in the Black Sea, 200 in the Sea of Azov, 28 in the Caspian and in Lake Aral (ZENKEVITCH 1963). Mediterranean forms invaded the Caspian Sea during the periods of its connection with the Black Sea. In the Pleistocene (Khwalyn stage) the fishes *Syngnathus nigrolineatus*, *Atherina mochon*, the bivalve *Cardium edule*, the Polychaete *Fabricia sabella*, the ell-grass *Zosteranana* reached the Caspian Sea. *Cardium edule (?lamarcki)* and *Zostera nana* entered Lake Aral. In recent times, that is in this century, the number of species in the Caspian and Lake Aral was increased through passive transport,

deliberate transfer in order to increase catch of fish, as well as by immigration through linking canals. Thus the following got into the Caspian Sea and, in parts, Lake Aral: the planktonic diatom *Rhizosolenia calcar-avis,* the bivalves *Mytilaster lineatus, Syndosmya alba,* the Polychaete *Nereis diversicolor,* the planktonic Cladocera *Podon polyphemoides,* the barnacles *Balanus improvisus* and *B. eburneus,* the Decapoda *Leander adspersus* and *L. squilla,* the fishes *Mugil auratus, M. saliens, Pleuronectes (Platichthys) flesus.* In the same way euryhaline brackish-water species which have been widely distributed through passive transport, have reached these brackish seas: e. g. the crab *Rithopanopaeus harrisi,* the Polychaete *Merceriella enigmatica,* the Bryozoon *Membranipora crustulenta* etc. All these animals are, in the main, Mediterranean only in so far as they have been transported from the Mediterranean or else through the Mediterranean to the Black Sea.

In spite of the different chemical composition of the water in the Caspian or Aral, these new immigrants have rapidly multiplied such as *Rhizosolenia, Nereis, Balanus, Mytilaster, Leander, Mugil* (see ZENKEVITCH 1963).

Though the Mediterranean immigrants invaded the Black Sea at a later geological period some isolated special subspecies (Cumaceae) or species, have already evolved. Among the Turbellaria P. Ax (1959) lists the following species from the Black Sea which are very closely related to species from the Mediterranean: *Macrostomum ermini, Otoplana bosporana, Postbursoplana pontica, Paramesostoma pulcherrimum, Trigonostomum mirabilis, Rogneda tripalmata, Baltoplana valkanovi* etc.

4) Fresh-water animals and plants were able to invade the Pontocaspian brackish seas at any time. Numerous euryhaline fresh-water species occur in them. There is a remarkably high number of special subspecies or even species which have evolved here, presumably as a result of frequently protracted periods of brackish water in these seas. Among the fishes the following should be mentioned: *Lucioperca marina, Percaria demidoffi,* the Cyprinidae *Barbus brachycephalus caspius, Rutilus rutilus caspius, Aspius aspius taeniatus, Cobites caspia,* probably also the Acipenser *nudiventris, A. güldenstädti* etc. These species ascend the rivers for spawning. Among Gastropoda the Pulmonate *Planorbia eichwaldi* is endemic in the Caspian Sea, the Pontocaspian species of *Theodoxus* may also be of fresh-water origin. This is certainly the case for the Crustacean *Astacus pachypus* and probably for the Polyphemidae of the genera *Apagis* and *Cercopagis.* Among the Turbellaria *Gieysztorie knipovici* (Caspian Sea) and *G. bergi* (Lake Aral) are species of their own of fresh-water origin (P. Ax 1959). Among the Trichada there is even a special genus *(Caspioplana pharyngosa)* which is of fresh-water origin.

Hudson Bay. If the boundary sea-brackish water is placed at $30^0/_{00}$ S, then the Hudson Bay also belongs to the brackish seas. Its salinity is $26^0/_{00}$ and even lower in its southern bays.

d) Former brackish seas

In the most recent geological past the Arctic Ocean may perhaps have been a brackish-water area at times; this view is held by ZENKEVITCH. During the Ice Age there were considerable eustatic fluctuations of sea level; during glacial periods the water was bound in the mass of glaciers with a resulting lowering of the sea level by about 100 m, during interglacial periods there was a rise due to an increase of melt-

waters. In consequence, the communication between the Pacific and the Arctic
Ocean over the Behring Straits was partly interrupted and the submarine ridge from
the Shetland Isles over Faroer-Iceland to Greenland lay nearer the water level so
that the exchange of water with the Atlantic became reduced.

„Nach der Meinung des Verfassers mußte dabei ein bedeutendes Süßwerden
eintreten. Dieses bald brackiges, bald süßes Wasser enthaltende Becken konnte den
Charakter ganzer, miteinander verbundener Wasserbeckensysteme annehmen. In
dem komplizierten System von Transgressionen und Regressionen konnten diese
Systeme der halb geschlossenen, eventuell auch ganz geschlossenen Wasserbecken
ihre Verbindung bald mit dem Atlantischen, bald mit dem Stillen Ozean erhalten,
erneut salzig werden, einen entsprechenden Komplex der Meeresfauna aufnehmen
und später wieder Süßwasser enthalten." „Phasen von bedeutendem Süßwerden
mächtiger Teile des Polarbeckens haben im Verlaufe der Umgestaltung der Fauna
auch die für die Eiszeitrelikte als Ausgangsformen dienenden Brackwasserformen
wie auch die Eiszeit- und Meeresrelikte selbst herausgebildet. Die einen wie die
anderen lebten in der Phase der stärksten Versüßung, die die eurasiatischen und
nordamerikanischen Küstengebiete in Mitleidenschaft zog und waren den Zonen
verschiedenen Grades von Salzgehalt angepaßt, wie wir es gegenwärtig an den Rand-
gewässern des Polarbeckens und im Baltischen Meer beobachten.

Die darauffolgende Versalzung hat die gegenwärtigen Meeresreliktformen der
Eiszeit ins Süßwasser getrieben, die ihnen nahestehenden Ausgangsformen ins Brackwasser.

Einen großen Teil dieser Fauna erhielten seinerzeit das Baltische und das Kaspische Meer, desgleichen die Festlandgewässer Eurasiens und Amerikas."

„Das Baltische Meer bildete kein Zentrum der Akklimatisation arktischer Mee-
resformen und Umgestaltung derselben in Reliktformen, sondern hat dieselben aus
dem Nordosten fertig erhalten und lediglich bei der weiteren Entwicklung dieser
Fauna eine Rolle gespielt" (ZENKEVITCH 1933).

[In the author's opinion this would produce a considerable reduction in salinity.
The basin sometimes containing brackish and sometimes fresh water, could assume
the features of whole systems of interconnected water basins. In the complicated
system of transgressions and regressions these systems of partially closed, or even
totally closed, water basins may have been connected now with the Atlantic, now
with the Pacific Ocean, become salt again, receive a corresponding share of marine
fauna and, later on, contain fresh water once more. Considerable parts of the
Polar Sea have undergone phases of marked reduction in salinity; in the course of
the transformation of the fauna there evolved brackish-water forms which served
as the origin of the glacial relicts; but the glacial and marine relicts themselves were
also developed. Both groups lived in the phase of the greatest reduction of salinity
which also affected the coastal areas of Eurasia and North America; they were adapted
to zones of varying degrees of salinity as can be observed at the present time in the
marginal waters of the Polar Basin and the Baltic.

The subsequent increase in salinity has driven the present marine relict forms of
the Ice Age into fresh water, the closely related original forms into brackish water.

A large part of this fauna was at one time received by the Baltic and Caspian,
as well as the inland waters of Eurasia and America.

The Baltic Sea did not form a centre for the acclimatization of arctic marine forms and their transformation into relict forms, but has received those fully evolved from the Northeast; it has merely played its part in the further development of this fauna (ZENKEVITCH 1933).]

According to ZENKEVITCH the brackish phases of the polar regions were of great importance for the dispersal of the fishes from the families of Salmonidae and Gadidae. Of the 5 species of Gadidae of the Arctic *(Lota lota, Eleginus navaga, Boreogadus saida, Arctogadus borissowi* and *A. glacialis) Lota lota* lives entirely in fresh water, the next three migrate into fresh water for spawning, the behaviour of *A. glacialis* is not yet known. In addition to the arctic-Baltic brackish-water animals there are also brackish-water animals which are now specially restricted to the arctic region, e. g. the Amphipod *Pseudalibrotus birulai.*

SZIDAT (1956) is inclined to see in a former brackish sea of South America a region of similar significance for the transition from the sea to fresh water as has been displayed by the Tertiary brackish seas of the Pontocaspian area. It occupied the basins of the Amazon and the La Plata. SZIDAT explains the marine elements of the South American fresh-water fauna, e. g. the ray *Potamotrygon*, the Belonidae, but also the Characinidae, rich in species, and the Siluridae, in a wider sense, as fresh-water descendants of the fauna of this brackish sea. The argument is supported with numerous parasites of marine origin, e. g. of the Isopoda family Cymothidae and the Trematoda. The brackish marginal and residual seas of the Tertiary sarmatic-pontic basin have been carefully studied by geologists.

B. Lagoons

(Strandseen, Haffe, Étangs, Limane, Nör, Noore, Lake-Estuaries)

The coasts of the continents have large numbers of lakes, bays which are more or less brackish. This abundance of lagoons is characteristic of the present geological period. After the Ice Ages the sea level rose about 100 m, the sea has invaded terrestrial regions of varying surface structure: this has led to quite an irregular shore line. These irregularities are gradually being smoothed out by the currents on the sea shore and the sediments they transport; there is erosion on the parts of the coast which jut out and deposition of sand banks in front of the bays. The following process takes place: Marine bay — delimitation of the bay by sand banks from one or both "corners" of the bay — Cutting-off of the bay which, if this is complete, usually transforms the marine bay into a fresh-water lake. Lagoons which have arisen in this way may be termed "typical lagoons". They are almost invariably shallow basins. Their largest diameter usually runs parallel to the shore line.

A second way in which lagoons are formed is the transgression of sea water into existing fresh-water lakes (atypical lagoons). One transgression has usually catastrophical consequences for the living organisms. But several transgressions may create a permanent connection with the sea, that is, form a lagoon. The best example of such a transgression lagoon is the Zuider Zee; at one time a fresh-water lake isolated from the sea, Lacus flevus, which became the Zuider Zee; but nowadays, by human interference it has once again been cut off from the sea and after that it quickly became a fresh-water lake (Ijsselmeer). Thanks to the initiative of Dutch biol-

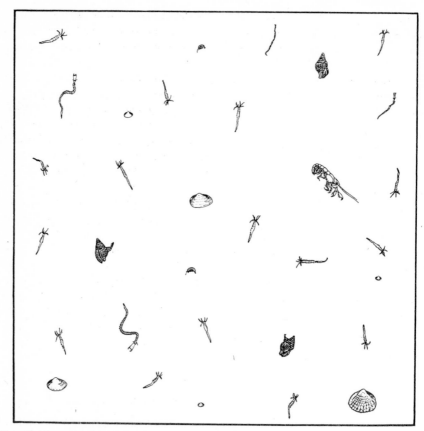

Fig. 68. Black Sea, Bay of Varna. *Zostera*-region in front of the Harbour = *Zostera* variation of the Upogebia-Melinna biocoenosis. 1 bottom sampler catch (1/10 qm) from station E 2, at 7 m depth, November 20th, 1939. 1 *Hippocampus hippocampus;* 5 *Nassa reticulata;* 85 *Bittium reticulatum;* 3 *Rissoa venusta;* 1 *Loripes lacteus;* 1 *Abra fragilis;* 1 *Corbula mediterranea; Mytilus*-clump; 1 *Portunus depurator;* 16 *Gammarus locusta,* 1 *Nototropis guttatus;* 11 *Idotea baltica; Balanus improvisus;* 2 *Nephthys hombergii,* 1 *Melinna palmata,* . 1 *Pectinaria koreni; Botryllus schlosseri; Membranipora.* From CASPERS 1951.

ogists the transformation of the brackish Zuider Zee into the fresh Ijsselmeer has been carefully studied (see REDEKE 1922b, BEAUFORT 1953); altogether, the Zuider Zee is the most thoroughly investigated brackish lagoon.

The morphology of atypical lagoons is far more variable than that of the typical lagoons since the structure of the fresh-water lakes into which the sea transgressed was very varied. They are frequently deeper, the bottom relief is often irregular, their shape is frequently elongated, extending in a direction perpendicular to the shore line. Atypical lagoons are chiefly found a) in the region of the last glaciation where there were numerous lakes. On the coast of northern Europe, for example, they form the large number of "Nooren" and "Förden" b) in the mountainous coastal zones in which sea water has penetrated into former glacial or river valleys.

Most lagoons with truly brackish water are found on the coasts of tideless seas

Fig. 69. Black Sea, Bay of Varna. Mud-community = Upogebia-Melinna biocoenosis.
Region of core. 1 bottom sampler catch (1/10 qm) from station H 11, at 24 m depth, July
29th, 1941. 1 *Upogebia litoralis;* 5 *Spisula subtruncata;* 1 *Venus ovata;* 3 *Nassa reticulata;*
16 *Melinna palmata;* 3 *Nephthys hombergii,* 1 *Capitella capitata* (2 *Balanus improvisus* on
living *Nassa*). From CASPERS 1951.

or seas with weak tides (Mediterranean, Black Sea, Baltic and Belt Sea) and on the
deltas of large rivers where the river itself produces head lands and sand banks from
its sedimentary material (Nile, Rhone, Mississippi).

Both hydrographically and biologically, there are, of course, transitions from
typical lagoons to Förden, Fjords, shallow marine basins which are bounded by
islands (Belt Sea) and to wadden seas.

Since lagoons are of importance to fisheries they have latterly been extensively
studied, also in the tropics. Lagoons are biologically markedly different from the
regions of the open sea, but they also differ from fresh-water lakes of similar morphology in the following features:

1) The flow of water is non-directional since the brackish lagoons show an "alternating current". In the connecting channels of the lagoon there is at times an inflow,
then an outflow. In the tidal seas inflow and outflow change daily with low and

high tide, if there is a permanent communication with the sea; it changes in a fortnightly rhythm if sea water can only enter with spring tides. In tideless seas or those with weak tides the change between inflow and outflow is chiefly produced by the weather; it is therefore irregular. If the sea-water level rises (mostly through heaping-up by shore winds) there is an influx of sea water; with offshore winds there is an outflow through the communicating channels. The influx of fresh water changes mainly in a yearly cycle, it is strong in the rainy season and weak in the dry season or, in cold climates, in winter. This alternating influx of water from both sides results in a complicated circulation of the water masses which varies from lagoon to lagoon (see PRITCHARD 1967, BOWDEN 1967, SAELEN 1967 and others).

As a result of the changing influx of water masses of different densities lagoons usually display a marked and changing stratification of water masses. It is much greater than in fresh-water lakes or in the ocean. It reaches similar values in the brackish seas, but here the bodies of water are far more extensive and more stable. Only in lagoons the waters of which are thoroughly mingled by the tides is it possible for the water to be nearly homohaline, that is without stratification in a vertical direction. This is undoubtedly a rare, exceptional case.

2) Variability of salt content and oxygen content. The frequent changes of influx of salt water and fresh water result in an intensive change of salinity at every point. In the normal case a brackish-water zone forms between the mouth, the region under marine influence, and the head, the influx of fresh water; this brackish zone is made up of one or more bodies of water with their fronts placed obliquely. Since they are frequently displaced, individual points show a variable salt content: In some lagoons, especially in the tropics, the change may range from 0—35⁰/₀₀ S. In the rainy season the marine basin gets completely filled with fresh water from the rivers, in the dry season it fills with sea water (see Lake Chilka in India, ANNANDALE). The bottom sediments usually retain a certain amount of salt. In arid regions evaportation may exceed the supply of fresh water; in that case hyperhaline areas of lagoons are produced, for example the lagoon Madre (HEDGPETH 1947). The oxygen content may undergo equally radical changes. In shallow lagoons a rich vegetation of diatoms, among others, may form on the floor. In such cases the water above the floor becomes supersaturated with oxygen. On the other hand, there is a stagnation of water in depressions and with abundant supply of organic detritus this can easily lead to oxygen deficiency and accumulation of H_2S. In these cases the lagoon gets into a "Black Sea" situation; this may be temporary in some places, but permanent in the deeper basins of atypical lagoons.

The variability of these factors in lagoons often leads to disastrous results in which there is a far-reaching extermination of animal life. There are various reasons for such disasters. Often they are due to oxygen deficiency in the deeper layers, caused by stagnation of the water in the depth and by intense processes of putrefaction. In lagoons with a high biomass of phytoplankton and little circulation, oxygen consumption during quiet, warm nights may suddenly rise to such an extent that mass extinction takes place. But sea water may suddenly transgress into brackish lagoons when the sand banks at the mouth are changed or broken through. Such disasters have, for example, occurred in Sandwich Bay (Southwest Africa) and St. Lucia Bay (South Africa); in the latter they even caused a rich population of birds to move off. As

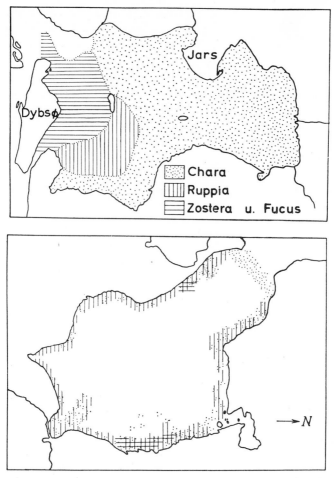

Fig. 70. Two brackish lagoons of similar morphology, but different extent of vegetation. Top: the Dybsöfjord, Denmark. After K. LARSEN. Bottom: The Jasmunder Bodden, its deeper regions (8 m) are covered with sapropel rich in H₂S, and lacking in higher plants. After TRAHMS (1939), the representation of plant forms has been simplified.

brackish-water areas the lagoons are undoubtedly of short duration. They probably disappear far more rapidly than many fresh-water lakes.

3) Animals and plants. The lagoons are eutrophic to hypertrophic so long as there is sufficient oxygen in the water. This also holds for tropical lagoons which are particularly rich in fishes and Crustacea. This high productivity is shown both by the plankton and the bottom plants and animals. The lagoons share this feature with river mouths and even the outer regions of brackish-water seas (Belt Sea at the entrance to the Baltic). However, it is specially marked in lagoons and similar marine bays (see GESSNER 1937, 1955). There are various reasons for this high productivity such as: Supply of nutrients, especially of phosphates, by the river and the "drainage water" from the land; the large number of dying planktonic organisms in the inflowing

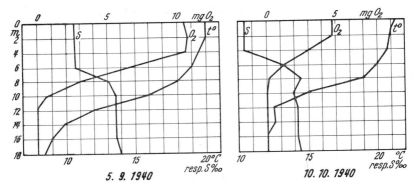

Fig. 71. Fluctuations of salinity, temperature and oxygen in a lagoon. Lake Varna, Bulgaria, Station IV. 5. After Paspalew from Caspers 1951.

	23. VII. 1924			2. IX. 1924 (after stormy weather)		
Depth	t	Cl	O$_2$	t	Cl	O$_2$
0	26.5	5.59	6.43	21.9	5.805	5.11
5	26.4.5	5.82	6.35	21.8	5.78	4.98
11	22.5	8.33	0.006	21.8	5.78	4.82

Change of conditions in the Sea of Azov in the same region. From Knipowitsch. Note oxygen content at a depth of 11 m.

sea and fresh water; the abundant supply of organic detritus; rapid circulation of nutrients as a result of the intense decomposition, especially in shallow lagoons,

In the plankton mass development is often shown by diatoms *(Chaetoceras, Coscinodiscus, Nitzschia, Skeletonema* etc.*);* up to 50% of the plankton may be diatoms. In addition Chlorophyceae play an important part and in areas where the effluent waters are particularly rich in nutrients, the Cyanophyceae.

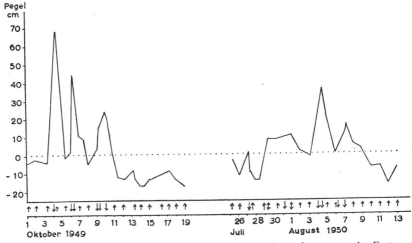

Fig. 72. Fluctuations of water level in the Sehlendorfer Lake, a lagoon on the East coast of Schleswig-Holstein, in October 1949 and August 1950. The arrows below the graph indicate the influx and outflow of water through the connection with the sea. After Lillelund.

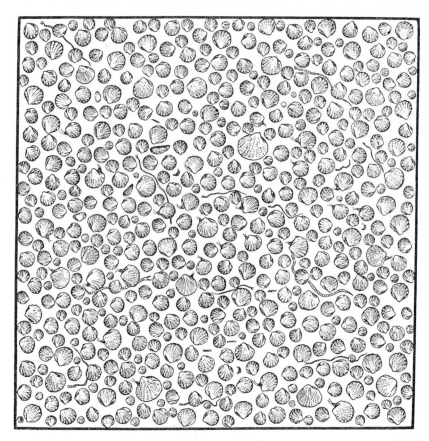

Fig. 73. Mass colonisation at the bottom of a lagoon. Lake Varna: Colonisation by *Cardium* (*endobious*). A bottom sampler catch (1/10 qm), Station IV, 1.2 m depth, October 26, 1940. 533 *Cardium edule;* 1 *Abra alba;* 2 *Mytilus galloprovincialis;* 15 *Hydrobia acuta;* 30 *Idotea baltica;* 2 *Sphaeroma pulchellum; Balanus* on living *Cardium;* 10 Amphipoda; 12 *Nereis succinea.* From CASPERS 1951 a.

In the zooplankton the Copepoda of the genera *Acartia (A. tonsa)*, *Eurytemora* and, locally, *Pseudodiaptomus* are important. The Rotifera of the genera *Synchaeta, Brachionus (B. picatilis), Trichocerca* are abundant.

The density of the plankton reduces the penetration of light. When production in the Schlei, a Förde of Kiel Bay, increased through effluents, the stand of bottom plants disappeared (NELLEN 1967).

The vegetation displays the following features: The quantities and numbers of species of Phaeophyceae and Rhodophyceae decrease more rapidly than would correspond to a normal reduction with lowered salinity. This is particularly the case with the large kelps. In parts this is due to the fact that in lagoons there is less rocky floor flooded over. Even if there is a rocky floor it is often covered with detritus and sand (DAY 1951). The Brown Algae are most common, *Fucus* (especially *F. vesiculosus*) and *Chorda filum,* in the South also species of *Sargassum.* The situation

is favourable for loose-lying masses of seaweed which are not fixed. They are formed by *Fucus, Polysiphonia* and *Phyllophora*. A formation of Red Algae, the Bostrychietum with *Bostrychia, Caloglossa*, species of *Catanella* is even a characteristic feature of the roots and stems of the mangroves.

The Chlorophyceae increase in quantity. Especially species of *Enteromorpha, Ulva, Monostroma* form large stands in places rich in nutrients; *Enteromorpha* produces floating layers in quiet places.

The Phanerogams of the families of Potamogetonaceae and Hydrocharididae also increase in numbers. In the more saline areas there are species of *Zostera, Halophila, Thalassia, Cymodocea* etc. They often form extensive meadows of sea grasses; in strongly brackish regions species of *Ruppia, Chara* and euryhaline fresh-water *Potamogeton* become dominant. This strong development of aquatic plants rooting in sand and unconsolidated floors results in a higher biomass compared with similar floors of the open sea.

The vegetation of the banks of lagoons habitually resembles that of fresh-water lakes. A belt of helophytes, consisting of species of reed *(Phragmites), Scirpus* and *Spartina* forms around the banks; it covers all the banks not exposed to strong wave action. It is this pounding of waves which more than the salt content keeps these plants away from the shores of the open sea. In areas of flooding salt meadows are common, with species of *Salicornia,* Chenopodiaceae *(Chenola, Obione)* and above all, grasses of the genera *Puccinella, Festuca,* in warmer regions *Sporobolus, Distichum* etc., as well as species of *Juncus* forming a short turf. In the tropics mangrove forests of *Avicennia, Brugueria, Ceriosa,* and others, occur in suitable places in the lagoons.

This mass development of vegetation results in a large production of vegetable detritus. In the lagoons this is mostly coarser than in the sea itself, as there is less grinding-up through surf action and the cell walls of land plants are more resistant. Thus in wide regions of the lagoons, even close to the banks, an admixture of organic, and in particular vegetable detritus in large quantities occurs in the sand.

In contrast to fresh-water lakes there is a complete lack of floating plants such as *Hydrocharis, Limnocharis, Pistia, Eichhornia* as well as Nymphaceae; only *Lemna* occurs in quiet places with low salinity.

The animals are represented by a number of species which have their optimal development in lagoons. DAY designated them as "estuarine species". They account for a large share of the biomass. GUNTER (1967) lists the following return for fisheries in lagoons in the Gulf of Mexico: Fresh-water species 9,3, salt-water species 24,4, estuarine species 1332,2 million pounds. The factors which lead to an optimal development of estuarine species in lagoons may be manifold. Abundance of nutrients in the lagoons is important. GUNTER calls the lagoons "nursery grounds". Very many motile species spawn in the sea outside the lagoons, enter the lagoons as young animals and leave them again as adults. This behaviour is displayed not only by many fishes, but also by Crustacea, e. g. species of *Penaeus, Callinectes* etc. The factors attracting the young animals into the lagoons is presumably lower salinity as they prefer this region. A large number of species of fishes occur in the lagoons of warm areas (see GUNTER, ANNANDALE and others). But the above-mentioned type of estuarine animals is rare in the lagoons of cooler regions. If marine fishes (excluding "guests") penetrate into brackish lagoons they are species which spawn in these

areas, e. g. forms of the herring *Clupeus harengus* which yields the relatively highest harvests in lagoons with a salintiy of $5^0/_{00}$ (see NELLEN 1965). There are other species which prefer the biotopes of lagoons and reproduce there: *Gobius (Pomatoschistus) microps, Nerophis, Syngnathus typhla* etc. Consequently there are predominantly euryhaline fresh-water fishes which spawn in fresh water, but appear in larger numbers in the brackish water of the lagoons. NELLEN (1965) lists 20 species occurring in the mesohalinikum, 8 species as far as the β mesohalinikum *(Cyprinus carpio, Tinca tinca, Scardinius erythrophthalmus, Alburnus alburnus, Abramis vimba, Blicca björkna, Esox lucius, Acerina cernua)*, 12 species are even found in the α mesohalinikum amongst others *Petromyzon fluviatilis, Coregonus lavaratus, Osmerus eperlanus, Rutilus rutilus, Idus idus, Abramis brama, Lucioperca luciolarca, Perca fluviatilis.* But in cooler latitudes there is a migration of marine animals between the sea and lagoons (MUUS 1967); this is a seasonal migration. In late summer or autumn several marine species migrate into the sea and into deeper water, returning to the lagoons in spring or summer, e. g. *Carcinus maenas, Crangon crangon, Pleuronectes (Platichthys) flesus, Gobius (Pomatoschistus) microps.* Temperature is evidently the decisive factor.

Furthermore, reduced competition and absence of predators may be important for estuarine species. The absence of Echinodermata *(Asterias, Ophiura)* which disappear rapidly in brackish water, is undoubtedly of significance for the large number of bivalves. Finally the biotopes of the lagoons determine the occurrence of many species in the lagoons. The extensive meadows of *Zostera, Ruppia, Chara* as well as the bottoms varying from fine sand to mud — rich in detritus — allow optimum development to many animals, e. g. Gastropoda *(Hydrobiae, Littorina* v. *tenebrosa)*, Oligochaeta *(Tubifex costatus, Peloscolex benedeni, Amphichaete sannio, Paranais litoraki, Rhizodrilus pilosus)*, Polychaeta *(Nereis diversicolor, Polydora ligni, Manayunkia aestuarina, Streblospio strubsoli)*, Crustacea *(Palaemonetes, Leander adspersus, Neomysis vulgaris (integer), Sphaeroma, Idotea viridis, Cyathura carinata, Gammarus duebeni, Melita palmata, Corophium volutator, C. insidiosum)*, many Harpacticidae *(Cletocamptus, Nitocra)*, Ostracoda *(Cyprideis litoralis, Loxocancha elliptica, Cytherois fischeri)*, *Protohydra*, numerous Ciliata.

Many data on the density of colonization can be found in MUUS's (1967) detailed study of lagoons in Denmark. He found up to 2 million Harpacticidae per square metre, up to 200,000 *Protohydra*, up to 387,000 Ostracoda, up to 80,000 Oligochaeta, 40,000 *Pygospio elegans*, 20,000 *Manayunkia aestuarina*, 4,000 *Neomysis*, 50,000 *Hydrobia*.

The polyhaline regions are mainly colonized by an impoverished marine fauna. Species with a preference for pure and moving water are absent. Such species, if they occur at all, are found in the belt of breaking waves (Turbellaria) and those species which prefer calm water, rich in detritus, are dominant. To them belong many ciliary feeders (bivalves, Spionidae, *Fabricia* etc.), and species resistant to oxygen deficiency *(Scoloplos, Capitella)*.

In the mesohaline area brackish-water species (see p. 19) are particularly rich in species and individuals. They extend into the oligohaline zone. There is a far greater percentage of brackish-water species than in the open regions of the sea with similar salinity. Species derived from fresh water are also relatively abundant. They occur among the Oligochaeta, but also among insects (larvae of *Chironomus sali-narius* and *Ch. halophilus)*, according to Muus up to 20,000 of *Acentropus niveus* on water plants up to 312 per square metre, all stages of the beetle *Haemonia mutica)*.

Muus (1967, p. 217) points out that in the pure *Macoma baltica* community and the "estuarine" lentic community there are numerous species which contribute varying proportions to a fixed total:

Macoma baltica community	*Cardium lm.* — *Hydrobia ventrosa* community
Cardium edule	*Cardium lamarcki*
Hydrobia ulvae	*Hydrobia ventrosa*
Littorina littorea	*Littorina tenebrosa*
Idotea baltica	*Idotea viridis*
Sphaeroma rugicauda	*Sphaeroma hookeri*
Nereis vireus	*Nereis diversicolor*
Pomatoschistus minutus	*Pomatoschistus microps*
Pleuronectes platessa	*Platichthys flesus*

The phytal zone is mainly colonized by the *Idotea viridis—Littorina rudis* community (Krüger and Meyer), the unconsolidated floors, rich in detritus, by the *Cyprideis-Manayunkia* community (Remane).

There are great regional differences in the fauna of the lagoons, for instance between Atlantic, American (Gunter, Hedgpeth) and South African lagoons (Day, Macnae). This applies in the first place to the marine and polyhaline regions. In the mesohaline region of tropical lagoons the greater abundance of Cyprinodontae, Decapoda and others is noticeable such as the numerous Penaidae, many Brachyura, both semiterrestrial on the shore *(Uca, Sesarma, Cardisoma* etc.) and in the water *(Callinectes, Cleistostoma, Hymenosoma* etc.). The Palaemonidae are also more abundant in warm lagoons and Decapoda which build tubes in the floor such as *Callianassa, Upogebia* and *Alpherus* penetrate further into brackish regions. But even in the least salin regions of the Kaysna lagoon (Old Drift summer 1.3⁰/₀₀, winter 16.5⁰/₀₀, Charlesford Rapids 0.3⁰/₀₀ and 10.7⁰/₀₀, Day 1962) there occur species of *Corophium* (c. *triaenonyx),* a *Melita (M. zeylanica),* *Cyathura carinata,* the Sphaeromidae *Sphaeroma terebrans* and *Pseudosphaeroma barnardi, Leander pacificus,* among molluscs a *Littorina (L. knymaensis),* a *Rissoa (R. prima),* a *Haminea (H. alfredensis)* that is a close relative of the European brackish-water one, in addition to numerous widespread genera and species, in algae (Bostrychietum) both of the mangrove roots and the mangrove mud. In the Bostrychietum there are many Nematoda of which *Paracyatholaimus vitrinus* and *Parachromadorella paramucrodonta* are typical; furthermore Harpacticidae of the genera *Nitocra* and *Mesochra,* Tanaidae. In the mud there are Polychaeta of the genera *Manayunkia (M. brasiliense)* and *Polydora redeki.* The Nematoda are represented with several species which also occur in Europe, e. g. *Theristus flevensis, Th. setosus, Anoplostoma viviparum, Terschillingia communis* and others. The Oligochaeta belong to the genera *Rhizodrilus, Limnodrilus, Peloscolex, Paranais,* among the Ostracoda we find species of the genera *Cyprideis, Cytherura, Iliocythere,* among the Opisthobranchiata species of the genus *Alderia* and *Stiliger.* In the region of the lagoons there is a stand of genera, and particularly of species, of world-wide distribution.

Several species of the meiofauna produce large numbers in the lagoons, especially on the floors rich in detritus where they are mainly found. According to P. Ax the following Turbellaria are such organisms of still water: *Pseudograffilla arenicola, Placorhynchus octaculeatus, Promonotus schultzei, Baicalellia brevitubus, Acrorhynchus robustus, Proxenetes westbladi, P. karlingi, Provortex balticus, Macrostomum balticum, M. rubrocinctum, Promesostoma rostratum* etc. There are large numbers of the Ostracoda *Cyprideis litoralis, Loxoconcha elliptica, Xestoleberis aurantia, Cytherura gibba,*

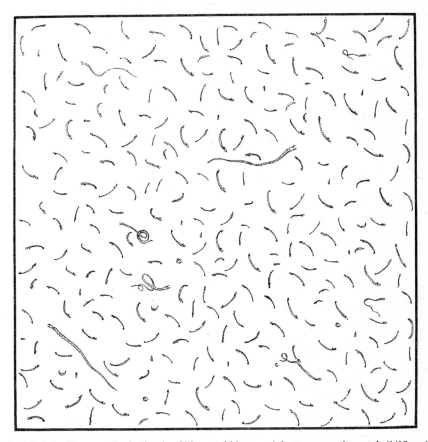

Fig. 74. Lake Varna: colonisation by Chironomid larvae. A bottom sampler catch (1/10 qm), Station X, 3.3 m depth, May 21, 1941. 240 Chironomid larvae (*Chir. salinarius*- and *halophilus*-Typus); 3 *Cardium edule;* 18 *Hydrobia acuta;* 8 *Nereis succinea;* 18 *Gammarus locusta* + *Corophium bonelli.*

Cytheris fischeri and species of *Leptocythere*. There is also considerable mass development of the Copepoda *Tachidius discipes, Mesochra Lilljeborgi, M. rapiens, Ectinosoma curticorne, Onychocamptus mohammed, Nitocra spinipes, N. lacustris, Cletocamptus confluens, Cl. retrogressus.* If we also take into consideration that the Ciliata are rich in species and individuals, then the floors of lagoons, rich in detritus, are among the most densely populated regions of the sea bottom.

A marked feature of lagoons is the presence of fresh-water species or those of fresh-water origin since — as already mentioned — the invasion of organisms from fresh water into marine regions occurs chiefly in the zones of calm water of the sea shore. Thus fresh-water snails (species of *Limnaea*), Planarians, insect larvae and Oligochaeta *(Stylaria lacustris, Nais)* are common in the phytal zone of these lagoons. *Chironomus* larvae also play an important part in the bottom zones (see fig. 74).

C. River mouths (estuaries)

All river mouths contain a zone of brackish water, as by its very nature, mixing of sea and fresh water takes place in it (see p. 9). The special features of river brackish waters are their directional currents. In the upper layers fresh water flows towards the sea, at the bottom a countercurrent runs upstream, carrying water rich in salt into the region of the mouth. How far the saline undercurrent penetrates the river mouth depends on the morphology of the latter. But since the mouth never represents an ideal gradient which widens and deepens regularly, the irregularities of structure of its banks and bottom result in considerable intermingling of salt and fresh water.

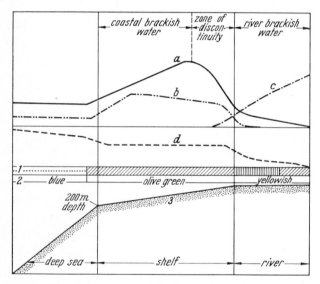

Fig. 75. Plankton, salt content, content of detritus, colour of water and depth of water in the region of the mouth of an ideal river. a, marine diatoms; b, Appendicularieae; c, fresh-water algae; d, salinity. 1, content of detritus; 2, colour of water. After THIEMANN 1934.

Thus a brackish intermediate zone is being inserted the salinity of which diminishes upstream; this zone consists of mixed water and in many regions there is but a slight stratification of salinity from the surface to the floor. These brackish bodies of water are constant as such, though there is a rapid exchange of the component water molecules. The consequence of this rapid change is that a true brackish-water plankton has far less chance of developing in river mouths than in calm bays and lagoons.

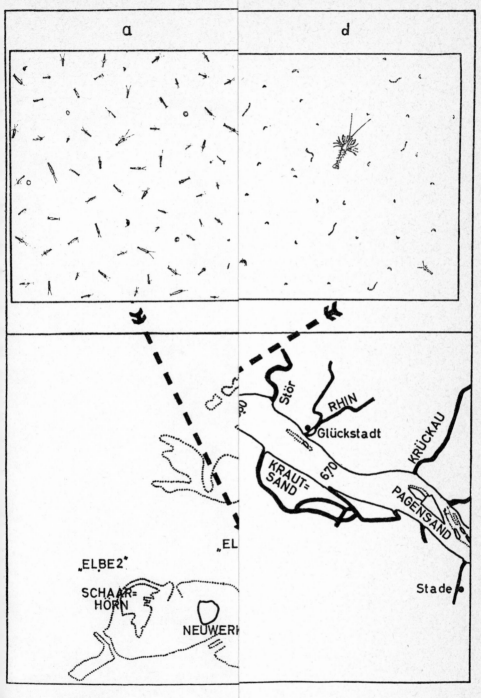

a) Bottom of fine sand at light ship "Elbe 3". ; 5 *Cardium edule;* 1 Spionid.
b) Colony of *Corophium* in soft mud off Cuxha*rcinus maenas;* 240 *Mytilus edulis;* 12 *Petricola pholadiformis;* 36 *Nereis succinea.*
c) Colony of *Mytilus* in the navigable channel
d) Fine sand with an admixture of loamy mud *us;* 1 *Gammarus zaddachi;* 1 *Neomysis vulgaris.*

Essentially it consists of euryhaline fresh-water organisms, though specific brackish-water planktonic organisms such as species of *Eurytemora* are not absent from river mouths.

Based on comparative investigations of many river mouths THIEMANN (1934) has formulated a basic scheme (see fig. 75). The marine region which is affected by river water, in front of the river mouth proper, is called coastal brackish water. How far it extends into the sea depends, of course, on the size and water masses of the river. In the river there is a region of river brackish water. Both areas are separated by a discontinuity layer. The coastal brackish water in front of the rivers shows a changing salinity, but is polyhaline throughout. Coastal brackish water in front of rivers with a wide mouth is characterized by a low content of detritus, but a strong development of diatoms which increases towards the zone of discontinuity. „Die häufigsten Diatomeen, die sich in allen Flußmündungen fast immer wieder finden, sind *Asterionella japonica, Biddulphia, Chaetoceros, Coscinodiscus, Laudera borealis, Leptocylindrus danicus, Lithodesmium undulatum, Rhizosolenia, Skeletonema costatum* und *Thalassiothrix.* Diese Diatomeen sind zum überwiegenden Teil Küstenformen, die schon wegen ihres Wohngebietes Schwankungen des Salzgehaltes ertragen müssen. Aber auch die Gattungen *Corethron, Rhizosolenia* und *Chaetoceros,* die im allgemeinen als Hochseeformen bezeichnet werden, finden sich im Küstengebiet, letztere vermag hier sogar gut zu gedeihen und eine hohe Bevölkerungsdichte zu entwickeln. Die schwach euryhalinen Diatomeen *Chaetoceros* und *Rhizosolenia* fehlen oder kommen nur vereinzelt vor im Flußbrackwasser der großen Ströme und grenzen dieses ab gegen die Küstenbrackwasser" (THIEMANN 1934, p. 248). [The commonest diatoms, almost invariably found in all river mouths are *Asterionella japonica, Biddulphia, Chaetoceros, Coscinodiscus, Laudera borealis, Leptocylindricus danicus, Lithodesmium undulatum, Rhizosolenia, Skeletonema costatum* and *Thalassiothrix.* These diatoms are predominantly coastal forms which must be able to tolerate salinity fluctuations in their habitat. But the genera *Corethron, Rhizosolenia* and *Chaetoceros* which are generally described as forms of the open sea, also occur in the coastal region; the last one flourishes and displays a high population density here. The weakly euryhaline *Chaetoceros* and *Rhizosolenia* are absent or have only isolated records in the brackish river water of large streams, delimiting it from coastal brackish water]. Colonization by animals shows essentially an increase of marine plankton, of the Tunicata *Oikopleura dioica* and *O. longicauda,* as has been established by LOHMANN (1896). Furthermore the Copepoda increase, but only in individuals, not in species; altogether coastal brackish water is much poorer in species than the oceanic region, in spite of its high productivity. According to DAHL (1894) *Euterpe acutifrons,* species of *Paracalanus, Corycaeus,* species of *Centropages* are widespread. Only in exceptional cases do fresh-water planktonic organisms occur here.

„Eine Erklärung für den Planktonreichtum in Brackwasser der Flußmündungen ist die, daß das Wasser hier reich an organischen Nährstoffen ist und eine starke Bevölkerungsdichte der euryhalinen Planktonformen ermöglicht. Im Küstengebiet vor den Flußmündungen sterben die Planktonformen des Süßwassers und die steno-halinen Formen des Salzwassers ab und sinken zu Boden. Diese abgestorbenen Organismen liefern bei ihrem Zerfall organische und anorganische Stoffe. Die für den Stoffwechsel des Phytoplankton notwendigen Nitrate und Phosphate werden hier beim Zerfall auf dem Boden neu gebildet. Diese Neubildung findet auf dem Boden des flachen Wassers intensiver statt als in den Tiefen des Ozeans mit niedriger Temperatur. Die auf dem Boden gebildeten Nitrate und Phosphate werden in den Schelfgebieten vor der Flußmündungen durch Vertikalströmungen, die durch die Tidenströme hervorgerufen werden, an die Wasseroberfläche gebracht. Weitere Nährsalze werden durch den Fluß in das Mündungsgebiet befördert. In diesem nährstoffreichen Gebiet liegt die größte Bevölkerungsdichte der schwach euryhalinen Formen (z. B. *Chaetoceros*) weiter außerhalb als die der stark euryhalinen Formen (z. B. *Coscinodiscus*)." (THIEMANN 1934, p. 248). [The wealth of plankton in the brackish water of the river mouths can be accounted for by the fact that the water here is rich in organic nutrients, allowing a high population density of euryhaline planktonic forms. In the coastal area in front of the river mouths the planktonic forms from fresh water and the stenohaline forms from salt water die and sink to the bottom. On decay these dead organisms supply organic and inorganic material. In this decomposition the nitrates and phosphates required for the metabolism of the phytoplankton are newly formed on the bottom. This re-forming proceeds with greater intensity on the bottom of shallow waters than in the ocean depth at low temperature. In the shelving areas in front of the river mouths the nitrates and phosphates which have formed on the bottom are carried to the water surface by means of vertical currents brought about by tidal currents. Further nutrient salts are carried by the river into the area of its mouth. In this productive region the greatest abundance of the slightly euryhaline forms (e. g. *Chaetoceros)* is attained further seawards than that of the strongly euryhaline ones *(e. g. Coscinodiscus)*].

The river water with its low salinity contains much detritus or fine detritus in suspension. Few forms of marine plankton occur, furthest the diatoms *Biddulphia* and *Coscinodiscus*. The plankton consists predominantly of euryhaline fresh-water forms, e. g. *Scenedesmus, Pediastrum, Conferva, Ulothrix, Merismopedia, Dictyosphaerium, Asterionella gracillima* and *Microcystis*. The Rotifera also show abundant development *(Keratella, Brachionus)*, as well as the Cyclopidae among the Copepoda. River brackish water appears to be extremely unfavourable to specific brackish-water species on account of the abrupt change of water. However, some of them appear here in large numbers. This applies for example to the diatom *Coscinodiscus rothii* which attains its maximum population density in the Elbe and Weser, especially in the fresh-water area. Among the Copepoda *Eurytemora (E. affinis),* species of *Acartia (A. tonsa, A. giesbrechti)* and, in the warmer regions, Pseudodiaptomidae should be mentioned here.

H. KÜHL and H. MANN (1961) have investigated the behaviour of numerous factors in the regions of river mouths for three rivers in North Germany (Elbe, Weser, Ems) (see fig. 81). The bottom regions of rivers are more favourable than the lagoons on account of the good supply of oxygen and rapid renewal of water which prevents the formation of the good supply of oxygen and rapid renewal of water which prevents the formation of regions of putrefaction and production of H_2S; in a biologically deleterious sense, there is a frequent restratification of the bottomiwith changes in the current; this also leads to a frequent change of the substratum. Once again, in front of the regions of river mouths mass colonization of euryhaline marine species occurs, e. g. *Cardium,* whereas the bottom regions of the river courses are poor in species and, over some stretches, also poor in individuals. In the Elbe CASPERS (fig. 76) found no macrofauna at all in the adjoining sandy, gravelly area of the fresh-water region. There are special features in the zone of the banks. Detritus is deposited on the slip-off slope and may here form a Watten area rich in sediments extending into the fresh-water region. According to CASPERS (1951 d) the picture in these regions is dominated by Tubificidae and *Chironomus* larvae in their mass colonization. On the undercut slope sandy floors are formed in which marine elements of the microfauna penetrate particularly far, even into the fresh-water regions (see. P. AX 1957). Altogether this is the area in which the sequence of species from salt to fresh water shows the greatest changes compared with marine regions. Actively migrating forms, e. g. *Neomysis vulgaris,* species of *Gammarus* and, of course, fishes *(Pleuronectes flesus)* as well as sessile animals reach the boundary of fresh water here, such as the barnacle *Balanus improvisus,* or else reach the fresh water of rivers like the polyp *Cordylophora caspia*. Other species are prevented from invading river mouths up to the limit of their salinity tolerance through the particular conditions at river mouths described above. The number of specific brackish-water species is also particularly low in the bottom fauna of the rivers, but *Cordylophora caspia* indicates that they may well occur here.

The publications by CASPERS and his students provide a detailed description of the fauna at the mouth of the river Elbe.

D. Pools

Small bodies of water (pools) with brackish water are very widespread on the sea shore as well as inland. Ecologically they are tremendously varied since the floor, the climate or weather and their position relative to the tides create an individual collection of factors for almost every pool. Nevertheless, colonization of these pools proves to be uniform to a large extent. The macrofauna of marine origin is reduced, but insects and insect larvae become prominent (Ephydridae, Nematocerae, Coleoptera, Sigara larvae). Among flowering plants *Ruppia, Zannichellia* and, in parts, brackish-water species of *Ranunculus* predominate. Microfauna and microflora are particularly well developed. The majority of species in pools obviously corresponds to those of the calm bays in lagoons, but there are a number of species which here attain their largest population density, e. g. the brackish-water prawn *Palaemonetes varians* among the macrofauna. The characteristic or differentiating species are more numerous in the microfauna. I mention the Ostracoda *Cypridopsis aculeata, Heterocypris salina, Candona angulata,* the Copepoda *Cletocamptus confluens, C. retrogressus, Nitocra spinipes, Onychocamptus mohamed,* species of *Horsiella* and of *Schizopera,* the Rotifera *Encentrum rousseleti, Testudinella clypeata*. Even in closely adjoining areas the composition of communities varies considerably, depending on whether or not the sea water has access to the pools through connecting channels. In the first case the fauna is much richer. It still contains the majority of species of the *Littorina rudis* — *Idotea viridis* community and of the *Cyprideis litoralis* — *Manayunkia* community which declines in the isolated pools. A large number of species of insects and insect larvae, above all beetles, occur in brackish pools (*Enochrus, Ochthebia* and others) as well as Diptera (Ephydridae, Chironomidae, Ceratopo-

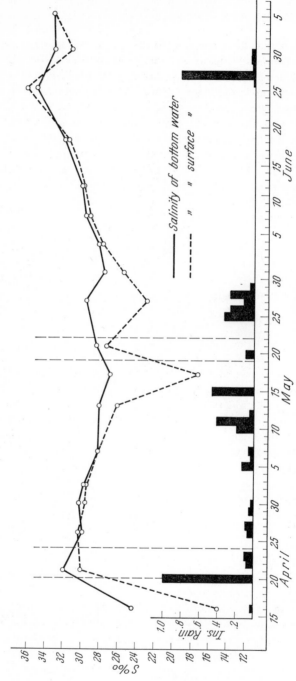

Fig. 77. Salinity fluctuations in a salt-marsh pool (III) in the spring of 1932. Salinity is affected by heavy rainfall and evaporation. The vertical, broken lines indicate the times when sea water enters with the tide. After NICOL 1935.

gonidae of the genera *Dasyhelia, Culicoides, Bezzia,* Stratomidae, Dolichopodidae (see
REMMERT 1955).

Typical kinds of brackish pools are 1. Rock pools. They arise in hollows in the
rocks which collect rain water on one hand, on the other hand are occasionally
reached by sea water, in the extreme case by spray water. They are found where-
ever impermeable rock forms a rocky coast. Some extremely euryhaline and eurythermic algae *(Enteromorpha)* and Crustacea *(Gammarus duebeni, Balanus improvisus)*
flourish in them. Rotifera are not uncommon. A typical species is the Copepod
Tigriopus fulvus (see fig. 37). 2. Pools of the salt marshes. In the firm clayey soil of
the salt meadows there arise smaller or larger pools with a loamy bottom which
receive salt water from the sea more or less frequently. Just as in rock pools, conditions
are subject to great fluctuations. On transgression of sea water there is a sudden
change of all the physical and chemical situation as shown in the graphs (figs. 77, 78)
after NICOL (1935).

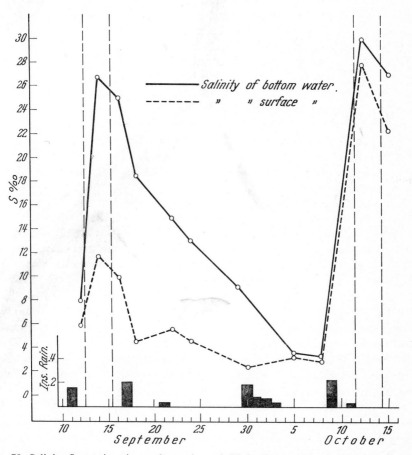

Fig. 78. Salinity fluctuations in a salt-marsh pool (X) in the autumn of 1931, showing the
effect of drainage water. The vertical lines again show the entry of the tides into the pool.
After NICOL 1935.

A number of marine animals of the macrofauna still penetrate into the brackish pools of the salt meadows, fishes e. g. *Gobius microps, Gasterosteus, Syngnathus*; molluscs e. g. species of *Hydrobia,* along the edges *Assiminea grayana,* Ellobiidea and in the algal carpets *Alderia modesta, Limapontia.* Crustacae are also represented here: *Gammarus, Jaera, Sphaeroma* and *Palaemonetes.* Depending on the conditions of the underlying floor the microfauna represents a variable selection from the microfauna of marine shore pools. In the frequent instances where diatoms form a dense layer on the floor, Rotifera occur in abundance.

A third type of marine shore pools develops on sandy shores. Here the waves frequently build up a beach ridge of sand. At high water level the hollows lying behind it are filled with water and soon turn brackish. These small bodies of water exist only intermittently as the water in them percolates into the sand floor after a shorter or longer period of time. They also frequently change position as the shore is constantly being altered. These small bodies of water may also be colonized by a microfauna and -flora. Here, too, a layer of diatoms may cover the floor, Rotifera and brackish-water Turbellaria may develop quickly. These pools are also important for the distribution of brackish-water animals.

As already mentioned, colonization of brackish-water pools quickly declines when the small bodies of water lose direct communication with sea water. This fact can be observed even if the sea water merely percolates through the soil. There are various reasons for this. The haline stratification is more pronounced and more stable; hence also the oxygen lack on the bottom. But the chief reason which accounts for the scarcity of species in inland brackish waters is probably connected with factors in the ecology of dispersal. Marine animals have their stages of dispersal almost exclusively in water, but fresh-water organisms are in the main dispersed through the air, either as adult stages capable of flight, such as insects, or as eggs or resistant stages capable of being transported passively by wind or by flying animals. This fundamental difference between marine and fresh-water organisms offers a much greater opportunity to the euryhaline fresh-water organisms and to the salt-water inhabitants of fresh-water origin, to colonize isolated and short-lived brackish-water pools. Such kinds of Rotifera, Phyllopoda and Protozoa are therefore inhabitants of inland saline waters, usually of world-wide distribution. The same does not apply to insects. In their case distribution is determined by factors relating to the ecology of their existence.

There are, however, a number of marine organisms which have colonized inland saline areas which do not represent relicts of former transgressions of the sea. To these belong in the first instance microorganisms, especially diatoms and Ciliata. There are a number of records of the rapid colonization of newly-formed inland saline places by marine diatoms, see KOLBE, BUDDE and others. No doubt the Ciliata show a similar behaviour: in any case it holds for the rich fauna of the Oldesloer saline (KAHL 1928) which shows considerable agreement with the fauna of the Ciliata from pools and lakes near the shore. It is interesting that even small Gastropoda have found their way into continental brackish water, e. g. *Hydrobia ventrosa* into the Central German salt waters near Halle; in the same area Foraminifera and acoelous Turbellaria were found (REMANE 1937a). More common in inland salines are Copepoda of marine origin, e. g. *Mesochra rapiens, Onychocamptus mohammed,* furthermore spe-

cies of the genera *Schizopera, Nitocra, Horsiella, Ectinosoma, Cletocamptus* assuming
the genus to be of marine origin. Among the Ostracoda *Cyprideis litoralis* and *Loxoconcha elliptica* in particular have been dispersed in this way. BEADLE (1931)
mentions the following marine forms in continental brackish waters: the planktonic
diatom *Chaetoceras*, the Green Alga *Enteromorpha intestinalis*, the prawn *Palaemo-
tes varians* and Foraminifera.

Transport of animal species by man shows how important the possibilities of
dispersal are in the colonization of continental brackish waters. Marine fishes were
transplanted into Lac Quarun and with these several marine bivalves, e. g. *Cardium
edule, Syndosmia ovata*, Gastropoda, *Balanus improvisus*, prawns and others were
carried into the lake where, in part, they show abundant development. The same
has been demonstrated by ZENKEVITCH (1940) in his successful experiments of transferring marine animals into the Caspian Sea.

E. Interstitial brackish water on the sea shore
(coastal ground water)

On the sea shore water will enter the soil of the coast — as a result of the tides
and waves, provided the soil can take up water. This is particularly the case in sandy
soils, but the interstices in the humous soils of temporarily flooded salt marshes
contain interstitial water as well. Here sea water mingles with the fresh water contained in the soil so that along the sea shore a subterranean strip of brackish water
develops. The region of the interstitial coastal water harbours a rich and peculiar
fauna which has only been investigated in recent times (REMANE and SCHULZ 1934,
DELAMARE-DEBOUTEVILLE, CHAPPUIS, GERLACH, JANSSON, SCHMIDT and others).

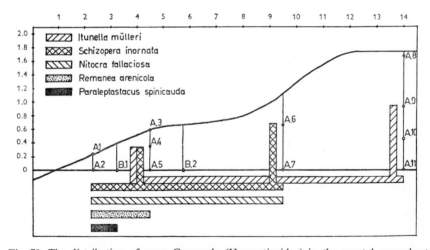

Fig. 79. The distribution of some Copepoda (Harpacticoidea) in the coastal groundwater
of Simrishamm on the SE coast of Sweden. The top row of figures shows the distance from
the shore in m, the one on the left the height above water level. In the region of 0 to 5 m
from the shore there is a rapid decline of salinity from $8^0/_{00}$ to about $2^0/_{00}$, up to 10 m to
about $1^0/_{00}$. After BRINCK, DAHL & WIESER 1955.

The fauna varies according to the soil and the tides. In tidal seas the sea water regularly infiltrates into the shore, during low tide interstitial water runs off again in a zone of small springs (microsprings). This region of circulating interstitial water has a salinity almost equal to that of sea water, thus it is euhaline to polyhaline. The region receives nutrients from the sea water, the oxygen content is good and is especially high in the region of run-off (JANSSON 1966). This region contains a rich fauna of Polychaeta (*Hesionides, Petitia*), Archiannelida (*Trilobodridus, Diurodrilus, Protodrilus*), Turbellaria (especially Otoplanidae, Kalyptorhinchia), Gastrotricha *(Paradasys, Turbanella, Xenotrichula)*, Kinorhyncha *(Carteria)*, very many Copepoda (*Psammotopa, Arenosetella, Paraleptastacus, Arenopontia, Paramesočhra* and other, see fig. 70). Only small numbers of Ostracoda (Polycopidae, *Microcythere*) are found, and few Foraminifera. There occur interesting genera of Isopoda (*Microcerberus, Angelierra* etc.) and of Amphipoda (*Bogidiella, Pseudoniphargus* and others.) This is also the habitat of the Mystacocarida (*Derocheilocaris* fig. 80); Ciliata only occur in zones rich in bacteria.

In tideless seas or seas with only weak tides this area is narrow and is displaced according to the water level on the shore (Otoplanen zone). In brackish seas this region is naturally also mesohaline or oligohaline, yet it still contains many charac-

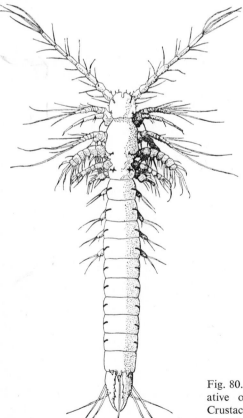

Fig. 80. *Derocheilocaris remanei*, a representative of the Mystacocarida, an order of Crustaceae. Enlarged. After CHAPPUIS and DELAMARE-DEBOUTTEVILLE

teristic species of the typical fauna, e. g. Turbellaria (*Bothriomolus balticus, Coelogynopora biarmata* and others), Archiannelida *(Protodrilus)*, Copepoda *(Remanea arenicola)* etc.

Landwards from this "marine coastal ground water" there follows a more stable region which is more or less brackish almost everywhere owing to admixture of fresh water. Since the supply of salt water (at high water level) and of fresh water (with heavy rains) varies, the salinity of this area also fluctuates. Often layers of different salinities are formed. This zone is the "coastal ground water proper" (GERLACH 1954) = high level subsoil region (GERLACH 1967) in seas without tides. If there is a strong influx of fresh water from the land the brackish zone may extend below the zone of circulation and drain into the sea. (DELAMARE-DEBOUTEVILLE 1955). On the other hand the influx of fresh water may be practically absent in arid regions and the water of the high level subsoil region may have the same or a higher salinity than the low level region or the sea (GERLACH 1967 on the Island of Sarso in the Red Sea). The high level subsoil region has a rich fauna with a number of characteristic species, e. g. Turbellaria: *Prognathorhynchus canaliculatus, Haplovejdovskya subterranea, Stygoplanellina halophila;* more euryoecious is *Coelogynopora schulzii*, Gastrotricha (*Paradasys subterraneus*, species of *Xenotrichula*), numerous Nematoda: *Dorylaimus obtusicaudatus, D. carteri, Plectus granulosus, Aegialospirina bulbosa, Mononchus rotundicaudatus, Adoncholaimus lepidus, Syringolaimus papallosus* etc., Copepoda: *Itunella mülleri*, species of *Schizopera* etc.

A number of fresh-water species or species of fresh-water origin enter this region, e. g. Rotifera: *Wierzejskia subterranea, Colurus colurus*, species of *Proales*, Oligochaeta: *Aeolosoma littorale*, Nematoda *(Dorylaimus, Plectus)*, Copepoda: *Parastenocaris vicesima, P. phyllura*.

In the coastal regions the habitat of most species is brackish water and many of them have a boundary of their occurrence both towards euhaline and polyhaline water, as well as fresh water. But experimental investigation of each species is required to find out whether they are true brackish-water species. GERLACH (1967) found on the Island Sarso in the Red Sea in this region interstitial water with excessive salinity. He found a rich fauna of Nematoda (37 species, among them 14 new ones) Copepoda and Annelida *(Protodrilus)*. Among the more thoroughly studied Nematoda which had so far been found in brackish coastal ground water where they have a world wide distribution he found e. g. *Synonchium obtusum, Latronema orcinum, Dolicholaimus oceanus, Haliplectus bickneri, Metachromadora chavat, Trefusia conica, Thalassolaimus brasiliensis, Diontolaimus tenuispiculum, Steineria parapolychaeta;* there are even isolated records of species known from the brackish water in Europe: *Theristus flevensis, Chromadora nudicapitata, Halalaimus gracilis*.

JANSSON (1966) examined the salinity tolerance of the Copepod *Paraleptastacus spinicauda*. It survived longest at 15⁰/₀₀ (in the ground water at 6⁰/₀₀ S); but populations have been found in much higher salinity so that here, too, the question arises whether the optimum may only be valid for the population Asko (Sweden, Baltic) and other populations may have much higher optima.

On sandy coasts air enters into the interstitial spaces at low water. A humid zone develops above the permanent ground water. In this the sand grains are covered with a film of water and there are air bubbles between them. In this moist zone there

are still numerous animals of the microfauna, as well as semiterrestrial animals (Enchytraeidae, Nematoda). But mites also invade this region (Rhodacaridae, the worm-shaped Nematolycidae and Oribatidae) with their own species. There are also numerous Collembola *(Archisotoma, Onychiurus* and others*)* (cf. DELAMARE-DE-BOUTEVILLE 1960) some other isolated Apterygota, e. g. Japygidae *(Parajapyx gerlachi)* and even Palpigrada *(Leptoloenenia gerlachi)*.

Both the upper and the lower coastal ground water (area of circulation) form a number of microhabitats — according to size of grains, volume of pores or water content, oxygen content, through-currents — for which certain species show a preference. Some species change their habitat to different regions of the shore in an annual cycle, e. g. the Polychaete *Hesionides arenarius* (WESTHEIDE 1966). Regional differences in the interstitial brackish-water fauna are mainly based on the soil type. The conditions described above apply to the situation in medium to coarse sands. The finer the material, the more circulation becomes reduced, but at grain size of about 0.2 mm capillary rise from the deeper ground water and evaporation at the surface become predominant. The Cyanophceae sand is such a facies (GERLACH 1954). It lies in the supralittoral and sea water only reaches it occasionally and irregularly. Near the surface a vegetation of Cyanophyceae develops which bind the sand grains. Salinity at the surface was very high (up to 120^0/$_{00}$) on account of evaporation, in the layers with living organisms about 2—10^0/$_{00}$ at 3—10 m depth. The water in the interstices was thus oligo- to mesohaline. Nematoda are the predominant group here. They attain numbers of 174 per cubic cm. They belong mainly to the euryhaline and brackish-water species. GERLACH (1954) records 61 species from the North Sea and Baltic regions. There are a few Turbellaria: *Macrostomum curvituba, M. pusillum, Vejdovskya ignava, Coronhelmis multispinosus, Placorhynchus octaculeatus, Coelogynopora schulzi,* several Copepoda, *Psammastacus confluens, Arenosetella tenuissima, Schizopera clandestina* etc.; several Oligochaeta from the shore etc. In addition there are a number of terrestrial Arthropoda, e. g. Collembola, Diptera larvae, mites and often mass colonization by the beetle *Bledius arenarius.*

Towards the sea black sand often forms below the layer of Cyanophyceae, that is a sapropel (putrefaction) zone poor oxygen, frequently also a "Farbstreifenwattsand" (see E. SCHULZ 1937), with a layer of purple bacteria between the stratum of the Cyanophyceae and the layer of black sand.

Interstitial brackish water with a fauna of an unexpected wealth of forms also occurs in the salt marshes as long as they are occasionally flooded by sea water that is in the region of *Puccinella maritima* and, partly, in the region of *Festuca rubra.* BILIO (1962—1966) has recently investigated this zone on the coasts of Germany, Holland and Finland. A surprisingly high number of species has been found.

The interstitial region in which brackish water may occur has here been chiefly produced by dead or living vegetable matter. There develop soils resembling humus or peat with different amounts of admixture of sand. Here, too, the Nematoda predominate; in the region of the North Sea and Baltic coasts of Schleswig-Holstein alone BILIO found 51 species. The number of species of Turbellaria is also very high (31; if the salt meadows of Holland and Finland are included, even 47), there follow Copepoda (19), Oligochaeta (20), Halacaridae (6) and a few species of other marine groups, among others Foraminifera. There are not only euryhaline species or wide-

spread species of the interstitial coastal ground water, but also typical species living only in this biotope. Among the characteristic species there are: Turbellaria: *Protoplana salsa, Proxenetes pratensis, Vejdovskya halilaimonia, Parautelga biloi, Proxenetes puccinellicola*, furthermore *Vejdovskya mesostyla, Minona baltica, Anthopharynx vaginatus, Westbladiella obliquipharynx* (see Ax 1960), the Triclad *Uteriporus vulgaris*. Nematoda: *Theristus spirus, Halalaimus terrestris, Microlaimus citrus, Leptolaimus puccinelliae, Haliplectus schulzi, Diplolaimella islandica*. Oligochaeta: *Lumbricillus (Pachydrilus) pagenstecheri, L. lenkingi* and others. Halacaridae: *Rhomobognathus prinicipes, Rh. uniscutatus*. Copepoda: *Schizopera pratensis, Sch. clandestina, Parepactophanes minuta, Sigmatidium minor, Horsiella trisetosa, H. ignava.*

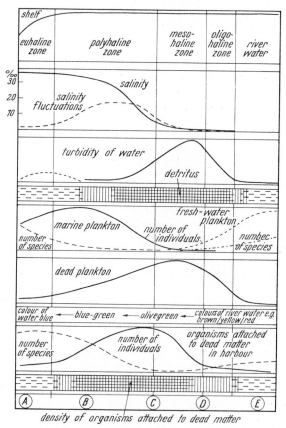

Fig. 81. Schematic illustration of physico-chemical and biological changes of different points of an estuary. After KÜHL 1963.

The ecological groups which colonize the brackish water of the salt marshes are very diverse. Besides the characteristic species we find "animals of the water edge" which are common in the boundary zones of the sea. They are allied to the "inhabitants of unconsolidated floors in the vicinity of the coast" such as occur in the lagoons and thus they partly belong to the *Cyprideis litoralis—Manayunkia* community.

Here, too, we find a number of species of the interstitial coastal ground water of sandy shores *(Coelogynopora schulzii, Halacarellus subterraneus* and others).

There is a very high percentage of brackish-water animals. BILIO (1966) records for the salt marshes of the North Sea 42% brackish-water animals, 51% euryhaline marine animals, 5% holeuryhaline animals, 2% fresh-water animals, for the salt marshes of the Baltic coast (Kiel Bay): 84% brackish-water animals, 8% euryhaline marine animals, 3% holeuryhaline animals, 5% euryhaline fresh water animals.

A problem of zoogeography has been solved by the interstitial fauna of the sea shore. In the subterranean fresh ground water a number of species have been found which were isolated in the fresh-water fauna, but closely related to marine species, e. g.: the Polychaete *Troglochaetus beranecki,* the Isopoda of the genera *Microcerberus, Microparasellus, Microcharon,* the Amphipoda of the genera *Ingolfiella, Bogidiella, Melita* and of the family of Niphargidae. Most of these species do not live in the waters of caves as had first been assumed, but interstitially in the ground water. Since new relations of most of these species, often belonging to the same genus, have been discovered in the subterranean water of the sea shore *(Bogidiella, Microcerberus, Ingolfiella)* it is certain that many species have migrated from the sea through the subterranean water of the coast directly into the subterranean fresh water.

References

AFANASHEV, G. (1938): Some data on the weight relationship of organs in the Black See Lamellibranchia. — Z. J. **17**.

ALEEM, A. & SCHULZ, E. (1952): Über die Zonierung von Algengemeinschaften. (Ökologische Untersuchungen im Nord-Ostsee-Kanal, I.) — Kieler Meresforsch. **9**: 70—76.

ALEXANDER, W. B., SOUTHGATE, B. A. & BASSINDALE, R. (1932): The salinity of the water retained in the muddy foreshore of an estuary. — J. Mar. Biol. Plymouth **18**: 297 to 298.

ALLEN, & TOPP (1900): The fauna of the Salcombe estuary. — J. Mar. biol. Ass. Plymouth **6**: 151—217.

ALTHAUS, B. (1955): Beitrag zur Kenntnis des Süßen Sees bei Mansfeld und seiner Fauna. — Wiss. Z. Univ. Greifswald **4**: 45—65.

— (1957): Faunistisch-ökologische Studien an Rotatorien salzhaltiger Gewässer Mitteldeutschlands. — Wiss. Z. Univ. Halle-Wittenberg, **6**, 1: 117—156.

D'ANCONA, U. (1931, 1932, 1934): Faune et Flore des eaux saumâtres I, II. Append. — Rapp. Proc. Ver. Comm. intern. Explor. Mer. Medit. **6, 7, 8, 75**.

ANDERSSON, K. A. (1938): A study of the rate of Growth of some fishes in the Baltic. — — Rapp. Proc. Verb. Cons. intern. Expl. Mer **108**: 67—87.

ANGELIER, E. (1950): Recherches sur la faune sables littoraux méditerranéens. — Vie et Milieu: 186—190.

ANNANDALE, N. (1928): Fauna of the Chilka Lake. — Mem. Indian Mus. **5**.

ANTIPA, GR. (1933): La vie dans la mer Noire. — Ann. Inst. océan. Monaco **13**, 2: 53—90.

— (1941): Marea negra. — Vol. I: Oceanografia, Bionomia si Biologia generala a Mariı negr. — Acad. Romana. Rubl. Fond. V. Adamachi **10**, 55 (Rum.): 1—313.

APSTEIN, C. (1906): Lebensgeschichte von *Mysis mixta.* — Wiss. Meeresunters. Kiel **9**: 239—308.

— (1906): Plankton in Nord- und Ostsee auf den deutschen Terminfahrten. — Wiss. Meeresunters. Kiel **9**: 1—23.

— (1908): Übersicht über das Plankton 1902—1907. — Beteil. Deutschl. Internat. Meeresforsch. **4—5**.

ARESCHOUG, J. E. (1847): Enumeratio Phycearum in maribus Scandinaviae crescentium. Sectio prior. — Nova Acta Reg. Soc. Sci. Ups. **13**.

ARLDT, TH. (1907): Die Entstehung der Kontinente. — Leipzig.

ARNDT, E. A. (1965a): Über das Vorkommen und die Morphologie von *Cordylophora caspia*, Hydrozoa in der Umgebung der Zoolog. Station Tvärminne, Finnland. — Comment. Biol. Soc. Sci. Fenn. **28**: 1—8.

— (1965b): Über die Fauna des sekundären Hartbodens der Martwa Wirla und ihres Mündungsgebietes (Danziger Bucht). — Wiss. Zeitschr. Univ. Rostock. **14**: 645 bis 653.

ARNDT, E. A., PANKOW, H. & KELL, H. (1966): Über das Phytoplankton der Wismarer-Bucht. — Internat. Rev. ges. Hydrobiol. **51**.

ARNDT, W. (1931/32): Die Tierwelt des Nord-Ostseekanals und ihr Lebensraum. — Der Naturforscher **8**.

AUGENER, H. (1928): Beitrag zur Polychätenfauna der Ostsee. — Z. Morph. Ökol. Tiere **11**: 102—104.

AURIVILLIUS, C. (1895): Littoralfaunens förrhallanden vid tiden för havets isläggning. — Övers. Kungl. Vetensk. Ak. Förhandl. **52**: 133—150.

— (1896): Das Plankton des Baltischen Meeres. — Bih. K. V. A. Handl. **21**, 8 Stockholm: 1—83.

AX, P. (1951a): Eine Brackwasserlebensgemeinschaft an Holzpfählen des Nord-Ostsee-Kanals. — Kieler Meeresforsch. **8**: 229—243.

— (1951b): Die Turbellarien des Eulitorals der Kieler Bucht. — Zool. Jb. Syst. **80**: 277—378.

— (1953): Eine Brackwasser-Lebensgemeinschaft an Holzpfählen des Nord-Ostsee-Kanals. — Kieler Meeresforsch. **8**, 2: 229—243.

— (1954a): Die Turbellarienfauna des Küstengrundwassers am Finnischen Meerbusen. — Acta Zool. Fenn. **81**: 1—34.

— (1954b): Marine Turbellaria *Dalyellioda* vor den deutschen Küsten. Die Gattungen *Baicalellia*, *Hangethellia* und *Canetellia*. — Zool. Jb. Syst. **82**: 481—496.

— (1956a): Les Turbellaries des Etangs côitiers du littoral Méditerranéen de la France meridionale. — Vie et Milieu Suppl. **5**: 1—215.

— (1956b): Das ökologische Verhalten der Turbellarien in Brackwassergebieten. — Proc. XIV. Intern. Congr. of Zool. Kopenhagen 1953: 462—464.

— (1957): Die Einwanderung mariner Elemente der Mikrofauna in das limnische Mesopsammal der Elbe. — Verh. Dt. Zool. Ges. Hamburg 1956: 428—434.

— (1959): Zur Systematik, Ökologie und Tiergeographie der Turbellarienfauna in den ponto-kaspischen Brackwassermeeren. — Zool. Abt. Syst. **87**: 43—184.

— (1960): Turbellarien aus salzdurchtränkten Wiesenböden der deutschen Meeresküsten. — Z. wiss. Zool. **163**.

AX, P, & AX, L. (1960): Experimentelle Untersuchungen über die Salzgehaltstoleranz von Ciliaten aus dem Brackwasser und Süßwasser. — Biol. Zbl. **79**: 7—31.

BACCI, G. (1954a): Alcuni relievi sulle faune die acque salmastre. — Pubbl. Staz. Zool. Napoli, **25**: 380—396.

— (1954b): Gradienti di salinita e distributione di Molluschi nel lago di Patria. — Boll. Zool. **21**.

BAČESCU, M. (1940): Les Mysidacés des eaux roumaines. — Ann. Sci. Univ. Jassy **26**: 453—804.

— (1948): Quelques observations sur la faune benthonique du défilé roumain de la Danube. — Ann. Sci. Univ. Jassy **31**: 240—253.

— (1949): Données sur la faune carcinologique de la Mer noire le long de la côte bulgare. — Trav. Stat. biol. marit. Varna **14**.

— (1951): Cumacea (Fauna Republicii Populare Romane). — Acad. Repl. Popul. Rom. Crustacea **4**, 1: 94 pp.

— (1954): Mysidacea (Fauna Republicii Populare Romane). — Acad. Repl. Rom. Crustacea **4**, 3: 126 pp.

— (1966): Die kaspische Reliktfauna im pontokaspischen Becken und in anderen Gewässern. — Kieler Meeresforsch. **22**: 176—188.

BAČESCU, M. et al. (1967): Eléments pour la caraktérisation de la zone sédimentaire medio-littorale de la Mer noire. — Trav. Mus. Hist. Nat. Antipa **7**.

BAČESCU, M., GOMOIO, X., BODEANU, X. et al. (1965): Recherches écologiques sur les fonds sablonneux de la Mer Noire. (Cote Roumanie.) — Trav. Mus. Hist. Nat. Antipa **5**: 33—81.

BALAZUC, I. & ANGELIER, E. (1951): Sur la capture à Banyuls-sur-Mer de *Pseudoniphargus africanus* CHEVREUX 1901. — Extr. Bull. Soc. France **76**: 309—312.

BANSE, K. (1956): Beiträge zur Kenntnis der Gattung *Fabricia, Manayunkia* und *Fabriciola*. — Zool. Jb. Syst. **84**: 415—438.

BARTENSTEIN, H. (1938): Foraminiferen der meerischen und brackigen Bezirke des Jade-gebietes. — Senckenbergiana **20**.

BASSINDALE, R. (1938): The intertidal fauna of the Mersey-estuary. — J. Mar. Ass. Plymouth **23**: 83—98.

— (1943): A comparison of the varying salinity conditions of the Tees and Severn estu-aries. — J. Anim. Ecol. **12**: 1—10.

BEADLE, L. C. (1931): The effect of salinity changes on the water content and respiration of marine invertebrates. — J. exp. Biol. **8**: 211—227.

— (1943): Osmotic regulation and the faunas of inland waters. — Biol. Rev. **18**: 172—183.

BEADLE, L. C. & CRAGG, J. B. (1940a): Studies on adaption to salinity in *Gammarus* spp. — J. exp. Biol. **17**: 153—163.

— The intertidalzone of two streams and the occurrence of *Gammarus* ssp. on south Rona and Raasey. — J. Anim. Ecol. **9**.

BEAUCHAMP, P. M. de (1914): Aperçu sur la répartition des êtres dans les zones des marées à Roscoff. — Bull. Sco. Zool. France **39**: 29—159.

BEAUFORT, L. F. de (1953): Veranderingen in de Flora en Fauna von de Zuiderzee na de Afsluiting in 1932: 1—359.

BEKLEMISHEW, W. N. (1922): Nouvelles contributions à la faune du lac Aral. — Russ., Hydrobiol. Z. **1**: 276—289.

BENTHEM JUTTING, T. van (1933): Mollusca I. — Gastropoda-Fauna van Nederland. **7** Leiden: 1—387.

— (1943): Mollusca: Lamellibranchia-Fauna van Nederland. **12** Leiden: 1—477.

BERG, L. S. (1928): Sur l'origine des éléments septentrionaux dans la faune de la mer Cas-pienne. — C. R. Acad. Sci. Leningrad 1928.

— (1933): Übersicht der Verbreitung der Süßwasserfische Europas. — Zoogeographica **I**: 107—208.

BEUDANT, F. S. (1816): Sur la possibilité de faire vivre des mollusques fluviatiles dans les eaux salées et des mollusques marins dans les eaux douces. — Ann. Chem. et Phys. **2**: 32—41.

BEYER, A. (1939): Morphologische, ökologische und physiologische Studien an den Larven der Fliegen *Ephydra riparia* FALLAN, *E. micans* HALIDAY und *Caenia fumosa*. — Kieler Meeresforsch. **3**: 265—320.

BILIO, M. (1962): Die aquatische Bodenfauna des Andelrasens *(Puccinellietum maritimae)*, eine vergleichend-ökologische Studie. — Diss. Univ. Kiel: 1—252.

— (1963): Die Zonierung der aquatischen Bodenfauna in den Küstensalzwiesen Schleswig-Holsteins. — Zool. Anz. **171**: 328—337.

— (1964a): Die biozönotische Stellung der Salzwiesen unter den Strandbiotopen. — Verh. Dt. Zool. Ges. 1963: 417—425.

— (1964b): Die aquatische Bodenfauna von Salzwiesen der Nord- und Ostsee. I. — Internat. Rev. ges. Hydrobiol. **49**: 509—562.

— (1966a): dto. II. — Internat. Rev. ges. Hydrobiol. **51**: 147—195.

— (1966b): Die Verteilung der aquatischen Bodenfauna und die Gliederung der Vege-tation im Strandbereich der deutschen Nord- und Ostseeküste. — Botanica Gotho-burgensis 3: 25—42.

— (1966c): Charakteristische Unterschiede in der Besiedlung finnischer, deutscher und holländischer Küstensalzwiesen durch Turbellarien. — Veröffentl. Inst. Meeresforsch. Bremerhaven. Sonderbd. II: 305—317.

BIRSTEIN, J. A. (1946): Some observations on the geographical distribution of Ponto-Caspian Amphipoda. — Bull. Soc. Nat. Moscou. S. Biol. Tl. **51** (3).

BLACK, V. (1951): Some aspects of the physiology of fish. II. Osmotic regulation in teleost fishes. — Univ. Toronto Bio. Ser. 59, Publ. Ontario Fish. Res. Lab. No. **71**: 53—89.

BLANCHARD, R. (1891): Résultats d'une excursion Zoologique en Algérie. — Mem. Soc. Zool. France **4**.

BLEGVAD, H. (1928): Quantitative investigations of bottom invertebrates in the Limfjord. — Rep. Danish Biol. Stat. **34**: 33—52.

— (1932): Investigations of the bottom Fauna at outfall of drains in the Sound. — Rep. Danish Biol. Stat. **37**: 1—20.

BOCK, J. (1950): Über die Bryozoen und Kamptozoen der Kieler Bucht. — Kieler Meeresforsch. **7**: 161—166.

— (1952): Zur Ökologie der Ciliaten des marinen Sandgrundes der Kieler Bucht I. — Kieler Meeresforsch. **9**: 252—270.

BOETTGER, C. (1932): Über die Ausbreitung der Muschel *Congeria cochleata* NYST in europäischen Gewässern und ihr Auftreten im Nord-Ostsee-Kanal. — Zool. Anz. **101**: 43—48.

BONDESEN, P. (1941): Preliminary investigations into the development of *Neritina fluviatilis* in brackish and freshwater. — Vidensk. Medd. Dansk. naturh. Foren. **104**: 283—318.

BONDESEN, P & KAISER, E. W. (1949): *Hydrobia (Potamopyrgus) jenkinsi* SMITH. — Oikos **1**: 252—281.

BORCEA, J. (1924a): Faune survivante de type caspien dans les limans d'eau douce de Roumanie (Note préliminaire). — Ann. Sci. Univ. Jassy **12**.

— (1924b): Faune survivante de type caspienne dans lacs limans d'eau douce de Roumaine. — Ann. Sci. Univ. Jassy **13**.

— (1925): Observations sur la faune des lacs Razelm. — Ibid. **14**, 1/2.

— (1927): Données sommaires sur la faune de la mer noire (littoral roumain). — Ibid. **14**: 536—581.

— (1930a): Quelques éléments de la faune de pénétration dans les eaux douces sur le littoral roumain de la mer Noire. — Arch. Zool. ital. **16**.

— (1930b): Faune des limans roumains en relation avec le problème de l'adaptation des êtres marins à l'eau douce et données sommaires sur la faune de la mer Noire (Littoral de Roumanie). — X. Congr. internat. Zool. Budapest: 1447—1451.

— (1934): Liste des animaux marins récoltés jusqu' à présent dans la région de la Station d'Agigéa (mer Noire). — Ibid. **19**, 3/4.

BORG, F. (1931): On some Species of Membranipora. — Arkiv Zool. Stockholm **22**.

— (1936): Einige Bemerkungen über Brackwasserbryozoen. — Zool. Anz. **113**: 188—193.

— (1947): Zur Kenntnis der Ökologie und des Lebenszyklus von *Electra crustulenta* (Bryozoa cheilostomata). — Zool. Bidrag Uppsala **25**: 344—377.

BOULOS, G. R. (1959): Ökologische Untersuchungen im Lake Qarun (Ägypten). — Diss. Kiel: 1—77.

BOWDEN, K. F. (1967): Circulation and Diffusion. — Estuaries Washington: 15—16.

BOZIC, B. (1956): Recherches taxonomiques sur des formes du genre *Tigriopus*. — Coll. internat. Biol. Mar. Roscoff: 269—282.

— (1965): Copepodes de quelques petits estuoires méditerranéens. — Bull. Mus. Hist. Nat. Ser. 2, **37**.

BRADSHAW, J. S. (1957): Laboratory studies on the rate of growth of the Foraminifer *Streblus beccari* (LINNÉ) var. *tepida* (CUSHMAN). — J. Paleont. **31**: 1138—1147.

BRAEM, F. (1951): Über Victorella und einige ihrer nächsten Verwandten sowie über die Bryozoenfauna des Ryck bei Greifswald. — Zoologica (Stuttgart) **37**.

BRANDES, C. H. (1939): Über die räumlichen und zeitlichen Unterschiede in der Zusammensetzung des Ostseeplanktons. — Mitt. aus d. Hamb. Zoolog. Mus. u. Inst. in Hamburg **48**: 1—47

BRANDHORST, W. (1955): Hydrographie des Nord-Ostsee-Kanals. — Kieler Meeresforsch. **11**, 2: 174—187.

184 Ecology of brackish water

BRANDT, K. (1890): Die mit der Kurre oder der Dredge auf der Expedition mit dem Dampfer
„Holsatia" vom 15. bis 25. September 1887 gesammelten Tiere. — VI. Ber. d. Komm.
z. Unters. d. dt. Meeres, Kiel: 141—148.
— (1896): Die Tierwelt der Kieler Bucht und des Kaiser-Wilhelm-Kanals. — Kiel Ein-
richtungen f. Gesundheitspflege und Unterricht: 1—8.
— (1897): Das Vordringen mariner Tiere in den Kaiser-Wilhelm-Kanal. — Zool. Jb.
Syst. 9: 387—408.
BRANDTNER, P. (1935): Eine neue marine Triclade. — Z. Morph. Ökol. Tiere 29: 422—480.
BRATTSTRÖM, H. (1941): Studien über die Echinodermen des Gebietes zwischen Skagerrak
und Ostsee, besonders des Öresundes usw. — Undersökn. över Öresund 27, Lund:
1—133.
— (1946): On the epifauna of an anti-submarine net in the northern part of the Sound.
— Smärre Unders. över Öresund 13. K. Fysiogr. sällsk. i Lund Förh. 16, 6, Lund:
1—7.
BREMEN, J. P. van (1905): Plankton van Noord- and Zuiderzee. — T. Nederland Dierkd.
Vereen, Sec. 2, 9: 1—329.
BRINCK, P., DAHL, E. & WIESER, W. (1955): On the littoral subsoil fauna of the Simrisham
Beach in Eastern Scania. — Kungl. Fysiograf. Sällskapets J. Lund Förhandl. 25:
1—21.
BRINKHURST, R. O. (1964): A taxonomic revision of the Alluroididae (Oligochaeta). —
Proc. Zool. Soc. London 142: 527—536.
BROCKMANN, CHR. (1908): Das Plankton im Brackwasser der Wesermündung. — Beitr.
Naturkde. Nordwestdeutschlands, N. F., 1.
— (1940): Diatomeen als Leitfossilien in Küstenablagerungen. — Westküste 2: 150
bis 181.
BRODNIEWICZ, J. (1965): Recent and some holocene Foraminifera of the southern Baltic
Sea. — Acta Palaeont. Polon. 10, 2: 131—248.
BROEKHUYSEN, G. J. (1956): On development growth and distribution of Carcinides maenas.
— Arch. Néerl. zool. 2: 257—399.
BROGMUS, W. (1952): Eine Revision des Wasserhaushaltes der Ostsee. — Kieler Meeres-
forsch. 9: 15—42.
BRUNELLI, G. (1929): Limnologia e ricerche lagunari. — Verh. IV. Congr. internat. Limnol.
249—251.
— (1933): Ricerche sugli stagni litoranei. — Rend. Ass. Lincei. 17: 246—249.
— (1934): La caratteristiche biologiche dell'ambiente lagunare e deglo stagni salmastri.
— Boll. Pesca Piscic. e Idrobiol. 10.
BUCH, K. (1949): Über den biochemischen Stoffwechsel in der Ostsee. — Kieler Meeres-
forsch. 6: 31—44.
BUCH, K. & HALME, E. (1947): Investigations on the Production of Life in the Baltic. —
Terra 59, Helsingfors: 1—8.
BUCHHOLZ, H. (1952): Das Brackwasserzooplankton an der schleswig-holsteinischen Ost-
seeküste. — Diss. Kiel: 1—69.
BUDDE, W. (1931, 1933): Limnologische Untersuchungen der rheinisch- und westfälischen
Gewässer, der Algenflora westfälischer Salinen. — Arch. Hydrobiol. 23, 24.
BÜLOW, TH. von (1955): Oligochaeten aus den Endgebieten der Schlei. — Kieler Meeres-
forsch. 11: 258—263.
— (1957): Systematisch-autökologische Studien an eulitoralen Oligochaeten der Kimb-
rischen Halbinsel. — Kieler Meeresforsch 13: 69—116.
BURSA, A., WOJTUSIAK, H. & R. J. (1947): Investigations of the bottom fauna and flora
in the Gulf of Gdansk made by using a diving helmet. — Part II Bull. intern. de
l'Acad. Polon. des Sci. et des Lettres. Ser. B Sci. naturelles. 9—10, B II. Cracovie.
CARAUSU, A. (1955): Contributii la studiul isopodelor (Crustacea Malacostraca) Marii
Negre (Litoralul ruminesc si regiunile invecinate). III. Familia Idoteidae. — Ann.
Scientifice ale Univ. Al. I. Cuza T. I.: 137—216.
CARAUSU, S. (1943): Amphipodes de Roumanie. I. Gammaridés de Type Caspien. — Mo-
nographia 1. Inst. Carcet. Piscic. Roman.: 1—293.

—   (1956): Introducere la managrafia amfipodelor Marii Negre (litoralul rominesc.).
    — Analele Scientifice Univ. Al. I. Cuza din Iase, I. Fasc. **1**: 66—83.

CARAUSU, S., DOBREANU, E. & MONOLACHE, C. (1955): Crustacea. — Fauna Republ. Po-
    pulare Romine, **4**: 4.

CASPERS, H. (1948): Ökologische Untersuchungen über die Wattentierwelt im Elbe—Ästuar.
    — Verh. Dt. Zool. Kiel: 350—359.

—   (1949 a): Biologie eines Limans an der bulgarischen Küste des Schwarzen Meeres
    (Varnaer See). — Verh. Dt. Zool. in Mainz: 288—294.

—   (1949 b): Die tierische Lebensgemeinschaft in einem Röhricht der Unterelbe. — Verh.
    Ver. naturwiss. Heimatforsch. Hamburg **30**: 41—49.

—   (1951 a): Quantitative Untersuchungen über die Bodentierwelt des Schwarzen Meeres
    im bulgarischen Küstenbereich. — Arch. Hydrobiol. **45**: 1—192.

—   (1951 b): Biozönotische Untersuchungen über die Strandarthropoden im bulgarischen
    Küstenbereich des Schwarzen Meeres. — Hydrobiol. **3**, 2: 132—193.

—   (1951 c): Untersuchungen über die Tierwelt von Meeressalinen an der bulgarischen
    Küste des Schwarzen Meeres. — Zool. Beitr. **1**, 3: 243—259.

—   (1951 d): Bodengreiferuntersuchungen über die Tierwelt in der Fahrrinne der Unter-
    elbe und im Vormündungsgebiet der Nordsee. — Verh. Dt. Zool. Ges. Wilhelmshaven:
    404—418.

—   (1952): Der tierische Bewuchs an Helgoländer Seetonnen. — Helgol. Wiss. Meeres-
    unters. **4**, 2: 138—160.

—   (1953): Biologische Untersuchungen über die Lebensräume der Unterelbe und des
    Vormündungsgebietes der Nordsee. — Mitt. Geol. Staatsinst. Hamburg **23**: 76—85.

—   (1954): Die Biologie von Elbe und Alster. — Gas u. Wasserfach, H. 20: 1—6.

—   (1955): Limnologie des Elbeästuars. — Verh. internat. Ver. theor. angew. Limnol.
    12: 613—619.

—   (1958): Biologie der Brackwasserzonen im Elbeästuar. — Verh. internat. Ver. Limnol.
    13: 687—698.

CHAPPUIS, P. A. & DELAMARE-DEBOUTTEVILLE, C. (1954): Recherches sur les Crustacés
    souterrains. — Arch. Zool. et Génér. **91**: 103—138.

CLAUS, A. (1937): Vergleichend-physiologische Untersuchungen zur Ökologie der Wasser-
    wanzen mit besonderer Berücksichtigung der Brackwasserwanze *Sigara lugubris*
    FIEB. — Zool. Jb. Physiol. **58**: 365—432.

COLLIER, A. & HEDGPETH, J. W. (1950): An introduction to the hydrography of tidal waters
    of Texas. — Publ. Inst. Mar. Sci. II, 2: 125—194.

CRAWFORD, G. J. (1937): The fauna of certain estuaries in West-England and South Wales.
    — J. Mar. Biol. Ass. Plymouth **21**: 647—662.

DAHL, E. (1944): The Swedish brackish water Malacostraca. — K. Fysiogr. Sällsk. Lund
    Förhl. **14**, 9, Lund: 1—17.

—   (1946): The amphipoda of the sound part II. Aquatic amphipoda with notes on
    changes in the hydrography and fauna of the area. — Lunds Univ. Arsskr. N. F.,
    Avd. 42. Kungl. Fysiolgr. Sällsk. Handl. N. F. **57**: 3—49.

—   (1948): On the smaller arthropoda of marine algas especially in the polyhaline waters
    of the Swedish West-Coast. — Undersökn. Oeresund 35, Lund: 1—193.

—   (1956): Ecological Salinity boundaries in poikilohaline waters. — Oikos. **7**: 1—21.

DAHL, F. (1893): Untersuchungen über die Tierwelt der Unterelbe. — Ber. Komm. Unters.
    Dt. Meere **6**: 1—24.

—   (1894): Die Copepodenfauna des unteren Amazonas. — Ber. Naturforsch. Ges.
    Freiburg i. Br. **8**: 1—14.

DAY, J. H. (1951): The Ecology of South African Estuaries. P. I. — Trans. R. Soc. South
    Africa **33**: 53—91.

—   (1959): The Biology of Langeboar Lagoon. — Trans. R. Soc. South Africa, **35**: 475
    to 547.

—   (1967): The Biology of Knysna Estuary, South Africa. — Estuaries (Washington):
    397—407.

DAY, J. H., MILLARD, N. A. & HARRISON, A. D. (1951): The Ecology of South African
   Estuaries. P. III. — Trans. R. Soc. South Africa 33: 367—413.
DEMEL, K. (1933): Liste des invertébrés et des poissons des eaux polonaises de la Baltique.
   — Fragm. Faun. Mus. Zool. Polon. 2, 13: 121—136.
   — (1935): Etudes sur la faune bentique et sa répartition dans les eaux polonaises de la
   Baltique. — Arch. Hydrobiol. Rybact. Suwalki 9: 239—311.
DEMEL, K. & MANKOWSKI, W. (1950): Studia nad fauna denna Baltyku poludniowego. —
   Biol. Morsk. Inst. Ryb. v. Gdyni 5.
   — (1951): Ilosciowe studia nad fauna denna Baltyku poludniowego. — Morsk. Inst.
   Ryb. v. Gdyni 6.
DEMEL, K. & MULICKI, Z. (1954): Quantitative investigations on the biological bottom pro-
   ductivity of the South Baltic. — Rep. Sea Fisher. Inst. Gdynia 7: 75—126.
DELAMARE-DEBOUTTEVILLE, M. (1955): Sur le mélange des eaux d'infiltrations marines et
   des eaux phréatiques continentales sous une plage d'une mer sans marées. — C. R.
   Acad. Sci. Paris.
   — (1960): Biologie des eaux souterraines littorales et continentales. — Paris: 1—740.
DERJUGIN, K. (1925): Reliktensee Mogilnogie. — Trav. Inst. Sci. Nat. Peterhof 2: 1—111.
DIETRICH, G. (1950): Die natürlichen Regionen von Nord- und Ostsee auf hydrographischer
   Grundlage. — Kieler Meeresforsch. 7: 35—86.
   — (1956): Beitrag zu einer vergleichenden Ozeanographie des Weltmeeres. — Kieler
   Meeresforsch. 12: 3—24.
DIETRICH, G. & KALLE, K. (1957): Allgemeine Meereskunde. — Berlin: 1—492.
DUNCKER, G. (1937): Die Fische der Nordmark. — Kiel 1937.
EHRENBAUM, E. (1936): Naturgeschichte und wirtschaftliche Bedeutung der Seefische Nord-
   europas. — Handb. Seefischerei, Stuttgart: 1—337.
EISMA, D. (1965): Shell-charakteristics of Cardium edule as indicators of salinity. — Netherlands J. Sea Res. 2: 493—540.
EKMAN, S. (1914—1920): Studien über die marinen Relikte der nordeuropäischen Binnengewässer. — Internat. Rev. Hydrobiol. Hydrogr. 6, 8, 1914: 335—372, 1917—1920:
   327—337, 477—528, 589—593.
   — (1916): Systematische und tiergeographische Bemerkungen über einige glazialmarine
   Relikte des Kaspischen Meeres. — Zool. Anz. 47: 258—269.
   — (1932): Die biologische Geschichte der Nord- und Ostsee. — Tierwelt Nord- u.
   Ostsee I b: 1—40.
   — (1935): Tiergeographie des Meeres. — Leipzig: 1—542.
   — (1953): Zoogeography of the Sea. — London: 1—418.
ELIASSON, A. (1920): Biologisch-faunistische Untersuchungen aus dem Öresund. 5. Polychaeta. — Acta Univ. Lund (NS) 16, Lunds Univ. Arsskr. N. F. avd. 2, 16.
ELOFSON, O. (1941): Zur Kenntnis der marinen Ostracoden Schwedens mit besonderer Berücksichtigung des Skagerraks. — Zool. Bidrag Uppsala 19: 215—534.
   — (1943): Neuere Beobachtungen über die Verbreitung der Ostracoden an den skandi-
   navischen Küsten. — Ark. Zool. 35, 2, Stockholm: 1—26.
EMERY, K. O. & STEVENSON, R. E. (1957): Estuaries and lagoons. — Mem. Geol. Soc. Amer.
   67, 1: 679—749.
ESTUARIES ed. S. R. LAUFF (1967): Amer. Assoc. Adv. Sci. Washington: 1—757.
FAUVEL, P. (1933): Histoire de Merceriella énigmatica, Serpulien d'eau saumâtre. — Arch.
   Zool. exp. gener. 75.
FINLEY, H. (1930): Toleration of fresh water Protozoa to encreased Salinity. — Ecology 11:
   337—347.
FISCHER-PIETTE, E. (1933): Nouvelles observations sur l'ordre d'euryhalinité des espèces
   littorales. — Bull. Inst. Oceanogr. Monaco Nr. 619: 1—16.
   — (1937): Sur la biologie du serpulien d'eau saumâtre Merceriella énigmatica FAUVEL.
   — Bull. Soc. Zool. de France 62.
FLANNERY, W. L. (1956): Current Status of knowledge of Halophilic Bacteria. — Bacteriol.
   Rev. 20.

FLORENTIN, M. R. (1901): Description de deux Infusoires cilies nouveaux des Mares Salées etc. — Ann. Sci. nat. Zool. **12**.

FOCKE, E. (1961): Die Rotatoriengattung *Notholca* und ihr Verhalten in Salzwasser. — Kieler Meeresforsch. **17**: 190—205.

FORD, E. (1923): Animal communities of the level sea-bottom in the waters adjacent to Plymouth. — J. Mar. Biol. Ass. Plymouth **13**: 164—224.

FORSMAN, B. (1938): Untersuchungen über die Cumaceen des Skagerraks. — Zool. Bidr. Uppsala **18**: 1—162.

— (1940): Cumaseer fran Öresund. — K. Fysiogr. Sällsk. Lund Förh. **10**.

— (1951): Studies on *Gammarus duebeni* LILLB., with notes on some rockpool organisms in Sweden. — Zool. Bidr. Uppsala **29**: 215—238.

— (1956): Notes on the invertebrate fauna of the Baltic. — Ark. Zool. n. s. **9**: 389—419.

FOX, M. H. (1926): Cambridge Expedition to the Suez Canal 1924, I. Gen. Part. — Trans. Zool. Soc. London **22**.

FRASER, J. A. (1933): Observations on the fauna and constituents of an estuarine mud in a polluted area. — J. Mar. Biol. Ass. Plymouth **18**: 69—85.

— (1936): The occurrence, ecology and life history of *Tigriopus fulvus*. — J. Mar. Biol. Ass. Plymouth **20**: 523—536.

— (1936): The Distribution of rockpool Copepoda according to the tidal interaction between predators and prey. — J. Animal Ecol. **4**.

— (1938): The fauna of fixed and floating structures in the Mersey Estuary and Liverpool Bay. — Proc. Liverpool Biol. Soc. **51**: 1—21.

FRIEDRICH, H. (1937): Einige Beobachtungen über das Verhalten der *Alderia modesta* LOV. im Brackwasser. — Biol. Zbl. **57**: 101—165.

— (1940a): Einige neue Hoplonemertinen aus der Ostsee. — Kieler Meeresforsch. **3**: 233—251.

— (1940b): Polychaeten-Studien V—X. Zur Kenntnis einiger bekannter oder neuer Polychaeten aus der westlichen Ostsee. — Kieler Meeresforsch. **3**: 362—373.

FRISCH, J. A. (1940): The experimental adaptation of *Paramecium* to sea water. — Arch. Protistenkde. **93**: 38—71.

GALLIFORD, A. L. & WILLIAMS, E. G. (1948): Microscopic Organisms of some Brackish Pools at Leasowe, Wirral, Ceshire. — The North Western Naturalist **23**: 39—62.

— (1956): Notes on the Ecology of Pools in the Salt Marshes of the Dee Estuary. — Liverpool Natural Field Club Proc. 1955.

GALTSOFF, P. S. (1954): Gulf of Mexico, its origin, waters, and marine life. — Fish. Bull. Fish. Wildlife Serv. Washington **55**: 604 pp.

GAUTHIER, H. (1928): Recherches sur le Faune de eaux continentales de l'Algerie et de la Tunesie. — Alger.

DE GEER, G. (1910): Quaternary Sea-bottom in Western Sweden. — Geol. Fören. Stockholm Förhandl. **32**.

GERING (1913): Ostpreußische Nemertinen. — Schr. Phys.-ökonom. Ges. Königsberg (Pr.). **54**: 292—295.

GERLACH, S. A. (1952): Die Nematodenbesiedlung des Sandstrandes und des Küstengrundwassers an der italienischen Küste. — Arch. Zool. Ital. I. System. Teil. **37**: 517—640; 1954, II. Ökolog. Teil **39**: 311—359.

— (1954): Das Supralitoral der sandigen Meeresküsten als Lebensraum einer Mikrofauna. — Kieler Meeresforsch. **10**: 121—130.

— (1956): Das Supralitoral der Meeresküste als Lebensraum. — XIV. Internat. Congr. Zool.: 121—129.

— (1958): Die Mangroveregion tropischer Küsten als Lebensraum. — Z. Morph. Ökol. Tiere **46**: 636—730.

— (1967): Die Fauna des Küstengrundwassers am Strand der Insel Sarso (Rotes Meer). — Meteor-Forschungsergebn. D, 2: 7—18.

GESSNER, F. (1933a): Phosphat, Nitrat und Planktongehalt im Arkonabecken. — J. Cons. Internat. l'Explor. Mer. **8**: 181—200.

— (1933b): Die Produktionsbiologie der Ostsee. — Naturwiss. **21**: 649—653.

— (1933c): Die Planktonproduktion der Brackwässer in ihrer Beziehung zur Produktion der offenen See. — Verh. internat. Ver. Limnol. **6**: 154—162.

— (1937): Hydrographie und Hydrobiologie der Brackwässer Rügens und des Darß. — Kieler Meeresforsch. **2**: 1—80.

— (1940): Produktionsbiologische Untersuchungen im Arkonabecken und den Binnengewässern von Rügen. — Kieler Meeresforsch. **3**: 449—459.

— (1955): Das Plankton des Lago Maracaibo. — Ergebn. dt. limnol. Venezuela Exped. 1952.

GIETZEN, J. (1931): Untersuchungen über marine Thiorhodaceen. — Zbl. Bakter., II. Abt. **83**.

GILLET, S. & ARCHARD, G. (1947): Le problème de l'eurihalinité: Quelques études sur les Faunes des limans. — Rev. Scient. (Paris) **85**.

GILLHARD, G. (1950): Rotifères d'une eau saumatre à Nienport. — Bull. Inst. Sci. nat. Belg. **35**.

GISLEN, T. (1930): Epibioses of the Gullmar Fjord I and II. — Kristinebergs Zool. Stat. 1877—1927, Uppsala. **3**: 123 pp., **4**: 380 pp.

— (1940): The number of animal species in Sweden. — Lunds Univ. Arsskr. N. F. 2, **36**: 1—23.

GOBE, CHR. (1874): Die Brauntange (Phaesporeae und Fucaceae) des Finnischen Meerbusens. — Mém. Acad. Imp. Sci. Sér. **7**, **21** (St. Petersbourg).

— (1877a): Die Rothtange (Florideae) des Finnischen Meerbusens. — Ibid. **24**.

— (1877b): Über einige Phaeosporeen der Ostsee und des Finnischen Meerbusens. — Bot. Ztg., Jg. 35, **33**.

GOOR, A. S. V. van (1925): Die Eugleninae des holländischen Brackwassers. — Rec. trav. bot. néerl. **22**.

GRESENS, J. (1928): Versuche über die Widerstandsfähigkeit einiger Süßwassertiere gegenüber Salzlösungen. — Z. Morph. Ökol. Tiere **12**: 706—800.

GRUNER, H. E. (1962): *Jaera albifrons* und ihre Unterarten an den deutschen Küsten. — Abh. Naturw. Ver. Hamburg **6**.

GUEYLARD, F. (1923): Résistance des Épinoches aux variations de salinité. — C. R. Soc. Biol. Paris **89**.

GUNTER, G. (1950): Seasonal Population changes and distributions as related to salinity of certain invertebrates of the Texas coast. — Publ. Inst. Mar. Sci. Univ. Texas **1**: 7—51.

— (1955): Mortality of oysters and abundance of certain associates as related to salinity. — Ecology **36**.

— (1956): Some relations of faunal distributions to salinity in estuarine waters. — Ecology **37**: 616—619.

— (1957): Predominance of the young among marine fishes found in fresh water. — Copeia **1**.

— (1961a): Some relations of estuarine organismes to salinity. — Limnolog. Oceanogr. **6**.

— (1961b): Habitat of juvenile Skrimp (Fam. Penaeidae). — Ecology **42**.

— (1967): Some relationships of estuaries to the Fisheries of the Gulf of Mexico. — Estuaries: 621—638.

GUNTER, G. & HALL, G. E. (1965): A biological investigation of the Caloosahatchee Estuary of Florida. — Gulf Res. Rep. **2**: 1—72.

GUNTER, G. & SHELL, W. (1958): A study of an estuarine area with water-level control in the Lousiana-Marsh. — Proc. Lousiana Acad. Sci. **21**: 5—34.

HAAHTELA, L. (1962): Kilkin Biologiasta ja Pyydystävaisästa. — Eripainos Suoman Kalastuslati **2**.

— (1964): Die Verbreitung der wirbellosen Tiere im Bottnischen Meerbusen (Finn.). — Luonnon Tutkija **68**: 162—166.

— (1965): Morphology, habitats and distribution of Species of the *Jaera albifrons* group (Isopoda, Janiridae) in Finland. — Ann. Zool. Fenn. **2**: 309—314.

HAAS, H. & STRENZKE, K. (1957): Experimentelle Untersuchungen über den Einfluß der ionalen Zusammensetzung des Mediums auf die Entwicklung der Tubuli von *Chironomus thummi*. — Biol. Zbl. **76**: 513—528.

HAGEN, G. (1951): Vergleichende ökologische und systematische Untersuchungen der eulitoralen Oligochaetenfauna in Süßwasser- und Meeresgebieten Schleswig-Holsteins. — Diss. Kiel.

— (1954): Strukturelle Abweichungen mariner und euryhaliner Oligochaeten in Grenzbereichen ihres Vorkommens. — Kieler Meeresforsch. 10, 1: 77—80.

HALME, E. (1944): Planktologische Untersuchungen in der Pojo-Bucht und angrenzenden Gewässern. I. Milieu und Gesamtplankton. — Ann. Zool. Soc. Vanamo 10: 188 p.

HART, T. J. (1930): Preliminary notes on the bionomics of the Amphipod, Corophium volutator. (Pallas.) — J. Mar. Biol. Assoc. 16: 761—789.

HARTMANN, G. (1956): Zur Kenntnis des Mangrove-Estero-Gebietes von El Salvador und seiner Ostracoden-Fauna. — Kieler Meeresforsch. 12: 219—248.

— (1957a): Zur Kenntnis des Mangrove-Estero-Gebietes von El Salvador und seiner Ostracoden-Fauna. — Kieler Meeresforsch. 13: 134—159.

— (1957b): Zur Biologie der peruanischen Garneele Cryphiops caemenbarins (MOLINA). — Kieler Meeresforsch. 13, 1: 117—124.

— (1959): Zur Kenntnis der lotischen Lebensbereiche der pazifischen Küste von El Salvador unter besonderer Berücksichtigung der Ostracodenfauna. — Kieler Meeresforsch. 15: 187—241.

— (1964): Das Problem der Buckelbildung auf Schalen von Ostracoden in ökologischer und historischer Sicht. — Mitt. Hamburg. Zool. Mus. Inst.: 59—66.

— (1966): Ostracoden. — In BRONNS Klassen und Ordnungen d. Tierreichs. 5: 1—216.

HARTOG, C. DEN (1960): Comments on the Venice-System for the classification of Brackish waters. — Internat. Rev. ges. Hydrobiol. 45: 481—485.

— (1961): Die faunistische Gliederung im südwestniederländischen Deltagebiet. — Internat. Rev. ges. Hydrobiol. 46: 407—418.

— (1964): Proseriate Flatworms from the deltaic area of the rivers Rhine, Meuse and Scheldt. I., II. — Proc. Kon. Ned. Akad. Wetensch. Amsterdam Ser. C, 67, 1: 10—34.

HASE, A. (1926): Zur Kenntnis der Lebensgewohnheiten und der Umwelt des marinen Käfers Ochthebius quadricollis MULS. — Internat. Rev. Hydrobiol. Hydrogr. 16.

HASS, G. (1936): Variationsstatische Untersuchungen an Proben von Gobius microps KRÖYER aus der Kieler Bucht und der Schlei. — Schr. Naturw. Ver. Schlesw.—Holstein 21: 419—426.

— (1937): Variabilitätsstudien an Gobius niger L., Gobius minutus Pallas und Cottus scorpius L. — Kieler Meeresforsch. 1: 279—321.

HAUER, J. (1957): Rotatorien aus dem Plankton des Van-See. — Arch. Hydrobiol. 53: 23—29.

HAVINGA, B. (1941): De ontwikkeling van den vischstand. De Biologie van de Zuiderzee tijdens haar drooghegging. — Meded. van der Zuiderzee-Commissie Ned. Dierk. Ver. Afl. 5,14

HAYES, F. R. (1929): Contributions to the study of marine gastropods. III. Development, growth and behaviour of Littorina. — Contr. Can. Biol. Fish. 4: 413—430.

— (1930): The physiological response of Paramecium to sea water. — Z. vergl. Physiol. 13: 214—222.

HAYRÉN, E. (1931): Aus den Schären Südfinnlands. — Verh. Internat. Ver. Limnol. 5, Helsingfors: 488—507.

— (1940): Über die Meeresalgen der Insel Hogland im Finnischen Meerbusen. — Acta Phytogr. Suec. 13, Uppsala: 50—62.

— (1941): Alger och vattenvegetation i Sibbo Skärgård, Sydfinland. — Mem. Soc. Fauna Flora Fenn. 17 (Helsingfors): 76—79.

— (1950a): Botaniska anteckningar från Raumo Skärgård. — Bidr. kännes Finl. Nat. Folk. 93 (Helsingfors): 1—23.

— (1950b): Botaniska anteckningar från Nystads skärgård. — Bidr. kännes Finl. Nat. Folk. 93 (Helsingfors): 1—19.

HEDGPETH, J. W. (1947): The Laguna Madre of Texas. — Reprinted from Transact. of the Twelfth North Americ. Wildlife Conf.: 364—380.

— (1950): Notes on the marine invertebrate fauna of Salt flat areas in Aransas National Wildlife refuge, Texas. — Publ. Inst. of Mar. Sci. **1**: 103—119.
— (1951): The classification of estuarine and brackish waters and the hydrographic climate. — Rep. Comm. on Treatise on Mar. Ecol. Palecol. **11**: 49—54.
— (1953): An introduction to the zoogeography of the Northwestern Gulf of Mexico with reference to the invertebrate Fauna. — Publ. of the Inst. of Mar. Sci. **3**: 107—224.
— (1956): The population of hypersaline and relict lagoons. — XIV. Internat. Congr. Zool. Proc. Kopenhagen.
HEIDEN, H. (1900): Diatomeen des Conventer Sees bei Doberan etc. — Mitt. a. d. Großherz. Meckl.-Geol. Landesanst. **10**.
HEIDEN, H. & FRIEDRICH, P. (1912): Die Litorina- und Praelitorinabildungen unter dem Priwall bei Travemünde. — Mitt. Geol. Ges. Lübeck 2. R., **25**.
HENKING, H. (1929): Untersuchungen an Salmoniden. Mit besonderer Berücksichtigung der Art- und Rassefragen. — Rapp. proc. verb. Réun. Conseil Expl. Mer **61**: 1—99.
HENSCHEL, J. (1936): Wasserhaushalt und Osmoregulation von Scholle und Flunder. — Wiss. Meeresunters. Kiel **22**: 89—121.
HENSEN, V. (1890): Das Plankton der östlichen Ostsee und des Stettiner Haffs. — 6. Ber. Komm. wiss. Unters. dt. Meere Kiel: 103—137.
HENTSCHEL, E. (1923): Grundzüge der Hydrobiologie. — Jena: 1—221.
— (1939): Die Planktonarbeiten für die Deutsche wissenschaftliche Kommission für Meeresforschung in der Hydrobiol. Abt. des Hamburger Zoologischen Museums und Instituts. — Ber. Dt. wiss. Komm. Meeresforsch. N. F. **9**: 233—236.
HERBST, H. V. (1951): Ökologische Untersuchungen über die Crustaceenfauna südschleswigscher Kleingewässer mit besonderer Berücksichtigung der Copepoden. — Arch. Hydrobiol. **45**: 413—542.
HESSE, R. (1924): Tiergeographie auf ökologischer Grundlage. — Jena: 1—613.
HESSLAND, I. (1943): Marine Schalenablagerungen Nord-Bohusläns. — Bull. Geol. Inst. Uppsala **31**: 1—348.
HESSLE, CHR. (1923): Undersökningar rörande bottnen och bottenfaunan i farvattnen vid Öland och Gotland. — Medd. Kungl. Landbruksstyr Nr. 243, Stockholm.
— (1924): Bottenboniteringar i inre Östersjöen. — Medd. Kungl. Landbruksstyr Nr. 250, Stockholm.
— (1925): The Herrings along the Baltic Coast of Sweden. — Cons. Internat. Publ. de Circons. **89**.
HESSLE, CHR. & WALLIN, S. (1934): Undersökningar över plankton och dess växlingar i Östersjöen under aren 1925—27. — Svenska Hydrobiol. Komm. Skr. (N. S. Biol. 1): 1—94.
HEUTS, M. J. (1947): Experimental studies on adaptive evolution in Gasterosteus aculeatus L. **28**. — Evolution **1**: 89—102.
— (1956): Temperatur adaptation in Gasterosteus aculeatus. — Publ. Staz. Zool. Napoli.
HILTERMANN, H. (1949): Klassifikation der natürlichen Brackwässer. — Erdöl u. Kohle **2**: 4—8.
— (1966): Klassifikation rezenter Brack- und Salinar-Wässer in ihrer Anwendung für fossile Bildungen. — Z. Dt. Geol. Ges. **115**: 463—469.
HIRSCH, A. (1915): Salzgewässer und Salzfaunen. — Arch. Hydrobiol. **10**.
HÖGBOM, A. S. (1917): Über die arktischen Elemente in der arolokaspischen Fauna, ein tiergeographisches Problem. — Bull. Geol. Inst. Uppsala **14**.
HÖHNK, W. (1940): Ein Beitrag zur Kenntnis des Phycomyceten des Brackwassers. — Kieler Meeresforsch. **3**: 337—361.
— (1956): Studien zur Brack- und Seewassermykologie. I—VI. — Veröff. Inst. Meereskde. Bremerhaven: 52—108.
HÖHNK, W. & VALLIN, ST. (1953): Epidemisches Absterben von Eurytemora im Bottnischen Meerbusen, verursacht durch Leptolegnia baltica nov. spec. — Veröff. Inst. Meeresforsch. Bremerhaven **2**: 215—223.

HOFFMANN, C. (1928): Über eine in der Kieler Förde neu aufgetretene Rotalge: *Porphyra atropurpurea* (*Olivi*) DE TONI. — Wiss. Meeresunters. **21**: 1—9.
— (1929): Die Atmung der Meeresalgen und ihre Beziehung zum Salzgehalt. — Jb. wiss. Bot. **71**: 214—268.
— (1933a): Beiträge zur Algenflora der westlichen Ostsee. — Schr. Naturw. Ver. Schlesw.-Holst. **20**: 106—115.
— (1933b): Die Vegetation der Nord- und Ostsee. — Tierwelt Nord- u. Ostsee. Tl. I c: 1-32.
— (1942): Beiträge zur Vegetation des Farbstreifen-Sandwattes. — Kieler Meeresforsch. **4**: 85—108.
— (1943): Der Salzgehalt des Seewassers als Lebensfaktor mariner Pflanzen. — Kieler Bl. **3**: 160—176.
— (1950): Über das Vorkommen endemischer Algen in der Ostsee. — Kieler Meeresforsch. **7**: 38—52.
HOFKER, J. (1922): Foraminifera. — In: REDEKE, H. C. (1922a).
HOLMQUIST, CH. (1949): Über eventuelle intermediäre Formen zwischen *Mysis oculata* FABR. und *Mysis relicta* LOVEN. — Lunds Univ. Arsskr. N. F. Avd. **2**: 1—26.
— (1959): Problems on marine-gracial relics on account of investigation on the genus *Mysis*. — Lund: 1—270.
HOOGENRAAD, H. H. & DE GROOT, A. A. (1940): Zoetwaterrhizopoden en Heliozoen. — Fauna van Nederland **9**: 1—302.
HOOP, M. (1940): Vergleichende Untersuchungen über die Schalenstruktur von euryhalinen Muscheln. — Zool. Jb. Syst. **74**: 269—276.
HOWES, N. H. (1939): The ecology of a saline lagoon in south-east Essex. — J. Linn. Soc. London Zool. **40**: 383—445.
HRABE, S. (1964): On *Peloscolex sirenkoi* and some other species of the genus *Peloscolex*. — Publ. Fac- Sci. Univ. Brno **450**.
HUSTEDT, FR. (1925): Bacillariales aus den Salzgewässern bei Oldesloe in Holstein. — Mitt. Geogr. Ges. u. Nat. Mus. Lübeck.
JAECKEL, S. (1937a): Tintenfische in der westlichen Ostsee. — Arch. Mollusken **69**.
— (1937b): Die Mollusken der Schlei. — Schr. Naturw. Ver. Schlesw.-Holst. **22**: 225 bis 230.
— (1949): Die Molluskenfauna des postglacialen Quellkalkes an der mecklenburgischen Küste bei Meschendorf. — Arch. Molluskenkde. **77**: 91—97.
— (1950): Die Mollusken der Schlei. — Arch. Hydrobiol. **44**: 214—270.
— (1951): Zur Verbreitung und Lebensweise der Opisthobranchier in der Nordsee. — Kieler Meeresforsch. **8**: 249—259.
— (1952): Zur Ökologie der Molluskenfauna in der westlichen Ostsee. — Schr. Naturw. Ver. Schlesw.-Holst. **26**: 18—50.
— (1954): *Aculifera aplacophora* und *Aculifera placophora* im Gebiet der Nord- und Ostsee. — Kieler Meeresforsch. **10**: 261—271.
— (1962): Die Tierwelt der Schlei. — Schr. Naturwiss. Ver. Schleswig-Holstein **33**: 11—32.
— (1964): Beiträge über Mollusken im Brackwasser (Abänderungen an den Schalen). — Schr. Naturw. Ver. Schlesw.-Holst. **35**: 19—27.
— (1965): Über die Herausbildung von Brackwasserformen bei Mollusken. — Zool. Anz. **174**: 119—125.
JÄRNEFELDT, H. (1940): Beobachtungen über die Hydrologie einiger Schärentümpel. — Verh. Internat. Ver. Limnol. **9**: 79—101.
JÄRVEKULG, A. (1962): Materialien zur Bodenfauna der Pernauer Bucht. — Eesri NSV Teaduste Akademia: 189—199.
— (1963): Einiges über die detaillierte Untersuchung der Bodenfauna in Küstenzonen. — Loc. cit. Biol. Ser.: 224—230.
— (1965): Über die Verbreitungsgrenzen der Meeres- und Brackwasserwirbellosen in der Matsaln-Bucht. — Loc. cit.: 362—365.
JANSSON, B. O. (1966): Microdistribution of factors and fauna in marine sandy beaches. — Veröff. Inst. Meeresforsch. Bremerhaven. Sonderbd. **2**: 77—86.

JEBRAM, D. (1968): A Cultivation Method for Saltwater Bryozoa and an Example for Experimental Biology. — Atti Soc. It. Sci. Nat. e Museo Civ. St. Nat. Milano, **108**: 119—128.
— (1970): Stolonen-Entwicklung und natürliches System bei den Bryozoa Ctenostomata. — Diss. Kiel; 1̄—106.
JOHANSEN, A. C. (1914): Om forandringar i Ringköbing Fjords Fauna. — Mindeskr. Japetus-Steenstrup, Kopenhagen **2**: 1—144.
— (1918a): Om hydrografiske faktorers Indflydelse paa Molluskernes Udbredelse i Östersjöen. — Forh. Skand. Naturf. Möde **16** (Kristiania).
— (1918b): Randersfjords Naturhistorie. — Kopenhagen: 1—520.
JOHNSEN, P. (1945): The Rock-pools of Bornholm and their Fauna. — Vidensk. Medd. Kobenhavn **109**: 1—53.
KÄNDLER, R. (1941a): Die Fortpflanzung der Meeresfische in der Ostsee und ihre Beziehungen zum Fischereiertrag. — Mh. Fischerei **11**: 158—163.
— (1941b): Fortpflanzung, Wachstum und Variabilität der Arten des Sandaals in Ost- und Nordsee, mit besonderer Berücksichtigung der Saisonrassen von *Ammodytes tobianus* L. — Kieler Meeresforsch. **5**: 45—145.
— (1944): Über den Steinbutt der Ostsee. — Ber. Dt. wiss. Komm. Meeresforsch. N. F. **11**: 1—136.
KÄNDLER, R. & PIRWITZ, W. (1957): Über die Fruchtbarkeit der Plattfische im Nordsee-Ostsee-Raum. — Kieler Meeresforsch. **13**: 11—34.
KAHL, S. (1928): Ciliata der Oldesloer Salzwasserquellen. — Arch. Hydrobiol. **19**.
— (1933): Ciliata libera et ectocommensalia. — Tierwelt Nord- u. Ostsee: 1—146.
KALLE, K. (1943): Die große Wasserumschichtung im Gotlandtief vom Jahre 1933/34. — Ann. Hydrogr. u. marit. Meteor **71**: 142—146.
KANNENBERG, E. G. (1956): Extrem-Wasserstände an der deutschen Beltseeküste im Zeitraum 1901—1954. — Schr. Naturw. Ver. Schlesw.-Holst. **28**: 17—20.
KARAMAN, ST. L. (1933): Neue Isopoden aus unterirdischen Gewässern Jugoslawiens. — Zool. Anz. **102**: 16—22.
— (1934): Beiträge zur Kenntnis der Isopodenfamilie Microparasasellidae. — Mitt. Höhlen- u. Karstforsch.: 42—44.
— (1953a): Pontokaspische Amphipoden der Jugoslavischen Fauna. — Acta Maced. Sci. Nat. **1**, 2: 21—60.
— (1953b): Über die *Jaera*-Arten Jugoslaviens. — Acta Adriat. **5**, 5: 3—20.
— (1955a): Die Fische der Strumica (Struma-System). — Acta Mus. Maced. Sci. Nat. **3**, Skoplje: 181—208.
— (1955b): Über einige Amphipoden des Grundwassers der jugoslavischen Meeresküste. — Acta Mus. Maced. Sci. Nat. **2**: 223—241.
— (1955c): Über eine neue *Microcerberus*-Art aus dem Küstengrundwasser der Adria. — Fragmenta Balcanica Skoplje **1**, 16: 141—147.
KARLING, T. G. (1931): Untersuchungen über *Kalyptorhynchia (Turbell. Rhabdocoela)* aus dem Brackwasser des Finnischen Meerbusens. — Acta Zool. Fenn. **11**: 1—66.
— (1934): Ein Beitrag zur Kenntnis der Nemertinen des Finnischen Meerbusens. — Mem. Soc. Fauna Flora Fenn. **10**: 76—90.
— (1947): Studien über Kalyptorhynchien (Turbellaria). I. Die Familien Placorhynchidae und Gnathorhynchidae. — Acta Zool. Fenn. **50**.
— (1952): Studien über Kalyptorhynchien (Turbellaria). IV. Einige *Eukaloptorhynchia*. — Acta Zool. Fenn. **69**.
— (1954): Über einige Kleintiere des Meeressandes des Nordsee-Ostsee-Gebietes. — Ark. Zool. Ser. **2**, 7, 14: 241—249.
KARSINKIN, G. S. (1924): Le Plancton de l'angle du Sud-Ouest de la mer d'Aral. — Russ. Hydrobiol. Z. **3**.
KAY, H. (1954): Untersuchungen zur Menge und Verteilung der organischen Substanz im Meerwasser. — Kieler Meeresforsch. **10**: 26—36.
KERTESZ, G. (1955): Die Anostraca-Phyllopoden der Natrongewässer bei Farmos. — Acta Zool. Acad. Sci. Hung. **1**.

KINNE, O. (1952a): Zur Biologie und Physiologie von *Gammarus duebeni* LILLJ., III: Zahlenverhältnis der Geschlechter und Geschlechtsbestimmung. — Kieler Meeresforsch. **9**: 126—133.

— (1952b): Zur Biologie und Physiologie von *Gammarus duebeni* LILLJ., V: Untersuchungen über Blutkonzentration, Herzfrequenz und Atmung. — Kieler Meeresforsch. **9**: 126—133.

— (1952c): Zum Lebenszyklus von *Gammarus duebeni* LILLJ. nebst einigen Bemerkungen zur Biologie von *Gammarus zaddachi* SEXTON subsp. *zaddachi* SPOONER. — Veröff. Inst. Meeresforsch. Bremerhaven **1**: 187—203.

— (1953): Zur Biologie und Physiologie von *Gammarus duebeni* LILLJ., VI. Produktionsbiologische Studie. — Veröff. Inst. Meeresforsch. Bremerhaven II.: 187—203.

— (1954a): Eidonomie, Anatomie und Lebenszyklus von *Sphaeroma hookeri* LEACH (Isopoda). — Kieler Meeresforsch. **10**: 100—120.

— (1954b): Die *Gammarus*-Arten der Kieler Bucht. — Zool. Jb. **82**, 5: 405—424.

— (1955): *Neomysis vulgaris* THOMPSON, eine autökologisch-biologische Studie. — Biol. Zbl. **74**, 3—4: 160—201.

— (1956a): Zur Ökologie der Hydroidpolypen des Nordostseekanals. — Z. Morph. Ökol. Tiere **45**: 217—249.

— (1956b): Über Temperatur und Salzgehalt und ihre physiologisch-biologische Bedeutung. — Biol. Zbl. **75**: 314—327.

— (1957): Salinity. — Geol. Soc. Amer. Mem. **67**, 1: 129—158.

— (1963): Über den Einfluß des Salzgehaltes auf verschiedene Lebensprozesse des Knochenfisches *Cyprinodon macularius*. — Veröff. Inst. Meeresforsch. Bremerhaven, Sonderbd. Drittes Meeresbiol. Sympos.: 49—66.

— (1964a): Physiologische und ökologische Aspekte des Lebens in Ästuarien. — Helgol. Wiss. Meeresunters. **11**: 131—156.

— (1964b): The effects of temperature and salinity on marine and brackish water animals. II. — Oceanogr. Mar. Biol. Ann. Rev. **2**: 281—339.

— (1964c): Non genetic adaptation to temperature and salinity. — Helgol. Wiss. Meeresunters. **9**: 443—458.

— (1966): Physiological aspects of animal life in estuaries with special reference to salinity. — Netherlands J. Sea Res. **3**, 2: 223—244.

— (1967): Physiology of estuarine organisms with special reference to salinity and temperature. — Estuaries 1967. Amer. Ass. Adv. Sci.: 525—540.

KINNE, O. & E. (1962): Rates of development in embryos of an eyprinodont fish exposed different temperature-salinity-oxygen combinations. — Canad. J. Zool. **49**: 231.

KLEMM, J. (1914): Beiträge zu einer Algenflora der Umgebung von Greifswald. — Diss. Greifswald.

KLIE, W. (1923): Muschelkrebse des Brackwassers. — Schr. f. Süßwasser- u. Meereskde. Büsum, **12**.

— (1925): Die Entomostraken der Salzwässer von Oldesloe. — In A. THIENEMANN: Das Salzwasser von Oldesloe.

— (1929a): Beitrag zur Kenntnis der Ostracoden der südlichen und westlichen Ostsee, der festländischen Nordseeküste und der Insel Helgoland. — Z. wiss. Zool. **134**: 270—306.

— (1929b): Ostracoden. — Tierwelt Nord- und Ostsee, Teil Xb: 1—56.

— (1929c): Die *Copepoda Harpacticoidea* der südlichen und westlichen Ostsee mit besonderer Berücksichtigung der Sandfauna der Kieler Bucht. — Zool. Jb. Syst. **57**: 329—386.

— (1933a): Bericht über eine Nachuntersuchung der Crustaceenfauna des alten Hafens zu Bremerhaven. — Verh. Internat. Ver. Limnol. **6**: 308—313.

— (1933b): Neues zur Crustaceen-Fauna Nordwestdeutschlands. — Abh. Nat. Ver. Bremen **28**.

— (1933c): Zoologische Ergebnisse einer Reise nach Bonaire, Curacao und Aruba im Jahre 1930. Nr. 5: Süß- und Brackwasser-Ostracoden von Bonaire, Curacao und Aruba. — Zool. Jb. **64**: 369—390.

— (1934a): Entomostraken aus dem Süß- und Brackwasser von Helgoland. — Schr. Ver. Naturk. Unterweser, N. F. **7**:1—7.

— (1934b): Die Harpacticoiden des Küstengrundwassers bei Schilksee (Kieler Förde). — Schr. Naturw. Ver. Schlesw.-Holst. **20**: 409—421.

— (1938): Zwei neue Ostracoden aus der Ostsee. — Kieler Meeresforsch. **2**: 345—351.

— (1949): Harpacticoida (Cop.) aus dem Bereich von Helgoland und der Kieler Bucht. — Kiel. Meeresforsch. **6**: 90—128.

KLUGH, B. (1924): Factors controlling the biota of the tide-pools. — Ecology **5**: 192—196.

KLUYVER, A. J. & BAARS, J. K. (1932): On some physiological artefacts. — Proc. Konikl. Akad. Wetensch. Amsterdam **35**.

KNIPOWITSCH, N. W. (1924): Über die Verteilung des Lebens im Schwarzen Meer. — Russ. Hydrobiol. Z. **3**: 199—203.

— (1924—1926): Zur Hydrologie und Hydrobiologie des Schwarzen und Asowschen Meeres. — Internat. Rev. Ges. Hydrobiol. Hydrogr. **12**: 342—349, **13**: 4—20, **16**: 81—102.

— (1930): Vertikale Zirkulation und Verteilung des Sauerstoffs im Schwarzen und im Kaspischen Meer. — Bull. Inst. Hydrobiol. **31**: 23—27, 38—42.

— (1933): Hydrologische Untersuchungen im Schwarzen Meer. — Abh. Wiss. Fischerei-exped. im Asowschen u. Schwarzen Meer. Liefg. **5**: 272 pp.

KNÖLLNER, F. (1935a): Ökologische und systematische Untersuchungen über litorale und marine Oligochaeten der Kieler Bucht. — Zool. Jb. Syst. **66**: 427—512.

— (1935b): *Stygocapitella subterranea* n. g. n. sp. — Schr. Naturw. Ver. Schlesw.-Holst. **20**: 468—472.

— (1935c): Die Oligochaeten des Küstengrundwassers. — Schr. Naturw. Ver. Schlesw.-Holst. **21**: 135—139.

KÖNIG, D. (1948): *Spartina townsendi* an der Westküste von Schleswig-Holstein. — Planta **36**: 34—70.

(1956): *Membranipora crustulenta* (PALL.) auf der Insel Pelworm. — Faun. Mitt. **7**: 1—3.

KOLBE, R. W. (1927): Zur Ökologie, Morphologie und Systematik der Brackwasser-Diatomeen. Die Kieselalgen des Sperenberger Salzgebietes. — Pflanzenforsch. **7**: 143 pp.

— (1932): Grundlinien einer allgemeinen Ökologie der Diatomeen. — Ergebn. Biol. **8**: 222—238.

KOLBE, R. W. & TIEGS, E. (1929): Zur mesohaloben Diatomeenflora des Werragebietes. — Ber. Dt. Bot. Ges. **47**.

KOLDERUP ROSENVINGE, L. (1935): Distribution of the Rhodophyseae in the Danish waters. — D. Kgl. Danske Vid. Selsk. Skr. 9, Raekke, Nat.-vid. og Mathm. Abd. **6**, S: 1—43.

KOLI, L. (1961 a): Die Molluskenfauna des Brackwassergebietes bei Tvärmime. — Ann. Zool. Soc. Vanamo **22**: 1—22.

— (1961b): Über die Hirudineen des Brackwassers in der Umgebung von Tvärminne. — Arch. Soc. Vanamo **15**: 58—63.

KOLLER, G. (1927): Über die geschlechtliche Fortpflanzung der *Protohydra leuckarti* GREFF. — Zool. Anz. **73**: 97—100.

KORRINGA, P. (1956): On the pretended compulsory relation between oviparous oysters and waters of reduced salinity. — Internat. Conf. Mar. Biol. Roscoff: 109—116.

— (1958): On the supposed compulsory relation between oviparous oysters and waters of reduced salinity. — Biologie comparée des espèces marines (Colloque internat. Roscoff) Paris.

KOSKE, P. H. & SZEKIELDA, V. A. (1965): Zur Hydrographie des Nord-Ostsee Kanals. — Kieler Meeresforsch. **21**: 132—143.

KOSSWIG, C. (1942): Die Faunengeschichte des Mittel- und Schwarzen Meeres. — C. R. Ann. et Arch. Soc. Turque Sci. Phys. Nat. Fasc. 9: 37—52.

— (1956): Beitrag zur Faunengeschichte des Mittelmeeres. — Pubbl. Staz. Zool. Napoli **28**: 78—88.

KOTHE, P. (1968): *Hypania invalida* und *Jaera sarsi* erstmals in der deutschen Donau. — Arch. Hydrobiol. Suppl. **34**: 88—114.

KOWALSKI, R. (1955): Untersuchungen zur Biologie des Seesternes *Asterias rubens* L. im Brackwasser. — Kieler Meeresforsch. **11**: 201—213.

KRAMP, P. L. (1935): Polypdyr I. — Danmarks Fauna **41** (Kopenhagen): 1—223.

— (1961): Synopsis of the medusae of the world. — J. Mar. biol. A. U. K. **40**: 1—469.

KREY, J. (1956): Die Trophie küstennaher Meeresgebiete. — Kieler Meeresforsch. **12**: 46—71.

KRISS, A. (1958): Marine Microbiology. — Moskau.

KRIZENECKY (1916): Beitrag zum Studium der Bedeutung osmotischer Verhältnisse des Mediums für Organismen. Versuche an Enchytraeiden. — Pflügers Arch. **163**.

KROGH, A. & SPÄRCK, R. (1936): On a new bottom-sampler for investigation of the microfauna of the Sea bottom. — Kgl. Danske Vidensk. Selsk. Medd. **13**: 3—12.

KRÜGER, K. & MEYER, P. F. (1938): Biologische Untersuchungen der Wismarischen Bucht. — Z. Fischerei **35**: 665—703.

KRUSEMANN jr. G. (1933): Welche Arten von *Chironomus* s. l. sind Brackwassertiere? — Verh. internat. Ver. Limnol. **6**: 163—165.

KÜHL, H. & MANN, H. (1957): Beiträge zur Hydrochemie der unteren Weser. — Veröff. Inst. Meeresforsch. Bremerhaven **5**: 34—62.

— (1961): Vergleichende hydrochemische Untersuchungen an den Mündungen deutscher Flüsse. — Verh. Internat. Verein. Limnol. **14**.

— (1962): Über das Zooplankton der Unterelbe. — Veröff. Inst. Meeresforsch. Bremerhaven **8**: 53—70.

KUNZ, H. (1935): Zur Ökologie der Copepoden Schleswig-Holsteins und der Kieler Bucht. — Schr. Naturw. Ver. Schlesw.-Holst. **21**: 84—133.

— (1938): Zur Kenntnis der Harpacticioden des Küstengrundwassers der Kieler Förde. — Kieler Meeresforsch. **2**: 223—254.

— (1940): Harpacticoiden vom Sandstrand der Kurischen Nehrung (Studien an marinen Copepoden III). — Kieler Meeresforsch. **3**: 148—157.

KÜNNE, C. (1933): Zur Kenntnis der Anthomeduse *Bougainvillia macloviana* LESSON. — Zool. Anz. **101**: 248—254.

— (1934): Über die Leptomedusen *Helgicirrha schulzii* HARTLAUB und *Eirene viridula* (PERON et LESUEUR). — Zool. Anz. **106**: 27—34.

LADD, H. S. (1951): Brackish-water and marine assemblages of the Texas Coast, with special reference to mollusks. — Publ. Inst. Mar. Sc. **2**: 125—164.

LAKOWITZ, K. (1929): Die Algenflora der gesamten Ostsee. — Danzig: 1—474.

LANG, K. (1948): Monographie der Harpacticiden I u. II. — Lund: 1—1683.

LARSEN, K. (1936): The distribution of the invertebrates in the Dybsö fjord. — Rep. Danish. Biol. Stat. **41**: 3—36.

LASSERRE, P. (1966, 1967): Oligochètes marins des côtes de France I, II. — Cah. Biol. Mar. **7**: 295—314, **8**: 273—293.

LASSIG, J. (1964): Notes on the occurrence and reproduction of *Prostoma obscurum* in the inner Baltic. — Ann. Zool. Fenn. **1**: 146.

— (1965a): Notes on the invertebrate fauna of the northern Baltic area. — Comment. Biol. Soc. Fenn. **28**, 8: 1—6.

— (1965b): The distribution of marine and brackish-water lamellibranchia in the northern Baltic area. — Comment Biol. Soc. Sci. Fenn. **28**, 5: 1—41.

LE CALVEZ, J. & LE CALVEZ, Y. (1951): Contribution à l'étude des Foraminifères des Eaux saumâtres. I. Etangs de Canet et de Salses. — Vie et Milieu **2**: 237—254.

LEEGAARD, C. (1920): Microplankton from the Finnish waters during the month of May 1912. — Acta Soc. Sci. Fenn.

LELOUP, E. & KONIETHKO, B. (1956): Recherches biologiques sur les eaux saumâtres du Bas-Escaut. — Mem. Inst. Roy. Sci. nat. Belg. **132**: 1—100.

LENGERKEN, H. von (1929): Die Salzkäfer der Nord- und Ostseeküste. — Z. Wiss. Zool. **135**: 1—162.

LENZ, F. (1933): Untersuchungen zur Limnologie von Strandseen. — Verh. internat. Ver. Limnol. **6**: 166—177.

LEVANDER, K. M. (1894): Materialien zur Kenntnis der Wasserfauna in der Umgebung von Helsingfors, mit besonderer Berücksichtigung der Meeresfauna. I. Protozoa. — Acta Soc. Fauna Flora Fenn. **12**, 2: 1—115.

— (1899): Materialien zur Kenntnis der Wasserfauna in der Umgebung von Helsingfors. II. Rotatoria, III. Spongien, Coelenteraten, Bryozoen und Mollusken des Finnischen Meerbusens bei Helsingfors. — Acta Fauna Flora Fenn. **12**, **3**; 3—72, **17**, 4.

— (1900a): Über das Herbst- und Winterplankton im Finnischen Meerbusen und in der Alands-See. — Acta Fauna Flora Fenn. **18**: 1—25.

— (1900b): Zur Kenntnis des Lebens in den stehenden Kleingewässern auf den Skären-inseln. — Acta Fauna Flora Fenn. **18**: 3—107.

— (1901a): Zur Kenntnis des Planktons und der Bodenfauna einiger seichter Brack-wasserbuchten. — Acta Fauna Flora Fenn. **20**: 1—34.

— (1901b): Übersicht in der Umgebung von Esbö-Löfö im Meerwasser vorkommenden Tiere. — Acta Fauna Flora Fenn. **20**: 1—20.

— (1911): Rotatoria. — Bull. trimestr. etc. Cons. permanent Explor. de la mer, Kopen-hagen.

— (1916): Zur Kenntnis des Küstenplanktons im Weißen Meer. — Medd. Soc. Fauna Flora Fenn. **42**: 150—158.

LEVRING, T. (1935): Zur Kenntnis der Algenflora von Kullen an der schwedischen West-küste. — Lunds Univ. Arsskr. N. F. **2**, 31.

— (1940): Studien über die Algenvegetation von Bleckinge, Südschweden. — Diss. Lund: 1—178.

LIEBETANZ, B. (1925): Hydrobiologische Studien an kujawischen Brackwässern. — Bull. Acad. Polon. Cracovie, Ser. B.

LILLELUND, K. (1953): Hydrographie und Netzplankton des Sehlendorfer Binnensees, einem Strandgewässer an der deutschen Ostseeküste. — Diss. Hamburg: 1—22.

— (1955): Über Untersuchungen im Sehlendorfer Binnensee. — Arch. Hydrobiol. Suppl. **22**: 426—432.

LILLJEBORG, W. (1901): *Cladocera suecica*. — Nova Acta Soc. Sci. Uppsaliens. **19**.

LINDBERG, H. (1936): Halophile und halobionte Hemiptera. — Tierw. Nord- u. Ostsee **11**e: 1—124.

— (1948): Zur Kenntnis der Insektenfauna im Brackwasser des Baltischen Meeres. — Soc. Sci. Fenn. Comment. Biol. **10**: 1—206.

LINDER, F. (1941): Contribution to the morphology and taxonomy of the Branchiopoda Anostraca. — Zool. Bidr. Uppsala **20**: 101—302.

LINDQUIST, A. (1960): Salthalten och djurvärlden im Baltiska havet. — Fiskeritidskr. för Finland, Ny Ser. **4**: 90—94.

LINDSTEDT, A. (1943): Die Flora der Cyanophyceen der Schwedischen Westküste. — Diss. Lund: 1—121.

LINKE, O. (1934): Beiträge zur Sexualbiologie der Littorinen. — Z. Morph. Ökol. Tiere **28**: 170—177.

LÖFFLER, H. (1956): Limnologische Untersuchungen an Iranischen Binnengewässern. — Hydrologia **8**.

— (1959): Zur Limnologie, Entomostraken- und Rotatorienfauna des Seewinkelgebietes (Burgenland—Österreich). — Sitz.-Ber. Österr. Akad. Wiss. Math. Nat. Kl. **1**: 168.

— (1959/61): Beiträge zur Kenntnis der Iranischen Binnengewässer II. — Internat. Rev. ges. Hydrobiol. **46**: 309—406.

LÖNNBERG, E. (1918): Undersökningar rörande Öresunds djurliv. — Medd. Kgl. Laut-bruksstyr. **1** (Uppsala).

LOHMANN, H. (1896): Die Appendicularien der Plankton Expedition. — Ergebn. Plankton Exped. Humboldt-Stift. **2** E c: 1—148.

LOVEN, S. (1845): On the bathymetral distribution of submarine life on the Northern Shores of Skandinavia. — Brit. Ass. Adv. Sci. 1844. Rep. **14**. London.

— (1864): Om Östersjön. — Föredrag vid Skand. Naturf. Sällsk. **1**, Stockolm.

LUNDBECK, J. (1928): Studien über das Frische Haff III: Die Strömungen und ihre Beziehungen zu Wasserhaushalt und Wasserbeschaffenheit im Frischen Haff. — Schr. Phys.-ökonom. Ges. zu Königsberg i. Pr. **65**: 1—111.

— (1932): Beobachtungen über die Tierwelt austrocknender Salzwiesentümpel an der Holsteiner Ostseeküste. — Arch. Hydrobiol. **24**: 603—628.

— (1935): Über die Bodenbevölkerung, besonders die Chironomidenlarven, des Frischen und Kurischen Haffes. — Internat. Rev. Hydrobiol. Hydrogr. **32**: 266—284.

LUTHER, A. (1909): Über eine Litorina-Ablagerung bei Tvärminne nebst einigen Bemerkungen über die kalkauflösenden Eigenschaften der jetzigen Ostsee und des Litorina-Meeres. — Acta Soc. Fauna et Flora Fenn. **32**: 1—22.

— (1918): Vorläufiges Verzeichnis der rhabdocölen und alloeocölen Turbellarien Finnlands. — Med. Soc. Fauna Flora Fenn. **44**.

— (1927): Über das Vorkommen der Bryozoe *Victorella paida* S. KENT im Finnischen Meerbusen bei Tvärminne. — Mem. Soc. Fauna Flora Fenn. **1**: 7—9.

— (1955): Die Dalyelloiden. — Acta Zool. Fenn. **87**: 1—337.

LUTHER, H. (1949): Vorschlag zu einer ökologischen Grundeinteilung der Hydrophyten. — Acta Bot. Fenn. **44**: 1—15.

— (1950): Die Funde von *Zostera marina* L. in der nördlichen Ostsee. — Mem. Soc. Fauna Flora Fenn. **25**: 25—36.

— (1951): Verbreitung und Ökologie der höheren Wasserpflanzen im Brackwasser der Ekenäs-Gegend in Südfinnland. I, II. — Acta Bot. Fenn. **49**: 1—231, **50**: 1—370.

LUTZE, G. (1965): Zur Foraminiferen-Fauna der Ostsee. — Meyniana Kiel **15**: 75—142.

MACNAE, W. (1963): Mangrove swamps in South-Africa. — J. Ecol. **51**: 1—25.

— (1966): Mangroves in Eastern and Southern Australia. — Austral. J. Bot. **14**: 67—104.

— (1967): Zonation within Mangroves associated with Estuaries in North Queensland. — Estuaries (Washington): 432—441.

MADSEN, H. (1936): Investigations on the shore fauna of East Greenland with a survey of the shores of the other arctic regions. — Medd. Grönland **100**: 4—79.

MANKOWSKI, W. (1961): Polish macroplankton observations in 1959. — ICES, Ann. Biol. **16**: 1—83.

MARCUS, E. (1940): Mosdyr. — Danmarks Fauna **46**: 1—303.

— (1950): Turbellaria Brasileiros. — 8. Zoologica Sao Paulo **15**: 5—123.

MARCUS, E. & E. (1963): On Brasilian supralittoral and brackish water spails. — Bol. Inst. Oceanogr. Sao Paulo **13**: 41—52.

MARRE, G. (1931): Fischereiwissenschaftliche Untersuchungen über die Grundlagen der Stintfischerei im Kurischen Haff. — Z. Fischerei **19**.

MARS, P. (1949): Contributions à l'étude biologique des étangs méditerranéens. — Bull. Soc. Linn. Provence **17**.

— (1950): Euryhalinité de quelques Mollusques méditerranéens. — Vie et Milieu **1**: 441—448.

MARTINI, E. (1923): Über Beinflussung der Kiemenlänge von Aedeslarven durch das Wasser. — Verh. internat. Ver. Limnol. **1**: 235—259.

MARX, W. & HENSCHEL, J. (1941): Die Befruchtung und Entwicklung von Plattfischeiern in verdünntem Nordseewasser im Vergleich zu den Befunden in der freien Ostsee. — Helgoländer Wiss. Meeresunters. **2**: 226—243.

MAUCHER, W. D. (1961): Statistische Untersuchungen in den Körperproportionen zwischen der Nord- und Ostseeform von *Crangon crangon*. — Kieler Meeresforsch. **17**: 219—227.

MEIXNER, J. (1938): Turbellaria I. — Tierwelt Nord- u. Ostsee, Teil IV b: 1—146.

MEUCHE, A. (1939): Die Fauna im Algenbewuchs. Nach Untersuchungen im Litoral ostholsteinischer Seen. — Arch. Hydrobiol. **34**: 351—520.

MEYER, H. (1935): Die Atmung von *Asterias rubens* und ihre Abhängigkeit von verschiedenen Außenfaktoren. — Zool. Jb. Physiol. **55**: 349—398.

MEYER, H. & MÖBIUS, K. (1865, 1871): Fauna der Kieler Bucht. — **1**: 1—88, **2**: 1—139.

MEYER, P. F. (1932): *Palaemonetes varians* LEACH. *(Crust. Decap.)* in den Oldenburgischen Gewässern. — Senckenbergiana **14**.

198 Ecology of brackish water

MIELCK, W. & KÜNNE, C. (1935): Fischbrut und Plankton-Untersuchungen auf dem Reichs-forschungsdampfer „Poseidon" in der Ostsee, Mai-Juni 1931: 1—118.
MÖBIUS, K. (1871): Die Faunistischen Untersuchungen der Pommerania-Expedition. — Jber. Comm. Wiss. Unters. Dt. Meere, Kiel: 97—150.
— (1873): Die wirbellosen Tiere der Ostsee. — Ber. Komm. Dt. Meere Kiel 1: 97—144.
— (1877): Die Auster und die Austernwirtschaft. — Berlin: 1—126.
— (1893): Über die Tiere der Schleswig-Holsteinischen Austernbänke, ihre physikalischen und biologischen Lebensverhältnisse. — Sitz.-Ber. Akad. Wiss. Berlin 1.
MÖBIUS, K. & HEINCKE, F. (1883): Die Fische der Ostsee. — IV. Ber. Unters. Dt. Meere, Kiel: 194—296.
MÖLLER, L. (1928): ALFRED MERZ' hydrographische Untersuchungen im Bosporus und Dardanellen. — Veröff. Inst. Meereskde. Berlin. N. F. R. A. 18.
MORDUKHAI-BOLTOVSKOY, PH. D. (1937): Zusammensetzung und Verbreitung des Benthos in der Taganrog-Bucht. — Arb. Wiss. Don-Kuban-Station f. Fischwirtsch. 5: 3—67.
MORDUKHAI-BOLTOVSKOY, R. A. (1957): Die gegenwärtige Verbreitung der Kaspischen Fauna (Russ.): 173—183.
— (1960): Catalogue of free living Invertebrate Fauna of the Sea of Azow (Russ. with engl. Summary). — Zool. J. 39: 1455—1466.
— (1965): Polyphemidae of the Pontocaspian Basin. — Hydrobiologia 25: 213—220.
MOTAS, C. & BAČESCU, M. (1938): *Hypania invalida* (GRUBE) et *Hypaniola Kowalewskii* (GRIMM) en Roumanie. — Ann. Sci. Univ. Jassi 24.
MOTAS, C. & SOAREC-TANASACHI, J. (1943): Un Halacaride reliquat ponto-caspien dans le Danube. — Bull. Soc. Natur. Roman. Bucuresti 16.
MÜLLER-LIEBENAU, J. (1956): Die Besiedlung der Potamogeton-Zone ostholsteinischer Seen. — Arch. Hydrobiol. 52: 470—606.
MÜNCH, H. D. & PETZOLD, H. G. (1956): Zur Fauna des Küstengrundwassers der Insel Hiddensee. I. — Wiss. Z. Greifswald 516: 413—429.
MÜNZING, J. (1959): Biologie, Variabilität und Genetik von *Gasterosteus aculeatus* L. — Internat. Rev. ges. Hydrobiol. 44: 517—582.
— (1961): *Gasterosteus aculeatus* im Ostseeraum. — Mitt. Hamb. Zool. Mus. Inst. 59: 61—72.
— (1962a): Ein neuer semiarmatus-Typ von *Gasterosteus aculeatus* aus dem Isniksee. — Mitt. Hamb. Zool. Mus. Inst. 60: 181—194.
— (1962 b): Die Populationen der marinen Wanderform von *Gasterosteus aculeatus* an den holländischen und deutschen Nordseeküsten. — Netherlands J. Sea Res. 1: 509 bis 525.
— (1963): The evolution of variation and distributional patterns in European populations of the threespines stickleback *Gasterosteus aculeatus*. — Evolution 17: 320—332.
MULICKI, Z. (1938): Note of the quantitative Distribution of the bottom Fauna near the Polish coast of the Baltic. — Bull. Stat. Marit. Hel. 2.
MUNTHE, H. (1910): Studier over Gotlands senkvartära historia. — Sver. Geol. Under. 4.
MUUS, B. J. (1963): Some Danish Hydrobiidae with the description of a new species *Hydrobia neglecta*. — Proc. Malacol. Soc. London 35: 131—138.
— (1967): The Fauna of Danish estuaries and lagoons. — Medd. Danmarks Fisk. Havunders. 5: 1—315.
NAGUIB, M. (1958): Studies on the Ecology of Lake Quarun (Fayum-Egipt). — Kieler Meeresforsch. 14: 167—222.
NAYLOR, E. (1955): The ecological distribution of British species of *Idotea* (Isopoda). — J. Anim. Ecol. 24: 255—269.
NEEDHAM, J. (1930): On the penetration of marine organisms into fresh water. — Biol. Zbl. 50: 504—508.
NELLEN, W. (1965a): Neue Untersuchungen über den „Schleihering", eine lokale Brack-wasserform von *Clupea harengus* L. — Ber. Dt. Wiss. Komm. Meeresforsch. 18: 163—193.
— (1965b): Beiträge zur Brackwasserökologie der Fische im Ostseeraum. — Kieler Meeresforsch. 21: 192—198.

—   (1967): Ökologie und Fauna (Makroevertebraten) der brackigen und hypertrophen Ostseeförde Schlei. — Arch. Hydrobiol. **63**: 273—309.

—   (1968): Der Fischbestand und die Fischereiwirtschaft in der Schlei. — Schr. Naturw. Ver. Schlesw-.Holst. **38**: 5—50.

NELSON-SMITH, A. (1965): Marine Biology of Milford Haven: The Physical Environment. — Field Studies **2**.

NICHOLLS, A. G. (1935): Copepods from the interstitial fauna of a sandy beach. — J. Mar. Biol. Ass. **20**: 379—406.

NICHOLS, M. & ELLISON, R. L. (1967): Sedimentary patterns of microfauna in a coastal plain estuary. — Estuaries (Washington): 283—290.

NICOL, E. (1935): The ecology of a salt-marsh. — J. Mar. Biol. Ass. U. K. **20**: 203—262.

—   (1936): The brackish water lochs of North Uist. — Proc. Roy. Soc. Edinburgh **56**: 169—195.

—   (1938): The brackish water lochs of Orkney. — Proc. Roy. Soc. Edinburgh **56**: 181 to 190.

NIENBURG, W. (1925): Eine eigenartige Lebensgemeinschaft zwischen *Fucus* und *Mytilus*. — Ber. Dt. Bot. Ges.: 292—298.

NIKITIN, W. N. (1931): Die untere Planktongrenze und deren Verteilung im Schwarzen Meer. — Internat. Rev. ges. Hydrobiol. Hydrographie **25**: 102—130.

NOODT, W. (1952): Marine Harpacticoiden (Cop.) aus dem eulitoralen Sandstrand der Insel Sylt. — Abh. Math. Naturwiss. Kl. Akad. Wiss. Lit. **3**: 105—143.

—   (1953): Entomostracen aus dem Litoral und dem Küstengrundwasser des Finnischen Meerbusens. — Acta Soc. Fauna Flora Fenn. **72**: 3—12.

—   (1954): Copepoda Harpacticoidea aus dem limnischen Mesospammal der Türkei. — Istanbul Univ. Fak. Hidrobiol. **2**: 27—40.

—   (1955): Harpacticiden (Crust. Cop.) aus dem Sandstrand der französischen Biscaya-Küste. — Kieler Meeresforsch. **11**: 86—109.

—   (1956): Verzeichnis der im Eulitoral der schleswig-holsteinischen Küsten angetroffenen Copepoda Harpacticoidea. — Schr. Naturw. Ver. Schlesw.-Holst. **28**: 42—64.

—   (1957): Zur Ökologie der Harpacticoidea (Crust. Cop.) des Eulitorals der Deutschen Meeresküste und der angrenzenden Brackgewässer. — Z. Morph. Ökol. Tiere **46**: 149—242.

—   (1964): Natürliches System und Biogeographie der Syncarida. — Gewässer u. Abwässer, **37/38**: 77—186.

OBERTHÜR, K. (1937): Untersuchungen an *Frontonia marina* FABRE-DOM. aus einer Binnenland-Salzquelle unter besonderer Berücksichtigung der pulsierenden Vakuole. — Arch. Protistenkde. **88**: 387—420.

OHM, G. (1964): Die Besiedlung der *Fucus*-Zone der Kieler Bucht unter besonderer Berücksichtigung der Mikrofauna. — Kieler Meeresforsch. **20**: 30—64.

OSTENFELD, C. H. (1913—1916): De Danske Farvandes Plankton. — Kgl. Danske Vidensk. Selsk. Skrift. Naturv. Math. Afd. **7**, 8.

—   (1931): Concluding remarks on the Plankton on the quarterly cruises in the years 1902—1908. — Cons. perm. internat. Explor. de la Mer. Bull. trimestr. **4**: 601—672.

OSTROUMOW, A. (1893): Distribution verticale des mollusques dans la mer Noire. — Congrès internat. Zool. 2. Sess. à Moscou.

OTTO, G. (1936): Die Fauna der Enteromorpha-Zone der Kieler Bucht. — Kieler Meeresforsch. **1**: 1—48.

OTTO, J. P. & WIELINGA, D. T. (1933): Hydrobiologische Notizen vom Brackwassergebiet der Provinz Friesland. — T. Nederl. Dierk. Ver. (3) **3**: 49—74.

OYE, P. VAN (1920): Note sur les micro-organismes de l'eau sâumatre du Vieux Port de Batavia (Java). — Ann. Biol. Lacustre **10**: 207—216.

PALMHERT, H. (1933): Beiträge zum Problem der Osmoregulation einiger Hydroidpolypen. — Zool. Jg., Allg. Zool. Physiol. **53**: 212—220.

PASPALEW, G. W. (1933): Hydrobiologische Untersuchungen über den Golf von Varna. — Arb. Biol. Meeresstat. Varna **2**: 1—18.

—   (1939): Temperatur und Salzgehalt des Wassers im Golf von Varna während des
      Jahres 1938. — Ibid., **8**: 35—45.
—   (1941): Untersuchungen über das Fischsterben im Varnaer See am 24. bis 26. Juli.
      1940. — Ribarski Pregled (Sofia) **11**, 3.
PATRICK, R. (1967): Diatom Communities in Estuaries. — Estuaries (Washington): 311
      to 315.
PAX, F. (1934): Anthozoa. — Tierwelt Nord- u. Ostsee 3, e: 1—80.
PEARSE, A. S. & GUNTER, G., (1957): Salinity. Treatise on Marine Ecology and Palaeecology
      1. — Mem. Geol. Soc. Amer. **67**.
PENNAK, R. W. & ZINN, D. J. (1943): Mystacocarida, a new order of Crustacea from in-
      tertidal beaches in Massachusetts and Connecticut. — Smithson. Misc. Coll. **103**, 9:
      1—11.
PERCIVAL, E. (1929): Report on the fauna of the estuaries of River Tama and the River
      Lynher. — J. Mar. Biol. Ass. Plymouth **16**: 81—108.
PETERSEN, C. G. J. (1897): Plancton studies in the Limfjord. — Rep. Danish Biol. Stat. 1897.
—   (1924): A brief survey of the animal communities in Danish waters. — Amer. J. Sci.
      7: 343—354.
PETERSEN, C. G. J. & BOYSEN JENSEN, P. (1911): Valution of the sea. I. Animal life of the sea
      bottom, its food and quantity. — Rep. Danish Biol. Stat. **20**: 1—76.
PETIT, G. (1950a): *Cyathura carinata* (KRÖYER) dans l'étang de Salses (P.-O.). — Vie et Milieu
      1: 476—477.
—   (1950b): *Corophium insidiosus* CRAWFORD dans les étangs du Roussillon. — Vie et
      Milieu **1**.
—   (1954): Introduction à l'étude écologique des Étangs Méditerranéens. — Vie et Milieu
      4: 569—604.
PETIT, G. & RULLIER, F. (1952): Observations sur deux stations nouvelles du Littoral des
      Pyrénées-Orientales. — Vie et Milieu **3**: 1—19.
—   (1956): Encore *Mercerilla enigmatica* dans les eaux saumâtre du Roussilon et du
      Languedoc. — Vie et Milieu **7**: 27—37.
PETIT, G. & SCHACHTER, D. (1951): Le Problème des eaux saumâtres. — Ann. Biol. **27**:
      533—543.
—   (1954): La Camargue. Étude écologique et faunistique. — Ann. Biol. **30**.
PORA, E. (1938): Contributiuni fiziologice la studiul raspandiri geografice a species. *Pachy-*
      grapsus marmoratus STIMPS in Marea neagra. — Acad. Romana Mem. Sect. Stint-
      Ser. 3, **13**, 4: 145—162.
—   (1939): Sur le comportement des Crustacés brachiures de la Mer Noire aux variations
      de salinité du milieu ambiant. — Ann. Sc. Univ. Jassy **25**: 1—34.
—   (1946): Problèmes de physiologie animale dans la Mer Noire. — Bull. Inst. Océanogr.
      (Monaco) **93**: 1—43.
—   (1951): Comportarea da Variationi de Salinitate. — Bul. Sci. **3**: 1—10.
PORA, E. A. & BAČESCU, M. (1939): Sur la résistance du Myside *Gastrosaccus sanctus* (VAN
      BENEDEN) de la mer Noire, aux variations de salinité du milieu ambiant. — Ann.
      Sci. Univ. Jassy **25**: 259—271.
PORTIER, P. (1938): Physiologie des animaux marins. — Biblioth. Paris.
POSTMA, H. (1967): Sediment transport and sedimentation in the estuarine environment. —
      Estuaries (Washington): 158—179.
POULSEN, E. M. (1938a): On the Growth of the Baltic Plaice. — Rapp. Proc. Verb. Cons.
      internat. Expl. Mer. **108**, 9: 53—56.
—   (1938b): On the Growth of the Cod within the Transition Area. — Ibid. **108**, 8:
      49—51.
PRATJE, O. (1931): Einführung in die Geologie der Nord- und Ostsee. — Tierwelt Nord u.
      Ostsee Id: 1—44.
PRECHT, H. (1935): Epizoen der Kieler Bucht. — Nova Acta Leopold. N. F. **3**: 405—474.
PRICE, J. B. & GUNTER, G. (1964): Studies of the chemistry of fresh and low salinity waters
      in Missisippi and boundary between fresh and brackish water. — Internat. Rev. ges.
      Hydrobiol. **49**: 629—636.

Pritchard, D. W. (1967): Observation of Circulation in Coastal Plain Estuaries. — Estuaries (Washington): 37—44.

Purasjoki, K. J. (1945): Quantitative Untersuchungen über die Mikrofauna des Meeresbodens in der Umgebung der Zoologischen Station Tvärminne an der Südküste Finnlands. — Soc. Sci. Fenn., Comm. Biol. 9: 1—24.

— (1948): Pygospio elegans Claparède (Polychaeta) belongs to our fauna. — Arch. Soc. Zool. Bot. Fenn. Vanamo 1 (Helsingfors): 69.

— (1953): Beobachtungen über die Einwirkung gesteigerten Salzgehalts auf das Auftreten einiger mariner Zooplanktonarten außerhalb Helsinki. — Arch. Soc. Vanamo 8, 1: 101—104.

Ramner, W. (1930): Phyllopoda. — Tierwelt Nord- u. Ostsee 10, a: 1—32.

Redeke, H. C. (1922a): Zur Biologie der niederländischen Brackwassertypen. — Bijdr. Dierk. Amsterdam 22: 239—335.

— (1922b): Flora en Fauna der Zuiderzee. — Amsterdam: 1—458.

— (1929): Die Zuidersee und ihre Bewohner, ein Beitrag zur Brackwasserbiologie. — Verh. internat. Ver. Limnol. 4: 536—537.

— (1932): Abriß der regionalen Limnologie der Niederlande. — Amsterdam: 1—39.

— (1933): Über den jetzigen Stand unserer Kenntnisse der Flora und Fauna des Brackwassers. — Verh. internat. Ver. Limnol. 6: 46—61.

— (1935a): De Zuiderzee als brackwatergebied. — Handelingen 25. Nederlandel, Natur-Geneeskundig. Congr. Delft.

— (1935b): Essay Review. Some Recent Publications on the Fauna of Brackish waters in the Netherlands. — Cons. Int. pour l'exploration de la mer. 10, 3: 319—323.

— (1935c): Synopsis van het Nederlandsche zoet- en brackwater-Plankton. — Hydrobiol. Club 2: 1—104.

— (1935d): Acartia tonsa Dana, ein neuer Copepode des niederländischen Brackwassers. — Arch. Nederland Zool. 1: 315—329.

Reibisch, J. (1902): Wirbellose Bodentiere der Ostsee-Expedition 1901. — Abh. dt. Seefischereiver. 2.

— (1914): Die Bodenfauna von Nord- und Ostsee. — Verh. dt. Zool. Ges. 24: 221—234.

— (1923): Über Änderungen in der Fauna der Kieler Bucht. — Schr. Naturw. Ver. Schlesw.-Holst. 17: 227—232.

Reid, D. M. (1930): Salinity interchange between sea water in sand and overflowing freshwater at low tide. — J. Mar. Biol. Ass. Plymouth 16: 609—614.

— (1939): On the occurrence of Gammarus duebeni (Lillj.) in Ireland. — Proc. roy. Irish Acad. 45.

Reid, G. K. (1961): Ecology of inland waters and estuaries. — Reinbold, New York: 1—375.

Reinbold, Th. (1889—1891): Die Chlorophyceen der Kieler Förde. — Schr. Naturw. Ver. Schlesw.-Holst. 8: 163—186.

— (1893): Die Phaeophyceen etc. — Ibid. 10: 109—144.

Reinhard, L. W. (1885): Algological Investigations. I. Material to the morphology and taxonomy of the algae in the Black Sea. — Zapiski Novorossijsk Obscestva Jestestvoisp. (Odessa) 9.

Reinke, J. (1879): Algenflora der westlichen Ostsee deutschen Antheils. — 6. Ber. d. Komm. d. wiss. Unters. d. Meere in Kiel f. d. Jahre 1887—1891: 1—103.

— (1896): Zur Algenflora der westlichen Ostsee. — Wiss. Meeresunters. N. F. Abt. Kiel 1.

— (1897): Untersuchungen über den Pflanzenwuchs in der östlichen Ostsee. — Ibid. 2: 1—6.

— (1898): Notiz über die marine Vegetation des Kaiser-Wilhelm-Kanals im August 1896. — Ibid. 3: 33—34.

— (1901): Untersuchungen über den Pflanzenwuchs in der östlichen Ostsee. 4. — Ibid. 5: 1—8.

Reinke, J. & Darbishire, O. V. (1898): Untersuchungen über den Pflanzenwuchs in der östlichen Ostsee. 2. — Ibid. 3: 17—24.

REMANE, A. (1933): Verteilung und Organisation der benthonischen Mikrofauna der Kieler
    Bucht. — Wiss. Meeresunters. Kiel **21**: 163—221.
—   (1934a): Die Brackwasserfauna. — Verh. Dt. Zool. Ges.: 34—74.
—   (1934b): Neue Gastrotrichen des Küstengrundwassers von Schilksee. — Schr. Naturw.
    Ver. Schlesw.-Holst. **20**: 473—478.
—   (1937a): Über eine marine Tierform (acöles Turbellar) in der Salzquelle von Artern.
    — Z. Naturwiss. Halle **91**, 2: 78—80.
—   (1937b): Die übrige Tierwelt (als 8. Beitrag in: Die Schlei und ihre Fischwirtschaft).
    — Schr. Naturwiss. Ver. Schleswig-Holstein **22**, 1: 209—224.
—   (1941): Einführung in die zoologische Ökologie der Nord- und Ostsee. — Tierwelt.
    Nord- u. Ostsee: 1—238.
—   (1949): Die psammobionten Rotatorien der Nord- und Ostsee. — Kieler Meeres-
    forsch. **6**: 3—11.
—   (1950): Das Vordringen limnischer Tierarten in das Meeresgebiet der Nord- und
    Ostsee. — Kieler Meeresforsch. **7**: 5—23.
—   (1952): Die Besiedlung des Sandbodens im Meere und die Bedeutung der Lebens-
    formentypen für die Ökologie. — Verh. Dt. Zool. Ges. Wilhelmshaven 1951: 327
    bis 359.
—   (1955): Die Brackwasser-Submergenz und die Umkomposition der Coenosen in
    Belt- und Ostsee. — Kieler Meeresforsch. **11**, 1: 59—73.
—   (1957): Das Farbstreifen-Watt. — Aus der Heimat **64**: 64—70.
—   (1959): Regionale Verschiedenheiten der Lebewesen gegenüber dem Salzgehalt und
    ihre Bedeutung für die Brackwasser-Einteilung. — Arch. Oceanogr. Limnol. Venezia
    11. Suppl.: 35—46.
—   (1963): Verschiedenheiten der biologischen Entwicklung in Meer und Süßwasser. —
    Veröff. Inst. Meeresforsch. Bremerhaven. Sonderbd.: 122—141.
REMANE, A. & LÄSSIG, G. (1960): Einwirkung des Salzgehaltes auf Vermehrung und Wachs-
    tum bei *Lemna minor*. — Kieler Meeresforsch. **16**: 221—228.
REMANE, A. & SCHULZ, E. (1934): Das Küstengrundwasser als Lebensraum. — Schr. Naturw.
    Ver. Schlesw.-Holst. **20**, 2: 399—408.
REMMERT, H. (1955a): Substratbeschaffenheit und Salzgehalt als ökologische Faktoren für
    Dipteren. — Zool. Jb. **83**: 453—474.
—   (1955b): Ökologische Untersuchungen über die Dipteren der Nord- und Ostsee. —
    Arch. Hydrobiol. **51**: 2—53.
RENTZ, G. (1940): Das Zooplankton der Hiddensee-Rügenschen Boddengewässer und seine
    Produktionsphasen. — Arch. Hydrobiol. **36**: 588—675.
REUTER, M. (1961): Untersuchungen über Rassenbildung bei *Gyratrix hernaphroditus* (Tur-
    bellarea Neorhabdocoela). — Acta Zool. Fenn. **100**: 1—32.
RHUMBLER, L. (1935): Foraminiferen der Kieler Bucht. I. — Schr. Naturwiss. Ver. Schlesw.-
    Holst. **21**: 143—195.
—   (1937): Foraminiferen der Kieler Bucht. II. — Kieler Meeresforsch. **1**: 179—242
—   (1938): Foraminiferen aus dem Meeressand von Helgoland. — Kieler Meeresforsch.
    2: 157—222.
RICHTER, R. (1927): „Sandkorallen"-Riffe in der Nordsee. — Natur u. Museum (Frank-
    furt) **57**.
RIECH, E. (1937): Systematische, anatomische, ökologische und tiergeographische Unter-
    suchungen über die Süßwassermollusken Papuasiens und Melanesiens. — Arch.
    Naturgesch. N. F. **6**: 37—153.
RIEMANN, F. (1965): Turbellaria Proseriata mariner Herkunft aus Sanden der Flußsohle im
    limnischen Bereich der Elbe. — Zool. Anz. **174**: 299—312.
RINGER, W. E. (1907): Over de eingenshappen van het Zuiderzeewater. In: REDEKE, Rapport
    over onderzoekingen betreffende de visscherig in de Zuiderzee. — Bigl. **4**: 3—55.
RIPPEL-BALDES, A. (1952): Grundriß der Mikrobiologie. 2. Aufl.: 1—404.
ROCH, F. (1924): Experimentelle Untersuchungen an *Cordylophora caspia*. — Z. Morph.
    Ökol. Tiere **2**: 350—426.

Röpke, H. (1957): Beitrag zum Chemismus der Unterweser unter besonderer Berücksichtigung der Erdalkalien und ihrer ökologischen Bedeutung. — Veröff. Inst. Meeresforsch. Bremerhaven 5: 103—123.

Rogenhofer, A. (1905): Über das relative Größenverhältnis der Nierenorgane bei Meeres- und Süßwassertieren. — Verh. zool. bot. Ges. Wien 55.

Rottgardt, D. (1952): Mikropaläontologisch wichtige Bestandteile rezenter brackischer Sedimente an den Küsten Schleswig-Holsteins. — Meyniana 1: 169—228.

Rozanska, Z. (1963): The Vistula Firth Zooplankton. (Poln.) — Zesz. Nauk. WSR 16: 278.

— (1964): The Firth of Vistula as a mixohaline body of water and its biological characteristic. — Zesz. Nauk. W. S. R. 17: 339.

— (1966): Plankton in der Bodenwasserschicht der Pucker Wieck. (Poln.). — Zesz. Nauk WSR 21.

Ruffo, S. (1954): Amfipodi di acque in terstiziali raccolte dal Dr. C. Delamare-Deboutteville in Francia, Spagna e Algeria. — Vie et Milieu 4: 669—681.

Ruinen, J. (1938): Notizen über Salzflagellaten. — Arch. Protistenkde. 90: 210—258.

Ruttner, F. (1937): Limnologische Studien an einigen Seen der Ostalpen. — Arch. Hydrobiol. 32.

Rylov, W. M. (1935): Das Zooplankton der Binnengewässer. — Die Binnengewässer 15: 1—272.

Saelen, O. H. (1967): Some features of Hydrography of Norwegian Fjords. — Estuaries (Washington): 63—70.

Sars, G. O. (1903): On the crustacean fauna of Central Asia. — Ann. Mus. St. Petersbourg 8.

— (1927): Notes on the crustacean fauna of the Caspian Sea. — Festschr. Knipowitsch, Leningrad: 315—319.

Sauramo, M. (1929): The quaternary geology of Finland. — Bull. Comm. Geol. Finland, 86.

Schachter, D. (1950): Contribution à l'étude écologique de la Camargue. — Ann. Inst· Oceanogr. 25: 1—108.

Schäfer, H. W. (1933): Unsere Kenntnis der Copepoden-Fauna des Brackwassers nebst Notizen über die Brackwasserfauna von Hiddensee. — Internat. Rev. Hydrobiol. Hydrogr. 29.

— (1936a): Harpacticoiden aus dem Brackwasser der Insel Hiddensee. — Zool. Jb. Syst. 68: 545—588.

— (1936b): Harpacticoiden aus dem Brackwasser der Insel Hiddensee. — Mitt. Naturw. Ver. Neuvorpommern u. Rügen 63: 16—19.

— (1953): Über Meeres- und Brackwasser-Ostracoden aus dem Deutschen Küstengebiet. — Hydrobiologia 5, 4.

Schellenberg, A. (1932): Neue Crustaceen der deutschen Küste. — Zool. Anz. 101: 61—65.

— (1934): Zur Amphipodenfauna der Kieler Bucht. — Schr. Naturw. Ver. Schlesw.-Holst. 20: 129—144.

Schlesch, H. (1937): Bemerkungen über die Verbreitung der Süßwasser- und Meeresmollusken im östlichen Ostseegebiet. — Loodusuurijate Seltsi truanded 43: 37—64.

Schlienz, W. (1923): Verbreitung und Verbreitungsbedingungen der höheren Krebse im Mündungsgebiet der Elbe. — Arch. Hydrobiol. 14: 429—452.

Schlieper, C. (1931): Über das Eindringen mariner Tiere in das Süßwasser. — Biol. Zbl. 51: 401—412.

— (1932): Die Brackwassertiere und ihre Lebensbedingungen vom physiologischen Standpunkt aus betrachtet. — Verh. internat. Ver. Limnol. 6: 113—146.

— (1933): Weitere Untersuchungen über die Beziehungen zwischen Bau und Funktion bei den Excretionsorganen decapoder Crustaceen. — Z. vergl. Physiol. 18: 255—257.

— (1955): Über die physiologischen Wirkungen des Brackwassers. (Nach Versuchen an der Miesmuschel Mytilus edulis.) — Kieler Meeresforsch. 11: 22—23.

Schlieper, C. & Herrmann, F. (1930): Beziehungen zwischen Bau und Funktion bei den Excretionsorganen decapoder Crustaceen. — Zool. Jb. 52: 624—634.

SCHMIDT, P. (1968): Die quantitative Verteilung und Populationsdynamik des Mesopsammons am Gezeitensandstrand der Norseeinsel Sylt. I., II. — Internat. Rev. ges. Hydrobiol. **53**: 723—779.

SCHMIDT, R. (1913): Die Salzwasserfauna Westfalens. — Diss. Münster: 1—69.

SCHMITZ, W. (1959): Zur Frage der Klassifikation der binnenländischen Brackwässer. — Arch. Oceanogr. Limnol. (Suppl.) **11**: 179—226.

SCHNAKENBECK, W. (1938): Heterosomata. — Tierwelt. d. Nord- u. Ostsee **12**, h: 1—60.

SCHULTZE, M. S. (1851): Beiträge zur Naturgeschichte der Turbellarien. — Greifswald, 1—78.

SCHULZ, B. (1932): Einführung in die Hydrographie der Nord- und Ostsee. — Tierwelt. Nord- u. Ostsee **Ic**: 1—88.

SCHULZ, E. (1937): Das Farbstreifenwatt und seine Fauna. — Kieler Meeresforsch. **1**: 359—378.

—   (1939): Über eine Mikrofauna im oberen Eulitoral auf Amrum. — Kieler Meeresforsch. **3**: 158—164.

—   (1950): Zur Ökologie von *Protohydra leuckarti* GREEF. — Kieler Meeresforsch. **7**: 53—57.

—   (1951): Über *Stygarctus bradypus* n. g. sp., einen Tardigraden aus dem Küstengrundwasser und seine phylogenetische Bedeutung. — Kieler Meeresforsch. **8**: 86—97.

SCHULZ, E. & MEYER, H. (1939): Weitere Untersuchungen über das Farbstreifen-Sandwatt. — Kieler Meeresforsch. **3**: 321—336.

SCHULZ, H. (1961): Qualitative und quantitative Planktonuntersuchungen im Elbe-Ästuar. — Arch. Hydrobiol. (Suppl.) **26**: 5—105.

SCHULZ, W. (1954): Zur Biologie von *Enchytraeus albidus* (HENLE). — Z. Wiss. Zool. **157**: 31—78.

SCHÜTZ, L. (1960): Die Hartbodenfauna des Nordostseekanals — ein Brackgewässer — im Jahre 1952/53. — Diss. Univ. Kiel: 1—504.

—   (1963a): Ökologische Untersuchungen über die Benthosfauna im Nordostseekanal. — Internat. Rev. ges. Hydrobiol. **48**: 301—418.

—   (1963b): Die Fauna der Fahrrinne des NO-Kanals. — Kieler Meeresforsch. **9:1** 104—115.

—   (1964): Die tierische Besiedlung der Hartböden in der Schwentinemündung. — Kieler Meeresforsch. **20**: 198—217.

SCHÜTZ, L. & KINNE, O. (1955): Über die Mikro- und Makrofauna der Holzpfähle des Nord-Ostsee-Kanals und der Kieler Förde. — Kieler Meeresforsch. **11**: 110—138.

SCHUURMANS-STEKHOVEN, J. H. (1931): Ökologische und morphologische Notizen über Zuiderseenematoden. — Z. Morph. Ökol. Tiere **20**: 613—677.

—   (1936): De oekologie der Zuiderzee Nematoden. — Med. Zuiderzee Comm. **4**.

SCHUURMANS-STEKHOVEN, J. H., ADAM, W. & PUNT, A. (1935): Ökologische Notizen über Zuiderseenematoden II. — Z. Morph. Ökol. Tiere **29**: 609—666.

SCHWARZ, S. (1959): Vergleichende Studien an tierischem Netz- und Vollplankton aus Brackwassergebieten der Ostsee. — Z. Fischerei Berlin **8** N. F.: 351—370.

—   (1960a): Hydrographisch-meteorologische und biologische Beobachtungen in den Brackwässern um Hiddensee. — Internat. Rev. ges. Hydrobiol. **45**: 327—338.

—   (1960b): Zur Crustaceenfauna der Brackwassergebiete Rügens und des Darß. — Hydrobiologia **16**: 293—300.

SCOTT, K. M. F., HARRISON, A. & MACNAE, W. (1952): The ecology of South African estuaries, t. 2. — Trans. R. Soc. S. Afr. **33**.

SEEMANN, W. (1960): Die Fischerei im Sehlendorfer Binnensee. In K. LILLELUND u. W. SEEMANN: Der Sehlendorfer Binnensee, Teil VI. — Z. Fischerei **9** NF: 604—658.

SEGERSTRÅLE, S. (1928): Quantitative Studien über den Tierbestand der Fucus-Vegetation in den Schären von Pellingen. — Comm. Biol. Soc. Sci. Fenn. **3**, 2: 1—14.

—   (1933): Studien über die Bodentierwelt in südfinnländischen Küstengewässern, I., II., III. — Comm. Biol. **4**, 8: 1—62, 9: 1—74.

— (1937): Studien über die Bodentierwelt in südfinnländischen Küstengewässern, IV. Bestandsschwankungen beim Amphipoden *Corophium volutator*. — Acta Soc. Fauna Flora Fenn. **60**: 245—255.

— (1937): Zur Morphologie und Biologie der Amphipoden *Pontoporeia affinis* nebst einer Revision der Pontoporeia-Systematik. — Comm. Biol. Soc. Sci. Fenn. **7**, 1: 1—183.

— (1939): Ein Planktonprofil Pellinge-Lill-Pernaviken (Südküste Finnlands) im August 1937. — Comm. Biol. Soc. Sci. Fenn. **7**, 10: 1—10.

— (1940): Studien über die Bodentierwelt in südfinnländischen Küstengewässern, VI. Zur Biologie des Amphipoden *Corophium volutator*, nebst Angaben über die Entwicklung und Rückbildung der Oostegitenborsten bei dieser Art. — Comm. Biol. Soc. Sci. Fenn. **7**, 1: 1—40.

— (1941): Einige Amphipodenfunde aus dem baltischen Meeresgebiet Finnlands. — Mem. Soc. Fauna Flora Fenn. **17**: 145—149.

— (1944a): Weitere Beobachtungen über das Vorkommen der Amphipoden *Leptocheirus pilosus* ZADDACH und *Corophium lacustre* VANHÖFFEN an der Südküste Finnlands. — Mem. Soc. Fauna Flora Fenn. **19**: 20—22.

— (1944b): Weitere Studien über die Tierwelt der *Fucus*-Vegetation an der Südküste Finnlands. — Comm. Biol. Soc. Sci. Fenn. **9**, 4: 1—28.

— (1947a): Neue Funde der Amphipoden *Calliopius l. laeviusculus* KRÖYER und *Bathyporeia pilosa* LINDSTRÖM aus dem baltischen Meeresgebiet Finnlands. — Comm. Biol. Soc. Sci. Fenn. **9**, 5: 1—4.

— (1947b): Über die Verbreitung der *Idotea*-Arten im baltischen Meeresgebiet Finnlands. — Comm. Biol. Soc. Sci. Fenn., **9**, 6: 1—6.

— (1947c): Über die Verbreitung der Mysiden in den Finnland umgebenden Meeresgewässern. — Comm. Biol. Soc. Sci. Fenn. **9**, 15: 1—15.

— (1947d): On the occurrence of the Amphipod *Gammarus duebeni* LILLJ. in Finnland, with notes on the ecology of the species. — Comm. Biol. Soc. Sci. Fenn. **9**, 18: 1—27.

— (1949): The brackish-water Fauna of Finnland. — Oikos **1**: 127—141.

— (1950): The amphipods on the coasts of Finland — some facts and problems. — Comm. Biol. Soc. Sci. Fenn. **10**: 1—28.

— (1951a): The recent increase in salinity of the coasts of Finland and its influence upon the fauna. — J. Cons. Internat. l'explor. Mer. **17**: 103—110.

— (1951b): The seasonal fluctuations in the salinity of the coasts of Finland and their biological significance. — Comm. Biol. Soc. Sci. Fenn. **13**, 5: 1—27.

— (1953): Further notes on the increase in salinity of the inner Baltic and its influence on the fauna. — Comm. Biol. Soc. Sci. Fenn. **13**, 15: 1—7.

— (1956): The distribution of glazial relicts in Finland and adjacent Russian areas. — Comm. Biol. Soc. Sci. Fenn. **15**, 18: 1—35.

— (1957a): Baltic Sea. — Mem. geol. Soc. Amer. **67**: 1—32.

— (1957b): On the immigration of the glacial relicts of Northern Europe, with remarks on their prehistory. — Comm. Biol. Soc. Sci. Fenn. **16**: 1—117.

— (1958a): On an isolated finnish Population of the relict Amphipod *Pallasea quadrispinola* exhibiting striking morphological reductions. — Comm. Biol. Soc. Sci. Fenn. **17**, 5: 1—33.

— (1958b): A quarter century of brackishwater research. — Verh. internat. Ver. Limnol. **13**: 646—671.

— (1959a): Synopsis of data on the crustaceans *Gammarus locusta*, *Gammarus oceanicus*, *Pontoporeia affinis* and *Corophium volutator*. — Comm. Biol. Soc. Sci. Fenn. **20**: 1—23.

— (1959b): Brackishwater classification, a historical survey. — Arch. Oceanogr. Limnol. **11** (Suppl.): 7—33.

— (1960a): Fluctuations in the abundance of benthic animals in the Baltic area. — Comm. Biol. Soc. Sci. Fenn. **23**: 1—19.

— (1960b): Investigations on Baltic populations of the bivalve *Macoma baltica* (L.). — Comm. Biol. Soc. Sci. Fenn. **23**, 2: 1—72.

— (1962): The immigration and prehistory of the glacial relicts of Eurasia and North America. A survey and discussion of modern views. — Internat. Rev. ges. Hydrobiol. **47**: 1—25.

— (1964a): Literature on marine biology in the Baltic area published in the years 1953 to 1962. — Comm. Biol. Soc. Sci. Fenn. **27**, 3: 1—44.

— (1964b): Marine Zoology in the Baltic Area in 1953—1964. — Oceanogr. Mar. Biol. Ann. Rev. **2**: 373—392.

— (1965a): On the salinity conditions of the south coast of Finland since 1950. — Comm. Biol. Soc. Sci. Fenn. **28**, 7: 1—28.

— (1965b): Biotic factors affecting the vertical distribution and abundance of the bivalve *Macoma baltica* in the Baltic Sea. — Bot. Gothoburg **3**: 195—204.

SEIFERT, R. (1933): Beiträge zur Kenntnis der Bodenfauna der Gewässer um Hiddensee. — Mitt. Naturw. Ver. Vorpommern-Rügen **60**: 1—21.

— (1938): Die Bodenfauna des Greifswalder Boddens. — Ein Beitrag zur Ökologie der Brackwasserfauna. — Z. Morph. Ökol. Tiere **34**: 132—156.

SERNOW, S. A. (1909—1911): Grundzüge zur Verbreitung der Tierwelt des Schwarzen Meeres bei Sebastopol. Abt. I. Benthos, Abt. II. Plankton. — Internat. Rev. ges. Hydrobiol. Hydrogr. **2**: 99—152, **3**: 299—305.

— (1913): Zur Frage der Untersuchung des Lebens im Schwarzen Meere. — Mém. Acad. imp. Sci. St. Petersbourg. Ser. III, **32**, 1 (russ.).

SICK, F. (1931): Die Fauna der Meerestümpel des Bottsandes (Kieler Bucht). — Arch. Naturgesch. N. F. **2**: 54—96.

SIDOROFF, S. A. (1929): Les mollusques du lac de Aral et de ses proches environs. — Russ. Hydrobiol. Z. **8**.

SIEWING, R. (1956): *Thermobathynella amyxi* nov. spec. aus dem Brackwasser der Amazonasmündung. — Kieler Meeresforsch. **12**: 114—119.

SILEN, L. (1943): Notes on Swedish marine Bryozoa. — Ark. Zool. **35 A**.

SJÖSTEDT, L. S. (1936): Beiträge zur Hydrographie des Sundes. — Undersökn. över Öresund **22**.

SKLOWER, A. (1930): Die Tierwelt vor der samländischen Ostseeküste und ihr Zusammenschluß zu Lebensgemeinschaften. — Zool. Anz. **92**: 254—366.

SMIDT, E. L. B. (1951): Animal Production in the danish Waddensea. — Medd. Komm. Danmarks Fiskeri Havundersög. Ser. Fiskeri **11**: 1—151.

SMITH, R. I. (1955 a): Comparison of the level of chlorid regulation by *Nereis diversicolor* in different parts of its geographical range. — Biol. Bull. **109**: 453—474.

— (1955b): Salinity variation in interstitial water of sand at Kames Bay, Millport with reference to the distribution of *Nereis diversicolor*. — J. Mar. Biol. Ass. Plymouth **34**: 33—46.

— (1956a): The ecology of the tamar estuary. VII. Observations on the interstitial salinity of intertidal muds in the estuarine habitat of *Nereis diversicolor*. — J. Mar. Biol. Ass. Plymouth **35**: 81—104.

— (1956 b): The tolerance of low salinities by Nereid Polychaetes and its relation to temperature and reproduction habit. — Internat. Confer. Mar. Bil. Roscoff: 93—107.

— (1964): On the early development of *Nereis diversicolor* in different salinities. — J. Morph. **114**: 437—451.

SPÄRCK, R. (1935): On the importance of quantitative investigations of the bottom fauna in marine biology. — J. Conseil internat. Explor. Mer **10**: 1—19.

— (1936): Bundfaunaen i Ringköbing Fjord i Brakvandsperioden 1915—1931. — Faunaen: Ringköbing Fjords sidste Brackvandsperiode sammenlignet med andre nordiske Brackvandsomrader. — Ringköbing Fjord Naturhist. i Brakvandsperioden 1915—1931. — Kopenhagen 1936.

— (1938): Über die Zoogeographische Bedeutung der Petersenschen Tiergemeinschaften. — Zoogeographica **3**: 132—144.

SPOONER, G. M. (1947): The distribution of *Gammarus* species in estuaries. — J. Mar. Biol. Ass. Plymouth **27**: 1—52.

SPRUIT, C. J. P. & PIJPER, A. (1952): An obligate halophilic bacterium from solar salt. — J. Microbiol. Serol. **18**.

STAMMER, H. J. (1928): Die Fauna der Ryckmündung, eine Brackwasserstudie. — Z. Morph. Ökol. Tiere **11**: 36—109.

— (1930): Eine neue Höhlenphäromide aus dem Karst, *Monolistra (Typhlopharoma) schottlaendri*, und die Verbreitung des Genus *Monolistra*. — Zool. Anz. **88**: 291—304.

— (1932): Die Fauna des Timavo. — Zool. Jb. **63**: 521—656.

— (1935): Ein neuer Höhlenschizopode, *Troglomysis vjetrenicensis* n. g. n. sp., zugleich eine Übersicht der bisher aus dem Brack- und Süßwasser bekannten Schizopoden. — Zool. Jb. Syst. **68**: 53—104.

STATHER, F. (1930): ,,Rote Verfärbung" und ,,Rote Erhitzung" auf gesalzenen Häuten. — Collegium **720**.

STEEMANN-NIELSEN, E. (1940): Die Produktionsbedingungen des Phytoplanktons im Übergangsgebiet zwischen der Nord- und Ostsee. — Medd. Komm. Danmarks Fisk. Havunders. Ser. Plankton **3**, 4: 1—55.

— (1951): The marine vegetation of the Isefjord. — A study on ecology and production. — Medd. Komm. Danm. Fisk. Havunders. Ser. Plankton **5**: 1—114.

STEPHEN, A. (1929): Studies on the Scottish marine Fauna: The fauna of the sandy and muddy areas of the tidalzone. — Trans. roy. Soc. Edinburgh **56**: 291—306.

— (1930): Studies on the Scottish marine fauna: Additional observations on the fauna of the sandy and muddy areas of the tidalzone. — Trans. roy. Soc. Edinburgh **56**: 521—535.

STEPHENSEN, K. (1910): Danmarks Fauna. Storkrebs I. Malacostraca: 1—193.

— (1928): Danmarks Fauna. Storkrebs II. Amphipoda: 1—399.

STEPHENSON, T. A. (1939): The constitution of the intertidal fauna and flora of South Africa I. — J. Linn. Soc. London: 487—536.

STERRER, W. (1965): Zur Ökologie der Turbellarien eines südfinnischen Sandstrandes. — Botanica Gothoburg **3**: 211—219.

STEUER, A. (1910): Planktonkunde. — Leipzig u. Berlin: 1—723.

— (1942): Ricerche idrobiologiche alle foci del Nilo. — Mem. Ist. Ital. Idrobiol. **1**: 85—106.

STOCK, J. H. (1952): Some notes on the taxonomy, the distribution and the ecology of four species of the amphipod genus *Corophium*. — Beaufortia **2**, 21: 1—10.

STRENZKE, K. (1951a): Notizen über die Milben und Collembolen der unterirdischen Feuchtzone des Nord- und Ostseestrandes. — Kieler Meeresforsch. **8**: 82—132.

— (1951b): Chironomiden von der bulgarischen Küste des Schwarzen Meeres. — Arch. Hydrobiol. Suppl. **18**: 678—691.

STUNDE, K. (1939): Limnologische Untersuchungen von Salzgewässern und Ziehbrunnen im Burgenland (Niederdonau). — Arch. Hydrobiol. **34**.

SUOMALAINEN, H. (1939): Beiträge zur Kenntnis der Verbreitung der marinen Bodentierwelt im östlichen Teil des finnischen Meerbusens. — Ann. Zool. Soc. Zool. Bot. Fenn. Vanamo **6**, 6: 1—13.

SVENONIUS, B. (1949): Über die Hydracarinenfauna im Bottenviken und im angrenzenden Küstengebiet. — Entomo. Tidskr. **70**, 4 (Uppsala): 253—262.

SZIDAT, L. (1956): Der marine Charakter der Parasitenfauna der Süßwasserfische des Stromsystems des Rio de la Plata und ihre Bedeutung als Reliktfauna des tertiären Tethys-Meeres. — XIV. Internat. Congr. Zool. Proc. Kopenhagen.

TATTERSALL, B. S. (1954): Report on a small collection of mysidacea from estuarine waters of South Africa. — Trans. R. Soc. South Africa **33**.

THAMDRUP, E. (1935a): Untersuchungen über die Grundwasserverhältnisse auf Skallingen. — Geogr. T. **38**, Kopenhagen.

— (1935b): Beiträge zur Ökologie der Wattenfauna auf experimenteller Grundlage. — Medd. Komm. Danmarks Fisk. Havunders. Ser. Fiskeri **10**: 1—125.

THIEMANN, K. (1934): Das Plankton der Flußmündungen. — Wiss. Ergebn. d. Dt. Atlant. Exped. a. d. Forschungs- u. Verm.-Schiff ,,Meteor" 1925—1927: 199—273.

THIENEMANN, A. (1913): Die Salzwassertierwelt Westfalens. — Verh. Dt. Zool. Ges. **23**: 56—67.

— (1915): Zur Kenntnis der Salzwasserchironomiden. — Arch. Hydrobiol. Suppl. **2**: 443—471.

— (1925a): *Mysis relicta*. — Z. Morph. Ökol. Tiere **3**: 389—440.

— (1925b): Das Salzwasser von Oldesloe. — Mitt. geogr. Ges. Lübeck **2** (1925): 53—195, (1926): 27—195.

— (1928): Die Reliktenkrebse *Mysis relicta, Pontoporeia affinis, Pallasea quadrispinosa* und die von ihnen bewohnten norddeutschen Seen. — Arch. Hydrobiol. **19**: 522 bis 582.

— (1937): Haffmücken und andere Salzwasser-Chironomiden. — Kieler Meeresforsch **1**: 167—178.

— (1950): Verbreitungsgeschichte der Süßwassertierwelt Europas. — Die Binnengewässer **18**: 1—809.

— (1954): Chironomus. — Die Binnengewässer **20**: 1—834.

THOR, S. (1928): Untersuchungen der von H. J. STAMMER im Brackwasser des Parkteiches und des Ryckflusses bei Greifswald gesammelten Acarina. — Z. Morph. Ökol. Tiere **11**: 104—114.

THORSON, G. (1946): Reproduction and larval development of Danish marine bottom invertebrates. — Medd. Komm. Danm. Fisk. Havunders. Ser. Plankton **4**, 1: 1—523.

THULIN, G. (1922): Bottenboniteringar i. S. Östersjön i samhand med fisktralningar. — Svensk. Hydrogr. Biol. Komm. Schr. **6**.

THUST, R. (1964): Zur Ökologie der Cladoceren und Copepoden in den Darsser Boddengewässern. — Limnologica **2**: 337—348.

TRAHMS, O. (1937): Zur Kenntnis der Salzverhältnisse und des Phytoplanktons der Hiddenseer und Rügenschen Boddengewässer. — Arch. Hydrobiol. **32**: 75—90.

— (1939a): Das Plankton des Großen Jasmunder Boddens. — Arch. Hydrobiol. **35**.

— (1939b): Beiträge zur Ökologie küstennaher Brackwässer. 2. Die Bodenfauna und Bodenflora des Großen Jasmunder Boddens. — Arch. Hydrobiol. **36**: 529—551.

— (1939c): Die Größen- und Kalkreduktion bei *Mytilus edulis* L. in Rügenschen Binnengewässern. — Z. Morph. Ökol. Tiere **35**: 246—250.

— (1941): Die Bedeutung des Untergrundes für die Nährstoffregeneration in stark eutrophen Brackwässern. — Z. Fischerei **39**.

TRAHMS, O. & STOLL, K. (1938): Hydrobiologische und hydrochemische Untersuchungen in den Rügenschen Boddengewässern während der Jahre 1936 und 1937. — Kieler Meeresforsch **3**: 61—98.

UCHIDA, T. (1951): A brackish-water medusa from Japan. — J. Fac. Sci., Hokkaido Univ. Ser. VI, Zool. **10**: 161—162.

ULRICH, W. (1926): Über das Vorkommen der *Victorella pavida* KENNT und einiger anderer Bryozoen im Brackwasser des Rostocker Hafens. — Z. Morph. Ökol. Tiere **5**: 559 bis 576.

VÄLIKANGAS, J. (1926): Planktologische Untersuchungen im Hafengebiet von Helsingfors 1. — Acta Zool. Fenn. **1**: 1—277.

— (1933): Über die Biologie der Ostsee als Brackwassergebiet. — Verh. internat. Ver. Limnol. **6**: 62—112.

VALKANOW, A. (1935): Notizen über die Brackwässer Bulgariens I. — Ann. Univ. Sofia, Phys.-math. Fac. **34**, 3: 249—303.

— (1936): Notizen über die Brackwässer Bulgariens II. Versuch einer hydrographischen und biologischen Erforschung derselben. — Ibid. **32**, 3: 209—341.

— (1937): Die Varnaseen. Beitrag zur Hydrographie und Biologie derselben. — Mitt. Bulg. Geogr. Ges. **4**: 118—139.

— (1938): Übersicht der Hydrozoenfamilie Moerisidae. — Jb. Univ. Sofia, Pys.-math. Fak. **34**: 251—320.

— (1955): Revision der Hydrozoenfamilie Moerisidae. — Arb. Biol. Meeresstat. Stalin. **18**: 33—47.

— (1957): Katalog unserer Schwarzmeerfauna. — Arb. Biol. Meeresstation Varna (Bulgarien) **19**: 1—27.

VANHÖFFEN, E. (1911): Beiträge zur Kenntnis der Brackwasserfauna im Frischen Haff. — Sitz.-Ber. Ges. naturf. Freunde Berlin: 399—405.

VIBE, CHR. (1939): Preliminary investigations on shallow water animal communities in the Upernavik and Thule districts. — Medd. Grönland **124**, 2: 1—42.

VIETS, K. (1925): In: THIENEMANN 1925 b.

VOGEL, R. (1927): Über drei an Salzwasser angepaßte Insektengattungen an der östlichen Mittelmeerküste. — Internat. Rev. Hydrobiol. Hydrogr. **17**.

VORSTMANN, A. G. (1933a): Über die Biologie von *Congeria cochleata* NYST. — Zool. Anz. **102**: 240—242.

— (1933b): Zur Biologie der Brackwassermuschel *Congeria cochleata* NYST. — Verh. internat. Ver. Limnol. **6**: 182—186.

— (1935): Biologische Notizen betreffs der sessilen Fauna im Hafen der Stadt Amsterdam. — Zool. Anz. **109**: 76—80.

WAERN, M. (1952): Rocky shore Algae in the Öregrund Archipelago. — Acta Phytogeogr. Suecica **30**: 1—298.

WATKIN, E. D. (1939): The pelagic phase in the life history of the amphipod genus *Bathyporeia*. — J. Mar. Biol. Ass. Plymouth **23**: 467—481.

WATTENBERG, H. (1935): Aufgaben der chemischen Meeresforschung. — Z. ges. Naturw. **6**: 220—230.

— (1949a): Die Salzgehaltsverteilung in der Kieler Bucht und ihre Abhängigkeit von Strom- und Wetterlage. — Kieler Meeresforsch. **6**: 17—30.

— (1949b): Entwurf einer natürlichen Einteilung der Ostsee. — Kieler Meeresforsch. **6**: 10—17.

WATTENBERG, H. & MEYER, H. (1936): Der jahreszeitliche Gang des Gehaltes des Meerwassers an Planktonnährstoffen in der Kieler Bucht im Jahre 1935. — Kieler Meeresforsch. **1**: 264—278.

WESTBLAD, E. (1930): Über die Geschlechtsorgane und die systematische Bedeutung von *Protohydra leuckarti* GREEF. — Arch. Zool. **21** A: 1—13.

WESTHEIDE, W. (1966): Zur Polychaetenfauna des Eulitorals der Nordseeinsel Sylt. — Helgol. wiss. Meeresunters. **13**: 207—213.

WIESER, W. (1964): Biotopstruktur und Besiedlungsstruktur. — Helgol. Wiss. Meeresunters. **10**: 359—376.

WIGGLESWORTH, B. (1933a): The effect of salts on the anal gills of the mosquito larva. — J. exper. Biol. **10**: 1—15.

— (1933b): The adaptation of mosquito larvae to salt water. — J. exp. Biol. **10**.

WILKE, U. (1954): Mediterrane Gastrotrichen. — Zool. Jb. **82**, 6: 497—550.

WILL, L. (1913): *Acaulis primarius* STIMPSON. — Ein neuer Ostseebewohner. — Abh. Naturf. Ges. Rostock **5**.

WILLER, A. (1925): Studien über das Frische Haff I. — Z. Fischerei **23**: 317—349.

— (1931): Vergleichende Untersuchungen an Standgewässern. — Verh. internat. Ver. Limnol. **5**.

WILLER, A. & QUEDNAU, W. (1934): Untersuchungen über den Lachs *(Salmo salar* L.*)* III. — Z. Fischerei **32**, 3: 533—555.

WILLMANN, C. (1935): Über eine eigenartige Milbenfauna im Küstengrundwasser der Kieler Bucht. — Schr. Naturw. Ver. Schlesw.-Holst. **20**: 422—434.

WILSON, D. C. (1935): Life of the shore and shallow sea. — London: 1—213.

WITTIG, H. (1940): Über die Verteilung des Kalziums und der Alkalinität in der Ostsee. — Kieler Meeresforsch. **3**: 460—496.

WOHLENBERG, E. (1934): Biologische Landgewinnungsarbeiten im Wattenmeer. — Der Biologe.

— (1937): Die Wattenmeer-Lebensgemeinschaften im Königshafen von Sylt. — Helgoländer Wiss. Meeresunters. **1**: 1—92.

— (1939): Die Nutzanwendung biologischer Erkenntnisse im Wattenmeer zugunsten der praktischen Landgewinnung an der deutschen Nordseeküste. — Cons. perm. internat. Expl. Mer Rapp. Proc. Verb. **109**, III, 24: 125—132.

— (1956): Die Versalzung im Gotteskoog (Nordfriesland) nach biologischen und chemischen Untersuchungen. — Die Küste, Heide **5**: 113—145.

Worobjew, V. P. (1949): Benthos of the Sea of Azov. — T. Azovo-Chernom. Inst. Rybn. Khoz. **13**: 5—193.

Wüst, G., Brogmus, W. & Noodt, E. (1954): Die zonale Verteilung von Salzgehalt, Niederschlag, Verdunstung, Temperatur und Dichte an der Oberfläche der Ozeane. — Kieler Meeresforsch. **10**: 137—161.

Wüst, G. & Brogmus, W. (1955): Ozeanographische Ergebnisse einer Untersuchungsfahrt mit dem Forschungskutter „Südfall" durch die Ostsee Juni und Juli 1954 (anläßlich der totalen Sonnenfinsternis auf Öland). — Kieler Meeresforsch. **11**: 3—21.

Wyrtki, K. (1954): Der große Salzeinbruch in die Ostsee im November und Dezember 1951. — Kieler Meeresforsch. **10**: 19—25.

Zacharias, O. (1889): Zur Kenntnis der Fauna des Süßen und Salzigen Sees bei Halle. — Z. wiss. Zool. **46**: 217—232.

Zenkevitch, L. (1933): Beiträge zur Zoogeographie des nördlichen Polarbassins im Zusammenhang mit der Frage über dessen paläogeographische Vergangenheit. — Zool. J. **12**: 17—34.

— (1935): Some observations on fouling in Ekaterininskaga Bay (Kolafjord, Barents Sea). — Bull. Soc. Nat. Moskau **44**.

— (1938): The influence of Caspian and Black Sea of different concentrations upon some commun Black-Sea invertebrates II. — Zool. J. **17** (Moskau) (Russ. with Engl. Summary): 976—1000.

— (1940): Sur l'aclimation dans la mer Caspienne de nouveaus invertébré alimentaires (pour les poissons) et sur les prémisses théoriques concernant cette aclimation. — Bull. MOIP, biol. **49**, 1.

— (1956): Neue Vertreter der Mittelmeerfauna im Kaspischen Meer. — XIV. Internat. Congr. Zool. Proc. Kopenhagen.

— (1959): The classification of brackish-water basins as exemplified by seas of the USSR. —Arch. Oceanogr. Limnol. **10** (Suppl.).

— (1963): Biology of the seas of the USSR. — London: 1—955.

Zmudzinski, L. (1963): The resources of Zoobenthos in the western part of Gdansk Bay. — Cons. Perm. Expl. Mer Ann. Biol. **20**: 99—100.

Zobell, C. E. & Rittenberg, S. C. (1948): Sulphate-reducing bacteria in marine sediments. — J. Mar. Res. **7**: 602—617.

Zuelzer, M. (1910): Über den Einfluß des Meerwassers auf die pulsierende Vakuole. — Arch. Entwicklungsmech. **29**: 632—640.

Part II

Physiology of brackish water
(Physiological features of life in brackish water)

by

Carl Schlieper

Kiel

With 84 Figures and 43 Tables in the text and on 1 Folder

1. Introduction

No aquatic organism can evade the influence of its surroundings. This applies not only to the inhabitants of sea and fresh water, but is also true for all species living in brackish water. Compared with fresh water, brackish water contains a large amount of inorganic salts. Consequently, this medium exerts a marked influence on the water economy and mineral economy of brackish-water animals.

In order to describe some of the fundamental correlations between the life processes of brackish-water inhabitants and the composition of their external medium it will be necessary to define and delimit brackish water; ecologists are confronted with the same problem (cf. part I, p. 4ff). According to GURNEY (1928/29) brackish water should simply be defined as dilute sea water. REDEKE (1933) states: "Brackwasser ist ein Gemisch von Süsswasser und Meerwasser s. str." [brackish water is a mixture of fresh water and sea water s. str.]. Like SCHMITZ (1959) I would consider as brackish waters in general all those natural bodies of water whose salt content lies between that of the oceanic sea water and that of pure fresh water. According to this definition brackish water is found in marine marginal areas (estuaries, marginal seas, lagoons, tide pools and splash water pools) as well as in land-locked or continental saline waters — provided their salt concentration is less than that of sea water. REDEKE puts the upper limit of salinity for brackish water at about $30^0/_{00}$ and the lower limit at about $0.2^0/_{00}$. REMANE (1940), however, suggests that the lower limit of salinity should be established at $0.5^0/_{00}$. At a Symposium in Venice in 1958 a small group of biologists agreed to propose a minimum salinity of $0.5^0/_{00}$ for marine brackish water and $30^0/_{00}$ as the upper limit against sea water; but it still remained debatable whether such more or less arbitrarily chosen limiting values are of equal validity in all circumstances and in all geographical latitudes (see also PRICE & GUNTER 1964).

From the point of view of the physiologist no clear-cut salinity boundaries can be established for the brackish-water region. For him the region of brackish water begins when marine species show the first signs of physiological reactions to a dilution of their external medium; and the region terminates when fresh-water organisms begin to exhibit the first signs of their metabolism being influenced by an increase in the salinity of the external medium. These physiological boundaries approximate to the ecological limits mentioned above! Three groups of organisms occur in brackish water: the marine immigrants, the specific brackish-water species and the immigrants from fresh water; the physical and chemical properties of their external medium have entirely different effects on each of these. But in all three these factors mainly influence metabolic processes. In the following chapters I shall try to describe this dependence and these correlations for representative examples; adaptations in physiological resistance and performance, vital for the success and survival of the species in brackish water, will be particularly stressed. But I have deliberately refrained from attempting a survey of all the literature on the subject in the form of a synopsis.

As an animal physiologist, my examples have been mainly drawn from the animal kingdom, but some important botanical and microbiological work has been included. In this connection I refer the reader to GESSNER (1959) whose second volume of "Hydrobotanik" contains a detailed exposition of the physiological basis of plant distribution in brackish water.

Some continental saline waters present a special physiological problem. As long as the ions are in the same proportions as in sea water, conditions found in estuaries for example, they will influence the metabolism of the animals living in the area in the same way. When the proportions are not the same, then effects that depend on which ion predominates will become evident.

Since the first edition of this work appeared, numerous new investigations have been published dealing with measurements and experiments on the physiology of brackish-water organisms. About 250 have been used in this revised edition. I have been helped by the reports of the conferences on brackish water and estuaries of Venice (Italy, 1958) and of Jekyll-Island (Georgia, U. S. A. 1964) and in particular by the reviews by PEARSE & GUNTER (1957), R. I. SMITH (1959), SCHWENKE (1960), SHAW (1961), SCHLIEPER (1960, 1964), POTTS & PARRY (1964), KINNE (1964, 1966) and VERNBERG (1967).

2. Salinity ranges and osmotic resistance

The problem whether a brackish-water species is able to exist in a small or in a larger range of salinities can be studied in two ways. On the one hand the biological limits of salinity can be established by studying the natural distribution of the species. I would call this the ecological method. In applying it it must be borne in mind that the occurrence of a species in a certain brackish-water biotope is determined not only by the salt content of the medium, but by other factors as well (temperature, pH, oxygen, ionic ratios). One must also take into account that the salinity range of a species may be less in biotopes where fluctuations are great than in biotopes where they are small. On the other hand the range of osmotic resistance of one or several populations of the species in question may be determined experimentally, in the laboratory, by producing more or less rapid changes of the salt content of the external medium. I would call this the physiological method. An extension of this method consists in long-term breeding experiments by which the reproductive limits of the species can also be established.

MARS (1950) among others, has used the "ecological method". By carrying out observations in the Etangs (lagoons) on the French coast of the Mediterranean he attempted to find out the extent of "euryhalinity" that is the biological salinity range of a number of marine molluscs (see table 1). MARS, however, is of the opinion that a wider range of salinities might be obtained if the osmotic resistance of the same species were determined by means of experiments in which the salt concentration of the medium is slowly changed. In the following paragraphs we shall see in how far this view has been supported by laboratory experiments.

The measure of osmotic resistance can be expressed as survival time in the experiment, depending on the salt content of the external medium, provided the other conditions (temperature, oxygen tension etc.) are either optimal or clearly defined.

Table 1. Biological salinity limits or "degree of euryhalinity" of some species of molluscs from the French coast of the Mediterranean. After P. MARS 1950.

Degree of euryhalinity	Species	Salinity range in ‰ S a b	Ratio a/b	
0	Theoretical degree: marine stenohaline species which cannot tolerate any change in salinity		∞	
1	Stenohaline species which do not penetrate into brackish water *Muricopsis blainvillei*	37 ± 2	18,5	12
2	*Rissoa ventricosa*	34 ± 4	8.5	8
3	*Cardium tuberculatum* *Murex trunculus*	32 ± 6	5.33	5
4	*Chlamys glabra* *Ostrea edulis* *Ocinebra erinaceus*	30 ± 9	3.33	3
5	*Tapes aureus* *Gibbula adamsoni*	27 ± 12	2.25	2
6	*Rissoa labiosa* *Mytilus galloprovincialis* *Loripes lacteus*	23 ± 13 25 ± 13 28 ± 15	1.77 1.92 1.86	1.5
7	*Brachydontes marioni*	24 ± 18	1.33	1.2
8	*Cardium edule*	32 ± 28	1.14	
9	Theoretical degree: marine species which tolerate maximal changes in salinity	30 ± 30	1.00	

Laboratory experiments by MANIGGAULT (1932) in Concarneau (Atlantic coast of France) may be quoted as an example of such measurements; they relate to the viability of the marine flat periwinkle *Littorina obtusata* in sea water of varying salinities. At room temperature individual snails from normal sea water were transferred into glass containers with dilute or with more concentrated sea water and their behaviour was observed. With a considerable change of salinity the snails promptly withdrew into their shell and remained in that condition until the moment of their death. At the moment of death the foot of the snail usually protruded somewhat from the shell. The snail was considered dead when a prick with a needle into the sole of the foot failed to produce any further contraction. By means of this method MANIGGAULT observed that in fresh water *Littorina obtusata* survives for about a day. Above 10‰ salinity (S) survival time increased rapidly. At 17‰ S the snail lived for an average of four days. Above 21‰ S the snails behaved normally during the whole period of observation which always lasted about six days. Thus 21‰ S represents the lower limit of osmotic resistance for the specimens of *Littorina obtusata* which had been collected at Concarneau, on the French coast of Atlantic. This does not correspond to the lower limit of occurrence of this species in other geographical

areas; according to SCHLESCH (1937) *L. obtusata* can tolerate a dilution of sea water up to 12—13⁰/₀₀ S in the Baltic. The reason for this rather considerable difference might be due to the fact in the experiments carried out by the French author the snails were transferred suddenly into the various concentrations to be tested. Had there been a gradual transfer with slowly increasing dilution of the experimental medium the lower limit of osmotic resistance — as determined experimentally — might have had a markedly lower value. The upper limit of osmotic resistance as established by MANIGGAULT was about 50⁰/₀₀ S (see fig. 1).

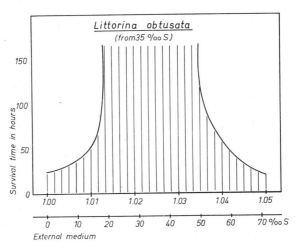

Fig. 1. Experimentally determined limits of osmotic resistance of the gastropod *Littorina obtusata* L. After MANIGGAULT 1932.

Experiments by MOTWANI (1955) on *Mytilus edulis* have established a lower limit of resistance which by no means coincides with the limit of distribution in the Baltic where the same species is still found in brackish water of 4—6⁰/₀₀ S. The mussels examined by MOTWANI came from the Blyth estuary (Northumberland, England) where salt concentrations of 23—33⁰/₀₀ S had been registered in the course of a year. After sudden changes in the salinity of the external medium the mussels first closed their shells and opened them again after 1—50 minutes. The greater the change in salinity, the longer was the period during which the temporary closing of the valves was maintained. In spite of this protective reaction that considerably slowed down an equilibration between internal and external medium, the mussels only tolerated salinities of 20—40⁰/₀₀. In brackish water of 10⁰/₀₀ S they only survived for a few days! In contrast, RANADE (1957) reports on laboratory experiments on *Tigriopus fulvus*, a harpacticid copepod which is widespread on European coasts. It showed normal survival times at salinities of 4.2—90⁰/₀₀. The experimental animals had been collected from rock pools where the naturally occurring salinity fluctuations were within the limits of resistance measured in the laboratory. WELLS (1961) studied the epifauna of the oyster banks in the estuary of Newport River; the distribution of 18 out of 20 species studied carefully appeared to coincide with the salinity which

laboratory experiments showed to be near the limits of tolerance. Only 2 species showed more marked discrepancies.

Identical results in such experiments can be expected only when the material is identical in age and time and place of capture and has been pretreated — as far as possible — in a similar manner.

Osmotic resistance can be specially affected by the salt content of the habitat water. It often happens that individuals of the same species which have grown up in brackish water of low salinity are able to tolerate greater dilutions than those coming from brackish water of higher salinity or from sea water. TOPPING & FULLER (1942) examined 14 marine species (*Gammarus* spec., *Nereis virens, Mya arenaria, Mytilus edulis* etc.) in the Narraguagus Bay (USA) from a biotope situated in the open sea and from brackish-water biotope (river mouth). Results showed that individuals coming from the brackish-water biotope were able to tolerate considerably larger dilutions though in the experiments with both groups the salinity of the external medium was reduced slowly and gradually. Similarly, ANDERSON & PROSSER (1953) found that individuals of the blue crab *Callinectes sapidus* obtained from brackish water of low salinity survived in sea water, diluted to 1/10, longer than those crabs which came from a higher salinity. If, however, both groups had been preadapted to sea water of normal salinity for a week before measurements of their osmotic resistance were taken, their behaviour was very similar.

Similar, non-genetic adaptations of resistance have been produced experimentally in *Zoothamnium hiketes* by PRECHT & LINDNER (1966). Colonies of this euryhaline peritrich ciliate were kept in sea and brackish water of about 30, 19 and $6^0/_{00}$ S for 7—11 days. The lower limits of resistance were established in the following way: Groups of preadapted animals were directly transferred into test media of about 6.3 and $0^0/_{00}$ S and after 30 minutes they were returned to the original media. After a further half hour the number of surviving specimens was determined. The values thus obtained showed conclusively that the position of the low limit of osmotic resistance depends on the salinity of the medium in which the period of preadaptation was spent.

NEUMANN (1961) was able to show and analyze differences in salinity range due to environmental conditions; he investigated two populations of the Nerite *Theodoxus fluviatilis* which occurs in fresh and in brackish water (up to $18^0/_{00}$ S). The home of one population was a fresh-water stream (Kossau, Holstein), while the other came from the brackish water of the Kiel Canal (Rendsburg $5—9^0/_{00}$ S). If the snails from both populations were preadapted to $0.2^0/_{00}$ S for 20 days, the upper salinity limit of the fresh-water animals was at $11^0/_{00}$ and that of the individuals from brackish water at $17^0/_{00}$ S. If, however, experimental animals from both populations were preadapted to $6.5^0/_{00}$ S, the upper salinity limit of the fresh water animals was displaced to $18^0/_{00}$ S and that of the brackish-water animals to $21^0/_{00}$ S. With abrupt changes the range of resistance depended on pretreatment and the origin of the animals. If, on the other hand, the salt content of the experimental medium was increased slowly and gradually (every four days by $2^0/_{00}$ S) the differences in resistance — as established in the experiments — disappeared: The animals from both populations attained the same upper salinity limit of $24^0/_{00}$ S. Thus no indication of a genetically controlled, physiological differentiation of the two populations proved.

Besides, the relationship between the osmotic resistance of a species and the salinity in its place of occurrence may be directly influenced by migratory or avoidance reactions of individuals of the species in question; this may happen both in nature and under experimental conditions. Thus diatoms from the tidal zone may withdraw into lower strata when the surface of the mud is uncovered by the tide. In this way they can avoid an increase in salinity caused by evaporation or a lowering produced by precipitation (GESSNER 1959). Corresponding migratory movements released by osmotic changes may assume the character of preferential reactions in brackish-water animals inhabiting a biotope with a salinity gradient. JANSSON (1962) found that two species of Oligochaeta from the moist sandy soil of the upper litoral always actively search for the salt content best suited to them. On the Baltic coast near Stockholm where there is a continous decrease of salinity from the shore towards the land *Aktedrilus monospermaticus* occurs in the interstitial soil water, closely above the water line at about $6^0/_{00}$ S, while *Marionina preclitellochaeta* prefers a

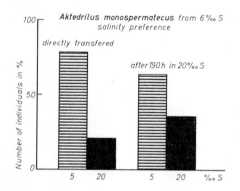

Fig. 2. Salinity preference of the Oligochaete *Aktedrilus monospermatecus* from brackish water of $6^0/_{00}$ S, before and after adaptation to $20^0/_{00}$ S. After B.-O. JANSSON 1962.

strip of interstitial water somewhat nearer the land with about $1.5^0/_{00}$ S. In survival experiments lasting several days both species tolerated much wider differential salinity ranges: *Aktedrilus* 1.25—20 and *Marionina* $2.5—10^0/_{00}$ S. If, however, in the experiment, both species were offered a choice of two different salinities, after a short time they always selected the salt content closest to their optimum. These experimentally established preferential optima (for *Aktedrilus* from $6^0/_{00}—1^0/_{00}$ S, for *Marionina* from $0—0.2—0.3^0/_{00}$ S) were relatively stable, although it was possible to change the extent of the reaction slightly by means of several days of readaptation to a different salinity (see fig. 2). LAGERSPETZ & MATTILA (1961) observed in alternative chamber experiments that *Gammarus oceanicus* shows a clear preferance for brackish water of $6^0/_{00}$ S against fresh water. But the reactions in choosing between 5 and $6^0/_{00}$ showed no significant deviation from a purely random distribution. The authors therefore conclude that the threshold of perception in this species is not low enough to ensure effective reactions of salinity preference under natural conditions.

If the distribution of a species is determined by the frequently lower resistance of the eggs or larvae or juvenile stages its ecological limits in brackish water cannot be established by means of experiments on resistance with adults. This is true, for instance, for the Indian brackish-water oyster *Ostrea madrasensis;* the adults survive in $20^0/_{00}$ S, while fertilized eggs will only develop in at least $22—26^0/_{00}$ S and the larvae require a minimum salinity of $28^0/_{00}$ for their development (RAO 1951; cf. RANSON 1948). — For the same reason BARNES (1953) would consider the differences in the osmotic resistance established by him for the nauplia of marine and brackish-water barnacles (*Balanus balanus* up to $21^0/_{00}$ S, *B. crenatus* up to $15^0/_{00}$ and *B. balanoides* up to $9^0/_{00}$ S) to be the decisive factor in the distribution of the three species. According to BOGUCKI (1954, 1963) the juvenils stages of the euryhaline Polychaete *Nereis diversicolor* are considerably more sensitive to dilute brackish water than the older stages; this applies particularly to the gastrulae, trochophora stages and the young larvae with only three segments bearing setae. However, this rule does not apply universally. In experiments on the gastropod *Nassarius obsoletus* which is widespread on the East coast of America SCHELTEMA (1965) found only slight differences between the lower lethal salinity limits of veliger larvae and of adults ($14-15.5^0/_{00}$ against 12.5 to $13.5^0/_{00}$ S). — In the decapod crustacean *Porcellana longicornis* the zoeae are even more resistant to dilute sea water than the subsequent post-larval stages (LANGE 1964). The same author observed that in a number of planktonic species of copepods inhabiting estuaries the adult males were generally less resistant to a dilution of the external medium than the females. In this connection it should be mentioned that in some species found in brackish water the adults are more susceptible to changes in salinity during their period of reproduction (KINNE 1964). According to my observations the same holds for the egg-bearing females of the Chinese mitten crab *Eriocheir sinensis* which is otherwise extremely euryhaline.

PLATEAU (1871) kept aquatic isopods *(Asellus aquaticus,* egg-carrying females*)* in a medium in which salinity was gradually increased. After a month all adults had died at $26^0/_{00}$ S, but there were some young isopods. When these young were transferred into pure sea water they survived for 108 hours, while adult isopods directly transferred from fresh water died within 5 hours.

In a paper of 1928 GRESENS demonstrated most elegantly that numerous factors may influence the extent of osmotic resistance of fresh-water organisms in brackish water. As a rule large individuals were more resistant than small ones. Deposition of gametes resulted in "subnormal resistance" while sex in itself did not seem to make any difference. In the middle range of temperature resistance (survival time)

Table 2. Experimentally determined limits of osmotic resistance ($^0/_{00}$ salinity) of some fresh-water species. After GRESENS 1928.

Species	Individual limits		Limits of reproduction
	sudden transfer	gradual transfer	
Glossosiphonia complanata	3.5	5.3	—
Herpobdella atomaria	5.3	5.3	3.5
Asellus aquaticus	5.3	15.0	5.3
Dendrocoelum lacteum	7.0	15.0	7.0
Pelmatohydra oligactis			
Pelmatohydra oligactis			
from fresh water	3.0	8.0	3.0
from brackish water	8.0	9.0	7.0

was doubled or trebled if the experimental temperature was lowered by 10°. But at 0° resistance was lower than at 5°C. There was often a very marked difference in the mean upper salinity limits, depending on whether salinity changed suddenly or gradually. If the salt content of the external medium was increased slowly the animals often tolerated higher salt concentrations than those in which the species occur in nature. Similarly, the limits of individual osmotic resistance, as measured in laboratory experiments did not correspond with the limits of reproduction which are usually considerably lower (see table 2).

The differences in osmotic resistance which PALMHERT (1933) found in the hydroid polyps *Pelmatohydra oligactis* (from fresh water) and *Clava multicornis* (from brackish water) are probably genetically controlled, that is species-specific. While the fresh-water species *(Pelmatohydra)* was able to survive up to $3^0/_{00}$ S, the brackish-water species *(Clava)* could be kept in concentrations between 8—10 and 30—$32^0/_{00}$ S (see fig. 3). In distilled water *Pelmatohydra* disintegrated within 24 hours, but was able to survive in a mixture of equal parts of aqua dest. and water from a pool. Populations of *Pelmatohydra* from brackish water — which PALMHERT investigated at the same time — tolerated higher than individuals coming from fresh water.

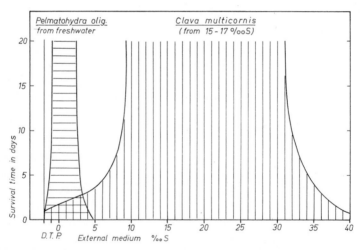

Fig. 3. Experimentally determined limits of osmotic resistance in the hydroid polyps *Pelmatohydra oligactis* and *Clava multicornis*. D = distilled water, T = tap water, P = pond water. After H. W. PALMHERT 1933.

Comparative measurements of the osmotic resistance of different species of the same genus may reveal remarkable differences; frequently similar variations of thermal and chemical resistance have been detected. SCHMITT (1955) found in my laboratory that the fresh-water Turbellarian *Planaria gonocephala* which is relatively stenothermic survives in brackish water of $3^0/_{00}$ S, while individuals of the more eurythermic *Planaria lugubris* are able to survive in $5^0/_{00}$ S (see fig. 4). LAGERSPETZ (1955) observed significant differences in osmotic resistance of *Daphnia magna, D. pulex* and *D. longispina*. In laboratory experiments *Daphnia magna* was

Table 3. The osmotic resistance of various species of *Daphnia* in brackish water. After
LAGERSPETZ 1955.

(Number of surviving individuals. 5 individuals were used for each separate experiment.
Times in hours from the beginning of the experiment).

Species	Salinity of the pool of origin in %0 S	Salinity in experimental dishes											
		0%0				2.9%0				5.8%0			
		hours				hours				hours			
		5	10	20	40	5	10	20	40	5	10	20	40
D. magna	0.10	5	5	5	5	5	5	5	5	5	5	5	5
	0.12	5	5	5	5	5	5	5	5	5	5	5	4
	0.17	5	5	3	1	5	5	5	4	5	5	5	5
	0.17	5	5	5	5	5	5	5	5	5	5	5	5
	0.23	5	5	5	5	5	5	5	5	5	5	5	5
	0.29	5	5	5	5	5	5	5	5	5	5	5	5
	1.54	5	5	5	4	5	5	5	5	5	5	5	5
D. pulex	0.06	5	5	5	5	5	5	5	1	5	4	–	–
	0.09	5	5	5	5	5	5	5	3	5	4	2	–
	0.09	5	5	5	5	5	5	5	3	5	2	1	–
	0.10	5	5	5	5	5	5	5	2	4	2	–	–
	0.20	5	5	5	5	5	5	4	3	5	3	–	–
	0.23	5	5	5	5	5	5	4	3	5	5	–	–
	1.54	5	5	5	5	5	5	4	4	5	3	–	–
D. longispina	0.05	5	5	5	5	5	3	3	–	2	1	–	–
	0.06	5	5	5	5	2	2	–	–	1	1	–	–
	0.12	5	5	5	5	4	3	1	–	3	2	1	–
	0.15	5	5	5	5	5	3	2	1	3	2	2	–

considerably more resistant than the other two species; this also finds expression in
their occurrence in brackish-water pools in Finland (see table 3).

In addition to the observations made on the osmotic resistance of invertebrates
similar results have been obtained for fishes. Here, too, results of observations in

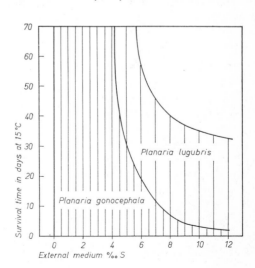

Fig. 4. Experimentally determined
limits of osmotic resistance of *Planaria gonocephala* and *Pl. lugubris*.
After E. SCHMITT 1955.

the laboratory by no means always coincide with findings made in nature. Thus the fresh-water teleost *Carassius carassius* — like the crayfish *Astacus fluviatilis* — can be kept in the laboratory in brackish water of 15%/oo S, though neither species occurs naturally in brackish water of that salinity. But there is agreement in the fact that of the 12 fresh-water fishes occurring in the Baltic in brackish water of 10—18%/oo S only the stickleback *Gasterosteus aculeatus* can reproduce at these salt concentrations (NELLEN 1965). — According to KEYS (1931) small specimens of the brackish-water teleost *Fundulus parvipinnis* tolerate greater dilutions of the external medium than larger individuals. — Results on planarians observed by SCHMITT (1955) have their parallel in differences in osmotic and thermal resistance in related flatfishes (WAEDE 1954): For the flounder *Pleuronectes flesus* MARX & HENSCHEL (1941) found differences within the species depending on the salinity of the water from which they came. While the eggs of the so-called "Bankflunder" of the Baltic are viable and capable of development in 7—8%/oo S, the eggs of the North-Sea flounder are said to be greatly damaged at these salinity grades.

In his beautiful experiments on acclimatisation McLEESE (1956) studied the influence of temperature and oxygen tension of the external medium on the osmotic resistance of *Homarus americanus*. According to his findings the lower lethal salinity limit of this poikilosmotic crustacean is about 6%/oo S at 5°C, about 11.2%/oo S at 15°C and about 16.4%/oo S at 35°C. In each case reduction of oxygen tension results in a lowering of osmotic resistance, that is a rise of the lower lethal salinity limit.

Some authors have discussed the problem whether differences in the osmotic resistance of two populations of a species from aquatic habitats of differing salinities can be explained only as a result of long-term adaptation or else in terms of genetic differences produced by selection. PROSSER (1955) is of the opinion that differences in such populations, as observed in experiments, mostly arise as a result of individual long-term adaptation. From my experience with *Mytilus edulis* I would hold the same view. Therefore I am not convinced that BINYON (1961) is correct in his findings that the North Sea and the Baltic populations of the starfish *Asterias rubens* represent different physiological races. He arrives at this conclusion as a result of experiments on adaptation which only lasted two weeks. He observed that the lower salinity limit of the starfish from the English coast (Whitstable, North Kent) lies at about 23%/oo S, while in the Baltic starfish occur near Rügen at salinities of 8%/oo S. I, too, have not succeeded in acclimatizing starfish from the North Sea within a few weeks to Baltic water from Kiel of 15%/oo S, in spite of careful, gradual dilution of the salinity of the external medium. But I would not like to infer from such relatively short-term experiments that there are different physiological races; they might well be modifying adaptations which, as REMANE (1959) has stressed, can be preserved for several generations. However, there is no doubt that in other instances differences in osmotic resistance have been brought about by true formation of physiological races. Thus, e. g. we know both a fresh-water form and a brackish-water form of *Gammarus duebeni*. According to BEADLE & CRAGG (1940) the brackish-water form soon dies in distilled water while the fresh-water form survives at least four days (see also HYNES 1954). Similarly the North European stickleback *(Gasterosteus aculeatus)* occurs in two morphologically and physiologically different races. While one race (subspecies *gymnurus*) lives predominantly in fresh water, the other

race (subspecies *trachurus*) occurs mainly in brackish and sea water. The race or subspecies *trachurus* is more resistant to an increase of salinity in the external medium. The differences in osmotic resistance of *Gasterosteus* which have been produced by natural selection of the surviving individuals are coupled with morphological characters such as number of vertebrae, body size and number of lateral plates (HEUTS 1947).

PORA (1946) holds an extreme view in this respect. According to his data the prawn *Palaemon squilla* is able to tolerate greatly differing salinity ranges depending on the salt content of the water of origin. Thus the osmotic resistance of specimens from Agigea on the Black Sea (where salinity is of an average of $18\pm5^0/_{00}$) is said to range from $14-25^0/_{00}$ S, while specimens from Naples range from $29-40^0/_{00}$ S and specimens living on the coast of Odessa on the Black Sea range from $2-13^0/_{00}$ S. From these and similar findings PORA deduces the existence of physiological races: „on peut dire qu'il y a autant d'espèces physiologiques que de milieux salins. Or, nous avons vu qu'il y a une infinité de milieux salins, un passage continu entre l'eau de mer et l'eau douce, ainsi on peut dire qu'il y a une infinité d'espèces physiologiques" (PORA 1946 p. 23, bottom). [It can be said that there are as many physiological races as there are saline media. As we have seen there exists an infinite number of saline media, a continuous transition from sea water to fresh water; so it may be said that there is an infinite number of physiological races.] Clearly, a decision on this point can only be made as a result of long-term experiments on adaptation and, if necessary, experiments on breeding and crossing.

R. I. SMITH (1955, 1963, 1964) has studied the Polychaete *Nereis diversicolor* from marine and brackish-water biotopes in Scotland, Southern England, Denmark and Finland; he thinks that the view that there are differently resistant, physiological races of this species has not been proven. But he emphasizes the fact that on the Finnish coast (Tvärminne) *Nereis diversicolor* still reproduces at suboptimal salinities which, on the Atlantic coast of Sweden (Kristineberg) are lethal for the larvae of the same species. In the Baltic *Nereis diversicolor* has a decreasing number of paragnatha with increasing distance from the North Sea; from this fact MUUS (1967) would draw the conclusion that there are a number of geographical morphological races ("isolated races or entities"). BATTAGLIA (1967) has made important advances in his analysis of the osmotic behaviour of three geographically separated populations of the harpactoid copepod *Tisbe furcata*. Populations derived from the English coast near Plymouth and two others from brackish-water biotopes in the Adriatic (lagoon of Venice; Lake Varano near Chioggia, Apulia) were cultivated for more than 60 generations, at constant salinity ($34^0/_{00}$ S), over a period exceeding two years. In the critical experiments groups of adults of the same age were transferred into brackish water of $18^0/_{00}$ S, this produced a shock from which the animals recovered within certain times. These times of recovery (in minutes) were always shorter for the brackish-water forms than for the marine ones: Plymouth 81.58 ± 1.86; Venice 54.42 ± 1.16 and Varano 43.19 ± 0.82. This result strongly indicates that differences in osmotic resistance which have been measured in this way represent genetically controlled characters. In accordance with this is the fact that the F[1]- hybrids Varano x Plymouth show greater tolerance of dilution than the parent population while F[2] displays increased variance.

Work on isolated organs or tissues may also throw light on the osmotic resistance of a brackish-water species. For many experiments with invertebrates dilute sea water can safely be used as external medium. For more accurate measurements, however, it is advisable to take note of any existing differences in the ionic composition of the external and the internal medium of the species chosen for the experiments (see also p. 265). WELLS & LEDINGHAM (1940) and WELLS, LEDINGHAM & GREGORY (1940) in England have worked with isolated tissue preparations from *Arenicola maritima, Nereis diversicolor, Perinereis cultrifera, Mytilus edulis* and *Pleurobrachia pileus*. They observed that after sudden transfer into dilute sea water the preparations absorbed water rapidly and became temporarily incapable of functioning ("shock-effect"). After a few hours some kind of balance was established and the tissues were once again capable of mechanical activity, either by muscular contraction or the beating of cilia. True, the muscles and cilia of specimens of *Arenicola marina* and *Mytilus edulis* coming from sea water ceased to function at such low salinities at which the same species are still viable in the Baltic. In the authors' opinion this points to the existence, in the Baltic, of brackish-water varieties (races) of the species in question or to the need for a longer period of adaptation. The latter seems to me to be the more probable explanation. With a slow, gradual reduction of the salt content these experiments with isolated organs revealed similar species-specific differences in osmotic resistance as had been obtained with whole individuals, that is isolated muscle preparations of *Perinereis cultrifera* were more susceptible than those from *Arenicola marina* and these again were less resistant than muscle preparations from *Nereis diversicolor*. The lower salinity limits which allowed spontaneous activity in isolated muscle preparations of the above-mentioned species were *Perinereis cultrifera* 20—25% sea water, for *Arenicola marina* 15—20% and *Nereis diversicolor* 5—10% sea water.

PILGRIM (1953a) conducted experiments with isolated, rhythmically contracting preparations of cardiac muscle from *Mytilus edulis, Ostrea edulis* and *Anodonta cygnea*. A sudden lowering of salinity caused a temporary shock-reaction (reduction of amplitude or inhibition of rhythmical contractions), followed by recovery (adaptation). The appearance of shock effects could be avoided when the salt content was slowly and gradually changed. The net result of the mechanical performance of the preparations was the same for brackish water of low salinity whether salinity had been reduced gradually or abruptly. The author holds the view that the osmotic resistance of the tissue preparations investigated by him is greater than would be expected from the observed range of the species in question (see table 4).

Table 4. The osmotic resistance of heart muscle preparations of bivalves, compared with the ecological salinity limits of the species concerned.
All salinity data in % sea water. 100% sea water = 35⁰/₀₀ S. After PILGRIM 1953 a.

Species	Heart muscle well active	Ecological salinity ranges
Mytilus edulis	40—160	45—100
Ostrea edulis	70—120	70—100
Anodonta cygnea	2— about 24	fresh water — about 14

In addition PILGRIM (1953 b) worked with isolated gill preparations from the same bivalves. In some cases shock-reactions could be seen after a sudden change of salinity in the external medium. Thus the rhythmical beating of the frontal gill cilia of *Mytilus edulis* was inhibited for 4—14 minutes when the sea water was diluted to 30% (3 parts of sea water +7 parts of fresh water). No inhibition occurred when the medium was diluted from 100% to only 40%. In 30% sea water the activity of the frontal cilia was reduced and in 20% sea water the cilia of only some pieces continued to beat. In *Ostrea edulis* the isolated gills were less resistant; this corresponds to the lower osmotic resistance of the whole oysters. In 20% sea water the cells of the gills immediately began to swell and soon after disintegrated. The gills of the fresh-water mussel *Anodonta cygnea* tolerated concentrations up to 8—12% sea water without any harmful effect. In 1% sea water a marked swelling of the gill cells could be observed. In tap water the swelling was greater, though in general the frontal cilia continued to beat with reduced activity.

Further experimental investigation of numerous other invertebrates has shown that there is generally a close correlation between the salinity range of their occurrence and their cellular osmotic resistance (SCHLIEPER et al. 1960, RESHÖFT 1961; VERNBERG et al. 1963; SCHLIEPER, FLÜGEL & THEEDE 1967). Thus isolated pieces of gills from euryhaline bivalves survive in much wider ranges salinity than e. g. those from

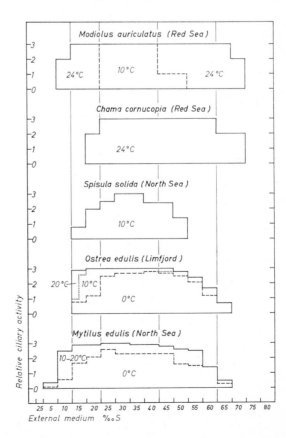

Fig. 5. Species-specific cellular salinity ranges for five marine bivalves. After H. THEEDE 1965b and C. SCHLIEPER et al. 1967, combined and simplified.

more stenohaline species (see fig. 5). Species from greater depths in sea water of constantly high salinity (e. g. *Spisula solida*) display particularly narrow cellular salinity ranges while the tissues of species penetrating into brackish water (such as *Mytilus edulis*) are also more viable in dilute sea water; species relatively resistant to desiccation (e. g. *Chama cornucopia*) from the upper littoral are characterized by a specially high cellular resistance in concentrated sea water. It is important to use optimum temperatures in all such experiments. A cold-water species has a considerably reduced salinity range at higher temperatures, while a tropical warm-water species will only reveal its full cellular osmotic resistance in warm, but not in cooler water.

Euryhaline brackish-water invertebrates also show non-genetic individual differences in their cellular osmotic resistance, depending on the salinity of the water they inhabit. Thus the tissues of *Mytilus galloprovincialis* from the brackish lagoons (Etangs) in the South of France are markedly more resistant to a dilution of the external medium than the specimens of the same species from normal sea water from the Mediterranean coast. In experiments in which the bivalves from sea water were transferred into brackish water the "resistance to dilution" of their tissues rose slowly. The first signs of adaptation of an altered resistance appeared after about two days, but a complete readaptation was only noticeable after 1—2 weeks (SCHLIEPER et al. 1960). As will be explained later (see chapter 5) a maximum of 24 hours is required for the poikilosmotic common mussel to adjust the concentration of its internal medium osmotically to a changed salinity of the external one, while keeping its shells open. From this it appears that the slow adaptation of osmotic resistance described above is probably a process taking place in the colloidal protoplast of the

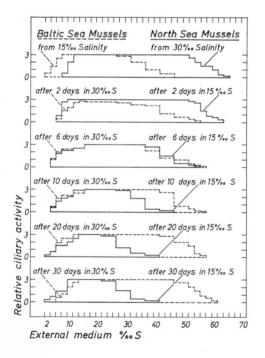

Fig. 6. Individual cellular salinity ranges for *Mytilus edulis* from the North Sea and from the Baltic, before and after adaptation to the other medium of 15 or 30‰ S respectively. After H. THEEDE 1965b, altered.

cell; it is by no means completed after isotonicity between internal and external medium has been established and it bears no relation to the process of attaining balance between the two media.

The investigations of THEEDE (1965b) on *Mytilus edulis* from the North Sea and from the Baltic provide a particularly fine example of such slow adaptation of osmotic resistance. Whole individuals of this species as well as surviving isolated gill pieces display differential salinity ranges. With reciprocal adaptation of the intact mussels to water from the North Sea and the Baltic respectively these differences disappear in about 20 days during a slow process of adaptation; this shows them to be conditioned by the environment to a large extent (see fig. 6).

We owe numerous experiments on the osmotic resistance of marine algae to BIEBL (1938, 1939, 1952). He found correlations between the extent of osmotic resistance of the algae and their occurrence in depth in the open coastal reaches (Plymouth, Heligoland, Naples). His results show clearly that for the algae which have been investigated the range of osmotic resistance decreases, the greater the depth at which they are found (see table 5). The algae of the tidal zone have an average resistance range of 10—3000% sea-water concentration, the algae from the zone

Table 5. Osmotic resistance of marine algae. The range of resistance is expressed in percentage sea water. 100% = normal, unaltered sea water. After BIEBL 1952.

Species	Range of resistance
a) Tidal zone	
Cladophora spinulosa	10—300
C. bertolinii	10—300
C. hamosa	10—300
C. laetevirens	10—280
Bangia fuscopurpurea	10—300
Porphyra leucosticta	10—300
Polysiphonia pulvinata	20—300
b) Low-water level	
Chaetomorpha linum	0—240
Cladophora utriculosa	10—240
Ceramium berneri	30—190
Callithamnion granulatum	40—260
Antithamnion cruciatum	40—190
c) Sublittoral zone	
Cladophora prolifera	10—200
C. ramellosa	50—170
Toania atomaria	80—150
Griffithsia furcellata	70—140
G. flosculosa (= setacea)	60—150
G. schousboei	60—140
Aglaeothamnion scopulorum	60—150
Ceramium diaphanum var. strictum	40—140
Pleonosporium borrerii	70—150
Plocamium coccineum	70—140
Nitophyllum punctatum	70—140
Acrosorium uncinatum	60—150

of low-water level and from tidal pools of 40—200% sea-water concentration and the algae from the sublittoral zone only 50—150% sea-water concentration.

Similar experiments on resistance of marine algae with graduated salt solutions were carried out by other workers (KNIEP 1907, HÖFLER 1931, KYLIN 1938); they also yielded a variety of species-specific values. Generally those species which occur only in normal sea water were less resistant than those which were also common in brackish water of low salinity. KNIEP studied the physiology of germination of *Fucus serratus* and in his experiments he found roughly the same limiting concentration as one obtains in nature. But for *Fucus vesiculosus* much narrower limits were found in the experiment than in natural brackish water. We owe to HOFFMANN (1943) a very lucid and illuminating discussion of the osmotic resistance in marine algae; he points out that in this case specimens of *Fucus vesiculosus* from normal sea water had been used while plants from areas of much lower salinity might show wider limits owing to acclimation, unless racial characters are the decisive factor.

Table 6. Salinity resistance of red algae living in the brackish water of Kiel Bay, as measured by the % numbers of surviving cells. Duration of experiment 24 hours, 5—10°C. After SCHWENKE 1958.

Medium %o S	*Delesseria sanguinea*	*Membranop-tera alata*	*Callithamnion corymbosum*	*Ceramium rubrum*	*Phycodris rubens*
0	0	0	0	10	0
1	75	50	50	90	25
2	75	95	50	100	25
3	100	100	75	100	50
4	100	100	75	100	100
8	100	100	100	100	100
30	100	100	100	100	100
40	100	100	100	100	100
50	75	75	100	100	100
55	25	25	100	75	100
60	0	0	100	75	100
65	0	0	95	50	100
70	0	0	95	50	100

In recent times SCHWENKE (1958, 1960) in particular, attempted to analyze the osmotic resistance of red algae living in brackish water. For the experimental determination of the lethal salinity boundaries he used the well-known phenomenon that during or shortly after death the red pigment of these algae exudes from the rhodoplasts, with resulting change of colour, e. g. in the case of *Delesseria* to a "minium red". As the main injury to the osmotic system occurs in less than 24 hours it is possible to measure species specific-differences by means of short-term experiments (see table 6). After longer periods of experimentation the salinity range established in this manner becomes narrower. Thus, in short-term experiments the lower osmotic limit for *Delesseria sanguinea* is about 3—4%o and at about 8—10%o in experiments lasting 2—3 months. Some characteristic red algae form the depths of the western Baltic have a greater resistance to dilution than the same algae on the European Atlantic coast (see table 7). But SCHWENKE also stresses that the ecology of the red algae in brackish water cannot be judged from the resistance behaviour of the adult vegetative stages alone. The reproductive stages of all the species which he investigated were more susceptible to dilution than the vegetative cells. This may be the reason why algae such as *Rhodomenia* and *Odonthalia* from the southern Kattegat do not occur in the Kiel Bay with its water of lower salinity in spite of the great resistance to dilution displayed by the fully-developed thalli.

Table 7. Lower salinity limits (minimum values in S $^{o}/_{oo}$) of different red algae in relation to the salinity of the sea water of their locality. Duration of experiment 24 hours, 14—16°C. After BIEBL 1958 and SCHWENKE 1960 b.

Locality	Salinity of locality	*Delesseria sanguinea*	*Phycodrys sinuosa*	*Membranipora alata*
Brittany	35	7	14	14
Heligoland	33	7	7	7
Kiel	15	3.5	3.5	3.5

We are indebted to HÖHNK (1952, 1953, 1956) for valuable studies on the osmotic resistance of brackish-water Phycomycetes. He considers the fungi of the brackish water as "transitional forms" which are clearly distinguished from the fungi of fresh water and of the sea by their salinity resistance. In general the specific brackish-water fungi (forms, races or species) are considered to be euryhaline, that is they tolerate larger salinity fluctuations than the fresh-water races of the same species, without being confined to a brackish water of restricted salinity range. HÖHNK grew, among others, a number of mycelia of *Pythium monospermum*, *P. salinum* and *P. undulatum* var. *litorale* from brackish water of different salinities. In these cultures the mycelia displayed a different range of resistance, depending on the salt content. Each form coming from brackish water of a certain salinity showed optimum development in a range corresponding to the salinity of the locality where they were found. From this the author deduced the existence of local races of the species of *Pythium* in brackish water, their resistance genotypically fixed. HÖHNKs further observations cannot be discussed in detail; but it is of special interest that the Phycomycetes inhabiting brackish water are less resistant during their sexual phase than during sporulation and the highest degree of resistance is displayd during the phase of vegetative growth.

ZOBELL (1941) made comparative studies of the osmotic resistance of marine bacteria collected some distance from the coast and of those collected near the land; he cultured them in nutrient solutions with a graded content of sea water. He found that bacteria from the open sea or from the mud bottom far from the coast showed optimum development in a nutrient medium prepared with 100% sea water. By contrast bacteria from the water and the mud of San Diego Bay and the Mission Bay (California) showed optimum growth in nutrient solutions with only 10—50% sea water. This result seems to indicate that bacteria living in the sea or in brackish water behave according to the salinity conditions of their external medium. Possibly mutations occurring within the same species and selection of the "suitable" mutants may play an important part. In this connection ZOBELL (1946, p. 120) draws attention to the fact that in older bacterial cultures individual cells are frequently found which are considerably euryhaline, while others behave as relatively stenohaline. The salinity range of stenohaline forms is said to be low, medium or high, after prolonged culturing, depending on the level of salinity of the nutrient solution. It should be noted that ZOBELL cultured marine bacteria over long periods of time (2—12 years). In contrast MacLEOD & ONOFREY (1956) maintain that marine bacteria cultured by them, had, even after two years, in no case acquired the ability to reproduce in fresh-water media (cf. LARSEN 1962 and MacLEOD 1965).

VENKATARAMAN & SREENIVASAN (1954) examined the osmotic resistance of numerous bacteria isolated from marine teleosts (mackerels) or from sea water. According to their data most species of the genus *Flavobacter,* e. g. are stenohaline. Some species in the genera *Achromobacter* and *Flavobacter, Pseudomonas* and *Paracolobactrum* completely fail to grow in the absence of salts. By contrast, species of the genera *Micrococcus, Sarcina, Bacillus* and *Bacterium* thrive in sea water as well as in fresh water.

More recent studies by RHEINHEIMER and co-workers show beyond any doubt that the distribution and composition of the bacterial flora in brackish water depends to a large extent on salt content (RHEINHEIMER 1966; KRUMM & RHEINHEIMER 1966; KOSKE, KRUMM, RHEINHEIMER & SZEKIELDA 1966). On the western Baltic, according to these authors, the proportion of halotolerant (that is maximum euryhaline) bacteria which grow in sea and fresh water is greatest near the coast and least in the open

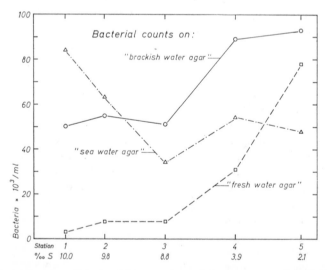

Fig. 7. Salt requirements of the brackish-water bacteria in the estuary of the Schlei. Bacterial counts in one ml Schlei water in relation to the salinity of the Schlei water and to the salt content of the medium.

"sea water agar" = ZoB.-Agar + sea water of 25⁰/₀₀ S. "brackish water agar" = ZoB.-Agar + brackish water of 8⁰/₀₀ S. "fresh water agar" = ZoB.-Agar + fresh water. After unpublished values by G. RHEINHEIMER.

sea. Only a few nautical miles away from the land, between 12 and $22^0/_{00}$ S, the percentage of halotolerants is fairly small. Similarly, in the Kiel Canal the number of "sea-water bacteria" thriving only on a nutrient medium prepared with sea water clearly diminishes from east to west with a decreasing salinity, whereas the number of "fresh-water bacteria" which will only thrive on nutrient media prepared with fresh water, rises almost continuously. Below about $8^0/_{00}$ S there is a rapid change in populations in favour of brackish-water and fresh-water bacteria (see fig. 7). In a similar manner the composition and salinity requirements of the bacterial flora in the lower reaches of the Elbe (Unterelbe) change with the salinity fluctuations of the tides.

3. Rate of activity and activity changes

The activity of a brackish-water animal can, amongst other criteria, be assessed by its rate of development, of growth and of locomotion. After changes in the salinity of the external medium an aquatic species may be considered to be adapted or undamaged as long as its activity, expressed in the above-mentioned functions, has remained unaltered. Unfortunately there are only few investigations available which allow us to make any statements about the relation between activity of brackish-water animals and the salt content of the external medium. A reduction in the activity of certain marine species has been observed when the salinity of the external medium was lowered. The rate of development of the eggs of some marine invertebrates decreases below a certain critical salt concentration (TSCHANG-SI 1930, MIYASAKI 1933, CLARK 1935). The fertilized eggs of the American oyster Ostrea virginica will only develop at a normal rate in a salinity range of 35 to $25^0/_{00}$ at 18—21.5° C (AMEMIYA 1926). In 23 and $21^0/_{00}$ S development is somewhat slowed down though it progresses normally in other respects. In brackish water of $19,3^0/_{00}$ S a small proportion of the eggs fail to develop. The rest of the eggs show uneven development which is markedly slowed down. The blastomeres are swollen. Nevertheless, many shell-bearing larvae developed after four days. In $17.5^0/_{00}$ S a large number of eggs fail to develop and an equal number of abnormal embryos are produced. Development is greatly retarded. A number of free-swimming larvae arise, but only very few shell-bearing ones. In $15.8^0/_{00}$ S development is very much retarded and hardly any shell-bearing larvae are produced. Similar observations on the dependence of developmental rate on the salinity of the external medium have been recorded by the same author for the Portuguese oyster Ostrea angulata and the British oyster Ostrea edulis. In the last two species a dilution of the external medium has an even more adverse effect on their rate of development. In contrast to the larvae the adults of the Ostreids are able to survive in a greater salinity range (RANSON 1948). The eggs of the Atlantic salmon Salmo salar develop at a normal rate in fresh water which is their natural medium, but in brackish water the rates of division and development are considerably retarded so that young fish in brackish water hatch much later than the controls in fresh water (BUSNEL, DRILHON & RAFFY 1946).

Except in the case of extreme euryhaline species growth is often more or less reduced in salinities outside the optimal range. LOEB (1904) already noticed this while investigating the effect of salt concentration of the surrounding medium on the growth processes of marine Tubulariae (hydroid polyps). He found that maximum growth in length occurred at about 25⁰/₀₀ S and that it decreased both in more concentrated and in more diluted sea water. Below 25⁰/₀₀ S the rate of growth decreased rapidly and a salinity of 13% allowed neither growth nor regeneration to take place. Consequently a reduction in size and even stunting of forms are found in numerous marine immigrants into brackish water, both invertebrates and some fishes. Reduction in size is mainly due to a slower growth rate. This also results in the annual spawning times of marine invertebrates being delayed in brackish water and in fishes reaching sexual maturity at a smaller size. KOWALSKI (1955) has made a comparative study of gonad development in the starfish *Asterias rubens* in populations from the western Baltic (15⁰/₀₀ S Kieler Förde) and from the North Sea (30⁰/₀₀ S, Island of Sylt). Whereas the starfish from the North Sea spawn on an average from January to March, those from the Baltic mature much more slowly and only spawn in May and June. This slower development of the gonads can be studied both from the changes in weight which they undergo in the course of a year, as well as from the increase in the diameter of the oocytes and eggs respectively. It is interesting to note that the mature eggs of the brackish-water starfish do not attain the size of the ripe eggs of the starfish from the North Sea. In spite of their smaller size and the delay in the deposition of the eggs by the starfish from brackish water, they are still capable of being fertilized and of developing so that the preservation of the species is safeguarded at least for this salinity (average of 15⁰/₀₀ S).

Some fresh-water invertebrates also show a reduced growth rate in brackish water; this decreases with increasing salinity of the medium. This is shown, among other studies, in a faunal investigation by BOETTGER (1950). He examined the coloniza-

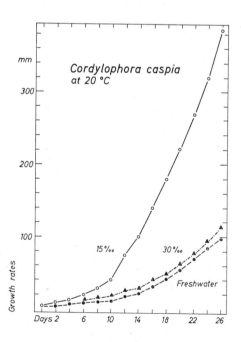

Fig. 8. Growth rates of the brackish-water polyp *Cordylophora caspia* in sea water, brackish water and fresh water. After O. KINNE 1956.

tion of molluscs in the coastal area of the central Baltic in brackish water of differing salinities; among other effects a marked reduction in size was observed in *Theodoxus fluviatilis, Bulimus tentaculatus* and *Galba palustris*. SCHMITT (1955) transferred egg cocoons of *Planaria gonocephala,* which had been deposited in fresh water, into brackish water of 2—4⁰/₀₀ S. After four to six weeks young animals hatched from all egg cocoons. These young planarians, 1—2 mm long, were then kept in the same solutions, regularly fed and measured. Reduction of growth proved to be more marked, the higher the salinity of the external medium. In brackish water of 3⁰/₀₀ S growth in length was reduced by about 30%.

By contrast, the typical brackish-water species show an entirely different relation between total activity and growth rate, and the salinity of the external medium. These species are adapted to an optimum of medium salt concentrations; therefore their greatest intensity of growth is displayed in brackish water of medium salinity, while both an increase and a reduction of salinity in the external medium result in a slower rate of growth. Such results e. g. were obtained by KINNE (1956, 1958) when experimenting with the brackish-water polyp *Cordylophora caspia* which occasionally occurs in running fresh water as a smaller-growing form. His original material came from the brackish water (17⁰/₀₀ S) of the western Baltic. Colonies which he reared in fresh water, at 15⁰/₀₀ and at 35⁰/₀₀ S, differed in several respects. With equal food supplies the individuals from brackish water grew far more rapidly (see fig. 8), they had longer hydranths with more tentacles and showed more lively movements of their tentacles. Their sexual reproduction, as measured by the number and size of the gonophores and the number of eggs per gonophore, was most vigorous in brackish water. The same author obtained corresponding results (1954) in his studies of the brackish-water amphipod *Gammarus duebeni*. For six months he cultured young animals at 18°C and they grew best at 10⁰/₀₀ S; their growth intensity

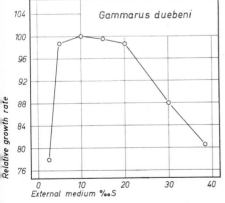

Fig. 9. Influence of the salinity of the external medium on the growth rate of *Gammarus duebeni* at 18°C, growth of young animals in the course of 6 months. 0⁰/₀₀ S = fresh water. After O. KINNE 1954.

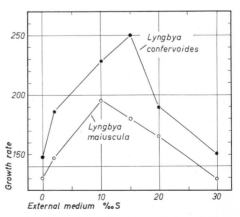

Fig. 10. Effect of salinity of the external medium on the growth rate of the blue-green algae *Lyngbya confervoides* and *Lyngbya maiuscula*, growth at 19°C. Initial length of filaments = 100; 0⁰/₀₀ S = fresh water. After E. M. KINNE-DIETRICH 1955.

decreased more rapidly with dilution of the medium and it decreased at a slower rate when the salt concentration was raised (see fig. 9). Similar observations were made by KINNE-DIETRICH (1955) with two blue-green algae from brackish-water pools (see fig. 10). The species *Lyngbya maisuscula* exhibited maximum growth at $10^0/_{00}$ S. Higher as well as lower concentrations inhibited growth to an increasing degree. In this case, too, it became evident that reduction of growth was greater with diminishing salinity than with increasing one.

BRAARUD (1951, 1961) examined the reproduction of some phytoplankton species in sea water of different salinities. The euryhaline Dinoflagellate *Exuviella baltica* showed relatively uniform rates of reproduction in a larger salinity range (10 to $40^0/_{00}$ S), with a slight maximum at $10^0/_{00}$ S. These experimental results agree with the fact that the same species is widespread in the brackish waters of the Baltic, in the North Sea, the North Atlantic and the Barents Sea. Its distribution is obviously only limited by extremely low salinities. *Peridineum triquetrum,* on the other hand, has a fairly good reproductive capacity even at $5^0/_{00}$ S. This accounts for the fact that this species also occurs in the low-salinity water ($3.5—5^0/_{00}$ S) of the Finnish coast. The planktonic diatom *Asterionella japonica,* however, has its reproductive optimum at $30^0/_{00}$ S and does not reproduce at all at salinities of $15^0/_{00}$ and below (KAIN & FOGG 1958). This result is in agreement with the widespread occurrence of *Asterionella* in the Atlantic and the Mediterranean at salinities above $35^0/_{00}$ and its absence from brackish water. No geographical races whose reproduction is variously dependent on salinity, have been found in widespread species of marine phytoplankton. BRAARUD's findings are in accordance with this; he reports that clones of *Peridineum trochoideum* from the Oslo fjord (with $20^0/_{00}$ S during the summer maximum there) and those from the Gulf of Naples (with $37^0/_{00}$ S) had practically identical curves of reproduction with a salinity optimum at $20^0/_{00}$ S.

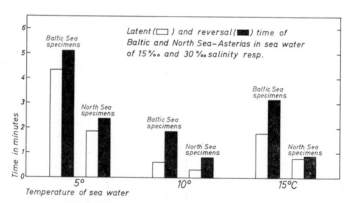

Fig. 11. Comparison of periods of latency and righting of *Asterias rubens* from the North Sea and from the Baltic (S = $30^0/_{00}$ and $15^0/_{00}$ resp.) After R. KOWALSKI 1955.

The interrelations between the activity of a species inhabiting brackish water and the salinity of the external medium can suitably be studied by means of their rate of locomotion or their righting reaction. Here, too, more strongly euryhaline species are less affected in their activity by alterations of salt content. Thus the highly

euryhaline shore crab Carcinus maenas rights itself in brackish water of 15⁰/₀₀ S
with approximately the same speed as in sea water of 30⁰/₀₀ S. But the starfish *Asterias
rubens* which also occurs in 30 and 15⁰/₀₀ S, shows a different behaviour. KOWALSKI
(1955) compared the behaviour of specimens of *Asterias* from the Baltic (15⁰/₀₀ S)
and the North Sea (30⁰/₀₀ S). If a North Sea starfish of average size was placed on
its back in sea water of 30⁰/₀₀ S and at 5°C, it began to right itself after a latent
period of about 2 minutes and after a further 2 minutes it had returned to its normal
position. Starfish of similar size from the Baltic had periods of latency and righting
which were more than double in brackish water of 15⁰/₀₀ S and 5°C (see fig. 11). When
the water temperature was raised to 10°C both starfish reacted faster, but the relative
differences between the brackish-water and the marine animals were maintained.
When the temperature of the medium was raised to 15°C the speed of righting showed
no further increase. This is an expression of sensitivity to temperature as displayed
by many cold-stenothermic marine animals. But it is interesting to note that the
starfish from brackish water were damaged to a greater extent by
this increase in temperature than those from sea water. Nonetheless, even at
15°C the differences between the slower individuals from brackish water and the
faster ones from sea water were fully maintained.

SCHMITT (1955) found the same phenomenon with fresh-water planarians in
brackish water. At 15°C their rate of crawling was slower, the higher the salinity
of the external medium. SCHMITT worked with *Planaria gonocephala* occurring in
streams and *Planaria lugubris* which is more eurythermic and comes from fresh-water
pools. When crawling in fresh water *Pl. gonocephala* was slightly faster than *Pl.
lugubris,* while in brackish water of 3—5⁰/₀₀ S *Pl. gonocephala* was distinctly slower
than *Pl. lugubris* in the same medium. The species which was faster in fresh water
was thus affected to a greater extent by the increasing isalnity of the medium (see
fig. 12).

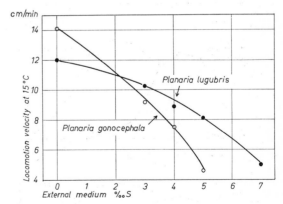

Fig. 12. The activity (speed of
crawling) of *Planaria gonocephala* and *Pl. lugubris* in fresh
water and in brackish water.
After E. SCHMITT 1955.

In recent years several investigations on marine bivalves have been carried out
in which the interrelations between the intensity of beating of their gill cilia
and the salinity of the external medium were assessed. These arose from earlier
studies on the pumping and filtration rate of oysters and common mussels in sea
water of different salinities. HOPKINS (1936) examined the current of respiratory

water of the Japanese oyster *Ostrea gigas*. He found that any marked change of salinity in the external medium resulted in an immediate slowing down or inhibition of the water current. With increased salinity the intensity of pumping returned to its original level within a few hours. But when salinity was diminished, complete recovery took several days. In 20⁰/₀₀ S, however, the same values as in higher salinities were not obtained even after several days. In 15 and 13⁰/₀₀ S the bivalves were extremely sensitive and the current of respiratory water was greatly reduced. But complete recovery took place after transfer into normal sea water.

Von HARANGHY (1942) examined the filtering performance of mussels by adding 30 drops of Indian ink per litre to the surrounding sea water and determining the reduction of turbidity every half hour by means of a PULFRICH photometer (Zeiss). A mussel of about 25 grammes (weight of soft parts 6.5 g) was able to clear 0.5 litre of sea water containing Indian ink completely in 90 minutes But this filtering performance declined rapidly if the salinity of the experimental solution was reduced. In sea water of about 17⁰/₀₀ S the same mussel took 5 times as long to clear the same amount of water. THEEDE (1963) used a similar, but partly improved method for investigating the filtration rate of *Mytilus edulis* from the North Sea and from the Baltic in relation to the salinity of the medium. If measured in the water from the locality where the animals are found the filtration performance of his mussels from the Baltic was but slightly less than that of mussels from the North Sea, having the same length of shell. But a drastic reduction of the filtration performance was produced in both forms after direct transfer into water of higher or lower salinity respectively (see fig. 13). This reduction was diminished if the experimental animals had a longer period of adaptation to the changed salt content. But in no case was a period of 7—10 days' adaptation sufficient to obliterate the differences in salinity dependence between the filtration performance of the two populations. Since the

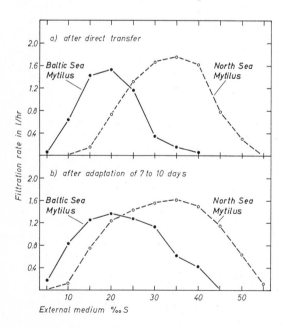

Fig. 13. Filtration rate of common mussels *(Mytilus edulis)* from the North Sea and from the Baltic, depending on the salinity of the external medium. After H. THEEDE 1963.

current of respiratory water of the common mussel is essentially produced by the action of the beating gill cilia, these results indicate that the activity of the ciliary beat in *Mytilus* depends on the salinity of the external medium. These conclusions are confirmed by direct measurements of ciliary activity of common mussels which have developed into maturity in sea or brackish water of different salinity.

If a mussel is opened by cutting the large adductor muscle and the valves are spread out, the large gill lamellae are exposed. If a small piece of tinfoil is put on the horizontally placed gill surface, the beating of the frontal cilia will always slowly transport it in the direction of the free edge of the gill. In this way the beating intensity or mechanical activity of the frontal cilia can easily be measured.

SCHLIEPER (1955) and SCHLIEPER & KOWALSKI (1957) have used this method to measure the ciliary activity of common mussels from the North Sea (30⁰/₀₀ S, List), the western Baltic (15⁰/₀₀ S, Kieler Förde) and from the eastern Baltic (5—6⁰/₀₀ S, Tvärminne). At 20°C ciliary activity of North Sea mussels (List, shell length 6—7 cm) at 30⁰/₀₀ S is on an average 41 mm/min., of the same size Baltic mussels (Kiel) in brackish water of 15⁰/₀₀ S only 30 mm/min. This means that for mussels which have had maximum adaptation the activity of the frontal gill cilia in brackish water of 15⁰/₀₀ S is about 25% less than in sea water of 30⁰/₀₀ S. The effect of the salinity of the external medium is even more pronounced in a comparison of mussels from Tvärminne (eastern Baltic) and from Kiel (western Baltic). In this case mussels of about 3 cm shell-length were examined at 17—18°C in the water of the locality where they were found. For the mussels from Tvärminne (6⁰/₀₀ S) the ciliary activity was 14 mm/min., while for those from Kiel (16⁰/₀₀ S) it amounted to 34 mm/min. that is more than double. If the salinity of the external medium for the mussels from Tvärminne is raised from 6 to 16⁰/₀₀ S, that is if they are transferred into Baltic water from Kiel, their ciliary activity is raised to an average of 25 mm/min. within a short time. To some extent these findings are confirmed by a comparison of the frequency of heart beat of the same mussels. If otherwise normal external conditions are maintained, the heart frequency of the mussel can also be taken as a measure of activity. We found an average of the following values (experiments carried out in August, shell-length about 10 mm, average temperature 18°C):

North Sea *Mytilus* in 30⁰/₀₀ S = 57 heart beats/min.
Kiel *Mytilus*   in 15⁰/₀₀ S = 50 heart beats/min.
Kiel *Mytilus*   after transfer and several days' adaptation
                     in 8⁰/₀₀ S = 36 heart beats/min.
Tvärminne *Mytilus* in 6⁰/₀₀ S = 36 heart beats/min.
Tvärminne *Mytilus* after transfer and several days' adaptation
                     in 16⁰/₀₀ S = 43 heart beats/min.

As will be shown later, the common mussel is poikilosmotic and so the osmotic concentration of its external and internal medium are the same; the observations just described show that the salinity of the body fluids has a very strong influence on the activity of the mussels.

In some species production of lime is reduced in brackish water. The reason for this particular reduction of the activity of the lime-secreting tissues in brackish water is not yet understood. But some quantitative data may be of interest here. Such data might form the basis for an experimental approach to this

problem by means of long-term breeding experiments; this has not been attempted
so far. The reduction in shell-weight of numerous marine bivalves
in the brackish water of the Baltic is particularly noticeable. The shells of
bivalves of equal size, coming from brackish water of different salt content, are
lighter, the lower the salinity of the external medium. Furthermore, an analysis
of the shells of the brackish-water common mussel *Mytilus edulis* shows that not

Table 8. Comparison of shells of the common mussel (*Mytilus edulis*) from the North Sea,
the western Baltic and the eastern Baltic.

	Length of shell mm	Dry weight total in mg	Calcareous content (soluble in HCl) mg	Organic substance mg
	a) Shells from the North Sea (List, 30°/₀₀ S)			
	30	1436.4	1399.2	37.2
	30	1589.4	1505.6	38.8
	28	1160.2	1134.2	26.0
	31	1819.9	1768.3	51.6
	32	1461.5	1416.6	44.9
	32	1789.3	1727.2	62.1
	29	1292.4	1253.6	38.8
	30	1448.5	1415.6	32.9
Ma	30	1500 = 100%	1458 = 100%	42 = 100%
	b) Shells from the western Baltic (Kiel, 15°/₀₀ S)			
	29	567.4	538.8	28.6
	32	1114.6	1062.9	51.7
	31	638.0	603.2	32.8
	30	626.4	593.0	33.4
	29	567.0	538.6	28.4
	29	670.2	629.5	40.7
	31	684.8	652.7	32.1
	29	839.8	801.3	38.5
	31	763.8	722.5	41.3
Mb	30	719 = 48%	683 = 47%	36 = 86%
	c) Shells from the eastern Baltic (Tvärminne, 6°/₀₀ S)			
	28	514.3	495.9	18.4
	28	488.3	468.6	19.7
	30	515.2	493.3	21.9
	27	468.3	441.5	26.8
	29	476.2	454.6	21.6
	32	634.3	604.5	29.8
	31	549.9	526.9	23.0
	30	457.6	435.6	22.0
Mc	29	513 = 34%	490 = 34%	23 = 55%

only the calcium content, but also the amount of organic substance (conchiolin, periostracum) is reduced (see table 8). However, the amount of organic matter does not diminish to the same degree, so that the percentage of organic substance in the formation of the shell increases. But from the analytical values reproduced in the table it becomes evident that in brackish-water mussels the performance (activity) of the shell-producing tissues as a whole is reduced. This rule holds in spite of some exceptional features as, e. g. represented by the thick-shelled mussels (dwarf forms) from water moved by surf waves; these are found in both sea and brackish water (see also TRAHMS 1939).

KOWALSKI (1955) made a comparison of the chemical composition of starfish *(Asterias rubens)* from the North Sea and from brackish water of the western Baltic from which it can be seen that the specimens from the North Sea have a significantly higher ash-content (8.8 against 5.9%, see fig. 14). The calcium content of the ash also differs. In the *Asterias* from the North Sea it amounts to 4.1% of live weight (= 17.7% of dry weight) as against the Baltic *Asterias* with 2.6% of live weight (= 15.4% of dry weight). A determination of the proportion of calcium carbonate in the skeleton for both groups of starfish gives the following results KOWALSKI (1955): for *Asterias* from the North Sea 57.2% of dry weight and for those from the Baltic 48.4% of dry weight.

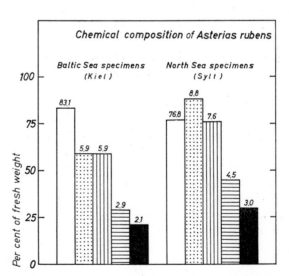

Fig. 14. Comparison of the chemical composition of *Asterias rubens* from the western Baltic (15⁰/₀₀ S) and from the North Sea (30⁰/₀₀ S). After R. KOWALSKI 1955.

The relatively lower calcium-content of the brackish water may have a direct bearing on the reduced production of calcium by the brackish-water invertebrates discussed above. But the question can only be solved by long-term breeding experiments in artificially prepared brackish water of varying calcium content.

Observations made by KINNE & KINNE (1962) on the teleost *Cyprinodon maculatus* have taken us a step further in the analysis of relations between the salinity of the

external medium and the activity of euryhaline fishes. This small fish (maximal
length 45 mm) is widespread in shore pools, brooks and springs in the region of
Lake Salton in southern California; it can survive in the very wide salinity range
from fresh water to 70⁰/₀₀ S. Its fertilized eggs develop at about the same rate in
fresh and in sea water at temperatures below 25° C. At higher temperatures development in sea water is somewhat slower. With unlimited food supplies the growing
fishes consume a maximum of food in sea water of 35⁰/₀₀ S and 30° C. But the

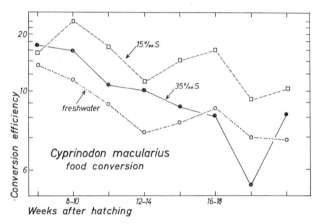

Fig. 15. Utilisation of food by the euryhaline teleost *Cyprinodon macularis* in sea, brackish
and fresh water. After O. KINNE & E. M. KINNE 1955.

efficiency of the conversion of food into body substance is greatest at the same
temperature, in brackish water of 15⁰/₀₀ S (see fig. 15). It is highly desirable that
this complex relationship, which is influenced by temperature as well, should be
further investigated.

Finally I would like to mention that a relation between the *activity of the thyroid*
and the salinity of the external medium has been observed in teleosts (LELOUP 1948).
Thus marine teleosts *(Muraena helena, Labrus bergylta)* which were transferred
into brackish water of 24⁰/₀₀ S displayed a higher activity of the thyroid in the first
days following the change of medium (OLIVEREAU 1948). With euryhaline teleosts
which spawn in fresh water a particularly intensive activity of the thyroid was
established at the moment when they entered fresh water (FONTAINE, LACHIVER,
LELOUP & OLIVEREAU 1948). All this points to the fact that in some ways the thyroid
gland exerts an influence on the osmoregulation of the teleosts in brackish and
fresh water. FONTAINE, the greatest expert on these problems remarks (1953): «La
glande thyroide intervient dans la régulation du metabolisme hydrominérale chez
les poissons». (The thyroid gland influences the regulation of the water and mineral
metabolism of fishes.) (See also FONTAINE 1956, C. W. SMITH 1956, HOAR 1958.)

It may be that the *activity of the nervous system* of teleosts in brackish water
also has an influence on regulating the water and mineral economy. If the spinal
cord is destroyed the resistance of fresh-water fishes *(Cyprinus)* to a small increase
in salinity of the external medium is markedly reduced (DRILHON 1942).

Relationships existing in homoiosmotic brackish-water inhabitants between the activity of the osmoregulatory organs and tissues proper and the salinity of the brackish water will be discussed in the two chapters on osmotic and ionic regulation of brackish-water animals.

4. Volume and regulation of volume

In analyzing the physiological effects of brackish water on the organisms inhabiting it or those which have been transferred into it one frequently observes processes of swelling and osmotic water movements. These processes which affect volume and water content are most conveniently studied in the eggs of stenohaline marine invertebrates, especially echinoderms and annelids. Since their permeability to salts is low, short-term experiments produce curves of swelling and shrinkage which correspond very closely to those theoretical graphs which have been calculated

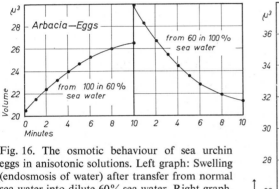

Fig. 16. The osmotic behaviour of sea urchin eggs in anisotonic solutions. Left graph: Swelling (endosmosis of water) after transfer from normal sea water into dilute 60% sea water. Right graph. Shrinking (exosmosis of water). After 25 minutes in 60% sea water the sea urchin cells were returned into normal sea water. After B. LUCKÉ & M. McCUTCHEON 1932.

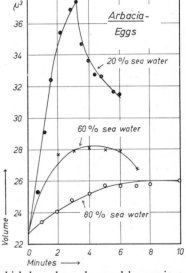

Fig. 17. The osmotic behaviour of sea urchin eggs which have been damaged by previous heating. After B. LUCKÉ & M. McCUTCHEON 1932.

on the basis of the gas laws (see fig. 16). This is particularly the case when the osmotically inert space, caused by the cell structures (it has also been called non-solvent volume) is taken into account. In the unfertilized egg of *Arbacia,* for example, this osmotically inert space amounts to about 7.3% of the total volume of the cell. The rate of water movement into or out of such an unfertilized sea urchin egg is proportional to the egg surface, the permeability of the cell membrane to water and the difference between the osmotic concentration of internal and external medium. If sea urchin eggs are transferred into very dilute sea water, e. g. 20% sea water (= 2 parts sea water + 8 parts aqua dest.) they rapidly swell and finally burst when the elasticity limit of their walls has been exceeded. No regulation

takes place which would produce a return to the normal volume. But curves of
swelling which fall off can be obtained by damaging the sea urching eggs before
the beginning of the experiment by heating them to 39°C for a short time, thus
interfering with their original relative semipermeability. When such heatdamaged
cells are transferred into 60% sea water they start swelling, like normal cells, for
the first few minutes, but then they shrink again (see fig. 17). This secondary passive
reduction in volume must not be mistaken for an active regulation.

RESÜHR (1935) has described similar, osmotically conditioned changes in volume
for the eggs of marine brown algae (Fucaceae from the Baltic), following changes
in the salt content of the external medium.

Similar observations can be made with isolated tissues of marine invertebrates
or with brackish-water Protozoa when the salinity of the medium is considerably
altered for short periods. If, on the other hand, smaller changes in salinity are
carefully applied it can be observed that in some specially suitable euryhaline Protozoa the cell volume remains more or less unchanged, owing to the regulatory
activity of contractile vacuoles. According to our present views the contractile
vacuoles of fresh-water Protozoa act as volume regulators, their
rhythmic pulsations continuously eliminating the water which has entered osmotically
through the outer surface. Numerous marine Protozoa have no contractile vacuoles
at all. But where contractile vacuoles occur in marine Protozoa their rate of pulsating
is usually less than that in fresh-water Infusoria. The big difference in pulsation
rate points to the fact that in marine Protozoa vacuolar activity only serves to
remove water which has been actively taken up and, presumably, metabolic products
as well. Several authors have investigated the activity of contractile vacuoles in
marine and fresh-water Protozoa after changes in the salinity of the external medium.
As regards short-term experiments considerable concurrent changes in vacuolar
activity have nearly always been observed. Some experiments by MÜLLER (1936)
may be quoted as an example. If *Amoeba proteus* which lives in fresh water is taken
through a series of brackish-water solutions with a step-wise increase in salinity,
the amount of water pumped out by the vacuoles in unit time increased up to a
salinity of $1.4^0/_{00}$ in the experimental medium. With further raising of salinity the
vacuolar performance decreased very considerably and already at $3^0/_{00}$ S it dropped

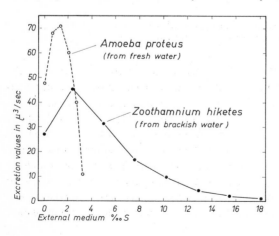

Fig. 18. The activity of the contractile vacuoles of *Amoeba proteus* and *Zoothamnium hiketes*
in fresh and in brackish water.
After R. MÜLLER 1936, altered.

to about ¼ of the original value (see fig. 18). Similar behaviour was shown by the peritrich Infusorium *Zoothamnium hiketes* which inhabits brackish water. Its vacuoles displayed maximum activity at about 2⁰/₀₀ S and worked more slowly both after an increase or reduction of salinity in the external medium (see fig. 18). Müller accounts for this initial increase of vacuolar performance on dilution of the medium by postulating that changes in the permeability and state of swelling of the cell membrane take place at the same time. In his opinion osmotic influx of water is greatest not in fresh water, but in brackish water of low salinity (1—2⁰/₀₀). We owe to Kitching (1954) a number of valuable studies on the physiology of contractile vacuoles; he takes the view that they react primarily to changes in body volume. His simultaneous observations of cell volume and vacuolar elimination in the marine peritrich Infusorium *Cothurnia* spec. are a model of careful microscopic-physiological analysis (see fig. 19). Reduction of the salt content of the external medium is said

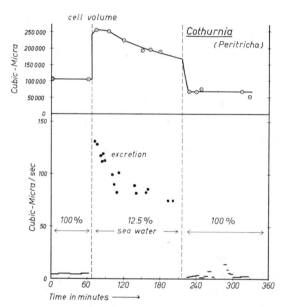

Fig. 19. The cell volume and vacuolar activity of a euryhaline peritrich marine Protozoon *(Cothurnia)* in sea and brackish water. After J. A. Kitching 1954.

to produce at first an increase of osmotic water influx and increase in cell volume and secondarily, triggered by this, to result in greater activity of the vacuoles. Some observations, however, seem to indicate that the contractile vacuoles do not merely function as regulators of volume. If the salinity of brackish water is altered gradually and euryhaline Protozoa are adapted to the new external concentrations for fairly long periods of time, no changes in vacuolar activity have been detected in many instances, or only very small ones. Oberthür (1937) stresses the fact that the holotrich ciliate *Frontonia marina* (from an inland salt spring of 44⁰/₀₀ S) directly transferred into either a more concentrated or more dilute medium, displays a depression or an increase in vacuolar elimination immediately after the transfer; the more marked the difference in concentration, the greater the change in vacular elimination. In the course of the next 48—120 hours a new regulation sets in, as a result of which the original pulsation frequency is regained and maintained thereafter. The fresh-water form *Frontonia leucas* shows the same behaviour in long-term experiments on adaptation, after transfer into brackish water up to $8^0/_{00}$ salinity.

Multicellular animals can also display considerable fluctuations in volume and water content in brackish water when its salt content has been changed. This can be observed in nature in a population of starfish *(Asterias rubens)* which lives at 10—12 m depth on the bottom of the Kiel Aussenförde (western Baltic). Depending on the direction of the prevailing wind this population either finds itself in a more saline bottom current from the open sea, or in less saline water. At a low salt content these forms look plump and fat, that is swollen through osmotic uptake of water, while at a high salinity they appear considerably smaller and their arms are more slender. This osmotic water shift which can easily be checked on live weight, can be demonstrated without difficulty in short-term experiments (see fig. 20). An active volume regulation does not seem to be possible, or only to a small degree. Starfish which have been directly transferred from North Sea water ($30^0/_{00}$ S) into Baltic water ($15^0/_{00}$ S) die as a rule within a few hours in a state of great swelling. Similarly

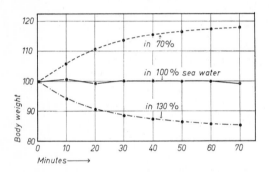

Fig. 20. Changes in volume (weight) of *Asterias rubens* after transfer into hypo-osmotic and into hyperosmotic sea water. After C. SCHLIEPER 1936.

starfish which have shrunk after transfer from Baltic into North Sea water are not able to regain their original water content. BINYON (1961) who experimented with starfish from the English coast (Whitstable) concludes from his experiments "*A. rubens* is incapable of weight or volume regulation". However, the differences obtained are only slight if comparative measurements are made of the volume of starfish from the western Baltic and starfish from the North Sea, of similar diameter and at the mean salinity of their respective biotopes (15 and $30^0/_{00}$ S respectively). This means that in the course of its immigration the brackish-water *Asterias* of the Baltic has been able to adjust its body volume to a great extent to the new external osmotic conditions. According to KOWALSKI (1955) the entire Baltic starfish has, on an average, a water content of 83% and North Sea starfish of the same size an average of 77% (cf. fig. 14, p. 237). This higher water content of the brackish-water individuals is brought about not only by a larger quantity of coelomic fluid, but by a greater swelling of the tissues as well (see table 9). But there is a limit to the increase of water content in the tissues. Among the adapted species of brackish-water inhabitants it hardly ever exceeds 10% of the total volume provided the species in question have not exceeded the lower limit of their osmotic resistance. These ecological observations have been confirmed by laboratory experiments which LANGE (1964) performed on *Strongylocentrotus droebachiensis*. In the next chapter a detailed

Table 9. Composition of ripe ovaries of *Asterias rubens* L. from the North Sea (30 °/₀₀ S) and from the western Baltic (15 °/₀₀ S). After KOWALSKI 1955. (All values in % of fresh weight).

Ovaries	North Sea *Asterias*	Baltic *Asterias*
Water content	80.5	84.0
Dry weight	19.5	16.0
Ash content	1.4	1.1
Organic substance	18.4	15.3

discussion will be found of the intracellular mechanism which operates in preventing an excessive water uptake and increase in volume in the tissues of euryhaline species in brackish water.

There are many species among marine invertebrates which almost behave as osmometers when transferred into brackish water as long as the reduction in the salinity of the external medium is not excessive; that is they swell through osmotic water uptake and appear unable to regain their original volume. If, however, the salinity of the external medium is suddenly and drastically reduced, these species initially show a considerable increase in volume, often followed by a decrease in volume which may lead to a return to the original volume or even beyond it. This may, e. g. be observed in the Nudibranch *Aplysia punctata* (see fig. 21). At the same time a more or less complete adjustment of the internal concentration and, especially of the Cl content, to the external medium can be seen. From this it follows that in the phase of volume reduction the individuals eliminate both water and salts. Since the same phenomenon can be observed when the intestinal tract has been blocked, this is presumably a case of passive elimination (filtration under pressure?) of water and salts through the body surfaces. In my opinion there can be no question of an active elimination of water and salts by these, mostly much damaged individuals which do not move and are probably often dead. It appears to be the same kind of passive adjustment of volume which has been observed when heat-damaged sea urchin eggs were transferred into brackish water and showed an increase in volume at first.

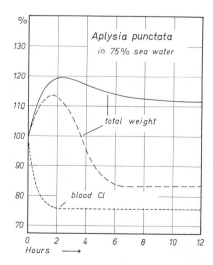

Fig. 21. Changes in total weight and in Cl content of the blood of the sea hare *Aplysia punctata* (Prosobranchia) after transfer from normal sea water into ¾ sea water. Two experimental animals displaying different responses. After A. BETHE 1934.

In other marine invertebrates which are more strongly euryhaline a transfer into brackish water also results at first in a more or less pronounced increase in volume. This is followed by a slow return to the original volume during which the individuals are able to move, being undamaged or only slightly injured. In many cases this kind of restoration of volume is an active regulation of volume. This difference between species with or without volume regulation can, among others, be very well seen in marine Polychaeta and Crustacea. Thus the more stenohaline species *Nereis pelagica* and *Nereis cultrifera* swell rather markedly after transfer into brackish water, become weak and die quickly if the external medium becomes more diluted. In contrast, the euryhaline species *Nereis diversicolor* shows only a slight or temporary uptake of water after transfer into brackish water; during the whole experimental period the individuals of this species have a normal appearance (SCHLIEPER 1929, BEADLE 1931, ELLIS 1939, BOGUCKI & WOJTCZAK 1964). *Nereis virens,* too, hardly takes up any water and remains active after transfer into brackish water up to $16^0/_{00}$ S. But at greater dilutions its weight rises rapidly and the animals become immobile (TOPPING & FULLER 1942). It is conceivable that in *Nereis diversicolor* and *Nereis virens* the nephridia assume a volume-regulating function in brackish water. More detailed analyses are still outstanding. KOLLER (1939) however, has proved that in the Pacific Gephyrea *Physcosoma japonicum* the nephridia influence its volume. As the two thin nephridial tubes, of 10—20 mm length, can easily be extirpated it has been possible to compare normal worms and those without nephridia. If *Physcosoma* is transferred from sea water ($34^0/_{00}$ S) into dilute water of only $22^0/_{00}$ S, both the normal individuals and those without nephridia show a considerable increase in weight, owing to osmotic water uptake during the first few hours. After that the weight of the normal animals rapidly decreases; within 24 hours their nephridia pump the water which has entered, out of their body. But the animals without nephridia are unable to regain their original weight, even after a long experimental period. Their maximum increase in weight is over 40% and after 48 hours they still show an increase in weight of about 25% (see fig. 20). But from the behaviour of the animals without nephridia it can be concluded that a certain amount of water and salts can be eliminated by a different path (skin and intestine) (cf. ADOLPH 1936 and GROSS 1954).

BEADLE (1934) has made a detailed study of the volume changes in the triclad Turbellarian *Gunda ulvae* in brackish water. If *Gunda* (from sea water) is transferred

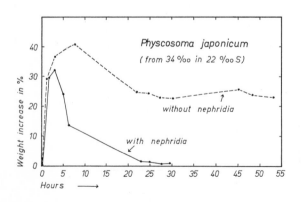

Fig. 22. Changes in volume and volume regulations of *Physcosoma japonicum* after transfer from sea water into brackish water. After G. KOLLER 1939.

into $1/10$ sea water a swelling occurs at first because of an osmotic influx of water through the ectoderm into the parenchyma; at the same time there is a marked and temporary reduction in the activity of the Turbellarian. Then the water which had been taken up also disappears from the parenchyma while at the same time the volume of the whole animal decreases and intracellular, water-holding vacuoles form in the cells of the gut (see fig. 23). Such vacuoles can be seen in the gut all the time while the Turbellarian, which has regained its activity, is being kept in the dilute medium. Elimination of the water which had been taken up from the intestinal cells, into the lumen of the gut and to the exterior, has not been demonstrated so far, but it is feasible.

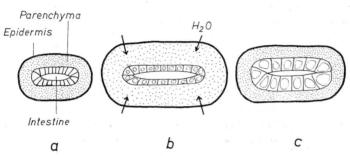

Fig. 23. Diagrammatic representation of the osmotic behaviour of *Gunda ulva* after transfer from sea water into 1/10 sea water.
a) Behaviour in normal sea water. b) Behaviour during the first hours after transfer into dilute sea water. Water enters osmotically through the ectoderm, causing the parenchyma to swell. The cells of the gut begin to collect this water in the form of intracellular vacuoles. c) Behaviour of the adapted animal, about 12 hours after transfer into 1/10 sea water. The water which had been taken up by the parenchyma has been removed by the cells of the gut which, in consequence, appear greatly swollen. The total volume of the animal has decreased again. After L. C. BEADLE 1934.

Recently, an experimental comparison between the capability for volume regulation of internal tissues and the degree of euryhalinity has been undertaken in some bivalves (LANGE 1968a). The volume of the muscle tissue of the euryhaline species *Mya arenaria* and *Cardium edule* seems to be independent within the natural salinity range. In contrast, the volume of the muscle tissue of more stenohaline species, as *Venerupis rhomboides, Pecten septemradiatus, Dosina exoleta,* and also to some extent *Modiolus modiolus,* apparently increases with decreasing salinity of the experimental medium. The possibility thus exists that the capacity for cellular volume regulation (or the capacity for preventing osmotically induced volume changes in tissues) might be one determining factor for euryhalinity in brackish water species.

If marine steno- and euryhaline Brachyuria are placed in brackish water the vital importance of volume regulation can be clearly demonstrated (see fig. 24). Though brachyurians are crustaceans with a hard shell, some species, in particular stenohaline or less euryhaline ones, swell to some extent in brackish water and then die. Only those individuals will survive which can reverse the original water intake and decrease in volume (see fig. 24, graph for *Maja,* c). In contrast to the more labile species the euryhaline shore crab *Carcinus maenas* does not change

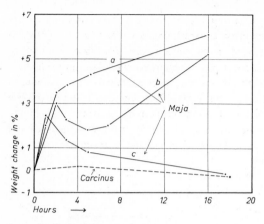

Fig. 24. Changes in volume (changes in weight) of *Maja verrucosa* (stenohaline) and *Carcinus maenas* (euryhaline) after transfer into dilute sea water (*Maja* a and b and *Carcinus* in 60% sea water, *Maja* c in 80% sea water).

its weight at all after transfer into brackish water. Only those individuals of *Carcinus* which had been somehow damaged before the start of the experiment (by heat or oxygen deficiency) increase their weight and die in the course of the experiment (see fig. 25). This does not mean that there is no osmotic influx of water when *Carcinus* is transferred into brackish water; but it is being compensated for by an increased production of urine which serves to regulate the volume (NAGEL 1934;

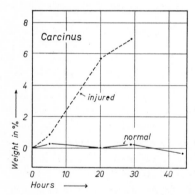

Fig. 25. Changes in volume of *Carcinus maenas* after transfer into brackish water (1 part sea water, 4 parts fresh water). After C. SCHLIEPER 1929.

BETHE, von HOLST & HUF 1935). The above authors studied the urine production of *Carcinus* by measuring the increase in weight of the crustaceans after blocking the excretory pores while SHAW (1961) based his calculations of urine production on the fact that *Carcinus* takes up radioactive sulphate from the external medium at a constant rate and excretes it just as continually by means of the antennary glands. The daily quantity of urine as measured in this manner amounts to about 3.6% of the body weight for individuals in normal sea water. After transfer of the crustaceans into brackish water of 14⁰/₀₀ S it increases more than eightfold (see fig. 26).

The changes in volume described above which some marine invertebrates undergo on transfer into brackish water are the expression of disturbed water economy. It must be assumed that all brackish-water inhabitants, like all marine and fresh-water organisms, constantly take up a certain amount of water and in the same way expel it to the exterior. Any permanent disturbance of this regular passage of water would make life in brackish water impossible for such individuals. The truth of this statement can be well demonstrated by examining the behaviour of fresh-water animals in brackish water. All fresh-water animals are subject to a constant "passive" osmotic influx of water from the hypo-osmotic external medium. Thus in fresh water they do not need to take in water by mouth in order to meet their water requirements. On the contrary, they have to safeguard themselves from a flooding of their tissues and too great a dilution of their body fluids by water entering osmotically. In the higher forms the water entering osmotically is continually being eliminated by the excretory organs (nephridia, antennary glands, kidneys etc.); it takes the form of a dilute hypo-osmotic (less commonly blood-iso-osmotic) urine. The mechanisms regulating water economy function in such a way as to ensure a constant water content and an equally regular passage of water. But the water economy of the fresh water animals is geared to a constant osmotic influx of water. They seem to be incapable

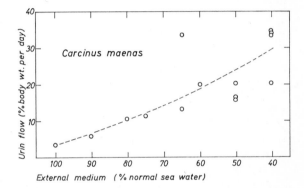

Fig. 26. Urine production of the euryhaline shore crab *Carcinus maenas*, depending on the salinity of the external medium. After J. SHAW 1961.

of actively taking up water if the "passive" osmotic uptake of water is diminished in hypotonic brackish water or if it ceases in iso- or hyper-osmotic brackish water. Only those species which survive both in fresh and salt water are able to maintain the vital passage of water in both media.

HERRMANN (1931) has shown that the crayfish *Astacus fluviatilis* loses weight in brackish water of over 20⁰/₀₀ S, owing to osmotic removal of water. At the same time the amount of circulating blood is reduced; the animals are then no longer able to

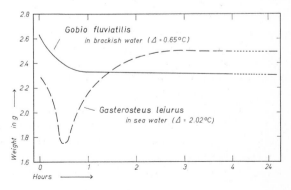

Fig. 27. Changes in weight in the gudgeon *(Gobio fluviatilis)* and the stickleback *(Gasterosteus leiurus)* after transfer into brackish and sea water, respectively. After F. GUEYLARD 1925.

compensate for the loss of blood by drinking. In this way crayfish which have been transferred into sea water become dehydrated and at the time of their death they contain but a few drops of blood or none at all. Similar behaviour is shown by steno-haline fresh-water fishes, e. g. *Gobio fluviatilis*, after transfer into hyperosmotic brackish water. Owing to osmotic removal of water they lose weight and die without being able to regain their normal water content. Euryhaline teleosts, however, such as the stickleback *Gasterosteus* are capable of compensating for initial water loss after transfer into hyperosmotic brackish water or sea water (GUEYLARD 1925); this enables them to survive the sudden salinity change in the external medium (see fig. 27).

The same holds for stenohaline marine teleosts. If transferred into brackish or fresh water they rapidly take up water osmotically, become immobile through the swelling of their tissues and die after some time. Euryhaline species of fish, e. g. the Flatfish *Pleuronectes platessa* and *Pleuronectes flesus,* can keep their volume constant, independently of the osmotic forces acting on them. They can do this over the whole range of concentrations in which they live, by producing and eliminating smaller or larger amounts of urine, as required. In contrast to the fresh-water fishes the marine teleosts drink continuously and take up water through the gut. They have to do this as they continuously pass out water osmotically to the outside, owing to the hypotonicity of their internal medium (see p. 260). As they suffer, as it were, from water shortage, their kidneys can only eliminate small amounts of urine.

We owe to HENSCHEL (1936) a fine analysis of the volume regulation and water economy of *Pleuronectes platessa* in brackish water. *Pleuronectes platessa* occurs in brackish water up to about $5^0/_{00}$ S. Its internal medium is approximately isosmotic with brackish water of $10^0/_{00}$ S. By taking continuous records of the weight of a starving specimen which has had a long-term adaptation to $16^0/_{00}$ S it will be found that weight decreases slowly as a result of hunger alone (at first about 1% a day). After transfer into brackish water of $8^0/_{00}$ S there is a slight increase in weight for the first three days (by about 2%) followed by a continuous decrease as

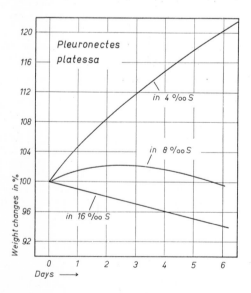

Fig. 28. Changes in weight in the plaice *(Pleuronectes platessa)* after transfer into brackish water of varying salinities. After J. HENSCHEL 1936.

a result of starvation. The initial increase in weight is presumably due to a slight osmotic influx of water which is counterbalanced after some time. In brackish water of 4⁰/₀₀ S, however, the water influx is considerably greater, resulting in an increase in volume which persists unchanged for several days; this cannot be compensated for since the regulating mechanisms no longer function (see fig. 28). HENSCHEL has attempted to analyze these regulating factors by ligaturing the pharynx and the urinary papilla. Like all other marine teleosts the plaice continuously takes up water by drinking and eliminates it partly as urine through the kidneys (while the salts of the sea water are in part eliminated by a different path). If drinking is inhibited by ligaturing the pharynx the plaice soon loses weight in sea water and in hyperosmotic brackish water (e. g. in 16⁰/₀₀ S) as the osmotic water loss referred to above still persists and the kidneys continue to eliminate urine (see fig. 29). If the urinary papilla is ligatured the weight of the fish increases markedly since it continues to drink but is no longer able to eliminate the urine. Only if (in 16⁰/₀₀ S) the pharynx and the urinary papilla are ligatured at the same time the weight remains fairly constant or else decreases slowly owing to exosmosis and the effects of starvation. But a plaice operated on in the same way behaves quite differently in brackish water of 8 and 4⁰/₀₀ S. Osmotic influx of water is now being added to the uptake of water through the gut. It appears that in brackish water of 8⁰/₀₀ S the fish is just able to regulate water intake in such a way that the kidneys can eliminate the total amount of water which has been taken up. Only if pharynx and urinary papilla are ligatured simultaneously will the weight increase slowly owing to osmotic water uptake. But in brackish water of low salinity (4⁰/₀₀ S) the osmotic water uptake is so great that the kidneys function is no longer adequate and the weight rises, without compensation, even if the pharynx has been ligatured (see fig. 29).

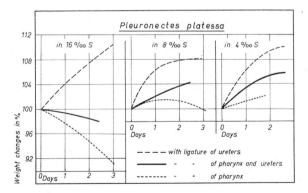

Fig. 29. Changes in weight of the plaice *(Pleuronectes platessa)* in brackish water of varying salinities. They are due to osmotic movement of water as well as active uptake and loss of water. After J. HENSCHEL 1936.

According to WIKGREN (1953) the strongly euryhaline freshwater lamprey *Lampetra fluviatilis* produces at 16—18°C 359 ml urine/kg per day in fresh water; in a medium of 100 mM corresponding to 17.4% sea water only 30—80 ml urine/kg per day; in 200 mM, corresponding to 35% sea water only 2—13 ml urine/kg per day.

Morris (1956) found amounts of urine which were lower on the whole, but showed similar variations, after the same experimental animals had been carefully adapted to the changed salinity of the external medium. The rainbow trout *Salmo gairdneri* is also euryhaline; in fresh water it produces 70—90 ml urine and after adaptation to sea water only 0.5—1 ml of urine per kg of body weight and day (Holmes 1961).

5. Osmotic concentration of the internal medium and osmoregulation

Survival of brackish-water organisms depends primarily on the salt content of the external medium and its effect on the internal medium. The salinity or osmotic concentration respectively of brackish water is solely determined by its content of inorganic salts or their ions. But in the internal medium of brackish-water organisms, that is their body fluids and tissue fluids, osmotic concentration is frequently determined by inorganic salts together with dissolved organic substances of low molecular weight.

Table 10. Physicochemical values of brackish and sea water.

Salinity $^o/_{oo}$	Specific gravity at 0° compared with that o distilled water at 4°C	Freezing point depression $\Delta °C$	Electrical conductivity at 25°C mS cm^{-1}
5.0	1.0040	0.267	8.83
10.0	1.0080	0.534	16.97
15.0	1.0120	0.802	24.68
20.0	1.0161	1.074	32.05
25.0	1.0201	1.349	39.23
30.0	1.0241	1.627	46.21
35.0	1.0281	1.910	53.01

A comparison of osmotic concentration in the external and internal media of brackish-water organisms points to a division into two large groups. In the one, the "poikilosmotic" group (osmotic conformers) there is equilibrium (isosmy) or approximate balance between the internal and external concentrations. The other "homoiosmotic" group (osmotic regulators) has the ability to maintain an internal concentration more or less independent of the external medium, thus to some extent avoiding the direct effects of the salinity of the brackish water (Schlieper 1929, 1932). For practical reasons it is best to record the osmotic concentrations of brackish and sea water, like those of animal and plant fluids, in freezing point values (see table 10) or in Milliosmol. These units have the advantage of being independent of chemical composition, molecular size and dissociation of the dissolved substances; they are proportional to the total number of dissolved particles. For instance sea water of 34.1 $^o/_{oo}$ S has an osmotic concentration of 1000 m Osmol/kg water (= 1129 mMol/kg water); it freezes at —1.86°C and could produce an osmotic pressure of 22.4 atmospheres in an osmometer with a membrane ideally semipermeable against distilled water. The same freezing point depression — exactly 1.858°C — is displayed by a solution of an ideal non-electrolyte containing a gram molecular weight of the nonelectrolyte in a kg of water. As a comparative measure of an effective osmotic concentration it is also possible to use the concentration of an isosmotic NaCl solution in per cent or in Millimol per kg of water or litre (see table 11).

Table 11. Freezing point depression of NaCl solutions.

% NaCl	0.58	1.17	1.75	2.92	4.08
mMol/kg H₂O	100	200	300	500	1000
Δ°C	0.34	0.68	1.02	1.69	2.38

Before we consider in detail the relationship between osmotic concentrations of the external and internal medium in brackish-water animals, the general importance of the osmotic concentration in the internal medium of animals will be discussed (see fig. 30). In the marine invertebrates which are closest to the origin of life the

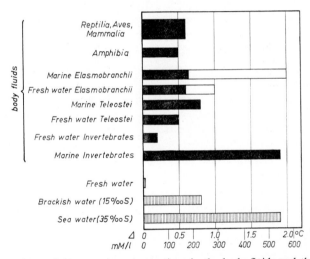

Fig. 30. Comparison of the osmotic concentrations in the body fluids and the natural external media of water living animals. (White segments of blocks for Elasmobranchia = the proportion of total concentration due to urea). After C. SCHLIEPER 1955.

Table 12. Comparison of freezing point depressions (Δ °C) of sea water, blood and cells of marine invertebrates. After POTTS 1952. Individual measurements, no average values.

a) *Psammechinus miliaris*				
Sea water	1.790	1.860	1.920	1.900
Egg cells	1.790	1.870	1.900	1.910
b) *Mytilus edulis*				
Sea water	2.070	2.080	2.080	2.080
Blood	2.080	2.090	2.080	2.075
Muscle cells	2.070	2.080	2.080	2.085

osmotic concentration of the body fluids still roughly corresponds with that of the surrounding sea water. The "blood" of many species contains inorganic salts in almost the same proportions. The cells and tissues of these animals are in equilibrium with the blood and the sea water (see table 12).

For these species sea water or a salt solution resembling sea water are the ideal fluid to substitute for blood. The fresh-water animals (invertebrates and fishes) behave differently. Even hard fresh water is totally unsuitable as a medium for their tissues. In order to maintain a constant salt concentration of their body fluids all fresh-water organisms must protect themselves from flooding by water entering osmotically from outside. Consequently many species continuously eliminate large amounts of a urine rich in water and strongly hypo-osmotic to blood. (Only a few slightly permeable species such as *Eriocheir sinensis, Telphusia fluviatilis* etc. produce small quantities of a blood-isosmotic urine.) The store of vital salts found in their cells and body fluids is a valuable capital for them and has to be managed carefully. But in contrast to the marine invertebrates their metabolism is adjusted to a considerably lower salt content in their tissues. Many species are capable of actively removing physiologically important ions from the surrounding fresh water and to transport them to the inside against the concentration gradient (KROGH 1939). The position is particularly complex for fishes living in the sea which may be considered as having returned from fresh water. They are descendants of ancestors which lived in fresh water and in memory of this — as it were — have maintained a salt concentration in their blood which is but ⅓ to ½ of that of the surrounding sea water. Consequently they are subjected to a constant exosmosis of water and an equally continuous intake of salts by diffusion from the hyperosmotic external medium. They compensate for the osmotic water losses by drinking sea water. It is inevitable that they also absorb in their intestine the univalent ions (especially Cl^-, Na^- and K^+) of the sea water which they take up. They then eliminate the excess of these ions through their gills (KEYS 1933). Their kidneys produce only small quantities of urine which is about blood-isosmotic or weakly hyperosmotic. Only in the marine sharks and rays (Elasmobranchia) does the osmotically effective total concentration

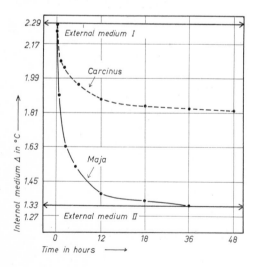

Fig. 31. Differential changes of blood concentration in *Maja verrucosa* (stenohaline) and *Carcinus maenas* (euryhaline) after transfer into dilute sea water. After E. SCHWABE 1933.

Table 13. Relationship between the freezing point depressions of the internal and the external medium in some aquatic invertebrates (examples).

Species	Internal medium	External medium	Difference Δ°C	Author
Mytilus edulis	2.00	1.97	+ 0.03	G. M. BELIAEV
	1.51	1.46	+ 0.05	,,
	1.36	1.28	+ 0.08	,,
	1.19	1.05	+ 0.14	,,
	0.43	0.30	+ 0.13	,,
Macoma baltica	2.02	1.96	+ 0.06	,,
	1.47	1.35	+ 0.12	,,
	0.38	0.30	+ 0.08	,,
Mya arenaria	2.04	2.02	+ 0.02	,,
	1.41	1.33	+ 0.08	,,
	1.13	1.04	+ 0.09	,,
	0.46	0.30	+ 0.16	,,
Carcinus maenas	2.00	1.91	+ 0.09	C. SCHLIEPER
	1.60	1.16	+ 0.44	,,
	1.52	1.08	+ 0.44	,,
	1.49	0.90	+ 0.59	,,
	1.43	0.63	+ 0.80	,,
Eriocheir sinensis	2.12	2.23	− 0.11	W. SCHOLLES
	1.69	1.79	− 0.10	C. SCHLIEPER
	1.66	1.72	− 0.06	,,
	1.27	0.87	+ 0.40	,,
	1.22	0.02	+ 1.20	,,
	1.18	0.02	+ 1.16	W. SCHOLLES
Nereis pelagica	2.06	1.96	+ 0.10	G. M. BELIAEV
	0.46	0.42	+ 0.04	C. SCHLIEPER
Nereis diversicolor	1.14	0.95	+ 0.19	,,
	1.10	0.88	+ 0.22	,,
	0.86	0.45	+ 0.41	,,
	0.70	0.21	+ 0.49	,,
	0.50	0.04	+ 0.46	,,
Palaemonetes varians	1.34	1.99	− 0.65	N. K. PANIKKAR
	1.37	1.69	− 0.32	,,
	1.18	1.24	− 0.06	,,
	1.28	1.09	+ 0.19	,,
	1.19	0.63	+ 0.56	,,
	1.24	0.28	+ 0.96	,,
	1.24	0.10	+ 1.14	,,
Astacus fluviatilis	0.80	0.02	+ 0.78	F. HERRMANN
	0.95	0.55	+ 0.40	,,
	1.00	0.80	+ 0.20	,,
	1.14	1.02	+ 0.12	,,

of the blood correspond to that of the surrounding sea water; they compensate for the difference by a high urea content of the blood (corresponding to the white parts of the blocks in fig. 30).

Let us consider the relationship between the osmotic concentrations of the internal and external medium of invertebrates occurring in brackish water. If two marine species, *Maja verrucosa* and *Carcinus maenas,* one more stenohaline and the other more euryhaline, are transferred from sea into brackish water (see fig. 31) the internal concentration of the stenohaline species will passively adjust itself to the changed concentration of the external medium within a few hours (poikilosmotic behaviour). By contrast, the internal concentration of the euryhaline species will be far less reduced and maintained above that of the external medium. This is done by means of active osmoregulation (homoiosmotic behaviour). Corresponding differences in osmotic behaviour can be observed in numerous invertebrates which occur in brackish water (see table 13 and fig. 32). As a rule the stenohaline or partially euryhaline species are passively poikilosmotic in brackish water of low salinity. If some of these species (e. g. *Asterias rubens, Mytilus edulis* etc.) nevertheless manage to exist in brackish water of low salinity, their dependence on the salt concentration of the external medium is often shown by a considerable reduction of body size and activity under these circumstances. Thus only those species may be called truly euryhaline

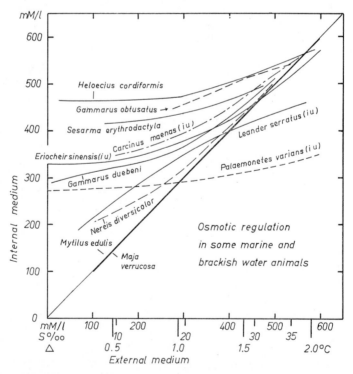

Fig. 32. Relationship between the osmotic concentrations of the internal and external media in some marine and brackish-water species (iu = isosmotic urine). After different authors from L. C. BEADLE 1943.

which exist in brackish water without suffering any marked reduction in size or loss of activity and these are only those species which are homoiosmotic in brackish water of low salinity, that is possess osmoregulatory ability. Among them the marine immigrants have an internal and external medium still approximately isosmotic while they live in the sea, but the farther they penetrate into brackish water the more hyperosmotic does the internal medium become. In this case the osmoregulatory functions are carried out by the body surfaces (e. g. gills) or, more rarely, the excretory organs which mostly produce an approximately blood-isosmotic urine, even in brackish water (see table 14).

Table 14. Comparison of freezing point depressions in the external medium, blood and urine of *Carcinus maenas*. After SCHLIEPER 1929 b.

External medium (°C)	Blood (°C)	Urine (°C)	Number of animals tested
1.91	2.00	1.97	1
1.90	1.94	1.94	2
1.16	1.60	1.60	4
1.08	1.50	1.48	4
1.08	1.53	1.54	4
1.00	1.47	1.48	2
0.98	1.56	1.58	3
0.82	1.48	1.55	3
0.64	1.25	1.27	3

Working in my laboratory NAGEL (1934) was the first to show that the gills of *Carcinus maenas* are able to remove ions (chlorides) from the external medium in brackish water and transport them to the inside against the osmotic gradient (see table 15). We owe to SHAW (1961) an exact analysis of this process. He investigated the active uptake of ions by *Carcinus* in brackish water by means of radioactive sodium chloride (^{24}NaCl). He found that the effective osmoregulatory mechanism provides specifically less for the maintenance of a certain osmotic blood concentration, but rather prevents a dropping of this concentration below a minimum level (see fig. 33). KOCH and co-workers (1954) have previously shown that isolated surviv-

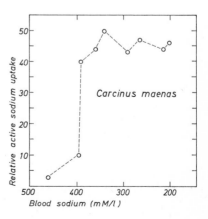

Fig. 33. Active uptake of sodium from the external medium by *Carcinus maenas*, depending on the sodium content of the internal medium. After J. SHAW 1961.

ing gills of the mitten crab *Eriocheir sinensis* in fresh or diluted brackish water absorb Na- and K-ions (together with Cl-ions) actively from the external medium (cf. p. 272). The difference in concentration inside and outside an animal varies considerably according to the species and to the concentration of the medium. The comparative study of 4 species of *Gammarus (G. obtusatus, locusta, duebeni* and *pulex)* by BEADLE & CRAGG (1940) provides a good example. It may be that physiological races are formed

Table 15. Chloride transport through the gills of *Carcinus maenas* from outside to the inside in hypotonic brackish water.

(The experimental animals were first long-adapted in external medium I, then transferred into the more saline, but still hypotonic medium II and examined after 24 hours' adaptation). After NAGEL 1934.

Medium	Salinity °/oo	Cl mg/ml	Δ°C
External medium I	14.9	8.57	0.89
Internal medium I..........	—	12.1	1.31
External medium II	20.1	11.45	1.18
Internal medium II	—	14.5	1.55

in some species. As mentioned before (see p. 220) we know a brackish-water race and a fresh-water race of *Gammarus duebeni* which possess different osmoregulatory abilities. According to BELIAEV & BIRSTEIN (1944) *Dikerogammarus* forms two physiological races in the Volga and in the Caspian Sea (at 13°/oo S) (see fig. 34).

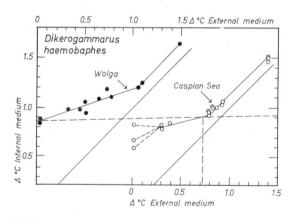

Fig. 34. The relationship of osmotic concentration in the external and internal medium in specimens of *Dikerogammarus haemobaphes* from the Volga and the Caspian, after adaptation to Caspian water of varying salinities. After G. M. BELIAEV & J. A. BIRSTEIN 1944.

Exact proof through long-term experiments in adaptation is still outstanding. Even the populations of *Carcinus maenas* from the North Sea (30°/oo S) and from the western Baltic (15°/oo S) show differences in osmoregulatory performance in brackish water of low salinity (see fig. 35); these persist even after several weeks of re-adaptation (THEEDE 1969).

The Polychaete *Nereis diversicolor* also appears to show different behaviour in the Baltic and the Black Sea. The populations occurring in the Baltic, like many others, are able to survive indefinitely in brackish water of 4°/oo S. While the internal

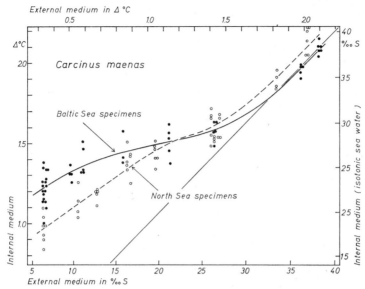

Fig. 35. The osmoregulatory performance of North Sea and of Baltic specimens of *Carcinus maenas* in brackish water. After H. THEEDE 1969.

medium of the worms in sea water is isosmotic to the external medium, the brackish-water individuals display active osmoregulation and have an internal medium which is hyperosmotic to the external one (SCHLIEPER 1929, SMITH 1955). By contrast the same species is less euryhaline in the Black Sea and does not appear to possess any osmoregulatory abilities (PORA & ROSCA 1944). "*Nereis diversicolor* de la Mer Noire n'a pas la possibilité de garder un milieu intérieur indépendant de l'extérieur" (p. 6). [*Nereis diversicolor* from the Black Sea is unable to maintain an internal medium independent of the external one]. There *Nereis* occurs at a salinity of 19 to $20^0/_{00}$ S in the *Zostera*-region of Agigea and tolerates e. g. $13^0/_{00}$ S for only about 100 hours and $4^0/_{00}$ S for only about 4 hours. At the same time the worms in these dilute media swell so much that the investigators speak of a "gigantisme endosmotique". The question arises whether the peculiar features of the behaviour of *Nereis diversicolor* from the Black Sea should be accounted for as formation of physiological races or by these worms suffering from some degree of hydrogen-sulphide poisoning. The water in Agigea where they occur is said to contain an average of 2 ccm H_2S per litre. In contrast to the marine immigrants among the brackish-water inhabitants which have been discussed so far, the brackish-water species proper are able to maintain the osmotic concentration of their internal medium constant to a much greater extent. In brackish water of low salinity their internal medium is hyperosmotic in respect of the external one and in more saline brackish and sea water their internal medium is hypo-osmotic. PANIKKAR (1941) was the first to demonstrate these astonishing osmoregulatory powers for the brackish-water prawn *Palaemonetes varians* (see fig. 36). This ability of a hypo-hyperosmotic regulation has also been developed to varying degrees in some tropical brachyurians from the littoral or from brackish water (JONES 1941, PROSSER et al. 1955, GROSS 1964). The greatest ability in this respect is shown by the small brine shrimp *Artemia salina* which exhibits hypo-osmotic regulation already in brackish water above 10⁰/₀₀ S (CROGHAN 1958). Finally the same type of osmoregulation is displayed by some Diptera larvae which can live in brackish water (BEADLE 1939, RAMSAY 1950—53, SUTCLIFFE 1960). Thus, in dilute brackish water all these species are able to transport inorganic ions actively from outside inwards, while in more concentrated media they eliminate salts equally actively and, by drinking, they satisfy their water requirements which have increased owing to exosmosis.

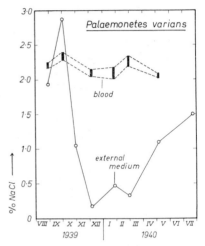

Fig. 36. Seasonal changes in osmotic concentration of the internal medium of *Palaemonetes varians* in relation to the concentration of the external medium. After N. K. PANIKKAR 1941.

Only rarely do the excretory organs proper take part in the osmoregulation of the brackish-water invertebrates discussed in the preceding paragraph. As a rule the urine excreted by the antennary glands of the brackish-water brachyurians and prawns is isosmotic to the blood. Only in the endemic brackish-water amphipod *Gammarus duebeni* is the urine hypo-osmotic to the blood in brackish water of low salinity (that is sea water more than half diluted) (LOCKWOOD 1961). In some Diptera larvae *(Aedes detritus, Coelopa frigida etc.)* which are able to live in brackish water the fluid secreted by the rectum plays an even more important part in osmoregulation. The "primary urine liquid" originally produced by the Malpighian tubules is still isosmotic to the haemolymph. But later the activity of certain cells of the rectum makes it hypo- or hyperosmotic to the internal medium, depending on the salt content of the external medium. (Compare also new results by SMITH 1970 in *Nereis*.)

However, fresh-water animals (invertebrates and fishes) behave quite differently on immigrating into brackish water. In brackish water they are no longer able to retain their original internal concentration. Even in brackish water of low salinity, hypo-osmotic and isosmotic brackish water, the osmotic concentration of their internal medium rises, while in brackish water of higher concentration it displays, to a large extent, a passive adaptation to the concentration of the external medium (see fig. 37). This process of a fresh-water animal adjusting to brackish water has been analyzed in detail by HERRMANN (1931) for the crayfish *Astacus fluviatilis*. If the crayfish (Δ = 0.8°C) is transferred into blood-isosmotic brackish-water of 15⁰/₀₀ S the internal concentration rises slowly, until after one or two weeks a new equilibrium between

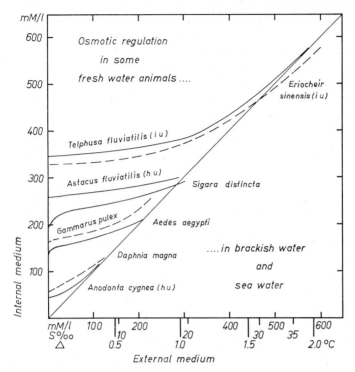

Fig. 37. Relationship between the osmotic concentrations of the internal and external medium of some fresh-water invertebrates after transfer into brackish and sea water (iu = isosmotic urine, hu = hypo-osmotic urine). After various authors from L. C. BEADLE 1943.

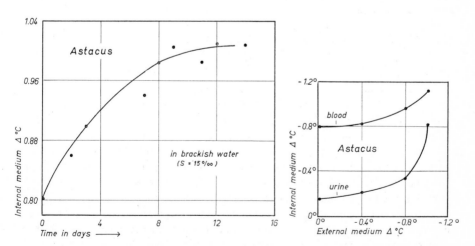

Fig. 38. The change in osmotic concentration in the blood of *Astacus fluviatilis* after transfer into isosmotic brackish water. After HERRMANN 1931.

Fig. 39. The osmotic concentrations of blood and urine of *Astacus fluviatilis* after adaptation to brackish water of varying salinities. After F. HERRMANN 1931.

internal and external concentration has been established (see fig. 38). If in brackish water of higher salinity the fresh-water animal completely loses its osmotic independence and the excretory organs cease to function, then the limits of osmotic resistance of the species have been exceeded (see fig. 39).

If a stenohaline fresh-water fish is gradually adapted to increased salinities of the external medium, the osmotic concentration of the internal medium also rises slowly (see fig. 40). From a certain salinity of the external medium onwards there is isosmy between internal and external medium. If e. g. a carp *Cyprinus carpio* (internal medium $\Delta = 0.50°C$) is transferred into hyperosmotic brackish water of $18.6^0/_{00}$ S ($\Delta = 1.0°C$) the concentration of the urine produced rises from $\Delta 0.07°C$ to a maximum of $\Delta 0.88°C$ (MARTRET 1939). At the same time the quantity of urine is greatly reduced. Individual carp may live for weeks or months in this concentration of the external medium. If, however, a carp is immediately transferred into brackish water of high salinity a strong osmotic removal of water sets in and no adjustment between the concentrations of the internal and external media takes place (see fig. 40).

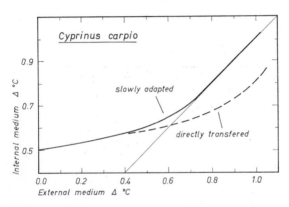

Fig. 40. The relationship between the freezing-point depressions of the internal and the external medium in the carp *(Cyprinus carpio)* after slow adaptation and after direct transfer into brackish and sea water. From P. PORTIER 1938.

After such considerable increase of the salt concentration of the external medium within a short time the condition of the fish quickly deteriorates and no recovery takes place when a certain salt concentration of the external medium has been exceeded. Obviously an entirely different behaviour is shown by euryhaline teleosts which like the eel *Anguilla vulgaris* are able to survive in fresh and in salt water (brackish and sea water). After a sudden change of the salinity of the external medium they show some parallel change in their internal concentration; but it is far less

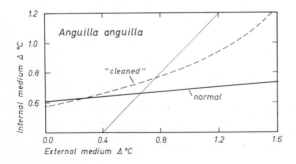

Fig. 41. Relationship between the freezing-point depressions of the internal and external medium in the eel *(Anguilla anguilla)*. Comparison between normal individuals and those in which there was no osmotic protection of the skin after it had been rubbed with a cloth. From P. PORTIER 1938.

pronounced. After a few days a new equilibrium is attained and finally the fish is capable of establishing its species-specific internal concentration more or less independently of the salt content of the external medium (see fig. 41). A different reaction is obtained for such a euryhaline teleost if, before such an experiment of transfer, the fish is somehow damaged or if its mucous covering and its sensitive epidermis have been mechanically injured (by rubbing with a rough cloth). In that case the fluctuations in the concentration of the internal medium are much greater (see fig. 41). If changes in concentration occurring in such specimens exceed a certain amount they become lethal. If marine teleosts are transferred into brackish water of low salinity it can be demonstrated that they only remain viable as long as they succeed in keeping the osmotic concentration of their internal medium fairly constant.

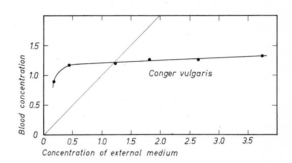

Fig. 42. The osmotic concentrations of the blood of *Conger vulgaris* in relation to the salt content of the external medium. The concentrations are expressed in % data of isosmotic sodium chloride solutions. After R. MARGARIA 1931.

Thus the conger eel *Conger vulgaris,* e. g., tolerates a dilution of sea water up to about a tenth. Up to this external concentration the blood concentration of the fish decreases only slightly. If, however, the external medium is further diluted, the osmotic concentration of the blood also diminishes considerably; at the same time the experimental animals swell up and die (see fig. 42). From such transfer experiments it may be concluded that, in the last resort, it is the magnitude of the volume- and osmoregulatory capacity which determines the degree of independence from the concentration of the external medium, both for homoiosmotic invertebrates and fishes as well.

The tissue fluids and cell fluids of invertebrates and fishes are generally isosmotic with the coelomic fluid and the circulating body fluids (see also table 9). While osmotic concentration of the primary body fluids is often determined chiefly by their content of inorganic ions (especially Na- and Cl-ions), in the cell fluid proper it is, to a considerable degree, brought about by their content of organic compounds of low molecular weight. Thus the blood of the common mussel contains about

Table 16. Comparison of chloride concentrations in the external medium, blood and foot muscle of the common mussel *Mytilus edulis*. After KROGH 1938. All data in mMol/kg water.

Duration of adaptation	External medium Cl mMol	Blood Cl mMol	Foot muscle Cl mMol
	532	526	189
24 hours	284	283	100
48 hours	229	224	55

as much chloride as the external medium. But in specimens from the sea and from
brackish water the chloride content of the tissue water is always less than 50% of
that of the external medium (see table 16). In the amphipod *Gammarus duebeni*,
too, the chloride content of the tissues reaches only a fraction of the chloride content
of the blood, whether this crustacean is adapted to fresh or sea water (see fig. 43).
The differences are even greater in a comparison of the cation content of the external
medium, of the blood and the tissues of brackish-water species. A detailed account
will be given in the next chapter (ionic content and ionic regulation).

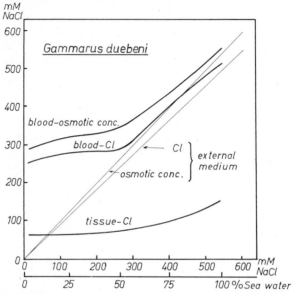

Fig. 43. Relationship between the chloride concentrations in the external medium, blood
and tissues of *Gammarus duebeni*. After L. C. BEADLE & J. B. CRAGG 1940.

Only brief reference will here be made to a problem which has been investigated
in recent years, that is the part played by free amino acids and other organic compounds in the osmotic concentration of tissue fluids of many aquatic animals. Among
others, large quantities of free glycine, proline, arginine, glutamic acid and alanine
have been found intracellularly (DUCHÂTEAU, SARLET, CAMIEN & FLORKIN 1952).
In individuals of euryhaline species adapted to brackish water the amount of amino
acids not bound to protein in the muscle tissue is always considerably smaller than
in individuals of the same species in normal sea water (DUCHÂTEAU & FLORKIN 1956).
The intracellular free amino acids etc. can be determined quantitatively as ninhydrin-
positive substances. In each case the decrease in their concentration is greater than
can be accounted for by the simultaneous increase of the water content in their
tissues (see fig. 44).

Conversely it has been observed that in the tooth carp *Platypoecilus maculatus*
which comes from fresh water the content of free amino acids rises within a few
hours after transfer into brackish water (see fig. 45); it reaches a maximum after
about 7 days and then it evidently attains a constant level (ANDERS et al. 1962).

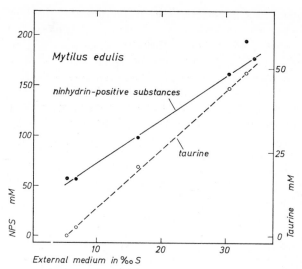

Fig. 44. Relationship between the salinity of the external medium and the concentrations of NPS (ninhydrin-positive substances, determined as taurine equivalents) and taurine in the tissue of the common mussel *Mytilus edulis*. After R. LANGE 1963.

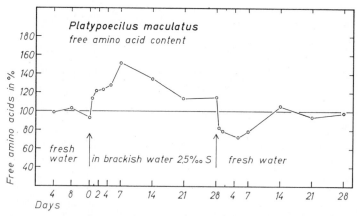

Fig 45. The effect of sudden salinity changes in the external medium on the free amino acid content in the toothed carp *Platypoecilus maculatus*. After F. ANDERS et al. 1962, altered.

After the fish had been returned to fresh water the amino acid level dropped, stayed at a minimum for about 7 days and finally regained its normal value. All these observations indicate that the free amino acids and other active organic compounds have a special significance for the intracellular osmotic adaptation and the osmotic resistance of the tissues. This problem will be dealt with in more detail in a discussion of ionic regulation in the tissues of brackish-water animals.

The osmotic conditions of brackish-water algae have not been studied so extensively. The cell fluids of the algae, both in sea and brackish water are known to be strongly hyperosmotic to the external medium (HOFFMANN 1932, OSTERHOUT

1933, COLLANDER 1936, MOSEBACH 1936, BIEBL 1952). In their cells there usually seems to be a complex equilibrium between the osmotic forces on one hand and the forces of imbibition of the protoplasm and cell membranes on the other (HOFF-MANN 1943). In general the osmotic value of the cell fluids in euryhaline species is at the same level above that of the external medium, irrespective of the salt content of the water in which they occur.

In this way the marine algae which occur in sea or brackish water of greatly varying salinity, are able to maintain their vital cell turgor, that is their intracellular hydrostatic pressure, always at approximately the same level, as the difference between the osmotic potentials of internal and external medium. In experiments with the green alga *Chaetomorpha linum* the salinity of the external medium was altered; according to KESSELER's measurements (1959) it took about 24—28 hours until the turgor had regained its original value or else a new final value had been established (see table 17). After a salinity change in the external medium the rate of

Table 17. Relation between the osmotic concentration of the cell fluid of the green alga *Chaetomorpha linum* and the salinity of the external medium. Medium of the locality of its occurrence 14—16°/oo S. Duration of adaptation 5 days in each case. After KESSELER 1959.

| External medium | | Cell fluid | Difference |
°/oo S	$\Delta°C$	$\Delta°C$	$\Delta°C$
0	0.02	1.35	1.33
5	0.27	1.75	1.48
10	0.53	1.83	1.30
15	0.80	2.25	1.35
20	1.08	2.28	1.20
25	1.35	2.62	1.27
35	1.91	3.13	1.22

cellular regulation is very high at first, but it declines rapidly. The new value of the osmotic potential of the cell sap is achieved mainly by K- and Cl-ions being expelled into the external medium or by active uptake of these ions from the external medium. According to KESSELER (1964) sodium appears to be of no importance for the regulation of cell turgor in the marine and brackish-water algae.

6. Ionic content and ionic regulation

The content of inorganic ions in brackish-water animals is related to the composition of the external medium and variable. In living organisms there is a continuous change in the total amount both of water and inorganic ions which proceeds at varying rates. Inorganic ions or salts are constantly taken up, either with food by way of the intestine or else directly through the integument. In the same way electrolytes are constantly eliminated to the outside, by excretory organs, the gut or else directly through certain cells of the integument. At any given time the concentration of the ions in the body fluids and in the tissues is determined, among other things, by the permeability of the bounding membranes, the transport performance of certain cells and organs as well as by the relative diffusion rates of the individual ions.

As mentioned before (see p. 252) the osmotic concentration of the coelomic fluid and the circulating body fluids is mainly determined by inorganic ions, (especially Na- and Cl-ions), while inside the living cells organic compounds of low molecular weight are also osmotically active to a considerable degree. In addition, the relative proportions of inorganic ions in the body fluids and in the tissues are quite different.

In some marine invertebrates (especially in certain poikilosmotic echinoderms and annelids) the composition of the coelomic fluid shows far-reaching agreement with that of the surrounding sea or brackish water. This applies e. g. for the euryhaline starfish *Asterias rubens* whose perivisceral fluid contains but the smallest amounts of organic compounds, while the content of inorganic ions shows no or only slight differences from that of the external medium (see table 18). Nevertheless, in this

Table 18. Composition of the perivisceral coelomic fluid of *Asterias rubens*, compared with that of the surrounding medium. All concentrations in mMol/l.

a) *Asterias rubens* in sea water (about 31 °/oo S). After BINYON 1962.

Ions	Na+	K+	Ca++	Mg++	Cl−	SO₄−−
Sea water.................	429	9.5	10.8	49.0	494	25.4
Coelomic fluid	428	9.5	11.7	49.2	487	26.7

b) *Asterias rubens* in brackish water (about 16 o/oo S). After SECK 1958.

Ions	Na+	K+	Ca++	Mg++	Cl−	SO₄−−
Brackish water	215	5.0	5.6	24.1	253	13.1
Coelomic fluid	216	5.4	5.6	24.2	255	13.1

species and in other echinoderms the first signs of an active ionic regulation of the body fluids can be detected. BINYON (1962) found in North Sea specimens of *Asterias rubens* a small accumulation of calcium in the coelomic fluid. But the starfish from brackish water of the western Baltic which SECK (1958) examined lacked this calcium regulation, while the potassium content of its perivisceral fluid was slightly higher

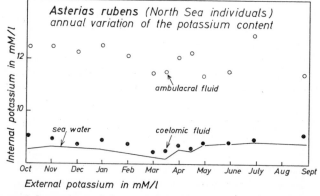

Fig. 46. Seasonal fluctuations of the potassium content in the coelomic and ambulacral fluids of *Asterias rubens* (North Sea specimens). After J. BINYON 1962.

than that of the external medium. There is a more marked potassium regulation in the ambulacral fluid of some echinoderms. BETHE & BERGER (1931) were the first to point this out for *Echinus esculentus* and later BINYON (1962) made a detailed study of it in *Asterias rubens* (see fig. 46). Brackish-water specimens of *Asterias rubens* from the western Baltic which we investigated in Kiel also displayed the same potassium regulation in the ambulacral fluid, both in the water where they occurred and after five days' adaptation to sea water of lower or higher salinity (see fig. 47). It may be assumed that the high potassium content of the ambulacral fluid is of importance for the movements of the tube feet of the starfish.

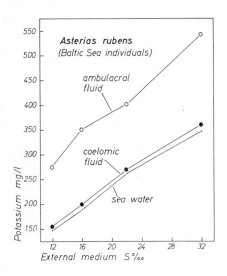

Fig. 47. Relation between the potassium content in the coelomic and ambulacral fluids of *Asterias rubens* (Baltic specimens) and the salinity of the external medium. After C. SCHLIEPER & U. RUNGE, unpublished observations.

As in all aquatic animals the cells and tissues of the echinoderms contain little sodium and chloride in comparison with the extracellular body fluids and sea water. As examples the values for the most important inorganic cell components in mMol/kg water may be quoted which ROTHSCHILD & BARNES (1953) have established for the unfertilized eggs of the sea urchin *Paracentrotus lividus:* Sodium 52 (485), potassium 210 (10), calcium 4 (11), magnesium 11 (55), chloride 80 (566), sulphate 6 (29), as well as PO_4- and acid-soluble P-compounds 68. The figures in brackets represent the corresponding values in the surrounding sea water (35⁰/₀₀ S). A comparison shows that in the sea urchin egg the inorganic components undoubtedly produce less than half the osmotic concentration of the cell sap. The rest is accounted for by dissolved organic compounds of low molecular weight, such as free amino acids etc. How do the cells and tissues of the euryhaline echinoderms behave in brackish-water? It should be remembered that on the whole echinoderms are poikilosmotic, that is their body and cell fluids are isosmotic to the surrounding external medium:

As has been mentioned before, after transfer into brackish water the egg cells of a stenohaline sea urchin behave virtually as osmometers, provided allowance is made for about 7—30% of the total cell volume; this is made up of cell structures consisting of proteins and lipids which are osmotically inert (nonsolvent space).

After transfer into half-diluted sea water they swell considerably through osmotic uptake of water and in this way they become again isosmotic to the new external medium (cf. also p. 239). But they soon perish because of this excessive and damaging water uptake. They are no longer able to survive in half-diluted sea water. But that does not apply for the cells and tissues of euryhaline echinoderms living in brackish water of the same salt content. The water content of their tissues is only slightly raised. Thus, for instance, the water content of the ovaries of specimens of *Asterias rubens* living in the western Baltic at 15⁰/₀₀ S is only 4% higher than that of individuals living in the North Sea in sea water of 30⁰/₀₀ S (KOWALSKI 1955). In an experiment in which starfish of the same species from 100% sea water are slowly and step by step transferred into 60% sea water (6 parts sea water + 4 parts aqua dest.), the water content of the tissue of the pyloric caeca has similarly increased by only 4% after an adaptation lasting for several days. This is only possible as "an intracellular isosmotic regulation" becomes effective; the term has been coined by the Belgian physiologist FLORKIN and the mechanism appears to be present in all euryhaline marine invertebrates. This regulation begins to act when, in the course of adaptation to brackish water, the osmotic concentration of the extracellular body fluids drops. It prevents a more marked osmotic water movement between the body fluids and the cells; it brings about an adjustment of the concentration of the cell fluids by reducing the quantity of the osmotically effective cell components, especially the free non-essential amino acids. Consequently, in the above mentioned adaptation experiment (transfer of *Asterias rubens* into 60% sea water) the content of free amino acids and taurine in the pyloric caeca drops from 320 to 170 mMol/l water. (JEUNIAUX, BRICTEUX-GRÉGOIRE & FLORKIN 1962). As LANGE (1964) has stressed, this organic "osmotic buffer" enables the cell to maintain more effectively the concentration of its physiologically important inorganic ions. If, on the other hand, the salt content of the external medium increases either in nature or in the experiment the reversible intracellular isosmotic regulation of *Asterias rubens* operates in the opposite direction; in accordance with the rise in the osmotic values of the extracellular body fluids it raises the quantities of free amino acids and other organic cell components until the intracellular osmotic concentration is once more in equilibrium with the extracellular one.

Beginnings of an ionic regulation (ROBERTSON 1949) as well as an efficient isosmotic cell adaptation (DUCHÂTEAU-BOSSON, JEUNIAUX & FLORKIN 1961) can also be detected in euryhaline and poikilosmotic Polychaeta, such as *Arenicola marina,* which are able to live in brackish water. Even in normal sea water the potassium content of the coelomic fluid of *Arenicola* is about 4% higher than that of the external medium. On adaptation to half-diluted sea water this difference is increased by a further 14—16% (ROBERTSON 1953). This relatively high potassium concentration of the coelomic fluid is probably the reason why isolated tissue fragments of *Arenicola* and other related species do not survive in unaltered brackish water as well as they do in sea water. Experiments by WELLS & LEDINGHAM (1940) with isolated proboscis preparations of *Arenicola* illustrate this. In normal sea water these preparations exhibited — very much like intact individuals — alternating periods of strong rhythmical activity and rest. But they soon lost their activity in dilute sea water. The contractions became increasingly smaller, even in salinities considerably above the lower limit of distribution of *Arenicola*. This rapid loss of activity can be prevented or at least slowed down by suitable addition of potassium chloride to the brackish water. ZAKS & SOKOLOVA (1960) arrived at the same result, taking as measure of survival the degree to which the isolated tissue may be dyed with neutral red (see fig. 48). The better a tissue survives in a certain medium the less neutral red will be taken up. According to their findings an isolated strip of muscular body wall of *Arenicola* survives best in dilute sea water if its potassium content has been raised to about that of normal sea water (400—500 mg/l).

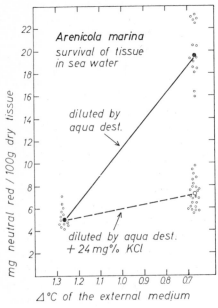

Fig. 48. Ability to survive of the muscular body wall of *Arenicola marina* in dilute sea water, depending on the potassium content of the external medium (further explanations in the text). After M. G. ZAKS & M. M. SOKOLOVA 1960.

In euryhaline poikilosmotic molluscs the difference between the ionic composition of the circulating body fluid and that of the external medium are even more pronounced, both in sea water and particularly in brackish water. To illustrate this I would like to refer to the common mussel *Mytilus edulis*. Though the freezing points of the internal and external medium are always the same, both in sea water and in half-diluted sea water (brackish water of 15—20⁰/₀₀ S), the potassium content of the blood is, on an average, 30—40% higher (see table 19). The total CO_2 content of the mussel blood may also exceed that of the external medium to a marked degree. This is equally true for marine and brackish-water specimens. It is obvious that the relatively higher potassium content of the mussel blood is of definite physiological importance; there is no doubt that the potassium content of the internal medium exerts an influence on the activity as measured by ciliary activity, mobility and reactivity (see also p. 310). For a physiologist experimenting with the surviving organs of *Mytilus* it is preferable not to use unaltered sea or brackish water as experimental medium, but to add to it potassium chloride up to about the level of concentration found in the haemolymph.

Like the euryhaline *Asterias rubens* the marine poikilosmotic bivalves which have immigrated into brackish water can only survive there because their tissues

Table 19. Composition of the blood of *Mytilus edulis*, compared with that of the surrounding medium.

a) ionic content of the blood of *Mytilus* in sea water (35°/₀₀ S) expressed as Cl = 100.
After ROBERTSON 1953.

Ions	Na^+	K^+	Ca^{++}	Mg^{++}	SO_4^{--}
Sea water	55.5	2.01	2.12	6.7	14.0
Blood	52.0	2.70	2.10	6.6	3.6

b) relative concentrations of the blood of the common mussel, compared with
blood which has been dialysed against sea water. After ROBERTSON 1953.

Na^+	K^+	Ca^{++}	Mg^{++}	Cl^-	SO_4^{--}	Protein
100.0	134.7	99.5	99.5	100.5	98.2	0.3 mg/ml

c) ionic content of the blood of *Mytilus* in brackish water (about 16°/₀₀ S) in mMol/kg H_2O.
After SECK 1957.

Ions	Na^-	K^+	Ca^{++}	Mg^{++}	Cl^-	SO_4^{--}
Brackish water	215	5.08	5.69	24.2	253.1	13.18
Blood	213.4	7.51	5.77	24.46	252.6	13.22
Blood %	99.3	147.8	101.4	101.1	99.8	100.3

also function at lower osmotic concentration. In analyzing their osmotically effective
intracellular components one has to bear in mind that their tissues may also contain
considerable interstitial spaces filled with haemolymph. In common mussels in
brackish water 10—30% of the total water content of the adductor and foot muscles
must be considered as extracellular fluid (KROGH 1939, POTTS 1958). Even taking
this fact into consideration it is still possible to show relatively large amounts of
Na-, Cl- and in particular K-ions intracellularly in the muscle fibres. So far the
most comprehensive investigation of muscle fibres of marine bivalves which can
live in brackish water as well has been carried out by BRICTEUX-GRÉGOIRE, DUCHÂ-
TEAU-BOSSON, JEUNIAUX & FLORKIN (1964 a and b). They worked with European
and Portuguese oysters, *Ostrea edulis* and *Gryphaea angulata*. According to their
results, as reproduced in table 20, the inorganic ions account for less than half
(37—46%) of the intracellular osmotic concentration of the "quick" fibres of the
adductor muscle of *Gryphaea*. In spite of this the potassium content of these muscle
fibres in specimens from sea water amounts to about eleven times that of the external
medium and in individuals adapted to brackish water even sixteen times. Osmotic
equilibrium with the extracellular haemolymph is brought about by dialysable
N-compounds dissolved in the cell sap; a large proportion is made up of free amino
acids, in particular glycine, alanine, asparagine etc, as well as betaine and taurine.
In those oysters which were gradually transferred into half-diluted sea water and
had been adapted, the water content of the same muscles increased only from 74
to 81%, in spite of a decrease of the interstitial portion. But the proportion of inorganic and organic components decreased much more than can be explained by

Table 20. Composition and osmotically active constituents in mOsmol/kg water of muscle
fibres (yellow "rapid" part of the adductor muscle) of the Portuguese oyster *Gryphaea angulata*
in sea water (M) and half-dilute sea water (M/2). After BRICTEUX-GRÉGOIRE, DUCHÂTEAU-
BOSSON, JEUNIAUX & FLORKIN 1964.

	Sea water animals		Brackish-water animals	
	tissue	M	tissue	M/2
Total water content % .	73.7		81.2	
Extracellular water con-				
tent as % of total water	9.6		5.9	
Cl	162.3	672.7	44.7	336.3
K	157.5	14.8	105.3	7.4
Na	188.4	492.2	59.0	246.1
Ca	9.2	8.1	4.3	4.1
Free amino acids	141.4		81.7	
Betaine	174.4		98.4	
Taurine	125.0		71.0	
Dialysable				
N-compounds, total ...	601.8		357.1	
Inorganic components				
(Cl, K, Na, Ca)	517.4	1087.8	213.3	593.9
Total intracellular con-				
centration	1119.2		570.4	
Calculated intracellular				
freezing point depression				
Δ°C	2.09		1.07	
External medium Δ°C .		2.20		1.02
External medium °/oo S .		41		21

this "dilution". It is the same process, already mentioned for *Asterias,* of an active
isosmotic intracellular regulation which prevents an excessive hydration of the
tissues in brackish water.

For the same reason the common mussel which is widespread in brackish water
displays a linear correlation between the intracellular amounts of ninhydrin-positive
substances (free amino acids and taurine) and the salinity of the external medium.
LANGE (1963) made a detailed study of these relationships in populations of *Mytilus*
which had reached maturity on the coasts of Norway, Sweden, Denmark and Finland,
in sea and brackish water of very different salinities. He found that the — probably
excretory taurine, especially, increases very much intracellularly with rising external
salinity (cf. fig. 44); in the haemolymph, however, it is found only in insignificant
concentrations.

Other brackish-water invertebrates maintain ionic concentrations in their body
fluids which differ even more from those of the external medium; this is done by
direct, controlled uptake of single ions and (or) by means of their excretory organs.
It appears that greater ionic regulation is specially developed in active, highly
organised species in which the blood has a high content of protein (haemocyanin).
This is true especially of the decapod brachyurians (cf. among others ROBERTSON

1953). Thus for instance the potassium and calcium content in the blood of the shore crab *Carcinus maenas* is relatively much increased, while the content of magnesium and sulphate is considerably smaller than that of the external medium (see table 21). In brackish-water specimens hyperosmy of the blood is produced

Table 21. Composition of the blood of the shore crab *Carcinus maenas* from sea water (32—34°/$_{oo}$ S) and from brackish water (about 16°/$_{oo}$ S). All values in % of the concentration of the external medium.

Ions	Na	K	Ca	Mg	Cl	SO$_4$	Author
a) Animals in winter							
From sea water	115	101	126	40	104	68	SECK 1957
From brackish water .	165	174	170	55	150	66	SECK 1957
b) Animals in summer							
From sea water	111	121	127	36	100	57	WEBB 1940
From brackish water .	166	202	174	59	152	62	SECK 1957

by the regulatory activity of the gills which actively remove ions from the external medium and transport them inwards (NAGEL 1934, SCHMIDT-NIELSEN 1941). We owe to SHAW (1961) an accurate analysis of the sodium exchange of *Carcinus* in sea and brackish water; he used tracers in his studies of active transport and of passive exchange movements due to diffusion (see table 22). The strikingly low magnesium content of the blood of *Carcinus* both in sea and brackish water is due to the activity of the excretory antennary glands and seems to be related to the relatively high activity of the species. According to ROBERTSON (1953) among the decapod crustaceans in general the slower species have high concentrations of magnesium, the more lively and active species have a lower level of magnesium in their blood.

Table 22. Sodium exchanges of *Carcinus* in sea water and brackish water (from SHAW 1961).

Fluxes mM/kg./hr.	In sea water of 33°/$_{oo}$ S and 450 mM Na/l.	In brackish water of 16.5°/$_{oo}$ S and 225 mM Na/l.
Inward diffusion	24.1	12.0
Outward diffusion	24.6	16.9
Active influx	1.2	8.2
Urine loss	0.7	3.3
Total flux	25.3	20.2

In *Carcinus maenas* the tissues of the brackish-water individuals have adapted themselves to the reduced osmotic concentration of the blood by the process of "isosmotic intracellular regulation" in a manner similar to that adopted by other euryhaline poikilosmotic brackish-water invertebrates. During this process the water content of their muscles (individuals in half-diluted sea water) has risen from about 74% to 77—78%. On the other hand, the concentration of the intracellular free amino acids glycine, proline and glutamic acid has diminished by 40—50%; it is not yet known whether the amino acids which have disappeared have been excreted or transformed into protein within the cells or possibly stored in the hepatopancreas (DUCHÂTEAU, FLORKIN & JEUNIAUX 1959).

From the results just described it can be seen that survival of the species in brackish water depends not only on any osmoregulatory abilities they may possess. Their ability for selective regulation of the ionic content of their body fluids and tissues is equally important for the degree of their osmotic resistance and the maintenance of their normal species-specific activity. This applies not only for the marine immigrants into brackish water, but equally for the brackish-water species proper and the immigrants from fresh water. Among the numerous investigations of such brackish-water inhabitants I want do draw particular attention to the papers by RAMSAY (1950) on Diptera larvae, by KOCH et al. (1954) on the mitten crab *Eriocheir sinensis*, by LILLY (1955) on *Pelmatohydra oligactis*, by FRETTER (1955) and JORGENSEN & DALES (1957) on *Nereis diversicolor*, by CROGHAN (1958) on *Artemia salina*, by SUTCLIFFE (1960) on Diptera larvae and lastly by POTTS & PARRY (1964) on *Palaemonetes varians*.

Because of their methodological importance the studies of the osmoregulatory ion pump of *Eriocheir sinensis* by means of radioactive tracers will be discussed in some detail. The mitten crab has eight pairs of gills. The first two are very small and the other six, situated farther back, are considerably larger. Only three pairs of gills situated farthest back contain large amounts of phosphatide; they alone are responsible for the intensive transport of ions which can be seen in whole animals in fresh water and in brackish water of low salinity. KOCH and co-workers (1954, 1956, 1965) were able to demonstrate that even isolated surviving gills absorb Cl^-, Na^+ and K^+ directly from the hypo-osmotic external medium, provided there is sufficient oxygen available. In an oxygen-free medium no ions whatever are taken up; they permeate rather quickly and passively to the outside. If the Na^+ of the external medium is only present in the form of $NaNO_3$ or Na_2SO_4 no active Na-absorption takes place, but only a slight exchange by diffusion. If, however, NH_4Cl or $CaCl_2$ are added to such an external medium, Na^+ is very rapidly absorbed. From this one might conclude that there is primarily an active absorption of Cl-ions. Just as Na^+, K^+ also is taken up together with Cl^-. If Na^+, K^+ and Li^+ are present in the external medium at the same time, all three cations are being absorbed. On addition of respiratory inhibitors and anticholin-esterases the absorption of ions suffers a reversible inhibition while, at the same time, considerable quantities of Na^+ diffuse outwards.

Studies of the ionic-regulatory function of the excretory organs in brackish-water invertebrates have been chiefly carried out in brachyurians, prawns, amphipods and insect larvae. As mentioned before, the antennary glands of the short-tailed crabs have, as a rule, no osmoregulatory function, even in the few fresh-water species since the urine they produce is usually approximately isotonic with the blood (SCHLIEPER 1929a: *Carcinus*, SCHLIEPER & HERMANN 1930: *Telphusa*, SCHOLLES 1933: *Eriocheir*). But in the marine and brackish-water species which have been examined the magnesium and sulphate content of the urine is always higher than that of the blood (BIALASCEWICZ 1933: *Maja*, SCHOLLES 1933: *Eriocheir*, ROBERTSON 1939: *Cancer*, WEBB 1940: *Carcinus*). In the mitten crab adapted to sea water the

Mg content of the urine is always more than double that found in the blood, while in fresh-water specimens it amounts to only about 20% of that in the blood. Also in the brackish-water prawn *Palaemon serratus* the urine produced by the antennary glands is isosmotic to the blood (PARRY 1954). The concentrations of Na^+, K^+ and Ca^{++} in the urine of marine and brackish-water individuals are always somewhat smaller, but the amounts of Mg^{++} and sulphate $^{--}$ quite considerably higher than those in the blood (see table 23). One might conclude that Mg^{++} and SO_4^- are being actively accumulated in the urine and excreted. PROSSER et al. (1955) made the first attempts to analyze this renal ionic regulation in *Pachygraspus crassipes*. In this brachyurous crustacean the concentrations of Mg and SO_4 in the urine increase quite considerably after transfer into hyperosmotic sea water, while sodium excretion diminishes. Further experimental analysis was carried out by GROSS & MARSHALL (1960) and GROSS & CARPEN (1966); this revealed that in *Pachygrapsus* the Mg concentration of the urine is independent not only of the Mg concentration of the external medium, but altogether of the Mg influx; but it is a function of the total salt content in the external medium and, in particular, of the magnitude of the water passage through the crabs and the length of time the urine remains in the bladder.

Table 23. Comparison of the composition of blood and urine in the prawn *Palaemon serratus* in sea water and brackish water. After PARRY 1954. All concentrations in mEqu/l.

Ions	Na^+	K^+	Ca^{++}	Mg^{++}	Cl^-	SO_4^{--}
a) Sea water	499	11.0	21.9	114.5	581	57.9
Blood	394	7.7	25.2	25.2	430	5.2
Urine	324	6.6	24.0	168.6	458	19.8
b) Brackish water (50% sea water)	249.5	5.5	10.95	57.3	290.5	28.95
Blood	257	6.7	28.3	22.2	300	2.3
Urine	206	7.4	25.0	67.1	330	3.6

Ionic regulation in brackish-water fishes is also complex. The mechanisms performing this function are various, depending on whether the internal medium to be regulated is hypo-osmotic, isosmotic or hyperosmotic to the surrounding brackish water. In so far as the internal medium of teleosts in brackish water of lower salinity is still hyperosmotic, the same osmo- and ionic regulatory mechanisms operate as in fresh water. Thus, on the one hand, these fishes constantly take up water osmotically, especially through their gills and the mucous membranes of the mouth, and, on the other hand, they excrete it again in corresponding amounts in the form of hypo-osmotic urine. Drinking of water does not take place, at least it is not necessary. The physiologically required ions are actively taken up from the external medium by means of certain cells in the gills and are transported into the blood. But teleosts capable of living in hyperosmotic brackish water have the same osmo- and ionic regulatory mechanisms as the marine teleosts. Since, to some extent, they constantly lose water osmotically they have to replace this water loss by drinking correspondingly. The sodium, potassium and chloride ions taken up with this "drinking water" are resorbed by the gut and to a lesser extent excreted by the kidneys (in the form of small amounts of a weakly hypo-osmotic urine). But the chloride content of the urine is low and the bulk of Cl- and Na-ions are expelled to the outside by certain cells in the gills, in a secretory process. In this the level of chloride secretion by the gills is solely and exclusively determined by the osmotic concentration of the blood (see table 24).

Table 24. Perfusion experiment to demonstrate the chloride secretion of the eel *Anguilla vulgaris* (heart-gill preparation). After SCHLIEPER 1933.

Experimental period	Perfusion media $\triangle^0$C		Change in mg Cl/hour kg	
	Internal medium	External medium	Internal medium	External medium
a	0.51	0.51	—22	+29
b	0.70	0.70	—84	+59
c	0.51	0.51	—39	+52

Even if in the experiment the concentration of the perfused internal medium is raised not by chloride, but by addition of glucose or sodium sulphate, more chloride is secreted by the gills (SCHLIEPER 1933). Thus the secretion of chloride represents a true osmotic regulation. But most of the calcium, magnesium and sulphate which have been taken into the gut with the drinking water are eliminated again with the faeces. We owe to the following authors more detailed quantitative investigations of the mineral economy of euryhaline teleosts capable of living in brackish water: MULLINS (1950), MOTAIS (1961), HOUSE (1963), as well as MOTAIS, ROMEU & MAETZ (1966).

According to MULLINS who measured by means of tracers the sodium exchange of the stickleback *Gasterosteus aculeatus* in brackish water of $17^0/_{00}$ S, it was 10.7 mMol per kilogram of live weight and hour, that is 15 times as much as in fresh water. The corresponding values for the potassium exchange are 0.40 mMol in brackish water and 0.15 mMol in fresh water. HOUSE who examined *Blennius pholis* in brackish water (10 and 40% sea water) and in undiluted sea water observed that after transfer of this fish from brackish into sea water the sodium exchange rose fivefold (= 100 mMol sodium per litre of blood and hour). According to experimental measurements which MOTAIS and co-workers carried out on the euryhaline flounder *Platichthys flesus* the ionic regulatory mechanism of this species changes by 90% within 20 seconds after an alteration of the salt content of the external medium. About 30 minutes later a further slow adaptation in the same direction takes place. The rapid reorientation of the sodium exchange is solely determined by the change of sodium concentration in the external medium. The Cl exchange is also said to be specifically dependent on the amount of Cl ions in the external medium. In stenohaline teleosts, e. g., the sea perch *Serranus scriba,* the adaptation of the sodium exchange after an alteration of the salt content of the external medium sets in far less vigorously and more slowly. The subsequent secondary phase of adaptation in ionic regulation is entirely lacking. Hence, for this species transfer into a strongly hypo-osmotic external medium results in considerable losses of electrolytes and finally in the death of the fish. ERCHINGER (1964) reports similar observations on the euryhaline teleost *Cottus bubalis*. If, after transfer into brackish water of $3^0/_{00}$ S this species exceeded its lower salinity limit, there occurs after a few days a drop in the high intracellular potassium content which is not compensated for and is presumably a major cause of death in this case (see fig. 49).

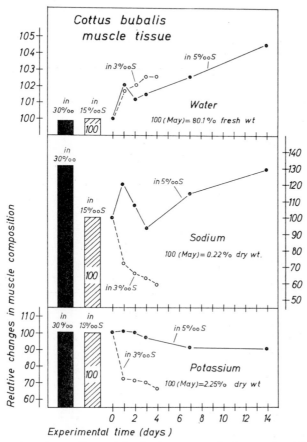

Fig. 49. Changes in water-, sodium- and potassium content in the muscles of *Cottus bubalis*, depending on the salinity of the medium. After H. ERCHINGER 1964.

In recent times attention has been drawn to the fact that some species of amphibians are capable of survival in brackish water. Thus the European green toad *Bufo viridis* penetrates as far as $20^0/_{00}$ S - brackish water in the Baltic region while the crab-eating frog *Rana cancrivora* in the mangrove zone on the coast of Thailand tolerates external salt concentrations up to $29^0/_{00}$ at least for some time. PORA & STOICOVICI (1955) were the first to experiment with *Bufo viridis* from the Black Sea; more recently GORDON (1962, 1965) has studied the green toad which among other places came from the Island of Saltholm near Copenhagen. In *Bufo* the osmotic concentration of the blood plasma rises in brackish water mainly through an increase in the concentration of NaCl. By contrast, the simultaneous increase in intracellular concentration in the muscles is brought about (to 47%) by inorganic ions (Cl, Na, K), to 33% by free amino acids and related compounds (in particular taurine, glycine, alanine) and to 22% by urea (see fig. 50). *Rana cancrivora* uses urea to an even greater extent as a means of osmotic adaptation in brackish water. After transfer from fresh into brackish water (80% sea water) the urea content of its blood plasma increases from 40 to 350 mMol/l. That means that in brackish-water individuals of this species 60% of the increase of the osmotic blood concentration is caused by urea (GORDON et al. 1961).

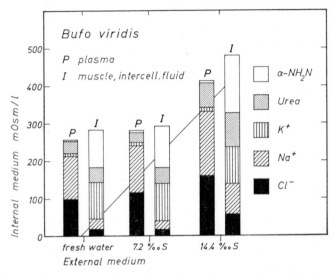

Fig. 50. Changes in the composition of the blood plasma and of the intracellular tissue fluid (muscles) of *Bufo viridis*, depending on the salinity of the external medium. After M. S. GORDON 1965.

Like animals living in brackish water, the plants which survive in brackish water are able to maintain the ionic content of their cell fluids more or less independently of the composition of the external medium. To illustrate this fact analytical values of the cell fluid of various plants from sea, brackish and fresh water are given (see table 25). All concentrations are in millimol; a Cl concentration of 580 mMol in

Table 25. Ionic concentrations in the cell fluid and external medium of different plants from sea, brackish and fresh water. (After OSTERHOUT 1933 and COLLANDER 1936 from KROGH 1939).

mMol	Sea water	*Valonia*	*Halicystis*	Brackish water	*Chara*	Fresh water	*Chara*
Cl	580	597	603	80	232	0.13	176
Na	498	90	557	68	148	0.21	84
K	12	500	6.4	1.4	69	0.04	77
Ca	12	1.7	8	7.6	22	6.6	26
Mg	57	Traces?	16.7				

sea water corresponds to a salt content of about $36^o/_{oo}$ and 80 mMol in brackish water to a salinity of $5^o/_{oo}$.

From the figures reproduced in table 25 it can be seen that in every case the total electrolyte concentration as well as the Cl content of the cell sap are higher than in the surrounding medium; this is particularly so for the Characeae found in brackish and fresh water. Just as in most living organisms the cell sap of *Valonia* shows a considerable increase in the concentration of potassium, compared with the surrounding medium; but that of sodium is reduced. Potassium is also concentrated in the cell sap of *Chara,* but not in *Halicystis.* The same holds for calcium and sodium in *Chara* occurring in brackish and fresh water. A comparison of *Chara* in brackish and fresh water in addition shows that the excess of the total concentration in the cell sap is practically the same in both media, while the proportion of the individual ionic concentration varies.

The euryhaline green alga *Chaetomorpha linum* maintains a constant hyperosmy almost independent of the salinity of the locality where it occurs (cf. p. 264); the following ionic relations roughly obtain between the cell sap and the external medium $(30^o/_{oo}$ S) in mMol/l for Na 44/410, K 743/9, Mg 17/47, Ca 4/9, Cl 762/478 and SO_4 23/23. In every case the adaptation of the cell sap to a changed salt content of the medium is brought about by a rapid uptake or giving off of K and Cl. Na has hardly any significance for the process of regulation (KESSELER 1964 a and b).

The problem of sodium and potassium balance in the cells of marine algae has recently been the subject of an admirable study by SCOTT & HAYWARD (1955) on the green alga *Ulva lactuca.* These authors succeeded in discovering separate mechanisms for the transport of sodium and potassium ,as well as an influence of the metabolism on the rate of exchange of potassium ions in *Ulva.* — We owe to MacROBBIE & DAINTY (1958) some quantitative measurements of ionic exchange in a brackish-water plant at constant salinity of the external medium. They investigated *Nitellopsis obtusa,* a Characea occurring in the brackish water of Finland at about $2—3^o/_{oo}$ S. The vacuolar sap of its giant internodal cells (length 4—10 cm, diameter $500—800 \mu$) has the following ionic proportions in mMol (internal medium/external medium): Na 54/30, K 113/0.65 and Cl 206/35. In the protoplasmic layer surrounding the cell sap vacuole Na and K are probably present in equal amounts, while Cl is found in smaller concentration (25 mMol). The ionic exchange rates (steady state fluxes) were determined by means of isotopic tracers. They were per cm^2 and second in moles $(10^{-12}$ moles) for Na 0.4, for K 0.25 and for Cl 0.5. These values mean that in this case the influx of the ions in question exactly corresponds to the efflux; in other words "the net flux is zero".

7. Oxygen requirements and respiration

An investigation of oxygen consumption as a measure of metabolic rate provides a valuable means of analyzing the physiological effects of the salinity of brackish water on the organisms inhabiting it. For a long time ecological observations have been known which indicate that there exists a correlation between the respiration of brackish-water inhabitants and the salt content of the brackish water. ROCH (1924) observed that the polyp *Cordylophora lacustris* can be found anywhere in brackish water, even if the medium is stagnant or shows but slight movement, whereas in fresh water it only occurs in oxygen rich, fast flowing water. From this ROCH concluded that the oxygen requirements of *Cordylophora* are higher the lower the salinity of the external medium. THIENEMANN (1928) drew attention to the fact that in the central Baltic the schizopod crustacean *Mysis relicta* occurs in large numbers even in deep high saline watters poor in oxygen, while in North Germany it is confined to such lakes in which the oxygen content of the deeper water layers does not drop below about 50% saturation. In THIENEMANN's opinion this ecological observation represents a "law of animal physiology" which he formulated as follows: "Die Atmung in Salzwasser ist leichter als in Süsswasser" [Respiration in salt water is easier than in fresh water].

SCHLIEPER (1929) then attempted an experimental investigation of the problem by analyzing the oxygen consumption of euryhaline invertebrates in brackish water of varying salinities. He found that a reduction of salinity in the external medium resulted in a permanent increase of oxygen consumption in *Carcinus maenas* and in a temporary one in *Nereis diversicolor* (see fig. 51). Respiration of isolated gill

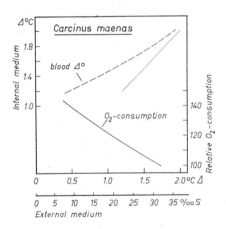

Fig. 51. Relationship between the freezing-point depressions of the blood and respiration in *Carcinus maenas* in brackish and sea water. After C. SCHLIEPER 1929.

tissue of *Mytilus edulis* also increased in brackish water of low salinity. On the other hand, intact common mussels and starfish *(Asterias rubens)* had reduced oxygen consumption under the same conditions. From these quite varied results it became clear at the time that the relationship between metabolic rate of the brackish-water animals and the salt content of the external medium can not be expressed by a simple formula. Numerous investigations during subsequent years led to the same conclusion. It became evident that a reduction or increase (respectively) of the salinity of the brackish water may have quite varying effects on the metabolism of the organisms inhabiting it. The following operative mechanisms have been discussed:

1. The oxygen consumption of a species may depend on the salinity of the external medium in so far as it always reaches its highest value in the region of the optimum osmotic concentration of the internal medium; in any transgression of the limits of the optimal region, whether upwards or downwards, this consumption will decrease.

2. The salt content of the external medium may have an indirect effect on oxygen consumption, by way of a stimulation or reduction of locomotory activity, of growth, of the activity of individual organs or of cell activity in general.

3. Parallel with a change in the salinity of the external medium the osmotic and (or) ionic regulatory performance of the species under investigation, linked with its energy requirements, may be raised or diminished, thus exerting their influence on oxygen consumption in the same direction.

4. An increase or a reduction in the salinity of the external medium may alter the water content of the tissues and through it secondarily, their oxygen consumption. Respiration might attain its highest value at a certain optimum water content of the tissues and diminish after any transgression of the limit of the optimum, both upwards or downwards.

A further factor which may play an essential part in an examination of the respiratory rate of an animal in water of varying salinity is, of course, the possibility that from a certain salt concentration onwards the experimental animal has been more or less seriously damaged and that its respiratory rate has declined for that reason.

Lastly it must be borne in mind whether the oxygen consumption of an animal has been examined immediately after a change of salinity in the external medium or after some period of adaptation to the changed salinity. Some species react to a sudden decrease or increase of salinity in the external medium at first by increased motor activity (unrest) which may give rise to violent attempts to escape and cause an immediate increase in oxygen consumption. Frequently shock reactions set in which may be coupled with a reduction in respiration, or even with a temporary stopping of respiration. In this way e. g. the sea anemone *Metridium marginatum* reacts to any change in the normal salt content of the sea water by contraction of the whole animal, together with a considerable lowering of the respiratory rate (SHOUP 1932).

Even when all the possible errors just mentioned are taken into account it must be admitted that the results of the first experiments reported have later been confirmed by numerous authors. Thus SCHWABE (1933) found increased oxygen consumption in the brachyurous crustaceans *Carcinus maenas* and *Eriphia spinifrons* after transfer into diluted sea water. Corresponding differences were observed by THEEDE (1964) with *Carcinus maenas* when he transferred crabs from the North Sea (30⁰/₀₀ S) and crabs from the Baltic (15⁰/₀₀ S) to the medium of the other. The same applies to *Hemigrapsus oregonensis* according to DEHNEL (1960) and to *Carcinus mediterraneus* and *Callinectes sapidus,* according to KING (1965). All these species which regulate hyperosmotically in brackish water show the lowest amount of respiration in isosmotic sea water and an increased metabolism in hypo-osmotic brackish water which, according to THEEDE, is also said to be reflected in a greater haemocyanin content in the blood of *Carcinus* adapted to brackish water.

MEYER (1935) and MALOEUF (1938) confirmed the reduction of oxygen consumption in the non-regulating *Asterias rubens* in diluted sea water. In addition BOCK & SCHLIEPER (1953) established the fact that respiration of North Sea starfish in sea water shows generally higher values than that of Baltic forms in brackish

water. The unfertilized eggs of *Asterias glacialis* provide an interesting exception in this respect; after reduction of the salinity of the external medium from 33.5 to 31.2°/$_{00}$ S their respiration increases by 8%, but then it slowly decreases on further dilution of the medium (BOREI 1936). BOUXIN (1931) studied the oxygen consumption of the common mussel *Mytilus edulis* var. *galloprovincialis* in normal, concentrated and diluted sea water. Apart from a slight increase in the respiratory rate, by a maximum of 5% on dilution of the external medium by 10—15%, oxygen consumption decreased, both after a raising or lowering of the salinity of the external medium; it was all the more pronounced, the greater the change in the salinity of the external medium (see fig. 52). BELIAEV & TSCHUGUNOVA (1952) have carried out similar measurements of the respiratory rate of common mussels, *Mytilus edulis,* from the Barents Sea and the Gulf of Finland in the Baltic (see fig. 53). The common mussels living in the Barent Sea in sea water of about 35°/$_{00}$ S show the highest oxygen con-

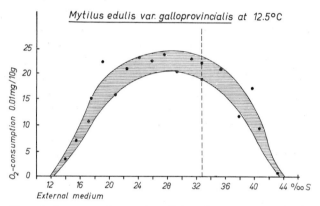

Fig. 52. The effect of short-term changes in salinity of sea water on the O$_2$-consumption of the common mussel *Mytilus edulis* var. *galloprovincialis*. After H. BOUXIN 1931.

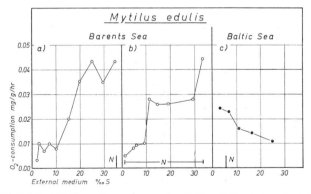

Fig. 53. Relation between O$_2$-consumption and salinity of the external medium in common mussels *(Mytilus edulis)* a) from the Barent Sea (about 35°/$_{00}$ S), b) from region of the Barent Sea with widely fluctuating salinity due to tides, and c) from the eastern Baltic (about 5.6°/$_{00}$ S). N is the salt content in the water of the place where the specimens were captured. After G. N. BELIAEV & M. N. TSCHUGUNOVA 1952.

sumption at that concentration which represents the optimum for them. If the external medium is diluted below $20^0/_{00}$ S they react with a considerable reduction of the respiratory rate. The same applies to common mussels from the coastal area of the Barent Sea with great salinity fluctuations. These mussels also have their most intensive respiration in normal, undiluted sea water. Conversely, the common mussels from the brackish water of the Gulf of Finland (about $5—6^0/_{00}$ S) have their greatest oxygen consumption at the low salt concentration of their locality to which they are specially adapted. After transfer into brackish water of higher salinity their respiratory rate declines; the higher the salinity of the external medium, the greater the decrease. (see also LAGERSPETZ & SIRKKA 1959.)

FRIEDRICH (1937) studied the respiration of the Opisthobranch *Alderia modesta* from the western Baltic, both in the brackish water where it occurs ($17^0/_{00}$ S) and in more dilute media. He noticed that a reduction of salinity from $17^0/_{00}$ S to $9^0/_{00}$ S resulted in an increase of the oxygen consumption by about 10%, but with a further lowering below $8^0/_{00}$ S it diminishes very rapidly. ,,Die bei $8^0/_{00}$ S beginnende ausserordentlich starke Abnahme der Atmungsgröße und der bei $5^0/_{00}$ beginnende osmotische Wassereinstrom begrenzen offenbar eine Zone, in welcher die Tiere der untersuchten Population bei 18—19°C noch gerade leben können, auch dann, wenn sie längere Zeit in diesem Medium gehalten werden" [The extraordinarily great reduction of respiratory rate which begins at $8^0/_{00}$ S and the osmotic water influx setting in at $5^0/_{00}$ appear to be delimiting a zone in which the animals of the population investigated are just able to survive at 18—19°C even if kept in this medium for some time].

There are numerous studies on the respiration of crustaceans in media of varying salinity. In addition to the euryhaline brachyurians already referred to, their stenohaline related species such as *Maja verrucosa, Hyas araneus* and *Libinia emarginata* should be mentioned here. All these non-regulating brachyurians have their highest oxygen consumption in normal sea water and their respiratory rates are more or less reduced after transfer into dilute sea water (SCHWABE 1933, KING 1965). A comparison of related species of crustaceans whose habitat is in water of differing salinities shows the oxygen consumption to be frequently greater, the lower the salinity of the water they inhabit. Thus, e. g. FOX & SIMMONDS (1933) and LÖWEN-STEIN (1935) found that *Gammarus marinus* in sea water consumes 562 mm³ O_2/g and h, while the related brackish-water species *G. chevreuxi* requires 648 mm³ O_2/g and h in brackish water and the fresh-water species *G. pulex* needs 1098 mm³ O_2/g and h. On examining one and the same euryhaline brackish-water species, *G. duebeni,* a constant respiratory rate is found in a medium salinity range which diminishes only when there is an increase to over $30^0/_{00}$ S in the external medium; it rises if the external medium is diluted below $10^0/_{00}$ S (see fig. 54). Parallel with this change in the respiratory rate *G. duebeni* shows a somewhat corresponding increase or reduction in heart frequency (KINNE 1952). This is evidence of a true change of total metabolic rate, depending on the salinity of the external medium. But it is not possible to generalize from these findings. Measurements of respiratory rates published by SUOMALAINEN (1956) show that the species of *Gammarus* from the brackish water in Finland *(G. oceanicus, G. zaddachi, G. duebeni)* have quite different reactions to changes in the salinity of the external medium. Thus *Gammarus*

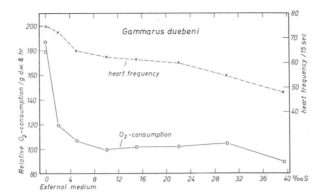

Fig. 54. The influence of salt content on the oxygen consumption and heart frequency of *Gammarus duebeni* (male individuals at 13 and 20° C respectively). After O. KINNE 1952.

oceanicus has its highest oxygen consumption at 7⁰/₀₀ S and its respiration diminishes, both in more dilute and in more concentrated brackish water.

LOFTS (1956) studied the effect of the salt content of the external medium on the oxygen consumption of two different populations of the brackish-water prawn *Palaemonetes varians,* var. *microgenitor,* which, as we have seen (see p. 258) is a hyper-hypo-osmotic regulator. The population from brackish water of higher salinity, the salt-marsh pool population, shows minimum oxygen consumption at 26⁰/₀₀ S. Oxygen consumption rose both with a dilution or a concentration of the medium. The greatest increase of respiration (about sixfold) occurred in brackish water of only 5⁰/₀₀ S. By contrast the other, sluice-pool, population coming from brackish water of very low salinity had its respiratory minimum at 6⁰/₀₀ S. In this case, too, oxygen consumption rose both with further dilution or concentration of the medium (see fig. 55).

The behaviour of the salt-marsh pool population is as might be expected since it displays the lowest respiratory rote when the external medium is isosmotic and

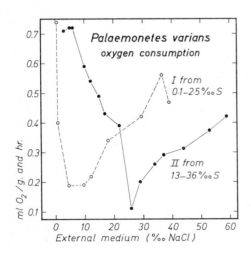

Fig. 55. Respiratory rate of *Palaemonetes varians* at varying salinities of the external medium, as investigated in two brackish-water populations. After B. LOFTS 1956.

the osmoregulatory work at its minimum. The sluice-pool population, on the other
hand, consumes least oxygen at 6⁰/₀₀ Na Cl, that is at a salt concentration which,
though not isosmotic, presumably represents its optimum.

We owe comparable observations to RAO (1956) who experimented with two
populations of the Indian prawn *Metapenaeus monoceros* which also exhibits hyper-
hypo-osmotic regulation. Individuals of this species coming from brackish water
consumed least oxygen in half-diluted isosmotic sea water; this increased in both
directions, in more concentrated or more dilute media. But individuals coming
from sea water showed the minimum respiratory rate at normal salinity of the sea
water and rising respiration with increasing dilution of the medium.

If fresh-water invertebrates are transferred into brackish water some
degree of reduction of the respiratory rate can be observed in most animals within a short time. It is interesting that in every case the respiration of species which
are more resistant in brackish water is lowered to a lesser degree (see fig. 56). Thus
in brackish water respiration of *Planaria alpina* is more reduced than that of *Pl.
gonocephala* and the latter again has a more diminished respiratory rate than *Pl.*

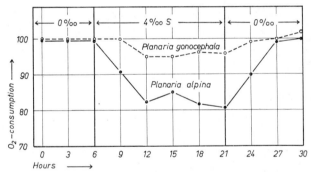

Fig. 56. Changes in metabolic rate of *Planaria gonocephala* and *Pl. alpina* after transfer into
brackish water (4⁰/₀₀ S). After C. SCHLIEPER 1952.

lugubris. As has been seen before, if the crayfish *Astacus fluviatilis* is transferred
into brackish water of 15⁰/₀₀ S, isosmotic with its blood, urine formation is reduced
and a slow osmotic adaptation takes place; the respiratory rate drops at a similar
speed and in the course of two or three weeks it reaches a new, lower level (see

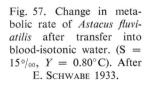

Fig. 57. Change in meta-
bolic rate of *Astacus fluviatilis* after transfer into
blood-isotonic water. (S =
15⁰/₀₀, $Y = 0.80°C$). After
E. SCHWABE 1933.

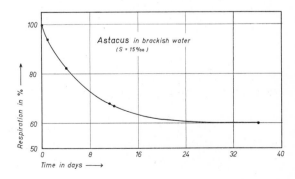

fig. 57). Especially oxygen consumption of the excretory organs, but also that of other tissues (muscles, hepatopancreas) is reduced (PETERS 1935). At the same time the water content of the tissues diminishes as the concentration of the external medium increases (see table 26).

Table 26. Water content of the muscles of *Astacus fluviatilis* in fresh water and in brackish water. After M. BOGUCKI 1934.

External medium	Water content of muscles in %
Fresh water	84
$^1/_5$ sea water	83.6
$^1/_2$ sea water	79.3
$^2/_3$ sea water	76.2

Thus the problem arises once more: What are the relations between the metabolism of isolated tissues from species viable in brackish water and the salt content of the medium? As mentioned before SCHLIEPER (1929) noticed an increase in the oxygen consumption of the isolated gill tissue of *Mytilus edulis* after lowering the salinity of the medium. In a later investigation (1955) the same author compared the respiratory rate of the gill tissue from a sea-water population (List, North Sea) and from a brackish-water population (Kiel, western Baltic). The following were the most important results: In brackish water of about $15^0/_{00}$ S respiration of the isolated gill tissues from the common mussel from Kiel, Baltic, is about 150 to 170% of the corresponding value for the North Sea mussels in sea water of $30^0/_{00}$ S. After transfer of North Sea mussels into the corresponding brackish water this increase in tissue respiration only develops slowly in the course of several weeks, in spite of a rapid osmotic adjustment (see fig. 58). Immediately after transfer the oxygen consumption of the tissues rises by about 20% within a few hours. Further increase takes place but slowly so that it requires four to seven weeks for the higher metabolic level of the Baltic mussels to be reached. The same slow adaptation of tissue metabolism in two steps (in this case a reduction of respiratory rate) can be observed when Baltic mussels from brackish water of $15^0/_{00}$ S are transferred into sea water of $30^0/_{00}$ S. In each case the rate of adaptation depends on the temperature of the external medium; it is greater in summer than in winter. If the salinity of the medium is further reduced from 15 to $10^0/_{00}$ S, no further increase in respiration occurred. This is in agreement with the findings of LAGERSPETZ & SIRKKA (1959) that oxygen consumption of isolated gill tissue of Finnish common mussels at a salinity of the water they inhabit ($= 5^0/_{00}$ S) resembles that of the mussels from Kiel in $15^0/_{00}$ S. If, however, the Finnish mussels were adapted to 15 or $30^0/_{00}$ S, the respiration of their gill tissue slowly dropped by 16 and 53% respectively. In sea water of $30^0/_{00}$ S respiration was about the same for the gill tissue of the three populations investigated (North Sea, western Baltic, Finnish coast).

Recently, LANGE (1968) investigated with improved methods the oxygen consumption of gill tissue, isolated from mussels collected near Oslo ($20—28^0/_{00}$ S). The animals were adapted to salinities of 30 to $10^0/_{00}$ for at least three weeks. Under

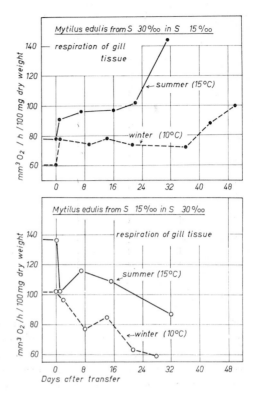

Fig. 58. Top: Changes in respiration of gill tissues of *Mytilus edulis* (North Sea 30⁰/₀₀ S) after transfer into brackish water (15⁰/₀₀ S). Bottom: Changes in respiration of the gill tissue of *Mytilus edulis* (Baltic, 15⁰/₀₀ S) after transfer into North Sea water (30⁰/₀₀ S). After C. SCHLIEPER 1955.

these conditions the respiration of the gill tissue showed a maximum at 20—23⁰/₀₀ S with decreasing values both with further dilution and concentration of the medium. The author concludes from his observations that the relation between the oxygen consumption of isolated tissue of the mussel and sea water is probably represented by a typical optimum curve. He believes that the results might be due to a salt effect on the respiratory enzymes. — At the same time WEBSTER (1968) measured the oxygen consumption of gill tissue from different bivalves of the American east coast. The animals were maintained for up to one week in artificial sea water of 5, 10, 15, 20, 25, and 30⁰/₀₀ S. In this case the oxygen consumption was relatively constant from 5—30⁰/₀₀ S for *Mytilus edulis* and *Crassostrea virginica,* but was greater at low salinities for *Modiolus demissus* and *Mercenaria mercenaria.* —

A study by ERMAN (1961) led a step further in the analysis of that phenomenon. First of all he demonstrated that only the gill tissue of the brackish-water mussels from Kiel showed about 50% increase in respiration, while the isolated mantle edge consumes about 12% more oxygen and the pericardial glands of brackish-water and sea-water mussels showed no difference at all in their respiration (see fig. 59). How can these results be interpreted? One possible explanation would be to assume that in *Mytilus* in brackish water only the cells of the epidermis adjoining the external medium react by means of increased respiration. The same author also found that the gill homogenate of the brackish-water individuals has an increased oxygen consumption independent of the osmotic concentration of the res

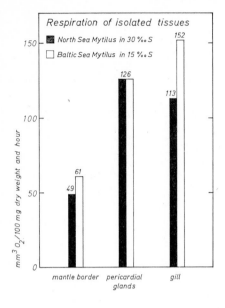

Fig. 59. Respiration of isolated tissues of *Mytilus edulis* from the North Sea and the Baltic depending on the salinity of the locality in which they occurred. After P. ERMAN 1961.

piratory medium. A similar behaviour was shown by the gill homogenate of North Sea mussels after seven days of adaptation to brackish water. This points to the possibility that during adaptation to brackish water the activity of some intracellular metabolic enzymes concerned rises. In agreement with this assumption it has been possible to demonstrate that the dehydrogenase activity in the gill homogenate of the brackish-water mussels is 60% greater. A slow increase in the intensity of dehydrogenase activity could also be seen in North Sea mussels transferred into brackish water (see fig. 60). Since *Mytilus* is poikilosmotic it is not possible to assume that the increased metabolism of the epidermal cells in brackish water is caused by increased energy requirement due to osmoregulatory work. But it might be conceivable that the energy requirements for ionic regulation in brackish water are greatly

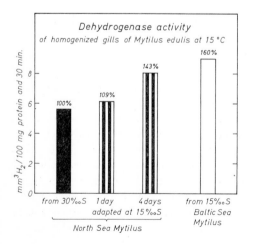

Fig. 60. Comparison of the activity of dehydrogenase in the gill tissue of *Mytilus edulis* from the North Sea and the Baltic. After P. ERMAN 1961.

increased, while the efficiency of the system of ionic transport is reduced under these conditions (see tables 19 and 20). It must also be borne in mind that both *Mytilus* and *Carcinus* exist in brackish water (of 15⁰/₀₀ S) in sub-optimal conditions to which they are but incompletely adapted, both genetically and individually.

The isolated gills of *Carcinus meanas* which is homoiosmotic in brackish water show similar behaviour. PIEH (1936) observed in green crabs from the North Sea (32⁰/₀₀ S) that in brackish water of 20⁰/₀₀ S the isolated gills reacted with a 53% increase in respiration, while the water content of the cells increased by about 4% at the same time. No additional effects were observed on further dilution. An increase in oxygen consumption was displayed by the same gills in sea-water-isosmotic NaCl solution. THEEDE (1964) continued these experiments by analyzing the respiration of gill tissues of *Carcinus* from the North Sea and from the western Baltic. He established the remarkable fact that the metabolic rate of gill tissue of the brackish-water individuals as well as of sea-water individuals with long-term adaptation to brackish water is increased, even independently of the osmotic concentration of the respiratory medium during the measurement (see fig. 61). A further con-

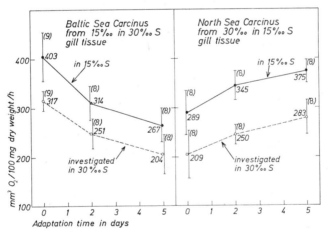

Fig. 61. Oxygen consumption of isolated gills of *Carcinus* depending on the salinity of the external medium. After H. THEEDE 1964.

tribution to the unravelling of these complex interrelations comes from the observation by DEHNEL & McCAUGHAN (1964) that in summer respiration of the gill tissues from *Hemigrapsus nudus* and *H. oregonensis* increases with a growing osmotic gradient between blood and external medium, while the corresponding values for animals in winter show only slight changes. In accordance with this, the authors assume that in summer the animals regulate their internal medium in brackish water through ionic absorption by the gills and with isosmotic urine, while in winter the animals maintain their blood concentration more by the production of a hypo-osmotic urine. A paper by KING (1965) published in the same year, reports that the isolated gill tissue of the poikilosmotic crustacean *Maja verrucosa* uses 6% more oxygen in half-diluted sea water. This is yet another proof that increased respiration need not necessarily be coupled with osmoregulatory activity. KING made a further revealing observation that gill tissue of *Callinectes sapidus* from the brackish water of an estuarine region (7—12⁰/₀₀ S) showed 30% greater respiration in a hypo-osmotic medium, where as the same tissue of *Callinectes* from sea water (34—36⁰/₀₀ S) showed only a 10% increase in respiration under identical conditions.

Mitochondria from the gills of *Carcinus, Callinectes* and two other brachyurians show corresponding metabolic reactions in hypo-osmotic media (KING 1966). The mitochondria functioning as centre for intracellular metabolism contain enzymes (cytochromoxidase, succinoxidase, fumarase, malicdehydrogenase) whose acticity increases by 200—300% on dilution of the fluid in which they were suspended from 1.6 to 0.16 Osmol (see fig. 62). After corresponding reduction of the osmotic concentration the oxygen consumption of the gill mitochondria of *Callinectes* from

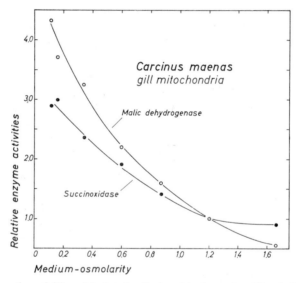

Fig. 62. Enzymatic activities of isolated mitochondria from the gills of *Carcinus maenas*, depending on the osmotic concentration of the medium. After values by E. N. KING 1966.

brackish water rose by 75% and in *Callinectes* from sea water by 35%. If concentrated sucrose solution was added to the hypo-osmotic suspension fluid the observed increase in enzymatic activity and in respiration were reversed. One may be allowed to assume that these remarkable results, produced in vitro and due to osmosis, are directly related to the changes in oxygen consumption within intact cells under brackish-water conditions. A well grounded detailed biochemical interpretation of the complex phenomenon is presented by FLORKIN & SCHOFFENIELS (1969).

In teleosts the correlations between respiratory rate and the salt content of the external medium are particularly complex since, on the one hand, their energy requirements for osmoregulation are higher, both in sea and in fresh water, than in brackish water isosmotic with their blood; on the other hand hormonal factors play an important part in the transition from one medium to another. In addition, a sudden change in the salinity of the external medium damages the osmotically susceptible gill tissue of some fishes; this impedes their gas exchange and the individuals concerned die quickly of asphyxia (P. BERT 1871). In other teleosts *(Sargus, Scorpaena)* the respiratory rate remains more or less unchanged up to the moment of their death (RAFFY 1932). Observations on teleosts which have been adapted over long periods appear to be more illuminating. Thus, respiratory rate of freshwater fishes drops in blood-isosmotic brackish water. This is true for the stenohaline goldfish *Carassius auratus* (VESELOV 1949) and the carp *Cyprinus carpio* (FONTAINE & RAFFY 1932), as well as for the more eurhyaline stickleback *Gasterosteus aculeatus* (GRAETZ 1931, WOHLSCHLAG 1957). Siimilarly in the marine lamprey *Petromyzon marinus* the oxygen consumption in half-diluted sea water is lower than in fresh water (FONTAINE & RAFFY 1935). On the other hand, the respiratory rate of the immigrant glass eels *Anguilla anguilla* (RAFFY & FONTAINE 1930) drops in fresh water as does that of smaller specimens of the starry flounder *Platichthys stellatus* (HICKMAN 1959).

At this point attention should be drawn to the fact that among invertebrates and vertebrates there are a few, specially euryhaline species, in which metabolism is more or less independent of the salt content of the external medium. To these belong, among others, the mitten crab *Eriocheir sinensis* whose respiratory rate in sea water of $32^0/_{00}$ S and in brackish water of $15^0/_{00}$ S, as well as in fresh water is always about the same (SCHWABE 1933). Similarly the gastropod *Theodoxus fluviatilis* which occurs in fresh and in brackish water (upper limit about $12—16^0/_{00}$ S) always has the same level of oxygen requirement in this range (LUMBYE 1958). A comparison of individuals of the same size and sex has shown the respiratory rate of the brine shrimp *Artemia salina* to be independent of the salinity of the external medium (GILCHRIST 1956). To the same group belong euryhaline teleosts such as the adult eel *Anguilla anguilla* (RAFFY 1933), the salmon *Salmo salar* (BUSNEL, DRILHON & RAFFY 1946) and within certain limits the small brackish-water teleost *Fundulus heteroclitus* (MALOEUF 1938) which has the same oxygen consumption in fresh water and in blood-isosmotic Ringer solution. There is strong indication that species which are to such large extent independent of the salt content of the external medium are, on the one hand, well protected from the surrounding medium by a relativ low permeability of the external membranes; on the other hand, their cellular mechanisms for ionic transport work with such efficiency that a change of the osmo-ionic regulatory performance is of no significance compared with the metabolism of the whole animal.

Similar observations have been made for algae occurring in brackish water. According to older data by LEGENDRE (1921) there is quite an extraordinary increase in the rate of assimilation in the marine algae *Fucus serratus* and *Ulva lactuca*, already in brackish water of $30^0/_{00}$ S and even more so in brackish water of $20^0/_{00}$ S. As a result of extensive studies HOFFMANN (1929, 1943) found that one group of algae *(Enteromorpha, Fucus vesiculosus* and *Porphyra)* are practically not affected by salinity changes of the external medium, while a second group *(Fucus serratus, Laminaria digitata, Ceramium rubrum)* experience an increasing rise in respiratory rate as the salinity of the external medium decreases (see fig. 63). He points out that the representatives of the first group are euryhaline to a marked degree. „*Enteromorpha* stellt den typischen Vertreter einer euryhalinen Alge dar. Sie kann im Brackwasser wie im Seewasser hoher Salinität ohne nachteilige Beeinflussung leben. Es ist also durchaus verständlich, wenn ihre Atmung von den Konzentrationsbedingungen des Substrates nicht beeinflußt wird. *Fucus vesiculosus* und *Porphyra,*

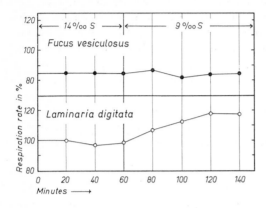

Fig. 63. Change in respiratory rate of *Fucus vesiculosus* and *Laminaria digitata* from the western Baltic after reduction of the salinity in the external medium. After C. HOFFMANN 1929.

beides Vertreter der typischen Gezeitenzone, sind dagegen durch Ebbe und Flut wiederholt starken Konzentrationsänderungen ausgesetzt. Es wird also auch hier verständlich, wenn sich die Atmung ziemlich unabhängig von der Konzentration des Außenmediums erweist" (p. 251) [*Enteromorpha* is a typical representative of a euryhaline alga. It can live in brackish water or sea water of high salinity without any adverse effects. Thus it is understandable that the concentration of the substrate does not affect its respiration. *Fucus vesiculosus* and *Porphyra*, by contrast, are representatives of the typical tidal zone, repeatedly exposed to strong fluctuations in concentration due to the tides. Here, too, it can be understood that respiration proves to be fairly independent of the concentration of the external medium]. Against that the same author stresses the fact that respiration in *Fucus serratus* and *Laminaria digitata* rises all the more with increasing dilution of the external medium „je näher wir dem für die Verbreitung der Alge gefundenen Minimum des Salzgehaltes kommen. Bei den Nordsee-Exemplaren, die direkt von $33^0/_{00}$ S einer Verdünnung von $11^0/_{00}$ ausgesetzt wurden, steigt schon bei dieser Konzentration die Atmung auffallend; bei den Ostsee-Exemplaren, die noch in stärkere Verdünnung gebracht wurden, die zum Teil die Konzentrationen der natürlichen Verbreitungsgrenze überschritten, wird die Atmungsintensität sogar weit über 100% gesteigert" (see p. 254) [the nearer we get to the minimum of salinity found for the distribution of the alga. In the North Sea specimens which were exposed directly from $33^0/_{00}$ S to a dilution of $11^0/_{00}$ S there is a marked increase in respiration even at this concentration; in Baltic specimens transferred into even greater dilutions — in part exceeding the concentrations of the natural limit of distribution — the respiratory rate is increased well over 100%]. HOFFMANN seems to have little doubt that there is a correlation between the water content of the algae and their respiratory rate, that is in the algae mentioned of the second group "mit zunehmendem Wassergehalt der Plasmakolloide die Atmungsintensität erheblich steigt" (see p. 250) [respiratory rate shows a considerable increase with a rising water content of the plasma colloids]. MONTFORT (1931) has followed this up with a study of carbon assimilation in some algae from the western Baltic in media of lower and higher salinity. According to his observations salinity of the external medium has little or no influence on the assimilation of highly resistant forms *(Enteromorpha, Chaetomorpha, Clado-phora)*. Other, somewhat less resistant forms show a considerable increase in assimilation after dilution of the external medium, followed by an irreversible decline; this takes some time in *Fucus vesiculosus*. In more susceptible forms *(Laminaria, Ceramium, Chorda)* there may be a transitory stimulation immediately followed by a steep drop. Fresh-water algae also display a reduction in assimilation after transfer into brackish water; but the limiting concentration at which the depression sets in varies for individual cases.

These findings have been confirmed and developed by recent investigations. KESSELER (1962) found transitory increase in respiration of the green alga *Chaeto-morpha linum* after transfer into sea water of either lowered or increased salt content. He would account for this by increased energy requirements due to restitution processes for the restoration of the normal structure of the protoplasm (after dilution of the external medium), and by positive turgor regulation (after concentration of the external medium). OGATA & MATSUI (1965) made manometric measurements of the photosynthesis of brackish-water algae (salinity of the water where they were growing $18.4^0/_{00}$). They found a maximum photosynthetic performance at $13.8^0/_{00}$ S, followed by a drop with increasing dilution or concentration. Only the tidal alga *Porphyra tenera* was an exception, showing almost uniform photosynthesis over the whole range under investigation ($0—37^0/_{00}$ S). According to NELLEN (1966) the average photosynthetic performance of marine forms (North Sea) of the brown alga *Fucus serratus* and the red alga *Delesseria sanguinea* is higher than that of their brackish-water forms (Kiel, western Baltic). The absolute values in mg oxygen pro 20 cm² and hour are an average of 0.81 for the North Sea form of *Fucus serratus* in $30^0/_{00}$ S and for the Baltic form of the same species an average of 0,67 in $15^0/_{00}$ S. In both species a short-term reduction or increase of the salt content provided a stimulation of the photosynthetic performance. If the algae remained in the changed medium the initial stimulation decreased again so long as the salt content of the medium was still within the limits of resistance of the species. If salinities were lower or higher a depression of the performance followed the initial stimulation. It is interesting that only *Fucus serratus* from the brackish water of the western Baltic reacted in the manner described, while the same species from pure sea water (littoral of Heligoland) was less susceptible: ,,Die Vermutung von HOFFMANN (1929), daß die Größenreduktion der Meeresalgen in Brackwasser nicht nur eine Folge der erhöhten Atmung, sondern auch der gleichzeitig erniedrigten Photosynthese und damit einer ungünstigeren Stoffbilanz sei, wird bestätigt (1) durch den Befund, daß die Normalleistung der Algen aus Gebieten mit geringerem Salzgehalt abnimmt, und (2) durch die nachgewiesene Leistungsdepression bei langer Einwirkung erniedrigter Konzentrationen" (quotation from NELLEN p. 306) [HOFFMANN's assumption that the reduction in size of marine algae in brackish water is not only the result of increased respiration, but also of simultaneously reduced photosynthesis and, with it, a less favourable metabolic balance, is confirmed (1) by the finding that the normal performance of algae from regions of lower salinity decreases and (2) the depression of performance which has been demonstrated when reduced concentrations have been acting for a long time].

8. Permeability of the integument

The permeability of the body surfaces of brackish-water organisms can be deduced from the exchange of material taking place passively, by diffusion. As with all living membranes their permeability depends on the size and shape of the molecules passing through them, on the number of ions and the distribution of their charges, on the solubility in the membrane, on the composition and, in particular the calcium content, of the external medium, on temperature etc. Biological factors, too, such as the physiological condition of the cells and active cell performance can exert an influence on permeability. Only after sudden, rather large changes in the external osmotic concentration may the surface of some species behave approximately as semipermeable membranes, through which water enters considerably faster than the dissolved substances, thus leading to swelling or shrinking.

Unfortunately there are but few quantitative studies of the surface permeability of species living in brackish water. KROGH (1939, 1946) has made a good summary of the available results which, in the main, are still valid to-day. He holds the view that the protoplasmatic surfaces of the animals are quite generally more or less permeable to water, dissolved gases, a number of organic substances and — with certain restrictions — to ions as well. BETHE (1929, 1930) is of the opinion that the body surfaces of all marine invertebrates are to some extent permeable to all ions of physiological importance. "The surfaces of all marine invertebrates which have been experimented upon are permeable for water and also for both the salts or their ions which are in solution in their blood and in sea water. The skin of these animals, save in the cases where special modifications have arisen, serves only as a protecting barrier preventing the loss of the body colloids" (p. 444). KOIZUMI (1932, 1935) examined the permeability of the body surfaces in the Holothurian *Caudina* and found that univalent ions permeate faster than divalent ones; this is probably of general validity:

$$K^+ > Na^+ > Ca^{++} > Mg^{++} \text{ and } Cl^- > So_4^{--}$$

In the same way the surface permeability is probably influenced by the protective layers of chitin or keratin overlying the outer epithelial cells. WEBB (1940) determined the following relative values of permeability in both directions for the chitinous cuticla in the foregut of *Homarus vulgaris*:

$$Na = 100, K = 169, Ca = 72, Mg = 42 \qquad Cl = 100, SO_4 = 52$$

In this case, too, the permeability is affected by the ionic equilibrium of the surrounding medium. Addition of calcium reduced the permeability of the membrane for sodium by about 50%. The permeability for water amounted to 0.4 μ^3 pro μ^2 surface, minute and difference in atmospheric pressure.

Some marine invertebrates die quickly after transfer from sea water into an isosmotic solution of NaCl. Within a short time they lose the physiologically important ions in their body fluids to such an extent that they are no longer able to live. But a comparison — made in this way — of the survival times of stenohaline and euryhaline marine invertebrates in isosmotic NaCl solution reveals remarkable differences. Medium-sized specimens of the stenohaline crab *Maja verrucosa* die within 1—3 hours in NaCl solution isosmotic with sea water, whereas specimens, of equal size, of the euryhaline shore crab *Carcinus maenas* survive for about 20 hours (SCHLIEPER 1932). Blood analyses show that the longer survival time of *Carcinus* in isosmotic, non-balanced salt solutions is due to a reduced permeability of its body surfaces (gill surfaces) (see fig. 64). It is hardly conceivable that the

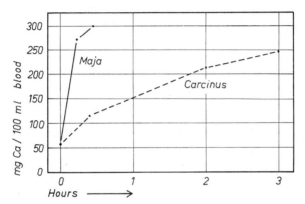

Fig. 64. Increase in the calcium content of the blood as a measure of the permeability of the outer membranes of *Maja verrucosa* (stenohaline) and of *Carcinus maenas* (euryhaline) after transfer into isotonic sea water of increased calcium content. After C. SCHLIEPER 1942.

permeability of the body surfaces of species inhabiting brackish water should be reduced for single ions only; in my opinion the rate of passage of one ion would permit general conclusions to be drawn regarding the permeability for ions and perhaps for water permeability as well. "The permeability of integuments (including gills) for electroneutral substances and ions seems to have some relation to the permeability for water" (KROGH 1939, p. 192). Based on such considerations NAGEL (1934) made comparative studies, in my laboratory, of the permeability of various marine, brackish-water and fresh-water crustaceans to sodium iodide. He added small, harmless quantities of sodium iodide (1.5 mg/ccm) to the external medium

Table 27. The rate of iodide uptake after addition of a little sodium iodide to the external medium as a measure of salt permeability of the outer membrane. Iodide content of the internal medium in relative values after 2.5 hours. After H. NAGEL 1934.

Medium	Species	Relative uptake of iodide
Sea water	*Portunus holsatus*	96
Sea water	*Hyas araneus*	76
Sea water	*Cancer pagurus*	14
Brackish water ..	*Carcinus maenas*	12
Fresh water	*Potamobius astacus*	2
Brackish water ..	*Eriocheir sinensis*	1

and analyzed the rate of penetration through the increase in the iodide content of the blood (see table 27 and fig. 65). These results show conclusively that the stenohaline species are more permeable then the euryhaline ones. The permeability of the integuments is specially low in species viable in fresh water. But the lowest

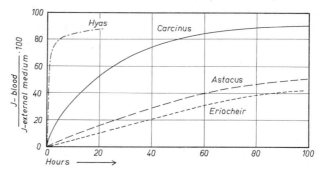

Fig. 65. The rate of iodide uptake as a measure of the permeability of the outer membranes of different crustaceans after transfer into sea water containing sodium iodide. After H. Nagel 1934.

level of permeability is found not in stenohaline fresh-water invertebrates, but in those particularly euryhaline species able to live both in fresh and in sea water, or else those largely independent of the salt content of the external medium. Consequently the mitten crab *Eriocheir sinensis* which can live both in sea and in fresh water takes up iodide more slowly than the fresh-water crayfish *Astacus fluviatilis*.

Corresponding relative differences in permeability were found by GROSS (1957) in the exoskeleton of decapod crustaceans. He made disks of the carapace of freshly killed animals whose hypodermis had been removed and used them as stoppers of

Table 28. Relative permeability to salt of the exoskeleton of various decapod crustaceans. (After W. J. GROSS 1957).

Habitat	Species	Relative permeability
Sea water	*Pugettia producta*	117
Sea water	*Cancer antennarius*	114
Sea water	*Cancer gracilis*	62
Brackish water ..	*Hemigrapsus oregonensis*	20
Brackish water ..	*Hemigrapsus nudus*	5.5
Brackish water ..	*Pachygrapsus crassipes*	5.0
Fresh water	*Cambarus clarkii*	3.5

glass tubes, each filled with 10 ml of sea water; these were placed into larger external containers each filled with 10 ml half-diluted sea water. After 24 hours the increase in the salt content of the external medium was determined by means of a conductivity bridge. In this case, too, the marine species unable to osmoregulate were most permeable for salts, while the regulating brackish-water species were least permeable, with the only fresh-water species investigated being least permeable (see table 28).

Measurements of quite a different kind, as carried out by SHAW (1961), also demonstrate that this must be a general rule. He determined the rate of sodium loss of various species of crustaceans after transfer into pure water and obtained quite comparable distinctions. The values in μ Mol per hour calculated for the same surface area and for 50 g live weight were as follows: for *Carcinus maenas* (from 40% sea water) 891, for *Eriocheir sinensis* (from 10% sea water) 153, for *Potamon niloticus* 27 and for *Astacus pallipes* 5.

Among earlier observations those of PANIKKAR (1941) will be mentioned at this point because of their special methodology. According to these the gills of *Palaemonetes varians* (though also of *Leander serratus*) are less permeable for organic dyes (methylene blue and vital red) than the gills of *Crangon vulgaris* and *Carcinus maenas*. According to USSING (see KROGH 1939, p. 98) the permeability for water — as determined by means of heavy water (D_2O) — of the brine shrimp *Artemia salina* is extraordinarily low.

Corresponding differences in permeability have also been measured in Polychaeta; these may be interpreted as being connected with a greater or lesser ability to survive in brackish or fresh water. Thus FRETTER (1955) found that the more stenohaline *Perinereis cultrifera* is about three times more permeable for radioactive ^{24}Na than the more euryhaline *Nereis diversicolor*.

Corresponding differences for radioactive chloride were found by JØRGENSEN & DALES (1957) in a comparison of *Nereis virens* and *Nereis diversicolor*. These authors also showed that after transfer into hyperosmotic media *N. virens* and *N. pelagica* lost water about twice or three times as fast as *N. diversicolor* under the same conditions. They also made it appear feasible that in *Nereis diversicolor* permeability for water and salts diminishes greatly after transfer into fresh water, thus possibly being actively reduced. Unfortunately far too little is as yet known about the occurrence of regulation of permeability in brackish-water animals.

Recently SMITH (1964) reported a possibly lowered D_2O (heavy water, deuterium oxide) influx when *Nereis succinea* and *Nereis limnicola* were tested in 5% sea water or fresh water, respectively, although the significance was doubted. — On the basis of the D_2O entry rates obtained under natural conditions of stress (10% and 70% sea water), SMITH (1967) concludes that also the small crab *Rithropanopeus harrisi* exhibits the capability of lowering its water-permeability as an adaptive response to a lowered external salinity. (Compare also SMITH 1970.)

WIKGREN (1953) draws attention to the fact that the water permeability of the euryhaline eel *Anguilla anguilla* — as calculated from the amounts of urine produced — is $1/5$ of the water permeability of the stenohaline goldfish *Carassius auratus* living in fresh water. The brook lamprey *Lampetra planeri* which occurs only in fresh water allows water to permeate in both directions considerably faster than the euryhaline, katadromous river lamprey *Lampetra fluviatilis* (HARDISTRY 1954). Several recent investigations of euryhaline teleosts indicate that, with an adaptation to a changed osmotic concentration of the medium, the permeability of their gills for certain ions or water respectively also changes (MULLINS 1950, GORDON 1962).

9. Effects of temperature and temperature resistance

It has long been known that numerous species, especially on tropical marine shores, have immigrated from the sea into brackish and fresh water (MARTENS 1858). Attempts have been made to explain this phenomenon in the following way:

1. The number of species in the marine littoral zone of the tropic is considerably higher than in the colder latitudes. This in itself may account for the absolute number of species penetrating into brackish and fresh water to be greater in tropical regions.

2. Immigration is facilitated by the relative constancy of water temperatures in the tropical latitudes. All the year long the invading species find themselves in temperatures which are optimal for them and so they only have to overcome the osmotic stress.

3. The higher water temperature in the tropics is in itself favourable to osmotic resistance and the osmoregulation of the immigrant species. Lower water temperatures impede the osmotic resistance and osmoregulation in brackish water.

The validity of the assertion made under 3) can be tested experimentally for relatively eurythermic species. Let us examine recent literature to see whether it is possible, in the laboratory, to influence osmotic resistance and osmoregulation through a change of water temperature. A short survey will show that this can, in fact, be achieved. But there is no indication that higher water temperatures in general raise the range of resistance and osmoregulatory performance of the brackish-water species or the species which can survive in brackish water. Obviously only these experiments offer a valid proof in which the experimental animals have not been exposed abruptly and simultaneously to a thermal and an osmotic stress. The importance of previous individual adaptation to one of the two environmental factors, before the other is being altered, is self-evident. It will be best to start changing one of the factors within the optimal range of the other and study its effects in this way. In establishing the limits of resistance of the American lobster *(Homarus americanus)* MCLEESE (1956) has taken the individual acclimatisation of his experimental animals into consideration in an admirable way. Before tests were carried out his crustaceans were acclimatized either to warmth or to cold, in sea water of $30^0/_{00}$ S. For individuals adapted to cold and examined at the same low temperature the lower lethal salinity limit was always considerably lower (see table 29). There was a similar, though somewhat smaller difference between animals acclimatized to $20^0/_{00}$ S at 15° and 25°C and tested in water of 15° and 25°C. In contrast specimens acclimatized to $20^0/_{00}$ S at 5°C had a lower limit of salt concentration which was above that of animals acclimatized at 15°C. This probably means that osmotic resistance may also be affected in very cold brackish water. As mentioned before (cf. p. 217) GRESENS (1928) has found the same phenomenon in his investigations into the resistance of fresh-water invertebrates in brackish water. Similarly the resistance to dilution is lowered in brackish water of low oxygen content. But the upper lethal temperature limits of *Homarus* in sea water of $30^0/_{00}$ S were always higher then in sea water of $20^0/_{00}$ S, provided that comparisons were made between individuals of the same pretreatment temperature (see table 29). ¾

While *Homarus vulgaris,* a temperate cold-water species is better able to tolerate lower salinities at 15°C than at 25°C, the blue crab *Callinectes sapidus,* widespread

Table 29. Upper lethal temperature limits and lower lethal salinity and oxygen limits of *Homarus americanus*. After MCLEESE 1956.

Conditions of acclimatization (during pretreatment)			Lethal limits		
Temperature °C	Salinity °/oo	Oxygen mg/l	Temperature °C	Salinity °/oo	Oxygen mg/l
5	20	2.9	20.6	11.0	0.72
		6.4	23.7	9.0	0.72
15	20	2.9	27.3	9.0	0.86
		6.4	27.8	8.2	1.20
25	20	2.9	28.5	11.5	1.72
		6.4	29.3	11.1	1.26
5	30	2.9	24.0	10.8	0.29
		6.4	25.7	6.0	0.20
15	30	2.9	27.8	10.6	0.66
		6.4	28.4	11.2	0.83
25	30	2.9	28.7	15.4	1.30
		6.4	30.5	16.4	1.17

Table 30. Survival rates in percent of megalops of *Callinectes sapidus* at 16 combinations of salinity and temperature. From J. D. COSTLOW 1967.

	°/oo S			
	5	10	20	30
15°C	—	0	14	42
20°C	0	70	95	90
25°C	22	79	85	100
30°C	41	95	90	86

in a fairly large area of the American Atlantic and Gulf coasts, shows the opposite behaviour. COSTLOW (1967) bred *Megalopa* larvae of this euryhaline species; their survival rate in brackish water of 15 °C is far lower than at 25° and 30 °C (see table 30).

In the same way an experimental comparison of marine bivalves from different latitudes shows that as a rule, their cellular resistance to low salinity is favoured by those temperatures closest to the species-specific optimal temperature (RESHÖFT 1961, VERNBERG et al. 1963, SCHLIEPER et al. 1967). Thus gill segments of the temperate cold-water species *Modiolus modiolus* (Kattegat) show much better survival in brackish water of low salinity at 10°C than at 20°C whereas the warm-temperate *Modiolus demissus* (North Carolina) has a slighlty greater cellular resistance to dilution at 25 °C than at 10 °C; the range of osmotic cellular resistance of the tropical species *Modiolus auriculatus* (Red Sea) is even more reduced by low temperatures.

Similar relations between temperature and osmotic resistance in brackish water have also been found in marine algae. Red algae from the Gulf of Naples have a reduced resistance to hypotony at 2°C (BIEBL 1939). By contrast, red algae from deep water of the western Baltic, with an ecological optimum temperature about 5 °C, are far more resistant to lower salinities at 0—5 °C than at 10—20 °C (SCHWENKE 1959, 1960 b).

Numerous marine species are more sensitive to extreme temperatures in brackish water than they are in sea water. This is shown, e. g. in the observations of GOMPEL & LEGENDRE (1928) on the Turbellarian *Convoluta roscoffensis* and by LOEB & WASTENEYS (1912) on the euryhaline brackish-water teleost *Fundulus heteroclitus;* according to their findings heat resistance of the species is diminished in dilute sea water. The same holds for the invertebrates *Nereis diversicolor, Gammarus duebeni* and *Sphaeroma hookeri* which occur in brackish water (KINNE 1954). In all three species a lowering of the natural external concentration produces an increased susceptibility to higher temperatures. This also holds for species which, in nature, are adapted to various salinities; this has been demonstrated by SCHLIEPER & KO-WALSKI (1957) in a comparative study of the cellular heat resistance of three populations of *Mytilus edulis* from the North Sea, the western Baltic and the Finnish coast (30, 15 and $6^0/_{00}$ S). In this case heat resistance was determined by observing the duration of beat of the terminal gill cilia at the lethal temperature of 35 °C. It was found that the cilia were beating on an average for 45 minutes in brackish water of $6^0/_{00}$ S, in $15^0/_{00}$ S, however, for 68 minutes and in $30^0/_{00}$ S even about 100 minutes (see table 31). A transfer of brackish-water mussels into sea water and, vice versa, of sea-water mussels into brackish-water changes heat resistance — after a short period of adaptation — in the same direction.

Table 31. Heat resistance of the gill tissues of *Mytilus edulis* in brackish water and in sea water, measured by the duration of beat of the terminal gill cilia at 35°C, in minutes (mean values). After SCHLIEPER & KOWALSKI 1957.

Experimental animal and salinity of medium	North Sea *Mytilus* (List)		Baltic *Mytilus* (Kiel)		Baltic *Mytilus* (Tvärminne) in $6^0/_{00}$ S
	in $30^0/_{00}$ S	after transfer into $15^0/_{00}$ S	in $15^0/_{00}$ S	after transfer into $30^0/_{00}$ S	
Ciliary beat					
vigorous	45	20	22	63	11
Weak	38	27	24	22	16
Very weak	16	17	22	23	18
Total time	99	64	68	108	45

Experiments by VOGEL (1966) make it seem probable that interrelations between heat resistance of a brackish-water species and the salinity of the external medium may also be influenced by the position of the salinity optimum of the species in question. He examined the heat resistance of the peritrich brackish-water ciliate *Zoothamnium hiketes* which at 8 °C and salinities of 3—$30^0/_{00}$ S is viable for at least several weeks. $40^0/_{00}$ S was only tolerated for a week. In the critical experiment the animals which had been adapted for 6—8 days to seven salinity grades between 3 and $30^0/_{00}$ S at 8 °C, were transferred directly into 36 °C and their survival time in seconds was determined. There was an increase in heat resistance up to $20^0/_{00}$ S, then the curve dropped again (see fig. 66). In this case the maximum of heat resistance appears to coincide with the salinity optimum for this limited euryhaline species.

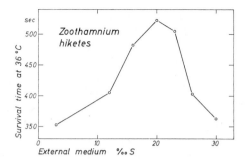

Fig. 66. Heat resistance of the per-itrich ciliate *Zoothamnium hiketes* depending on the salt content of the external medium. After W. VOGEL 1966.

Different interrelations between heat resistance and the salinity of the external medium are found if a maximal euryhaline species is used as experimental animal. As an example I would quote the copepod *Tigriopus fulvus* which RANADE (1957) has studied; it occurs on European coasts in pools at the high water border. In experiments it can withstand salt concentrations between 4.4 and $90^0/_{00}$ S. In order to determine heat resistance the experimental animals — in groups of 20 — were directly transferred from sea water of $34^0/_{00}$ into 11 different salinity grades between 0 and $90^0/_{00}$ S. In this way the lethal temperature in each case, at which about ¾ of the animals died immediately on transference could be determined accurately to 0.1 °C; the rest of the animals died within 3 minutes. The results show that in this case heat resistance rises continuously with increased salt content (see fig. 67). This is in good agreement with the behaviour of the species in nature as in the tidal pools salinity fluctuations are also correlated with those of water temperature.

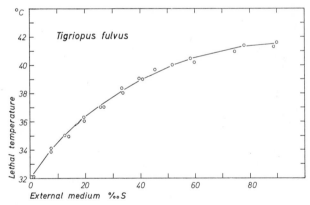

Fig. 67. Lethal temperature of the copepod *Tigriopus fulvus* depending on the salinity of the external medium. After M. R. RANADE 1957.

How can the mechanisms of the combined effects of salinity and temperature — described above — be interpreted in detail? A possible explanation is as follows: Heat death of living cells is often considered to be an inactivation of vital cellular enzymes through heat. This results from denaturation or destruction of the proto-plasm-protein complexes with deformation of the individual molecules and the disruption of intermolecular bonds (hydrogen bridges). In the context of this working hypothesis the degree of heat resistance would also provide a measure of the stability of the plasma proteins. It would be logical to conclude that the stability of the plasma proteins is all the more reduced in brackish water, the lower its salinity. As already mentioned (see p. 279) this reduction in the stability of the living structure with diminishing salinity of the external medium might be brought about in the tissues of the brackish-water inhabitants by increased water content and degree of swelling, through a rise in bound water at the expense of free water. In the context of this working hypothesis it might further be concluded that the more euryhaline species frequently maintain the water content of their tissues to a greater degree independently of the influence of the salinity of the surrounding medium. This might well be one of the reasons why euryhaline species, invertebrates and fishes, often have a higher heat resistance in brackish water than their more stenohaline relations.

In accordance with this the plaice e. g., *Pleuronectes platessa* is far more sensitive to heat in brackish water of 7.5⁰/₀₀ S than in sea water of 30⁰/₀₀ S, while the related, more euryhaline flounder *Pleuronectes (Paralichthys) flesus* is less affected in its heat resistance by the salt content of the medium, at least in the above-mentioned range (see fig. 68). In the same manner the heat resistance of fresh-water species

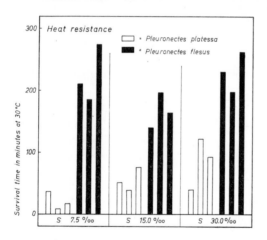

Fig. 68. Heat resistance (survival at 30° C) of *Pleuronectes platessa* and *Pl. flesus* in sea and in brackish water. After M. WAEDE 1954.

immigrated into brackish water or else that of fresh-water species experimentally adapted, to brackish water rises as SCHMITT (1955) demonstrated for fresh-water planarians and WAEDE (1954) for the golden orfe *Idus melanoticus* (see figs. 69 and 70).

Unfortunately no investigations are as yet available on the mechanism of resistance in poikilosmotic brackish-water species. But THEEDE (1965b) has determined the cellular freezing resistance of marine littoral bivalves in correlation to the salinity of the external medium. His method consisted of freezing isolated gill segments at —10°C for varying periods. After rapid thawing the degree of damage to the tissue was estimated by observing the activity of ciliary beats (3 normal, 2 somewhat reduced, 0 stoppage). The results show that the cellular resistance to freezing is a species-specific character which is affected by the salinity of the medium of the locality (see fig. 71). If, in an experiment, bivalves from the brackish water of the

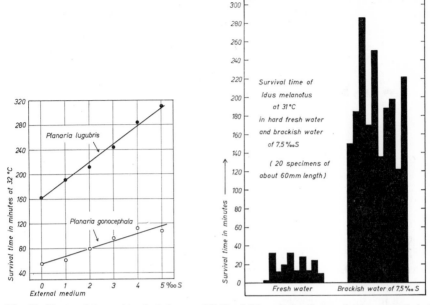

Fig. 69. Heat resistance (survival time at 32°C) of *Planaria lugubris* and *Pl. gonocephala* in fresh water and in brackish water. After E. SCHMITT 1955.

Fig. 70. Heat resistance (survival time at 31° C) of the golden orfe *(Idus melanotus)* in hard fresh water and brackish water of 7.5°/oo S. After M. WAEDE 1954.

western Baltic (15°/oo S) were transferred into North Sea water (32°/oo S) the cellular freezing resistance rose continuously in the course of a slow process of adaptation and after about 30 days it reached the higher degree of resistance of the sea-water population (see fig. 72). Like heat resistance, the individual cellular freezing resistance of the bivalves drops during their seasonal reproductive phase. THEEDE holds the view that the reduction in cellular freezing and heat resistance in brackish water

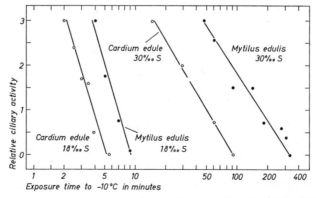

Fig. 71. Differences in the cellular resistance to freezing of two marine bivalves from the Baltic (18°/oo S) and from the North Sea (30°/oo S). After H. THEEDE 1964.

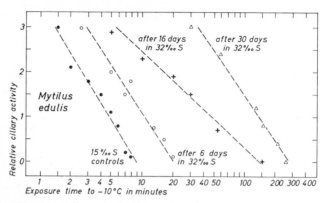

Fig. 72. The effect of salinity and of salinity adaptation on the cellular resistance to freezing of the isolated gill of *Mytilus edulis*. After H. THEEDE 1965.

is connected with a general, non-specific reduction of resistance caused by the lowered salinity content. He considers the mechanisms to be a decrease of intermolecular forces in the macromolecular region; these are responsible for the structural stability of the protoplasm.

What influence has temperature of the brackish water on the osmoregulatory performance of homoiosmotic species living in it? The functioning of the osmo-regulatory mechanisms of a brackish-water species always depends on a number of factors; of these, the permeability of the external membranes and the performance of osmotic work by the epidermal cells taking part are of special importance. In general the permeability of a living membrane increases with rising temperature. As corresponding measurements in brackish-water inhabitants are not available I would mention WIKGREN's (1953) fine studies in this connection; he examined the influence of temperature on the water permeability and urine production of the river lamprey *Petromyzon fluviatilis* in fresh water (see table 32).

Table 32. Urine formation and water permeability of *Petromyzon fluviatilis* in fresh water at various temperatures. After WIKGREN 1953.

Temperature °C	Urine formation ccm/kg and day	Water permeability days	Temperature quotient
2—3	60	715	$\dfrac{715}{266}$ 4.8
7—9	160	266	$\dfrac{266}{113}$ 2.6
16—18	359	113	

In *Petromyzon* the osmotic influx of water and consequently its urine production rise three — to fivefold if the temperature of the water is raised by 10 °C. This means that for homoiosmotic species the expenditure of osmoregulatory energy increases in brackish water to a degree corresponding with a rise in water temperature, that is a higher osmoregulatory performance is required in order to maintain the same differences in concentration in warmer brackish water. In addition, sudden changes in temperature may cause disturbances in the coordination of the osmoregulatory mechanism of teleosts; these are manifested by temporary volume changes of the experimental animals (RAFFY 1954).

The results of LOCKWOOD's (1960) investigations on the effects of temperature on the ionic regulation of the fresh-water isopod *Asellus aquaticus* are in accordance with the observations on vertebrates just referred to. In brackish water of low salinity and with a constant medium temperature this species maintains the osmotic concentration and sodium content of the hyperosmotic haemolymph in a steady state (constant level) by an active uptake of sodium from the external medium which is of the same magnitude as the losses by diffusion and excretion of urine. If in the experiment the water temperature is lowered the rate of active uptake of sodium diminishes to a greater extent than the rate of sodium loss: as a result of this the sodium concentration of the haemolymph goes down. This drop in the internal sodium concentration, however, immediately triggers off a certain increase in the rate of sodium uptake. So after some time an equilibrium between the rates of loss and of uptake is established and a new, constant, though somewhat lower level of internal concentration is set up.

Apparently not all temperate brackish-water species are able to perform the required higher osmoregulatory work in brackish water of low salinity and at a higher temperature. This is true e. g. for the brachyurous decapods *Heteropanope tridentatus* and *Eriocheir sinensis* as well as the brackish-water amphipod *Gammarus duebeni* (OTTO 1934 and 1937, KINNE & ROTTHAUWE 1952, and KINNE 1952.) Consequently the blood concentration of *Heteropanope* in winter animals (7°C) is markedly higher than in summer animals (20—21°C) and mortality of this crustacean is also particularly high in dilute media at higher temperatures (see fig. 73).

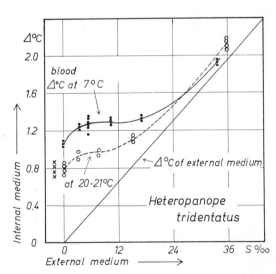

Fig. 73. The effect of temperature on the osmotic concentration of the internal medium of *Heteropanope tridentatus* in sea and brackish water. After O. KINNE & H. W. ROTTHAUWE 1952.

The shrimp *Crangon crangon* and the shore crab *Carcinus maenas* from the western Baltic (FLÜGEL 1960, 1963) behave very much like *Heteropanope*. In brackish water (below 20⁰/₀₀ S) and at 5°C both crustaceans can maintain their internal medium more strongly hyperosmotic against the external medium than at 15°C (see figs. 74 and 75). From a comparison of the lowering of the freezing point and the electrical conductivity of the blood it can be concluded that the essential difference in the osmoregulatory performance at 5° and 15° is brought about by a greater ionic regulation at the lower temperature. Below 5 °C, near zero, the ionic regulatory performance of *Crangon* drops once again. It is hardly necessary to emphasize that these results obtained with temperate species may not be simply applied to warm-water species. This is shown, among others, by the results of WILLIAMS's (1960) studies of the subtropical estuarine shrimps *Penaeus setiferus* and *P. aztecus*

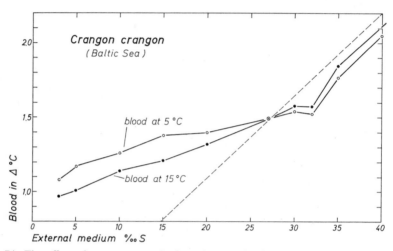

Fig. 74. The effect of temperature of adaptation on the freezing-point depression of the blood of the shrimp *Crangon crangon*. After H. FLÜGEL 1960.

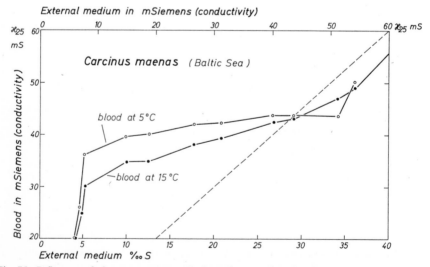

Fig. 75. Influence of the temperature of adaptation on the electrical conductivity of the blood of the shore crab *Carcinus maenas*. After H. FLÜGEL 1963.

which are common on the South east and the Gulf coasts of the United States. At a medium temperature (15—30 °C) their internal medium is hypo-osmotic towards sea water and hyperosmotic towards brackish water. At a lower temperature (about 9 °C) their osmoregulatory performance declines and the blood concentration becomes increasingly isosmotic to the external medium.

A very good indication of the importance of the temperature of the locality where the animals occur, for osmoregulation in brackish water is provided by the investigations of SEGAL & BURBANCK (1963), as well as FRANKENBERG & BURBANCK (1963) on the isopod *Cyatura polita* which lives in estuaries. A comparison was made between several populations, among others from the area of a river mouth near Cape Cod (1.5—50% sea water, 2—23 °C) and from the warm saline Silver Glen Springs in Florida (3.5% sea water, 20—23 °C). Both groups have a hyperosmotic internal medium in half- and more strongly diluted sea water. In the Cape Cod specimens the osmoregulatory performance is in no way affected over the whole range of temperatures from 5—32 °C. The Florida population, on the other hand, is unable to maintain corresponding hyperosmy at 5 °C.

HOHENDORF (1963) has followed in detail the breakdown of osmoregulation in the euryhaline Polychaete *Nereis diversicolor* at extreme temperatures. The combination of low salinity and temperatures round freezing point have a lethal effect on this species. Even at 1 °C in brackish water below 2⁰/₀₀ S the osmotic concentration of the internal medium shows a considerable drop (see fig. 76).

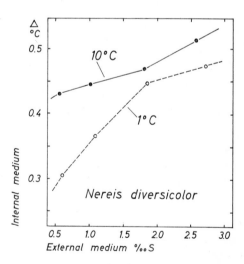

Fig. 76. Influence of temperatures near freezing point (+ 1 °C) on the osmoregulation of the Polychaete *Nereis diversicolor* in brackish water of low salinity. After K. HOHEN-DORF 1963.

A similar reduction of the osmoregulatory performance at very low temperatures has been observed for temperate teleosts in sea water and in particular in brackish water (ERCHINGER 1964). On the other hand, well-adapted species may also display a compensating increase in the osmotic internal concentration in winter which might be interpreted as protective reaction against freezing (GORDON et al. 1961, RASCHACK 1967) (see fig. 77).

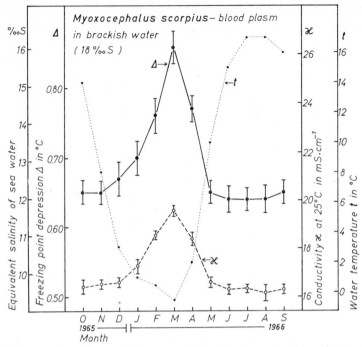

Fig. 77. Seasonal changes in osmotic concentration (freezing-point depression) and of the electrolyte content (conductivity) in the blood of the teleost *Myoxocephalus scorpius*. After M. RASCHACK 1968.

10. Ionic effects in balanced and non-balanced brackish water

Sea water and brackish water are balanced salt solutions in which the physiological effects of ions — which are partly antagonistic — cancel or balance each other. If brackish water has formed from sea water by influx of soft fresh water free of electrolytes or, at least, poor in electrolytes, the original ionic proportions of the sea water are maintained. This e. g. applies to the brackish water of the Norwegian fjords whose fresh-water supply comes from Precambrian rocks. In brackish water of this type not only the salt content, but alkalinity (the ability to neutralize acids) and the calcium content decrease in proportion to the dilution. If, however, the formation of brackish water from sea water is brought about by an influx of fresh water which is more or less hard (rich in calcium) a greater dilution may result in abnormally high values of alkalinity and, in particular, in the calcium content. Such a situation occurs e. g. in the southern Baltic where the incoming fresh water has a high content of calcium bicarbonate (see table 33). In this way brackish waters of low salinity arise which have an abnormally high calcium content. The proportions of the remaining cations and anions are less disturbed by influx of hard fresh water unless this fresh water has an extremely high content of magnesium or sulphate (cf. table 34).

Table 33. Alkalinity and calcium content in sea and brackish water of the Atlantic and the Baltic. After WITTIG 1940.

Station	Salinity %/oo S	Chlorine content gCl/l	Alkalinity mEqu/l	Calcium content		
				present mg/l	"normal" mg/l	relative %
Outside Oslo	33.89	18.76	2.308	409.0	409.0	100
Kieler Förde	18.26	10.11	2.160	232.5	225.5	103
Bornholm-Basin	17.32	9.58	2.005	221.3	213.9	104
Kieler Förde	15.75	8.71	2.175	206.5	194.5	106
Western Baltic	15.23	8.42	1.790	198.7	183.6	108
Kiel Bay	12.67	7.00	1.723	165.9	156.5	112
Arkona-Basin	7.34	4.05	1.510	107.4	90.7	116
Eastern Baltic	6.60	3.64	1.430	92.4	79.7	116
Schwentine-mouth ...	5.26	2.90	2.375	107.0	64.8	165
Schwentine-mouth ...	0.34	0.17	2.690	57.6	4.1	1405

Table 34. Relation of salinity components in sea water, brackish water and hard fresh water. From KALLE 1943 and BALDWIN 1949.

	A Sea water (Challeng. Exp) S = 35.19%/oo	B Baltic water between Öland and Gotland S = 7.22%/oo		C Hard fresh water	
	%	%	$\frac{B}{A}$	%	$\frac{C}{A}$
Na	30.593	30.47	0.996	6.98	0.23
K	1.106	0.96	0.868	5.32	4.79
Ca	1.197	1.67	1.40	21.6	18.0
Mg	3.725	3.53	0.95	4.65	1.25
Cl	55.292	55.01	0.995	13.62	0.25
Br	0.188	0.13	0.69	—	—
SO₄	7.692	8.00	1.04	8.30	1.08
CO₃	0.207	0.14	0.68	39.5	190
	100	100		100	

Numerous biological and physiological observations have shown that a high calcium content facilitates the penetration of marine species into brackish water of low salinity and into fresh water. PANTIN (1931) was the first to observe that a certain calcium content in the external medium was necessary for the functioning of osmoregulatory mechanisms in the euryhaline Turbellarian *Gunda (Procerodes) ulvae. Gunda* can tolerate fluctuations in the salinity concentration from sea water to pure fresh water. PANTIN transferred *Gunda* from sea water into brackish water, river water and distilled water. Even in river water the Turbellarian still behaved homoismotically while it reacted like a poikilosmotic species in distilled water swelling up through water uptake and giving off salts at the same time. An analysis of this phenomenon revealed that the calcium content of the river water was the factor which enabled *Gunda* to survive in this medium. Numerous studies of cellular physiology have shown that calcium ions reduce the permeability of animal membranes; therefore it seems likely that the effect of calcium on *Gunda* can also be explained in this way. Other workers have made similar observations. ELLIS (1937) was able to show that after transfer into $^1/_5$ sea water *Nereis diversicolor* begins to swell at first by osmotic water uptake, but within the next few hours it regains its original weight (volume) by means of active volume regulation. If, however, calcium--free brackish water is used in such an experiment nothing will prevent the worm from swelling and it will only be able to regulate its volume after the missing calcium ions have been added to the external medium. According to BREDER (1933) the calcium content of the external medium is of the same physiological importance to marine fishes immigrating into brackish and fresh water (cf. also HEUTS 1944).

SCHWENKE (1958, 1960) carried out simple experiments on red algae from the western Baltic and was able to demonstrate that, in a similar manner, calcium is of special significance for their survival in brackish water (see table 35).

Table 35. Relationship between the resistance to hypotony and the calcium content of the medium in the Red Alga *Delesseria sanguinea* (January, 7—8° C). After H. SCHWENKE 1960.

Ca content		% proportion of surviving cells after 24 hours in salinities of		
mg/l	Corresponding to dilution of oceanic water in $^0/_{00}$ S	1 $^0/_{00}$	5 $^0/_{00}$	10 $^0/_{00}$ S
12	1.0	50	70	90
60	5.0	75	90	100
120	10.0	80	90	100
180	15.0	90	100	100

SCHLIEPER & KOWALSKI (1956) have attempted to analyze the mode of action of the calcium content in brackish water by means of cell physiological experiments. They determined the osmotic and thermal resistence of the isolated gill tissue of the common mussel *Mytilus edulis* in relation to the alkalinity and calcium content of the medium. In every case the duration of beating by the gill cilia of Baltic mussels from Baltic water of 15$^0/_{00}$ S ($= 233$ mMol/l) was used as the measure of resistance. This Baltic water was diluted to 30 mMol/l with aqua destillata or with aqua destillata containing 200 mg Ca/l. If the water was diluted with aqua destillata alone the gill cells survived on an average for 66 minutes at 10 °C. If, however, the dilute solution containing calcium was used, the gill cells survived for 94 minutes, that is considerably longer. In the course of further experiments it was possible to demonstrate that it was not the calcium content of the external medium, but that of the cells themselves which was decisive for the level of the osmotic resistance: If the experimental animals had been pre-adapted to a Ballic water of 15$^0/_{00}$ S with addition of 200 mg Ca/l and the gills of these mussels were subsequently transferred into dilute Baltic water with a relatively increased calcium content their survival time was 145 minutes. The calcium content of the brackish water is of equal importance for the heat resistance of the gill tissue. As previously mentioned (see p. 298) the cellular heat resistance of the common mussel measured as survival time at 35°C is reduced to about $^2/_3$ when animals are transferred from sea water of 30⁰/₀₀ S into brackish water of 15⁰/₀₀ S. But the heat resistance of the gill tissues remains practically unchanged if, for a dilution of sea water to half its original salt concentration, if the calcium content is kept constant. In this case, too, it is the calcium content of the tissues rather than that of the external medium which is the decisive factor for the magnitude of resistance. For an interpretation of the effect of the calcium content of the brackish water on the level of osmotic and thermal resistance it might be pointed out that calcium ions increase the stability of the protein molecules in the protoplasm and with it plasmatic resistance. An increased stability of plasmatic structures might result in an increase of heat resistance and osmotic resistance. A reduction of permeability for water and dissolved substances caused by increased density of the cortical layer of the protoplasm would, however, might specially favour an increase in osmotic resistance.

The fact that the effect on heat resistance produced by a reduction of the total salt content in brackish water may, in certain cases, be counteracted simply by keeping the calcium content constant is of particular interest for the problem of the physiological effects of individual ions in brackish water. It is obvious that the physiological effects of the individual ions contained in brackish water can not be analyzed in experiments using isotonic solutions of pure salts. Isotonic solutions of the chlorides of sodium, potassium, calcium and magnesium will kill the brackish-water animals placed into them within a short time. However, there is the possibility of reducing the total salt content of the brackish water while keeping the concentration of individual ions constant. Solutions of this kind are tolerated without ill-effects by many brackish-water animals and they allow certain conclusions to be drawn on the effects of the ions in question. SCHLIEPER & KOWALSKI (1956) have used this method to compare the effects of a natural brackish water of 15⁰/₀₀ S with that of corresponding brackish water in which, in each case, the concentration of ionic sodium or potassium or calcium or magnesium had been doubled. Results showed that the heat resistance of the tissue — which was used as the measure of stability of the cell colloids — was increased by addition of calcium or magnesium, but lowered by addition of potassium. The mechanical performance of the gill cilia (transportation activity) — which was taken as a measure of activity — was, however, raised by addition of sodium or potassium and reduced by addition of calcium or magnesium. A comparison of the effects of the total osmotic concentration with that of individual ions in brackish water clearly brings out the opposing physiological effect of the total osmotic concentration on the one hand and that of the cation concentration (K, Ca, Mg) on the other (see fig. 78 a and b). From this it may perhaps be concluded that the observed effects are due to specific ions and that the natural brackish water, provided it is dilute sea water, exerts its physiological effects on the marine organisms inhabiting it mainly by its lower total osmotic concentration, unless single ions (calcium, potassium) fall below the species-specific minimum essential for survival.

The situation is different if the saline inland waters are included in the concept of "brackish water". Their ionic composition frequently differs from that of the sea water. Consequently the occurrence of certain species in them may be more limited than in marine brackish water, due to the effects of certain ions or else to

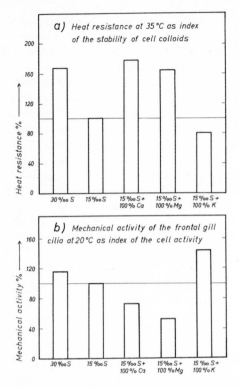

Fig. 78. Effect of salt concentration and of the ionic content on the stability of the plasma colloids and the mechanical activity of the gill tissue of *Mytilus edulis*. After C. SCHLIEPER & R. KO-WALSKI 1956.

anomalous ionic proportions. Such limiting effects of ions are to be expected in every case when an animal species is no longer capable to compensate — by means of its regulatory mechanisms — for a deficit or a surplus of any ion, and thus no longer able to maintain an adaequate ionic balance in its cells (NEUMANN 1962). We shall have to examine every inland brackish water in order to establish whether — both in nature and in the laboratory experiment — physiological effects can be traced which exceed those of marine brackish water of similar total osmotic concentration.

The brackish water of the Black Sea (18.6⁰/₀₀ S) has a relative KCl content which is 77% higher than that of sea water, though no marked adverse effects of

Table 36. Relative ionic content in % equivalents, compared with those of sea water. After BIRSTEIN & BELIAEV 1946 and KALLE

Ions	Lake Balkhash 7.5⁰/₀₀ S	Lake Aral 12⁰/₀₀ S	Caspian Sea 13⁰/₀₀ S	Sea water (ocean) 35⁰/₀₀ S
Na^+	23.35	28.27	31.95	38.69
K^+	2.96	0.61	0.49	0.84
Ca^{++}	1.67	7.60	3.96	1.74
Mg^{++}	19.54	13.68	13.78	8.64
Cl^-	19.78	29.17	34.62	45.16
SO_4^{--}	18.57	19.76	14.37	4.68
CO_3^{--}	14.13			0.20
HCO_3^-		0.91	0.83	

this have been demonstrated in the species inhabiting it. But the ionic relations in the Russian Caspian Sea, Lake Arčl and Lake Balkhash are disturbed to a greater extent (see table 36). In the Caspian Sea and Lake Aral the relative proportion of magnesium sulphate is three-and a half to four times that of the relative proportion in sea water. In addition the relative proportion of calcium sulphate in the Caspian Sea has risen by nearly double and that in the Lake Aral over three times. In the water of Lake Balkhash, in particular, the potassium-ion content of the water shows an extreme rise. BIRSTEIN & BELIAEV (1946) were the first to demonstrate that this produces limiting ionic effects. They investigated the viability of the amphipod *Dikerogammarus haemobaphis*, coming from the fresh water of the Volga, in water from Aral and Balkhash. They found that *Dikerogammarus* dies in Aral and specially quickly in Balkhash water (see fig. 79). Another example is provided by the Myside *Mesomysis kowalevskyi* (KARPEVICH 1958). This schizopod crustacean still tolerates a salinity range of 0—10⁰/₀₀ S in Aral water, but in Balkhash water only 0—1.5⁰/₀₀ S.

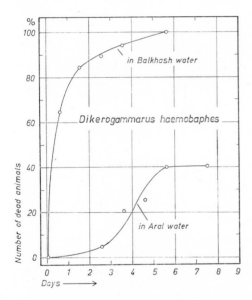

Fig. 79. Comparison of the effects of Balkhash water (9.5⁰/₀₀ S) and Aral water (12⁰/₀₀ S) on *Dikerogammarus haemobaphes* from the Volga delta. After I. A. BIRSTEIN & G. M. BELIAEV 1946.

In Germany there is an interesting case of an inland brackish running water in the lower Werra river which has become salty through effluents of the local potassium industry. In the particular region this river has mean salinities of 6—8⁰/₀₀ S, though with considerable fluctuations. A comparison, after SCHMITZ (1959) of the relative ionic concentrations in the Werra water (in val⁰/₀₀) with those obtaining in the sea gives the following values:

Ions	Na	Ka	Ca	Mg	Cl	SO₄	CO₃/HCO₃
Werra	33.8	2.8	3.7	19.5	43.3	4.6	2.3
Sea	38.7	0.8	1.7	8.7	45.2	4.7	0.2

NEUMANN (1959, 1960) analyzed — by means of breeding experiments — the behaviour of the fresh- and brackish-water gastropod *Theodoxus fluviatilis* from the Werra by placing individuals in marine brackish water and Werra water. He observed that the animals from this population showed no inhibition of their development and reproduction in marine brackish water up to $10^0/_{00}$ S. In water from the Werra, with its higher potassium content growth stopped even below $10^0/_{00}$ S. In other experiments in a medium with relatively increased magnesium concentration the formation of the shell was disturbed, though no other visible reduction in vitality could be noticed. In addition NEUMANN demonstrated in fine experiments that in the salinity range round $4^0/_{00}$ S the variability of the colour pattern of the shells is modified, owing to disturbance of the Ca/Mg ratio of the medium. Such changes in the ionic composition of the medium produce specific change of the pattern that is determined in the tissue of the mantle edge, causing a reversible alteration of the colour pattern in the new growth of the shell.

SCHMITZ (1956, 1959) reports that at one time the salinity of the Werra rose above the average of the highest value. When the $12^0/_{00}$ S limit was exceeded there was a mass death of the fresh-water fishes which were well adapted to the normal total salt content of the Werra ($6^0/_{00}$ S). Only a few, more strongly euryhaline species such as *Perca fluviatilis, Lucioperca sandra, Anguilla anguilla* and the Salmonideae survived. An experimental examination of the presumed limits of resistance confirmed that the death of the fishes was entirely due to the total concentration having increased too much. In experiments with marine brackish-water the fishes in question from the Werra population died at about the same total salt content. Thus the fishes were able to compensate for the ionic ratios in the Werra water as well as for that in marine brackish water. If, however, the ionic ratios of the external medium were markedly more altered in the experiment, the limits of resistance were soon exceeded. The fresh-water teleost *Abramis brama* represents a special case; in marine brackish water it survived salinities of $15—17^0/_{00}$ S, but when the potassium content was doubled, it only tolerated total salt content up to $13^0/_{00}$ S.

Similar studies have recently been carried out on the importance of single ion components in brackish water for resistance in algae. Calcium in brackish water of low salinity has already been mentioned as a decisive factor for the resistance to hypotony in algae (see p. 308). In accordance with this KESSELER (1964) found that calcium deficiency in the green alga *Chaetomorpha linum* leads to a strong discharge of KCl and a considerable loss of turgor. DROOP (1956) also refers to sodium as a possible factor for the hypotonicity limit of euryhaline algae. In agreement with this, SCHWENKE (1958) reports that red algae in brackish water require a certain minimum Na Cl content. As a rule red algae living in brackish water will tolerate a decrease in potassium and magnesium for several days. The algae of the *Delesseria*-group are more susceptible to a deficiency in magnesium, the *Ceramiaea*, however, to deprivation of potassium.

11. The food cycle in brackish water

There are several reasons why nutrient cycles in the sea and in brackish water differ. First of all, different species often predominate in brackish water on the bottom and in the open water. If these brackish-water species have been genetically and individually adapted to the prevailing salt content and no other differences in environmental conditions exist, then one may expect the processes of production and turnover to be the same as in normal sea water. Consequently, in an experiment, the planktonic algae occurring in brackish water may display, at a certain low salinity, an optimum in reproduction and photosynthetic activity (BRAARUD 1961, NAKANISHI & MONSI 1965). Nevertheless, values for productivity in the surface brackish-water rarely equal those in the open sea. Among possible reasons may be the fact that in brackish water, with its thermal and saline stratification, the concentrations of the limiting substances (P, N) may be lower in the illuminated layer; also, exchange between bottom water and sediment on the one hand, and with the surface water on the other hand, is more difficult. In order to illustrate these differences in detail, the example chosen will be from the relatively shallow (average depth = 55 m) marginal sea of the Baltic; its hydrography already has been discussed on p. 137 ff. of this book.

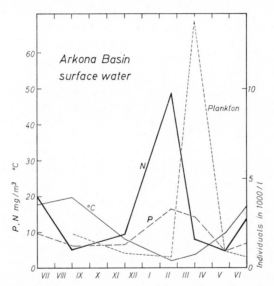

Fig. 80. Annual cycle of plankton production and inorganic nutrients in the surface water of the Arkona Basin (8—10⁰/oo S) of the central Baltic. After F. GESSNER 1933.

In principle there is, for instance, in the surface water of the central Baltic with a salinity of 8—10⁰/oo S (Arkona Basin) the same annual antagonistic cycle between the plankton and the inorganic limiting substances as in a comparable temperate coastal sea (s. fig. 80). In winter there is a minimum development of plankton and a steep rise in the curve of nutrients. But when diatoms start to multiply in spring the values for N and P drop abruptly, only to reach a second lower peak in the summer. But the total production is much lower. This is shown clearly in the diagram in fig. 81. The values reproduced for the open Atlantic have been taken from the plankton counts by HENTSCHEL (1930, 1933) and been contrasted with those found by GESSNER (1933) in the surface water of the Arkona Basin. Though this diagram is based only on rough values, it yet demonstrates that organic production in the brackish water of the central Baltic (55° northern latitude) is less than in the Atlantic at the same latitude.

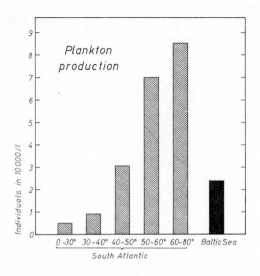

Fig. 81. Comparison of plankton
production of a stratified brackish-
water sea (Baltic, 55° N latitude)
and that from different latitudes of
the South Atlantic Ocean. After
F. Gessner 1933.

Relatively lower values have also been found in other regions of the Baltic by
Gilbricht (1952) in Kiel Bay and by Steemann-Nielsen (1958, 1964) in the Great
Belt and the Kattegat. The measurements of these authors, together with other productivity values for the Atlantic are reproduced in table 37. They are highest in
the temperate North Atlantic, with an oceanic salinity and, in winter, a vertical
convection down to the bottom; this always results again in a uniform distribution
of nutrients throughout the whole, enormous mass of water. In the North Sea, with
its slightly lower salt content (32⁰/₀₀ S) they are markedly less since this shallow

Table 37. Comparison of the primary production of the phytoplankton of different waters.

Water	Author	Primary production g C/m² and year
Atlantic		
Sargasso Sea	Steemann-Nielsen & Jensen 1957	18
Sargasso Sea	Menzel & Ryther 1960, 1961	69
North Atlantic	Berge 1958	300—400
North Atlantic — Island	Kändler 1962	300
North Sea	Fao 1958	100—200
North Sea	Kändler 1962	150
Baltic		
Kattegat	Steemann-Nielsen 1964	67
Belt	Steemann-Nielsen 1958, 1964	59
Kiel Bay	Gilbricht 1952	58
Freshwater lake		
Hyperthrophic	Hübel 1964	200—400
Eutrophic	Hübel 1964	70—200
Mesotrophic	Hübel 1964	30—70
Oligotrophic	Hübel 1964	10—30

epicontinental sea is, to some extent, cut off from the high nutrient concentrations of the North Atlantic deep water. Productivity values are lower still in the stratified Baltic which is even further impeded in its connection with the ocean; in its deeper parts it also lacks a vertical circulation extending to the bottom in winter. Nevertheless, the same biological laws apply to the total food cycle in the Baltic.

Here, too, the phosphate content of the surface water is of great importance for the production of phytoplankton. Based on his measurements on the light ship Flensburg in Kiel Bay GILBRICHT (1952) reports that in spring the total amount of P is rapidly used up by a diatom maximum. Lack of nutrients terminates this first seasonal plankton outburst unless the normal course of production has already been disturbed by vertical turbulence caused by wind (HICKEL 1967). Larger quantities of nutrient salts only become available after an increase in water temperature and, with it, in the rate of remineralisation. The diatoms multiply once more when, about August, sufficient products of remineralisation have reached the surface from the deep in this way. In the autumn when the diatoms have disappeared and there is an increased growth of Peridineae, and especially in winter, an increase in phosphorus in the upper layer is brought about by thermal convection and convection due to wind; this forms the basis for a renewed development of plankton in the following spring.

Apart from other, more secondary influences such as turbulence produced by wind, stability of the water column, stirring-up of sediment, consumption by zooplankton, the main limiting factor of primary production in the stratified water of the Baltic lies in the slow renewal of inorganic nutrient salts in the illuminated upper layer. The regeneration in situ in light surface water of low salinity is not sufficient to ensure a longer, optimum plankton bloom in the spring. Similarly, the supply of water richer in nutrients from the land plays but a subordinate role in the western and central open Baltic. The bulk of the remineralisation processes in dead and sinking plankton and detritus takes place in the deeper strata of the water and on the bottom. The density stratification of the water masses which become colder and more saline with increasing depth, impedes the transport upwards of the inorganic nutrient salts released in these regions to a greater extent than does the thermal stratification alone in the open temperate ocean. For this reason primary production is at its lowest in the shallow Danish Sound, only 7 m deep near Copenhagen, where the outflow of infertile surface water of the Baltic, with reduced salinity, is predominant in the photic layers. It is somewhat greater over the deep sections of the Kattegat and the Great Belt where a part of the North Sea water, rich in nutrients and of higher specific gravity comes in at the bottom and mixes with the low-saline Baltic water which flows out (see also part I, page 137, fig. 58). Nevertheless, even in the Belt the average annual productivity is relatively low as, here too, the density stratification — caused by the outflow of low-saline water from the Baltic — offers resistance to the rising of water from the deep.

The results of ten years of observations by KREY (1968) in the western Baltic (Kiel Bay, Boknis Eck, water 27 m deep) have been summarized in fig. 82. These observations are remarkable as they represent continuous measurements over the whole course of the year, of a large number of factors which are important for the course of production and the cycle of nutrients. The values may well be described as highly typical for a stratified temperate brackish sea (2—14° C, 17—20⁰/₀₀ S). The concentration of chlorophyll may be taken as the measure of primary production; it shows a wide maximum in March and April. The total inorganic and organic particulate matter (plankton, detritus, stirred-up sediment) which can be collected on a filter with a pore-diameter of 1 μ has been designated "Seston". The Seston-maximum corresponds to some extent with the chlorophyll maximum. The relatively high Seston values for December and Januar are presumably due to disturbed bottom sediment. Total P is the total phosphorus which can be determined in 100 ml water. This is the total of inorganic and organic, dissolved and undissolved phosphorus contained in a certain quantity of water and in the particulate material found in it. The

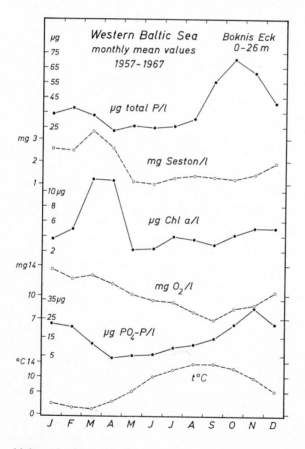

Fig. 82. Annual cycle of the factors essential for the course of productivity and nutrient cycle in the western Baltic (Boknis Eck, Kiel Bay). After J. KREY 1968.

high peak of total P observed in October indicates an intermediate stage of remineralisation in which quantities of organic phosphorus predominate which are as yet incompletely decomposed, probably still partly in colloidal solution. As expected, the inorganic phosphate attains its maximum about two months later, in November to January. It is characteristic that the oxygen minimum in September coincides with the temperature maximum, that is a period of intensive consumption and remineralisation.

Owing to reduced depth of the water in the western Baltic the dead plankton reaches the bottom of the sea relatively quickly. Therefore it only undergoes partial decomposition during sinking. ZEITZSCHEL (1965) has made more detailed studies in the Eckernförder Bay and reports that there about half the organic matter produced sinks to the bottom (40% not decomposed and 60% as organic detritus). The particulate material which has sunk is further decomposed on the surface of the sediment. Unless the sediment is stirred up again by gales, a part of the organic matter deposited in the lower sediment is irretrievably lost to the nutrient cycle. This is one of the contributory causes of the comparatively low food production in the Baltic.

The further we get into the Baltic, the more stable does the hyaline stratification become and the greater are the difficulties in the way of exchange taking place between the surface water, poor in salts, and the deep water which is rich in salts and

Baltic Sea, Gotland Basin *(longitudinal section, August 1956)*

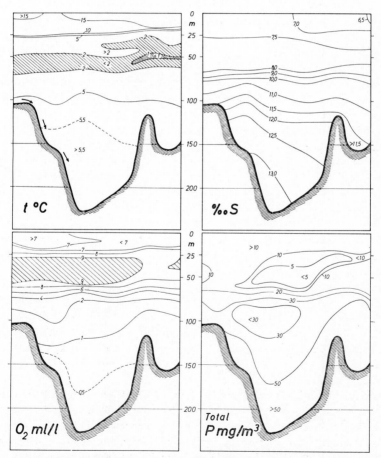

Fig. 83. The vertical distribution of temperature, salinity, oxygen and total phosphorus in a stratified brackish-water sea (Baltic, Gotland Basin).
After G. WÜST, K. BANSE & J. KREY 1957.

colder. In the central Baltic no bottom water from the North Sea occurs which is unmixed and rich in nutrients. The deep water found here usually comes from the western Belt Sea and Kiel Bay. In addition deep water rich in salts flows only occasionally, at irregular intervals, over the Darsser sill, situated on the eastern edge of the Bay of Mecklenburg (cf. part I, fig. 5) eastwards into the large troughs of the central Baltic (Basins of Arkona, Bornholm and Gotland). Subsequently larger amounts of nutrients may accumulate in the oxygen-deficient bottom water of these deep troughs; but they reach the upper layers in such small quantities that they are not able to raise the yearly productivity figures for long periods of time (see fig. 83).

Under quite exceptional meteorological conditions (persistent gales from the East) bottom currents of heavy, highly saline water from the Kattegat reach the central part of the Baltic. They may then produce longer periods of stagnation in the deep water of the Baltic troughs. According to FONSELIUS (1967) such a long-term period started in December 1951. In the subsequent ten years the chemical decomposition continuously taking place at the bottom and in the sediment brought about a slow reduction of the oxygen content of the bottom water. When all the oxygen had been used up hydrogen sulphide formed. At the same time large quantities of inorganic phosphates were produced, their release in the sediment apparently being favoured by a relatively low pH value. When in the middle of 1961 new water had replenished the Gotland Basin the phosphate which had accumulated reached the upper layers by vertical convection in the following winter. This caused an enormously increased primary production for several months in the following spring; this, in turn, led to a marked improvement in the stand of herrings in that part of the Baltic. The same author also gives a detailed account of a shorter period of stagnation which set in in 1963. When in the winter of 1965 the stagnant water masses began to move, larger amounts of phosphates again appeared in the surface water; these could in parts be traced as far as the Gulf of Finland.

Conditions in the shallow coastal areas of the Baltic, near the land, may take an entirely different course. Here a relative eutrophication from the land can be observed, frequently in the Forden (Northern German and Danish fjords) of the Baltic, especially in the harbour areas near towns, but also in the Bodden waters of Rügen enclosed by land, in the Frisches Haff (Vistula Firth) and even in the Gulf of Finland. Already in the Great Belt a supply of released nutrient salts from lateral areas of shallow water can be seen in summer. Here it results in a displacement of the maximum of productivity from spring into summer such as never occurs in the open Baltic at some distance from the land (see also ANDERSON & BANSE 1961, as well as BANSE & KREY 1962). In the shallow water of the inner Isefjord STEEMANN-NIELSEN (1951) even found a yearly primary production of 240 g C/m². GRØNTVED (1960) was able to record high productivity values of the microbenthic vegetation (mainly pennate diatoms) in water depths between 0,5 and 0,7 m in seven other Danish fjords, especially over floors covered by loam or mud, less over sandy bottoms. In the uppermost 10 metres of the Förden on the coast of Schleswig-Holstein into which enormous amounts of phosphates flow with the sewage of larger towns such as Flensburg and Kiel, one plankton outburst after the other could be seen during the whole of the summer — as KREY (1953), in particular pointed out; the delicate diatom *Skeletonemacostatum* was most abundant. This rich and persistent primary production is undoubtedly, among other things, responsible for the luxuriant development of *Mytilus edulis* in the Innenförde (inner fjord) of Kiel and of Flensburg (see also BOJE 1965).

The Rügener Binnenbodden (see fig. 84) first studied by GESSNER (1933, 1937), TRAHMS & STOLL (1938) and WASMUND (1939) are another brackish-water area of the central Baltic which is of particular interest for production biology. They are a series of communicating shallow brackish-water basins, penetrating deeply inland and, at their outer edges, issueing into the open Baltic. Their average depth is only 2.5 m. The Grosse Jasmunder Bodden is situated farthest inland, its greatest depth is 8—9 m. While in the open Baltic, in the surface water in front of Rügen salinities of 18—8⁰/₀₀ have been recorded, values of 10—7⁰/₀₀ S are found in the outer Bodden mouth, the Rassower River; farther inland these drop to 9—6⁰/₀₀ S towards the Grosse Jasmunder Bodden. In the adjoining Kleiner Jasmunder Bodden (not shown in fig. 84) the salt content drops to 2.2⁰/₀₀ S within a bare 100 m. A narrow sluice represents its communication with the Grosse Jasmunder Bodden. Thus in the Bodden waters the proportion of fresh water slowly increases inwards. The turbulence due to wind prevents any thermal or saline stratification in their shallow

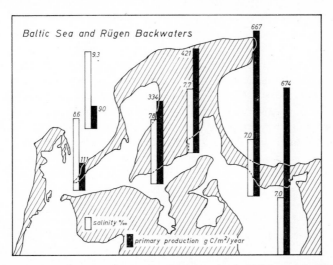

Fig. 84. The primary production of the phytoplankton in the Baltic and the Boddenwaters of Rügen in relation to the salt content. Original after values by H. HÜBEL 1964.

water and provides good aeration down to the bottom; this is sandy and, in parts has a dense vegetation. Only the deeper runnels and basins are filled with a black sapropel containing iron sulphide; according to WASMUND (1939) this is several metres thick in places. Just above the mud the oxygen content of the water may be somewhat reduced, but nowhere is there any free hydrogen sulphide in the water (OVERBECK 1964). In the mud there is, besides pyrite, markasite and admixtures of sand, detritus with plant and animal remains among which the numerous shells of diatoms are prominent.

The upper layers of the bottom sediment and the processes of exchange taking place between it and the water above are of great importance in supplying nutrients to the Bodden waters. Inorganic nitrogen is always present in sufficient amounts in the form of nitrate (40—100 mg NO_3/m^3). The remineralized phosphate supplied from the uppermost layers of sediment is of decisive importance for the primary production of the phytoplankton; according to Overbeck (1964) it may amount to about 7—14 mg P/m^3 in the Bodden water. With increasing distance from the sea this rises considerably and reaches its maximum value in the Grosse Jasmunder Bodden (with over 2 mg total phosphorus/l). The phytoplankton of the inner Bodden differs clearly from that of the open Baltic. While diatoms and Peridineae predominate in the open sea and the outer Bodden, most of the rich nannoplankton in the brackish water of the Grosse Jasmunder Bodden consists of Cyanophyceae *(Merismopedia, Microcystis)*. Hübel (1964) has made a thorough investigation of the annual net primary production. He describes a gradual increase in this value, beginning from the open Baltic near Libben, over the outer Bodden to the inner Bodden situated farthest landward (see fig. 84). Since the salt content of the water declines in the same direction, there is a regular relationship between the plankton production and the salinity; this had previously been pointed out by Gessner (1933, 1937), Trahms & Stoll (1938) and Overbeck (1962). This "relation" is so rigid that slight fluctuations in the salinity of the Binnenbodden are coupled with big changes in the production of phytoplankton. Thus, e. g. a considerably lower primary production was recorded in the Grosse Jasmunder Bodden in 1960, compared with 1961 and 1962. But in 1960 salinity there was 1—$2^0/_{00}$ higher than in the other two years. Comparable observations can be made by following, in the course of one summer, the changes in salinity and in the quantity of plankton in the water of the inner Bodden in the direction towards the open sea. A striking and considerable drop in unicellular Cyanophyceae will be found exactly in the zone in which the salt content rises from about 6.7 to $7.7^0/_{00}$ S. Hübel (1964) draws attention to the fact that the production in the inner Bodden further distant from the sea may be compared with conditions in a hypertrophic fresh-water lake (see also table 37). In the course of the seasons the outer Bodden have a spring, a summer and an autumn maximum, whereas in the inner Bodden there is only one fairly broad maximum in summer and if, at all, a second, lesser maximum in the autumn. Towards the depth production rises to a maximum, the position of which depends on the transparency of the water; it is followed by a decrease.

According to Overbeck (1964) the specially high level of plankton production in the inner Bodden is due to the fact that the phosphate, remineralized from covalent bonding in the sediment, and dissolved in water, is not coprecipitated with adsorptive binding to ferric hydroxide and fixed in the sediment. The hydrogen sulphide produced in the mud through bacterial sulphate-reduction precipitates the iron — which is also present — as sulphide. This prevents a fixation of the phosphate to colloidal ferric hydroxide which would be produced from $Fe(HCO_3)_2$ in the presence of oxygen; the phosphate ions remain in solution and from the sediment reach the water above it. Ohle (1954) had established this significance of sulphate in the water in the mobilisation of phosphate in respect of the fresh-water nutrient cycle. The relatively high sulphate content of brackish water, together with a rich microflora of the sediment, are in the last resort of decisive importance for the magnitude of phytoplankton production. As can be shown experimentally, no hydrogen sulphide is formed without microbial activity, the iron present is not fixed as sulphide and phosphorus not released from its combination with the iron.

Conditions in the northern and eastern Baltic on the Finnish coasts where salinities around $6^0/_{00}$ S and less are the rule, provide evidence that a high primary production need not per se be linked with a low salinity of the brackish water. Here, in the Gulf of Finland, a relatively rich plankton can be seen in summer whereas the Gulf of Bothnia is poor in organisms by comparison. The reason for this striking distinction lies in the difference in topography and hydrography of these two bays, as BUCH (1932) was able to show. The Gulf of Finland opening trough-like to the Southwest, is in direct and undisturbed communication with the adjoining central basin of the open Baltic. The low saline surface water, rich in plankton which flows outwards from it is continuously being compensated for by a bottom current in the opposite direction which consists of relatively fertile deep water, richer in salts. In this way in the Gulf of Finland the amounts of phosphate used up in the upper layer for primary production are replaced with equal regularity. By contrast the Gulf of Bothnia is cut off from the open Baltic by a submarine sill near the Aland Isles; while the surface water containing plankton can flow out there is no corresponding antagonistic bottom current which can replace the lost nutrients. Therefore primary production in the Gulf of Bothnia is completely dependent on the supply of nutrients from the land. But the supplies of fresh water coming from the Precambrian rocks of Finland and Sweden are extremely poor in phosphates. With approximately the same salinity in the two bays, the different conditions of currents are responsible, to some extent, for the different productivity and density of colonisation.

To end this short paragraph on problems of productivity in the brackish water of the Baltic a few remarks will be added on the interrelations between organic primary production and catches of fisheries in this minor sea. A comparison between the North Sea and the Baltic which KÄNDLER (1962) carried out, is particularly revealing in this respect. From reports of the FAO a mean catch of fisheries of 35 kg/ha in the North Sea can be calculated for the last few years. According to the same source the average annual catch for the Baltic is only 12 kg/ha. These differing figures confirm yet again that the Baltic, a stratified brackish sea, with an upper layer poor in nutrients and without a sufficient vertical circulation to the bottom, is at a considerable disadvantage in contrast to the North Sea, an epicontinental sea of the open ocean, while otherwise conditions are comparable. NELLEN (1965) recently pointed out that conditions for production in fisheries can also be unfavourable in the brackish water of the Baltic near the coast in spite of a high content of nutrients and a high level of food production on the bottom, as is the case in the Frisches Haff (Vistula Firth). In this case the low salinity of only $4—7^0/_{00}$ is the decisive factor; on the one hand it prevents immigration of economically important fishes (spawn herrings, *Clupea harengus* L.) and, on the other hand, it represents an unsurmountable obstacle to the reproduction of fresh-water fishes.

12. Tables of the ionic composition of sea and brackish water

For some experiments with brackish-water organisms it will be necessary to work with sea or brackish water of known composition. Some methods and tables for use in the laboratory are included here.

The salt content of the sea and brackish water ($^o/_{oo}$ S) is defined as the total amount of dissolved substances contained in 1 kg of sea water or brackish water respectively, assuming that all carbonates have been converted into oxides, the bromides and iodides been replaced by chlorides and the whole of the organic matter has been completely oxidized. The simplest way of determining salt content is hydrometrically, using the hydrographic tables by M. KNUDSEN (1901). Similarly, salt content may be calculated from the chloride content titrated after MOHR. Details about these methods for determining salinity are found, among others, in SCHLIEPER (1955a, 1968). In oceanography the chloride content of sea water in $^o/_{oo}$ is the weight of chlorine in g, which is contained in 1000 g sea water or brackish water. For titration a silver nitrate solution is used containing about 37.11 g $AgNO_3$ per litre and standardized to a normal sea water with a Cl content of 19.379$^o/_{oo}$ (obtainable from the Hydrographic Laboratory, Copenhagen). There is a fixed relation between the salt content and the chlorine content determined in this manner:

$$S = 0.03 + 1.805 \cdot n \text{ g Cl/kg sea water.}$$

When sea or brackish water have been titrated with n/10 silver nitrate solution — as is usually done in the chemical laboratory giving a result of g Cl/l sea water, then the values obtained can easily be converted to g Cl/kg sea water after KALLE:

$$\begin{array}{rcl}
\text{g Cl/l} - \text{k} &=& \text{g Cl/kg} \\
8.87 - 0.10 &=& 8.77 \\
12.56 - 0.20 &=& 12.36 \\
15.40 - 0.30 &=& 15.10 \\
17.80 - 0.40 &=& 17.40 \\
19.92 - 0.50 &=& 19.42 \\
21.85 - 0.60 &=& 21.25 \\
23.61 - 0.70 &=& 22.91
\end{array}$$

Intermediate values have to be worked out by interpolation.

The above-mentioned relation between chlorine content and salt content may occasionally be somewhat disturbed in brackish water of low salinity, depending on the composition of the inflowing fresh water. For this reason e. g. the additive constant of the above-mentioned relation for brackish water of the eastern Baltic (Gulf of Riga) is not 0.03, but 0.13 since in these parts the inflowing fresh water carries larger amounts of salts with lower proportion of chlorine.

In low-salinity brackish water it is important to determine alkalinity, that is the excess of bases chiefly bound to carbon dioxide. Alkalinity is measured by boiling the quantity of brackish water to be examined with an excess of n/10 hydrochloric acid, and after cooling, titrating it back with n/20 solution of caustic soda in the presence of a mixed indicator (WATTENBERG & WITTIG 1940). While in the open sea a linear relationship between alkalinity, calcium content and chlorine content exists, this is no longer the case in brackish water which has arisen from an influx of fresh water rich in carbonates. For this reason it is desirable to give

Table 38r litre (20º C) of sea water (molarity) for chlorinity

(S = 1.84—41.55 º/oo)

Cl‰	S‰	Cl	B₃	Ca	Mg	K	Sr	Na	H₃BO₃
1	1·84	0·9989	0·006	0·8892	2·830	0·5114	0·008008	24·15	0·02215
2	3·64	1·9979	0·008	1·396	5·585	1·024	0·01604	48·37	0·04435
3	5·45	2·9968	0·0	1·905	8·347	1·538	0·02409	72·65	0·06662
4	7·25	3·9958	0·0	2·414	11·12	2·054	0·03216	97·00	0·08894
5	9·06	4·9947	0·0	2·925	13·90	2·571	0·04026	121·4	0·1113
6	10·86	5·9936	0·0	3·438	16·68	3·089	0·04838	145·9	0·1338
7	12·67	6·9926	0·0	3·952	19·47	3·609	0·05652	170·4	0·1563
8	14·47	7·9915	0·0	4·467	22·27	4·130	0·06468	195·1	0·1789
9	16·28	8·9905	0·0	4·984	25·08	4·653	0·07286	219·7	0·2015
10	18·08	9·9894	0·0	5·502	27·90	5·177	0·08107	244·5	0·2242
11	19·89	10·9883	0·0	6·021	30·72	5·702	0·08929	269·3	0·2469
12	21·69	11·9873	0·0	6·542	33·55	6·229	0·09754	294·2	0·2697
13	23·50	12·9862	0·0	7·064	36·39	6·757	0·1058	319·1	0·2926
14	25·30	13·9852	0·0	7·588	39·23	7·287	0·1141	344·1	0·3155
15	27·11	14·9841	0·0	8·112	42·08	7·818	0·1224	369·2	0·3385
16	28·91	15·9830	0·0	8·639	44·94	8·350	0·1308	394·3	0·3616
17	30·72	16·9820	0·0	9·167	47·81	8·884	0·1391	419·6	0·3847
18	32·52	17·9809	0·0	9·696	50·69	9·419	0·1475	444·8	0·4079
19	34·33	18·9799	0·0	10·23	53·57	9·956	0·1559	470·2	0·4311
20	36·13	19·9788	0·0	10·76	56·46	10·49	0·1643	495·6	0·4544
21	37·94	20·9777	0·0	11·29	59·36	11·03	0·1728	521·1	0·4778

data on the alkalinity (and pH value) of the brackish water used in every case when experiments are conducted with brackish water of low salinity.

For making up artificial brackish water it is best to use the recipe given by LYMAN & FLEMING (1940) for sea water (p. p. m):

```
NaCl   . . . . . . . . . . . . . . . 23.477
KCl . . . . . . . . . . . . . . . . . 0.664
MgCl₂ . . . . . . . . . . . . . . . 4.981
Na₂SO₄ . . . . . . . . . . . . . . 3.917
CaCl₂ . . . . . . . . . . . . . . . 1.102
NaHCO₃ . . . . . . . . . . . . . . 0.192
KBr . . . . . . . . . . . . . . . . . 0.096
H₃BO₃ . . . . . . . . . . . . . . . 0.026
SrCl₂ . . . . . . . . . . . . . . . . 0.024
NaF . . . . . . . . . . . . . . . . . 0.003
```

Add H_2O to make up 1000 g. Sea water is obtained with $Cl = 19^0/_{00}$ and a salt content $= 34.5^0/_{00}$. In making up this water it should be noted whether the salts being used contain any water of crystallisation. In practice the preparation of artificial sea water from the above recipe is best done by first dissolving all the chlorides and potassium bromide in about 800 ml aqua dest.; the remaining salts are dissolved in about 100 ml aqua dest. and added slowly in a fine jet, while stirring. Then fill up to 1000 g with aqua dest. Leave to stand for a day and then filter in order to remove the small amount of turbid matter formed. After thorough aeration the pH value is 7.9—8.3.

There follow data on the ionic composition of sea and brackish water. These have been calculated by H. BARNES (1954), the wellknown English marine biologist, based on the values of LYMAN & FLEMING and the tables by KNUDSEN.

References

AEBI, H. (1950): Kationenmilieu und Gewebsatmung. — Helvet. Physiol. Acta 8: 525—543.

ADOLPH, E. F. (1936): Differential permeability to water and osmotic changes in the marine worm, *Phascolosoma*. — J. cell. comp. Physiol. 9: 117—135.

AHUJA, S. K. (1964): Salinity tolerance of *Gambusia affinis*. — J. Exp. Biol. 2: 9—11.

AMEMIYA, I. (1926): Notes on experiments on the early developmental stages of the Portuguese, American and English oysters with special reference to the effect of varying salinity. — J. Mar. Biol. Assoc. Plymouth 14: 161—175.

ANDERS, F. et al. (1962): Genetische und biochemische Untersuchungen über die Bedeutung der freien Aminosöuren bei Art- bzw. Gattungsbastarden lebendgebärender Zahn-karpfen (Poeciliidae). — Biol. Zbl. 81: 45—65.

ANDERSON, D. & PROSSER, C. L. (1953): Osmoregulating capacity in populations occurring in different salinities. — Biol. Bull. Woods Hole 105: 369.

ARAI, M., COX, E. T. & FRY, F. E. J. (1963): An effect of dilutions of sea-water on the lethal temperature of the guppy. — Canad. J. Zool. 41: 1011—1015.

ARNOLD, D. C. (1957): The response of the limpet, *Patella vulgata* L., to waters of different salinities. — J. Mar. Biol. Assoc. U. K. 36: 121—128.

ARVY, L., FONTAINE, M. & GABE, M. (1954): Action des solutions salines hypertoniques sur le système hypothalamo-hypophysaire, chez *Phoxinus laevis* AGASS. et chez *Anguilla anguilla* L. — C. r. Soc. Biol. (Paris) 148: 1759—1761.

ATHANASSOPOULOS, G. D. (1930): L'action de la salinité sur les formes planctoniques. — Bull. Soc. Zool. France LX: 472—474.

BAAS-BECKING, L. G. M. et al. (1957): Biological processes in the estuarine environment. X. The place of the estuarine environment within the aqueous milieu. — Proc. Koninkl. Ned. Akad. Wetenschap., Ser. B, 60: 88—102.

BACQ, Z. M. & FLORKIN, M. (1957): Accumulation of silver in the gills of *Potamon perlatus* (M. EDW.) in relation to osmoregulation. — Arch. internat. Physiol. Bioch. 65: 379—390.

BALDWIN, E. (1949): An introduction to comparative biochemistry. — Cambridge.

BANSE, K. (1957): Ergebnisse eines hydrographisch-produktionsbiologischen Längsschnittes durch die Ostsee im Sommer 1956. II. Die Verteilung von Sauerstoff, Phosphat und suspendierter Substanz. — Kieler Meeresforsch. 13: 186—201.

BANSE, K. & KREY, J. (1962): Quantitative Aspekte des Kreislaufes der organischen Substanz im Meere. — Kieler Meeresforsch. 18, 3. Sonderh.: 97—106.

BARNES, H. (1953): The effect of lowered salinity on some barnacle nauplii. — J. Anim. Ecol. 22: 328—330.

— (1954): Some tables for the ionic composition of sea water. — J. Exp. Biol. 31: 582—588.

BATEMAN, J. B. (1933): Osmotic and ionic regulation in the shore crab, *Carcinus maenas*, with notes on the blood concentrations of *Gammarus locusta* and *Ligia oceanica*. — J. Exp. Biol. 10: 355—371, London.

BATTAGLIA, B. (1967): Genetic aspects of benthic ecology in brackish waters. — In "Estuaries" ed. by G. H. LAUFF, pp. 574—577, Publ. No. 83, Amer. Assoc. Adv. Sci. Washington, D. C.

BATTAGLIA, B. & BRYAN, G. W. (1964): Some aspects of ionic and osmotic regulation in Harpacticoid Copepods of the genus *Tisbe* in relation to polymorphism and geographical distribution. — J. Mar. Biol. Assoc. U. K. 44: 17—31.

BEADLE, L. C. (1931): The effect of salinity changes on the water content and respiration of marine invertebrates. — J. Exp. Biol. 8: 211—227.

— (1934): Osmotic regulation in *Gunda ulvae*. — J. Exp. Biol. 11: 382—396.

— (1937): Adaptation to changes of salinity in the Polychaetes. I. Control of body volume and of body fluid concentration in *Nereis diversicolor*. — J. Exp. Biol. 14: 56—70.

— (1939): Regulation of the haemolymph in the saline water mosquito larva, *Aedes detritus*. — J. Exp. Biol. 16: 46—62.

— (1943): Osmotic regulation and the faunas of inland waters. Biol. Rev. **18**: 172—183.
— (1957): Osmotic and ionic regulation in aquatic animals. — Ann. Rev. Physiol. **19**:
    329—358.
— (1959): Osmotic and ionic regulation in relation to the classification of brackish and
    inland saline waters. Arch. Oceanograf. e Limnol. (Venezia) **11**, Suppl.: 143—151.
BEADLE, L. C. & CRAGG, J. B. (1940): Studies on adaptation to salinity in *Gammarus* sp.
    I. Regulation of blood and tissues and the problem of adaptation to fresh water. —
    J. Exp. Biol. **17**: 153—163.
— — (1940): Osmotic regulation in freshwater animals. — Nature **60**: 588, London.
BEADLE, L. C. & SHAW, J. (1950): The retention of salt and the regulation of the nonprotein
    nitrogen fraction in the blood of the aquatic larva, *Sialis lutaria*. — J. Exp. Biol.
    27: 96—109.
BELIAEV, G. M. (1951): Die osmotische Konzentration des Innenmediums der wasserleben-
    den Wirbellosen in Medien von verschiedenem Salzgehalt. — Akad. Wiss. UDSSR,
    Arb. hydrobiol. Ges., Ökologie und Physiologie der wasserlebenden Organismen. **3**:
    92—139 (Russ.).
BELIAEV, G. M. & BIRSTEIN, J. A. (1940): Osmoregulation of some Caspian Invertebrates.
    — Zool. J. **19**: 548—565 (Russ. mit engl. Zusammenf.).
— (1944): A comparison between the osmoregulatory ability in Volga river and Caspian
    Amphipods. — C. r. (Dokl.) Acad. Sci. URSS **45**, 7: 304—306.
BELIAEV, G. M. & TSCHUGUNOVA, M. N. (1952): Die physiologischen Unterschiede zwischen
    den Mytili *(Mytilus)* der Barentsee und der Ostsee. — Vortr. Akad. Wiss. UDSSR,
    Ökologie **85**, No. 1: 233—236 (Russ.).
BENAZZI, M. (1933): Influence of diluted sea water on rhythmic motions of Jelly-fish. —
    Boll. Zool. **4**: 211—218, Naples.
BERGER, E. (1931): Über die Anpassung eines Süßwasser- und eines Brackwasserkrebses
    an Medien von verschiedenem Salzgehalt. — Pflügers Arch. ges. Physiol. **228**:
    790—807.
BERT, P. (1871): Sur les phénomènes et les causes de la mort des animaux d'eau douce que
    l'on plonge dans l'eau de mer. — C. r. Acad. Sci. Paris **73**: 382—385 and 464—467.
BETHE, A. (1929): Die Salz- und Wasserpermeabilität der Körperoberflächen verschiedener
    Seetiere in ihrem gegenseitigen Verhältnis. — Pflügers Arch. ges. Physiol. **234**:
    629—644.
— (1930): The permeability of the surface of marine animals. — J. gen. Physiol. **13**:
    437—444.
BETHE, A. & BERGER, E. (1931): Variationen im Mineralbestand verschiedener Blutarten.
    — Pflügers Arch. ges. Physiol. **227**: 571—584.
BETHE, A., V. HOLST, E. & HUF, E. (1935): Die Bedeutung des mechanischen Innendrucks
    für die Anpassung gepanzerter Seetiere an Änderungen des osmotischen Außendrucks.
    — Pflügers Arch. ges. Physiol. **235**: 330—334.
BEUDANT, F. S. (1816): Sur la possibilité de faire vivre des Mollusques d'eau douce dans
    les eaux salées et des Mollusques marines dans les eaux douces. — Ann. chim. et
    phys. **2**: 32—41.
BIEBL, R. (1937): Ökologische und zellphysiologische Studien an Rotalgen der englischen
    Südküste. — Beih. Bot. Cbl. **57**, Abt. A: 381—424.
— (1938): Trockenresistenz und osmotische Empfindlichkeit der Meeresalgen verschie-
    den tiefer Standorte. — Jb. wiss. Bot. **86**: 350—386.
— (1938): Zur Frage der Salzpermeabilität bei Braunalgen. — Protoplasma **31**: 518—523
    (1939): Protoplasmatische Ökologie der Meeresalgen. — Ber. Dt. Bot. Ges. **57**: 78—90.
— (1952): Ecological and non-environmental constitutional resistance of the proto-
    plasma of marine algae. — J. Mar. Biol. Assoc. U. K. **31**: 307—315.
— (1958): Temperatur- und osmotische Resistenz von Meeresalgen der bretonischen
    Küste. — Protoplasma **50**: 217—242.
BIALASCEWICZ, K. (1933): Contribution à l'étude de la composition minérale des liquides
    nourriciers chez les animaux marins. — Arch. int. Physiol. **36**, 41—53.

BIELAWSKI, J. (1961): The influence of the salinity of the medium on respiration in isolated gills of the clam *Dreissena polymorpha* (PALL.). — Comp. Biochem. Physiol. 3: 250—260.

— (1964): Chloride transport and water intake into isolated gills of crayfish. — Comp. Biochem. Physiol. 13: 423—432.

BINYON, J. (1961): Salinity tolerance and permeability to water of the starfish *Asterias rubens* L. — J. Mar. Biol. Assoc. U. K. 41: 161—174.

— (1962): Ionic regulation and mode of adjustment to reduced salinity of the starfish *Asterias rubens* L. — J. Mar. Biol. Assoc. U. K. 42: 49—64.

BIRSTEIN, J. A. & BELIAEV, G. M. (1946): The action of the water of Balkhash lake on the Volga-Caspian invertebrates. — Zool. J. 25: 225—236 (Russ., engl. Zusammenf.).

BLACK, V. (1951): Some aspects of the physiology of fish. II. Osmotic regulation in teleost fishes. — Univ. Toronto Biol. Ser. 59, Publ. Ontario Fish. Res. Lab. No. 71: 53—89.

— (1951): Changes in body chlorids, density and water content of chum *(Oncorhynchus keta)* and coho *(O. kisutch)* salmon fry when transferred from fresh water to sea water. — J. Fish. Res. Bd. Canada 8: 164—176.

BOCK, K. J. & SCHLIEPER, C. (1953): Über den Einfluß des Salzgehaltes im Meerwasser auf den Grundumsatz des Seesternes *Asterias rubens* L. — Kieler Meeresforsch. 9, 2: 201—212.

BOETTGER, C. R. (1950): Ein Beitrag zur Frage des Ertragens von Brackwasser durch Molluskenpopulationen. — Hydrobiologia (Den Haag) 2: 360—379.

BOGEN, H. J. (1948): Untersuchungen über Hitzetod und Hitzeresistenz pflanzlicher Protoplaste. — Planta 36: 298—340.

BOGUCKI, M. (1930): Recherches sur la perméabilité des membranes et sur la pression osmotique des oeufs des Salmonides. — Protoplasma 9: 345—369.

— (1932): Recherches sur la régulation osmotique chez l'isopode marin, *Mesidotea entomon* L. — Arch. Internat. Physiol. 35: 197—213.

— (1934): Recherches sur la régulation de la composition minérale du sang chez l'écrevisse *(Astacus fluviatilis* L.). — Arch. Internat. Physiol. 38: 172—179.

— (1954): Adaptation of *Nereis diversicolor* to diluted Baltic water and fresh water. — Pol. Arch. Hydrobiol. 2: 237—251 (Poln. mit engl. Zusammenf.).

— (1963): The influence of salinity on the maturation of gametes of *Nereis diversicolor* O. F. MÜLLER. — Pol. Arch. Hydrobiol. 11: 343—347.

BOGUCKI, M. & WOJTCZAK, A. (1962): Contractility of isolated muscles of *Nereis diversicolor* cultured in hypotonic media. — Pol. Arch. Hydrobiol. 23, 10: 233—239.

— (1964): Content of body water in *Nereis diversicolor* O. F. M. in various medium concentrations. — Pol. Arch. Hydrobiol. 12: 125—143.

BOREI, H. (1936): Über die Einwirkung des Salzgehaltes auf den O_2-Verbrauch des Echinodermeneies. — Z. Morph. Ökol. Tiere 30: 97—98.

BORUSK, V. & KREPS, E. (1929): Untersuchungen über den respiratorischen Gaswechsel bei *Balanus crenatus* bei verschiedenem Salzgehalt des Außenmediums. — Pflügers Arch. ges. Physiol. 222: 371—380.

BOUXIN, H. (1931): Influence des variations rapides de la salinité sur la consommation d'oxygène chez *Mytilus edulis* var. *galloprovincialis* (LMK.). — Bull. Inst. Océanograph. No. 569: 1—11.

BOVBJERG, R. V. (1952): Comparative ecology and physiology of the crayfish *Orconectes propinquus* and *Cambarus fodiens*. — Physiol. Zool. 25: 34—56.

BRAARUD, T. (1951): Salinity as an ecological factor in marine Phytoplankton. — Physiol. Plantarum 4: 28—34.

— (1961): Cultivation of marine organisms as a means of understanding environmental influences on populations. In "Oceanography. Invited lectures at the Int. Oceanogr. Congress, New York 1959" 271—298. — Washington.

BRAARUD, T. & PAPPAS, I. (1951): Experimental studies on the dinoflagellate *Peridinium triquetrum* (EHRB.) LEBOUR. — Avhandl. Norske Videnskaps-Akad. I. Mat.-Naturv. Kl. 2: 1—23.

BREDER, C. M. (1933): The significance of calcium to marine fishes invading fresh-water. — Anat. Rec. 57 (Suppl.): 57.

BRICTEUX-GREGOIRE, S., DUCHATEAU-BOSSON, GH. & JEUNIAUX, CH. (1964): Constituants osmotiquement actifs des muscles adducteurs de Gryphaea Angulata adaptée à l'eau de mer ou à l'eau saumâtre. — Arch. internat. Physiol. Bioch. 72, 5: 835—842.

BRICTEUX-GREGOIRE, S., DUCHATEAU-BOSSON, GH., JEUNIAUX, CH. & FLORKIN, M. (1964): Constituants osmotiquement des muscles adducteurs d'Ostrea edulis adaptée à l'eau de mer ou à l'eau saumâtre. — Arch. internat. Physiol. Bioch. 72, 2: 267—275.

BROEKEMA, M. M. M. (1942): Seasonal movements and the osmotic behaviour of the shrimp, Crangon crangon L. — Arch. Néerland. Zool. 6: 1—100.

BROEKHUYSEN, G. J. (1937): On development, growth and distribution of Carcinides maenas L. — Arch. Néerland. Zool. 2: 257—399.

BRYAN, G. W. (1963): The accumulation of ^{137}CS by brackish water invertebrates and its relation to the regulation of potassium and sodium. — J. Mar. Biol. Assoc. U. K. 43: 541—565.

BUCH, K. (1932): Untersuchungen über gelöste Phosphate und Stickstoffverbindungen in den nordbaltischen Meeresgebieten. — Havsforskingsinst. skr. (Helsinki) Nr. 86.

VON BUDDENBROCK, W. (1954): Physiologie der Decapoden. Bronns Klassen und Ordnungen des Tierreiches. 5, 1 Abt., 7. Buch, Decapoda, pp. 863—1283, Leipzig, Akad. Verlagsges., Geest & Portig K. G.

BULL, H. O. (1938): Studies on conditioned responses in fishes. Part VIII. Discrimination of salinity changes by marine teleosts. — Dove Mar. Lab., Cullercoats, Rep. for the Year ending July 31th, 1937, 19—36, Newcastle-upon-Tyne.

BULLIVANT, J. S. (1961): The influence of salinity on the rate of oxygen consumption of young quinnat salmon (Oncorhynchus tschawytscha). — N. Z. J. Sci. 4: 381—391.

BURBANCK, W. D. (1967): Evolutionary and ecological implications of the zoogéography, physiology, and morphology of Cyathura (Isopoda). — In "Estuaries", ed. by G. H. LAUFF, pp. 564—573, Publ. No. 83, Amer. Assoc. Adv. Sci., Washington, D. C.

BUSNEL, R. G., DRILHON, A. & RAFFY, A. (1946): Recherches sur la physiology des Salmonides. — Bull. Inst. Océanogr. (Monaco) 893: 1—23.

CALLAMAND, O. (1943): L'anguille européenne (Anguilla anguilla L.) les bases physiologiques de sa migration. — Ann. Inst. Océanogr. (Monaco) 21: 361—440.

CALLAMAND, O., FONTAINE, M., OLIVEREAU, M. & RAFFY, A. (1951): Hypothèse et osmorégulation chez les poissons. — Bull. Inst. Océanogr. (Monaco) 984: 1—6.

CAMIEN, M. N., SARLET, H., DUCHATEAU, G. & FLORKIN, M. (1951): Nonprotein amino acids in muscle and blood of marine and fresh water crustacea. — J. Biol. Chem. 193: 881—885.

CARTER, L. (1957): Ionic regulation in the ciliate Spirostomum ambiguum. — J. Exp. Biol. 34: 71—84.

CASPERS, H. (1967): Estuaries: Analysis of definitions and biological considerations. — In "Estuaries" ed. by G. H. LAUFF, pp. 6—8, Publ. No. 83, Amer. Assoc. Adv. Sci., Washington, D. C.

CLARK, A. E. (1935): The effects of temperature and salinity on the early development of the Oyster. — Progr. Rep. Atlantic Biol. Stat., St. Andrews, N. B. 16: 10.

CLAUS, A. (1937): Vergleichend-physiologische Untersuchungen zur Ökologie der Wasserwanzen mit besonderer Berücksichtigung der Brackwasserwanze Sigara lugubris FIEB. — Zool. Jb., Abt. Allg. Zool. 58: 356—432.

CLEGG, I. S. (1964): The control of emergence and metabolism by external osmotic pressure and the role of free glycerol in developing cysts of Artemia salina. — J. Exp. Biol. 41: 879—892.

COLE, W. H. (1940): The composition of fluids and sera of some marine animals and of the sea water in which they live. — J. gen. Physiol. 23: 575—584.

COLLANDER, R. (1936): Der Zellsaft der Characeen. — Protoplasma 25: 201—210.

Conference on Estuaries (1964): Abstracts, pp. 75. March 31—April 4, Jekyll Island, Georgia, USA. [Amer. Assoc. Adv. Sci., Publ. No. 83, LAUFF, G. H. (Ed.), 757 pp., 1967.]

CONKLIN, R. E. & KROGH, A. (1939): A note on the osmotic behaviour of Eriocheir in concentrated and *Mytilus* in dilute sea water. — Z. vergl. Physiol. **26**: 239—241.

CONTE, F. P., WAGNER, H. H., FESSLER, J. & GNOSE, C. (1966): Development of osmotic and ionic regulation in juvenile coho salmon *Oncorhynchus kisutch*. — Comp. Biochem. Physiol. **18**: 1—15.

CONWAY, E. J. (1945): The physiological significance of inorganic levels in the internal medium of animals. — Biol. Rev. **20**: 56—72.

COPELAND, D. E. (1948): The cytological basis of chloride transfer in the gills of *Fundulus heteroclitus*. — J. Morph. **82**: 201—228.

COSTLOW, JR., J. D. (1967): The effect of salinity and temperature on survival and metamorphosis of megalops of the blue crab *Callinectes sapidus*. — Helgoländer wiss. Meeresunters. **15**: 84—97.

CRAIGIE, D. E. (1963): An effect of water hardness in the thermal resistance of the rainbow trout, *Salmo gairdnerii* RICHARDSON. — Canad. J. Zool. **41**: 825—830.

CROGHAN, P. C. (1958): The osmotic and ionic regulation of *Artemia salina*. — J. Exp. Biol. **35**: 219—233, 234—242, 243—249, 425—436.

DAKIN, W. J. (1935): The aquatic animal and its environment. — Linn. Soc. N. S. Wales Proc. **60**, Pts. 1, 2: 8—332.

DAKIN, W. J. & EDMONDS, E. (1931): The regulation of the blood of aquatic animals, and the problem of the permeability of the bounding membranes of aquatic invertebrates. — Austral. Exp. Biol. and Med. Sci **8**: 169—187.

DARNELL, R. M. (1967): Organic detritus in relation to the estuarine ecosystem. — In "Estuaries" ed. by G. H. LAUFF, pp. 376—382, 1967, Publ. No. 83, Amer. Assoc. Adv. Sci., Washington, D. C.

DEHNEL, P. A. (1960): Effect of temperature and salinity on the oxygen consumption of two intertidal crabs. — Biol. Bull. **118**: 215—249.

— (1962): Aspects of osmoregulation in two species of intertidal crabs. — Biol. Bull. **122**: 208—227.

— (1964): Ion regulation in two species of estuarine crabs. — Helgoländer wiss. Meeresunters. **9**: 474—475.

— (1967): Osmotic and ionic regulation in estuarine crabs. — In "Estuaries" ed. by G. H. LAUFF, pp. 541—547. Publ. No. 83, Amer. Assoc. Adv. Sci., Washington, D. C.

DEHNEL, P. A. & MCCAUGHRAN, D. A. (1964): Gill tissue respiration in two species of estuarine crabs. — Comp. Biochem. Physiol. **13**: 233—259.

DEHNEL, P. A. & STONE, D. (1964): Osmoregulatory role of the antennary gland in two species of estuarine crabs. — Biol. Bull. **126**: 354—372.

DÉPÈCHE, J. (1964): Osmoregulation embryonnaire et adaptation à l'eau de mer de *Lebistes reticulatus* (Cyprinodonte vivipare). — C. r. Acad. Sci. (Paris) **259**: 908—910.

DIAMOND, J. H. & SOLOMON, A. K. (1959): Experiments with the brackish water alga *Nitella* confirm that both sodium and chloride are actively transported between the cell sap and the exterior, the potassium moving passively. — J. gen. Physiol. **42**: 1105 to 1128.

DREVS, P. (1896): Über die Regulation des osmotischen Drucks in Meeresalgen bei Schwankungen des Salzgehaltes im Außenmedium. — Arch. Verein. Freunde d. Naturgesch. Mecklenburg (Güstrow) **49**: 91—135.

DRILHON, A. (1942): Destruction de la moelle et adaptation aux changements de salinité chez un poisson homéiosmotique (Carpe). — C. r. Acad. Sci. **214**: 575—577, Paris.

DRILHON, A. & PORA, E. A. (1936): Régulation du milieu intérieur chez les poissons sténohalins. — Ann. Physiol. (Paris) **12**: 139—168.

DROOP, M. R. (1956): Optimum, relative and actual ionic concentrations for growth of some euryhaline algae. — Verh. internat. Verein. Limnol. Helsinki.

DUCHATEAU, GH. & FLORKIN, M. (1956): Systèmes intracellulaires d'acides aminés libres et osmorégulation des crustacées. — J. Physiol. **48**: 520.

DUCHÂTEAU-BOSSON, GH. & FLORKIN, M. (1961): Change in intracellular concentration of free amino acids as a factor of euryhalinity in the crayfish *Astacus astacus*. — Comp. Biochem. Physiol. **3**: 245—249.

— (1962): Adaptation à l'eau de mer de crabes chinois *(Eriocheir sinensis)* présentant, dans l'eau douce, une valeur élevée la composante aminoacide des muscles. — Arch. internat. Physiol. Bioch. **70**, 3: 345—355.

— (1962): Régulation isosmotique intracellulaire chez *Eriocheir sinensis* après ablation des pédoncules oculaires. — Arch. internat. Physiol. Bioch. **70**, 3.

DUCHATEAU, GH., FLORKIN, M. & JEUNIAUX, CH. (1959): Composantes aminoacides des tissus chez les Crustacés. I. Composantes aminoacides des muscles de *Carcinus maenas* L. lors du passage de l'eau de mer à l'eau saumâtre et au cours de de la mue. — Arch. internat. Physiol. **67**: 489—500.

DUCHATEAU, GH., SARLET, H., CAMIEN, M. N. & FLORKIN, M. (1952): Acides aminés non-proteiniques des tissus chez les mollusques lamellibranches et chez les vers. Comparaison des formes marines et des formes dulcicoles. — Arch. internat. Physiol. **60**: 124—125.

DUVAL, M. (1925): Recherches physico-chimiques et physiologiques sur le milieu intérieur des animaux aquatiques. Modifications sous l'influence du milieu extérieur. — Ann. Inst. Océanogr. (Monaco) 2,232—407.

EDMONDS, E. (1935): The relations between the internal fluid of marine invertebrates and the water of the environment, with special reference to Australian crustacea. — Proc. Linn. Soc. N. S. Wales, LX: 233—247, Sydney.

ELIASSEN, E. (1953): The energy-metabolism of *Artemia salina* in relation to body size, seasonal rhythms, and different salinities. — Univ. Bergen, Årbok 1952. Nat. vit. bekke Nr. **11**: 3—17.

ELIASSEN, E., LEIVESTAD, H. & MØLLER, D. (1960): The effect of low temperatures on the freezing point of plasma and on the potassium/sodium ratio in the muscles of some boreal and subarctic fishes. — Arbok Univ. Bergen, Mat.-Nat. Ser. 14, 24 pp.

ELLIS, W. G. (1937): The water and electrolyte exchange of *Nereis diversicolor* (MUELLER). — J. Exp. Biol. **14**: 340—350, London.

— (1939): Comparative measurements of water and electrolyte exchange in a stenohaline and a euryhaline Polychaete. — J. Exp. Biol. **16**: 483—486.

— (1933): Calcium and the resistance of *Nereis* to brackish water. — Nature **132**: 748.

EPPLEY, R. W. & CYRUS, C. C. (1960): Cation regulation and survival of the red alga, *Porphyra perforata*, in diluted and concentrated sea water. — Biol. Bull. **118**: 55—65.

ERCHINGER, H. (1964): Zur Frage der Brackwasseranpassung mariner Knochenfische. Untersuchungen über die Frage der Ionenregulation im Muskelgewebe von *Cottus bubalis* EUPHR. (Pisces, Teleostei). — Internat. Rev. ges. Hydrobiol. **49**: 563—610.

ERMAN, P. (1961): Atmungsmessungen an Geweben und Gewebehomogenaten der Miesmuschel *(Mytilus edulis* L.) aus Brack- und Meerwasser. — Kieler Meeresforsch. **17**: 176—189.

FAGE, L. (1912): Essais d'acclimatation du saumon dans le bassin de la Mediterranée. — Bull. Inst. Océanogr. (Monaco) **225**: 1—13.

FAO (1958, 1962): Yearbook of Fishery Statistics.

FLEMISTER, L. J. & FLEMISTER, S. C. (1951): Chloride ion regulation and oxygen consumption in the crab *Ocypode albicans* (BOSQ.). — Biol. Bull. Woods Hole **101**: 259—273.

FLORKIN, M. (1961—1962): Régulation anisosmotique extracellulaire régulation isosmotique intracellulaire et euryhalinité. — Ann. Soc. Roy. Zool. Belg. **92**, 1: 183—186.

— (1962): La régulation isosmotique intracellulaire chez les invertébrés marins euryhalins. — Bull. de la Cl. Sci. T. XLVIII, 5e Sér.

— (1966): Nitrogenous compounds in osmotic régulation. — In "Physiology of Mollusca" ed. by K. M. WILBUR & C. M. YONGE. Vol. II, pp. 333—336. New York.

FLORKIN, M. & DUCHATEAU, GH. (1948): Sur l'osmorégulation de l'anodonte. — Physiol. comp. **1**: 29—45.

FLORKIN, M., DUCHATEAU-BOSSON, GH., JEUNIAUX, CH. & SCHOFFENIELS, E. (1964): Sur le mécanisme de la régulation de la concentration intracellulaire en acides aminés lebres, chez *Eriocheir sinensis*, au cours de l'adaptation osmotique. — Arch. internat. Physiol. Bioch. **72**, 5: 892—905.

FLORKIN, M. & SCHOFFENIELS, E. (1969): Molecular Approaches to Ecology. — Acad. Press, New York and London (Chapter VII, pp. 149—158).

FLÜGEL, H. (1960): Über den Einfluß der Temperatur auf die osmotische Resistenz und die Osmoregulation der decapoden Garnele Crangon crangon L. — Kieler Meeresforsch. 16: 186—200.

FONSELIUS, S. H. (1967): Hydrography of the Baltic deep basins II. — Fish. Board Sweden, Ser. Hydrograph. Rep. No. 20: 1—31.

FONTAINE, M. (1930): Sur le parallélisme existant chez les poissons, leur résistance aux variations de salinité et l'indépendence de leur milieu intérieur. — C. r. Acad. Sci. 191: 796—798, Paris.

— (1943): Les facteurs physiologiques des migrations reproductrices des cyclostomes et poissons potamotoques. — Bull. Inst. Océanogr. (Monaco) 40 (848): 1—8.

— (1953): Equilibre hydrominéral et quelques particularités de sa régulation chez les vertébrés. — Arch. Sci. Physiol. 7: 55—78.

— (1954: Du déterminisme physiologique des migrations. — Biol. Rev. 29: 390—418.

— (1956): The hormonal control of water and salt-electrolyte metabolism in fish. — Mem. Soc. Endocrinol., No. 5, Part II: 69—82, Cambridge.

FONTAINE, M. & BARADUC, M. M. (1954): Influence d'une thryoxinisation prolongée sur l'euryhalinité d'un salmonide, la truite arc-en-ciel (Salmo gairdnerii RICH.). — C. r. Soc. Biol. (Paris) 148: 1942—1944.

FONTAINE, M. & CALLAMAND, O. (1947): Influence d'une diminution de salinité sur le pH sanguin de quelques téléostomes marins. — Bull. Inst. Océanogr. (Monaco) 910.

FONTAINE, M., CALLAMAND, O. & OLIVEREAU, M. (1949): Hypophyse et euryhalinité chez l'anguille. — C. r. Acad. Sci. 228: 513—514, Paris.

FONTAINE, M., DELATTRE, S. & CALLAMAND, O. (1945): Influence des variations de salinité sur la teneur en hématies de deux téléostéen (Anguilla anguilla L. et Cyprinus carpio L.). — Bull. Inst. Océanogr. (Monaco) 42: 886.

FONTAINE, M. & FIRLY, S. B. (1933): Influence des variations de salinité sur la réserve alcaline du sang des poissons. — C. r. Soc. Biol. 113, 306—308, Paris.

FONTAINE, M. & KOCH, H. (1950): Les variations d'euryhalinité et d'osmorégulation chez les poissons. Leur rapport possible avec le déterminisme des migrations. — J. Physiol. 42: 287—318.

FONTAINE, M., LACHIVER, L., LELOUP, J. & OLIVEREAU, M. (1948): La fonction tryoidienne du Saumon (Salmo salar L.) au cours de sa migration reproductrice. — J. Physiol. Path. Gén. 40: 182—184.

FONTAINE, M. & RAFFY, A. (1932): Sur le mécanisme des modifications de la consommation d'oxygène observées chez la civelle au cours des changements de salinité. — C. r. Soc. Biol. 110: 538—540, Paris.

— — (1935): Sur la consommation d'oxygène de la lamproie marine (Petromyzon marinus L.) Influence de la salinité. — Assoc. Franc. Sci. 59: 330—333.

— — (1932): Recherches physiologiques et biologiques sur les Civelles. — Bull. Inst. Océanogr. (Monaco) 603: 1—19.

FORSMAN, B. (1951): Studies on Gammarus duebeni LILLI, with notes on some rock pool organisms in Sweden. — Zool. Bidrag. Uppsala 19: 215—237.

FOX, D. L. (1941): Changes in the tissue chloride of the California mussel in response to heterosmotic environments. — Biol. Bull. 80: 111—129.

FOX, H. M. & SIMMONDS, B. G. (1933): Metabolic rates of aquatic Arthropods from different habitats. — J. Exp. Biol. 10: 67—74.

FRANKENBERG, D. & BURBANCK, W. D. (1963): A comparison of the physiology and ecology of the estuarine isopod Cyathura polita in Massachusetts and Georgia. — Biol. Bull. 125: 81—95.

FRETTER, V. (1955): Uptake of radioactive sodium (^{24}Na) by Nereis diversicolor MUELLER and Perinereis cultrifera GRUBE. — J. Mar. Biol. Assoc. U. K. 34: 151—160.

FRIEDRICH, H. (1937): Einige Beobachtungen über das Verhalten der Alderia modesta LOV. im Brackwasser. — Biol. Zbl. 57: 101—104.

FROMAGEOT, C. (1923): Influence de la concentration en sels de l'eau de mer sur l'assimilation chlorophyllienne des algues. — C. r. Acad. Sci. 177: 779—780, Paris.

FROMAGEOT, C. & GAUSE, G. F. (1941): The effect of natural selection in the acclimatization of *Euplotes* to different salinities of the medium. — J. Exp. Zool. 87: 85—100.

FUGELLI, K. & LANGE, R. (1965): Osmotic response of erythrocytes in the euryhaline teleosts, the flounder, *Pleuronectes flesus* L., and the three-spined stickleback, *Gasterosteus aculeatus* L. — Rep. Fourth Scand. Congr. Cell. Res.: 44—45.

GAVRILESCU, N. (1937): La résistance globulaire chez quelques espèces de poissons euryhalins et sténohalins. — C. r. Acad. Sci. Roumanie II: 90—92.

GESSNER, F. (1933): Nitrat, Phosphat und Planktongehalt im Arkonabecken. — J. Cons. internat. pour l'exploration de la mer 8: 181—194.

— (1933): Die Produktionsbiologie der Ostsee. — Naturwiss. 21: 649—653.

— (1933): Die Planktonproduktion der Brackwasser in ihrer Beziehung zur Produktion der offenen See. — Verh. Internat. Verein. Limnol. 6, b.

— (1937): Hydrographie und Hydrobiologie der Brackwässer Rügens und des Darß. — Kieler Meeresforsch. 2: 1—180.

— (1940): Produktionsbiologische Untersuchungen im Arkonabecken und in den Binnengewässern von Rügen. — Kieler Meeresforsch. 3: 449—459.

— (1940): Untersuchungen über die Osmoregulation von Wasserpflanzen. — Protoplasma 34: 593—600.

— (1956): Meer und Strand. Die Lebensgemeinschaften im deutschen Meeresraum. 2. Aufl. — Leipzig.

— (1959): Hydrobotanik. Die physiologischen Grundlagen der Pflanzenverbreitung im Wasser. Bd. II. Stoffhaushalt. Kap. IV. Die salzreichen Gewässer der Erde. 341—470. — Berlin.

GESSNER, F. & HAMMER, L. (1960): Die Photosynthese von Meerespflanzen in ihrer Beziehung zum Salzgehalt. — Planta 55: 306—312.

GILBRICHT, M. (1952): Untersuchungen zur Produktionsbiologie des Planktons in der Kieler Bucht. II. Die Produktionsgröße. — Kieler Meeresforsch. 9: 51—61.

GILCHRIST, B. M. (1956): The oxygen consumption of *Artemia salina* (L) in different salinities. — Hydrobiologia 8: 54—63.

GOMPEL, M. & LEGENDRE, R. (1928): Limites de température et de salure supportées par *Convoluta roscoffensis*. — C. r. Soc. Biol. 98: 572—573, Paris.

GORDON, M. S. (1957): Observations on osmoregulation in the arctic char (*Salvelinus alpinus* L.). — Biol. Bull. 112: 28—33.

— (1959): Osmotic and ionic regulation in Scottish brown trout and sea trout (*Salmo trutta* L.). — J. Exp. Biol. 36: 253—260.

— (1959): Ionic regulation in the brown trout (*Salmo trutta* L.). — J. Exp. Biol. 36: 227—252.

— (1962): Physiological ecology of osmotic regulation among littoral vertebrates. — Pubbl. staz. zool. Napoli 32: 294—300.

— (1963): Osmotic regulation in the Green Toad *(Bufo viridis)*. — J. Exp. Biol. 39: 261—270.

— (1965): Intracellular osmoregulation in skeletal muscle during salinity adaptation in two species of toads. — Biol. Bull. 128: 218—229.

GORDON, M. S., SCHMIDT-NIELSEN, K. & KELLY, H. M. (1961): Osmotic regulation in the Crab-eating Frog *(Rana cancrivora)*. — J. Exp. Biol. 38: 659—678.

GRAETZ, E. (1931): Versuch einer exakten Analyse der zur Osmoregulation benötigten Kräfte in ihrer Beziehung zum Gesamtstoffwechsel von Süßwasserstichlingen in hypo- und hypertonischen Medien. — Zool. Jb., Abt. Physiol. 49: 37—58.

GRAFFLIN, A. L. (1937): A problem of adaptation to fresh and salt water in the teleosts, viewed from the standpoint of the structure of the renal tubules. — J. Cell. Comp. Physiol. 9: 469—475.

— (1938): The absorption of fluorescein from fresh water and salt water by *Fundulus heteroclitus,* as judged by a study of the kidney with the fluorescence microscope. — J. Cell. Comp. Physiol. 12: 167—170.

GRESENS, J. (1928): Versuche über die Widerstandsfähigkeit einiger Süßwassertiere gegen- über Salzlösungen. — Z. Morph. Ökol. Tiere **12**: 707—800.

GRØNTVED, J. (1958): Planktological Contributions. III. Investigations on the phytoplankton and the primary production in an Oyster culture in the Limfjord. — Medd. Danmarks Fisk. og Havunders. N. S. **2**, No. 17: 1—15.

— (1960): On the productivity of microbenthos and phytoplankton in some Danish fjords. — Medd. Danmarks Fisk. og Havunders., N. S. **3**: 55—92.

GROSS, W. J. (1954): Osmotic response in the sipunculid *Dendrostomum zostericolum*. — J. Exp. Biol. **31**: 402—423.

— (1957): An analysis of response to osmotic stress in selected decapod crustacea. — Biol. Bull. **112**: 43—62.

— (1964): Trends in water and salt regulation among aquatic and amphibious crabs. — Biol. Bull. **127**: 447—466.

GROSS, W. J. & CARPEN, R. L. (1966): Some functions of the urinary bladder in a crab. — Biol. Bull. **131**: 272—291.

GROSS, W. J. & MARSHALL, L. A. (1960): The influence of salinity on the magnesium and water fluxes of a crab. — Biol. Bull. **119**: 440—453.

GROSS, W. J., LASIEWSI, R. C., DENNIS, M. & RUDY, P. JR. (1966): Salt and water balance in selected crabs of Madagascar. — Comp. Biochem. Physiol. **17**: 641—660.

GUEYLARD, F. (1925): De l'adaptation aux changements de salinité. Recherches biologiques et physicochimiques sur l'épinoche (*Gasterosteus leiurus* CUR. et VAL.). — Arch. phys. Biol. **3**: 79—187.

GUNTER, G. (1945): Studies on marine fishes in Texas. — Publ. Inst. Mar. Sci. **1**: 1—190.

GURNEY, R. (1928—1929): The fresh-water crustacea of Norfolk. — Trans. Norf. Norw. Nat. Soc. **12**: 5.

HAAS, H. & STRENZKE, K. (1957): Experimentelle Untersuchungen über den Einfluß der ionalen Zusammensetzung des Mediums auf die Entwicklung der Analpapillen von *Chironomus thummi*. — Biol. Zbl. **76**: 513—528.

HAGEN, G. (1954): Strukturelle Abweichungen mariner und euryhaliner Oligochaeten in Grenzbereichen ihres Vorkommens. — Kieler Meeresforsch. **10**: 77—80.

VON HARANGHY, L. (1942): Die Muschelvergiftung als biologisches Problem. — Helgoländer wiss. Meeresunters. **2**: 279—353.

HARDER, R. (1915): Beiträge zur Kenntnis des Gaswechsels der Meeresalgen. — Jb. wiss. Bot. **56**: 254—298.

HARDISTRY, M. W. (1954): Permeability to water of the lamprey integument. — Nature. **174**: 360—361, London.

— (1956): Some aspects of osmotic regulation in lampreys. — J. Exp. Biol. **33**: 431—447.

— (1957): Osmotic conditions during the embryonic and early larval life of the brook lamprey *(Lampetra planeri)*. — J. Exp. Biol. **34**: 237—252.

HARMS, J. W. & DRAGENDORFF, O. (1933): Die Realisation von Genen und die consecutive Adaptation. 3. Mitt.: Osmotische Untersuchungen an *Physcosoma lurco* SEL. aus den Mangrove-Vorländern der Sundainseln. — Z. wiss. Zool. **143**: 263—322.

HARNISCH, O. (1934): Osmoregulation und osmoregulatorischer Mechanismus der Larve von *Chironomus thummi*. — Z. vergl. Physiol. **21**: 281—295.

HARVEY, H. W. (1955): The chemistry and fertility of sea waters. — Cambridge Univ. Press, 224 pp.

HAYES, R. F. (1930): The physiological response of *Paramaecium* to sea water. — Z. vergl. Physiol. **13**: 214—222.

HAYWOOD, C. & CLAPP, M. J. (1942): A note on the freezing points of the urines of two fresh-water fishes; the catfish *(Ameiurus nebulosus)* and the sucker *(Catostomus commersonii)*. — Biol. Bull. **83**: 363—366.

HEILBRUNN, L. V. (1953): An outline of general physiology. Third ed. — Philadelphia, W. B. Saunders Co.

HELLEBRUST, J. A. (1967): Excretion of organic compounds by cultured and natural popula- tions of marine phytoplankton. — In "Estuaries" ed. by G. H. LAUFF, pp. 361–366, Publ. No. 83, Amer. Assoc. Adv. Sci., Washington, D. C.

HENSCHEL, J. (1936): Wasserhaushalt und Osmoregulation von Scholle und Flunder. — Wiss. Meeresunters. **22**: 89—121, Kiel.

HENTSCHEL, E. (1933): Untersuchungen über das Kleinplankton an den Küsten von Island. — Ber. dt. wiss. Komm. Meeresforsch. N. F. **6**: 4.

HENTSCHEL, E. & WATTENBERG, H. (1930): Plankton und Phosphat in der Oberflächenschicht des Südatlantischen Ozeans. — Ann. Hydr. u. mar. Met. **58**: 8.

HERRMANN, F. (1931): Über den Wasserhaushalt des Flußkrebses. — Z. vergl. Physiol. **14**: 479—524.

HEUTS, M. J. (1943): La régulation osmotique chez l'épinochette (*Pygosteus pungitius* L.). — Ann. Soc. Roy. Zool. Belg. **74**: 99—105.

— (1944): Calcium-ionen en geografische verspreitung van *Gasterosteus aculeatus*. — Nat. T. **26**: 10—14.

— (1945): La régulation minérale en fonction de la température chez Gasterosteus aculeatus. Son importance au point de vue de la zoogeographie de l'espèce. — Ann. Soc. Roy. Zool. Belg. **76**: 88—99.

— (1946): Physiological isolating mechanisms and selection within the species *Gasterosteus aculeatus*. — Nature **158**: 839—840, London.

— (1947): Experimental studies on adaptive evolution in *Gasterosteus aculeatus*. — Evolution **1**: 89—102.

HICKEL, W. (1967): Untersuchungen über die Phytoplanktonblüte in der westlichen Ostsee. — Helgoländer wiss. Meeresunters. **16**: 3—66.

HICKMAN, C. P. (1959): The osmoregulatory role of the thyroid gland in the starry flounder, *Platichthys stellatus*. — Canad. J. Zool. **37**: 997—1060.

HISCOCK, I. D. (1953): Osmoregulation in Australian freshwater mussels. — Austral. J. Mar. Freshw. Res. **4**: 317—342.

HOAR, W. S. (1951): Hormones in fish. (Some aspects of the physiology of fish. I.). — Univ. Toronto Biol. Ser. No. 59. Publ. Ontario Fish. Res. Labor. **71**: 1—51.

— (1958): Endocrine factors in the ecological adaptation of fishes. In: Comparative Endocrinology, ed. GORBMAN, 1—23. — New York.

HÖBER, R. (1926): Physikalische Chemie der Zelle und der Gewebe. 6. Aufl. — Leipzig.

HÖFLER, K. (1931): Hypotonietod und osmotische Resistenz einiger Rotalgen. — Österreich. Bot. Z. Wien **80**: 51—71.

HÖFLER, K., URL, W. & DISKUS, A. (1956): Zellphysiologische Versuche und Beobachtungen an Algen der Lagune von Venedig. — Boll. Mus. civ. Storia Nat. Venezia **9**: 63—94.

HÖHNK, W. (1952): Studien zur Brack- und Seewassermykologie. II. — Veröff. Inst. Meeresforsch. Bremerhaven **1**: 247—278.

— (1953): III. Ibidem **2**: 52—108.

— (1956): IV. Ibidem **4**: 195—213.

— (1953): Mykologische Studien im Brack- und Meerwasser. — Atti Congr. Int. Microbiol. Roma **7**, Seg. 12: 374—378.

HOFFMANN, C. (1929): Die Atmung der Meeresalgen und ihre Beziehung zum Salzgehalt. — Jb. wiss. Bot. **71**: 214—268.

— (1932): Zur Bestimmung des osmotischen Druckes an Meeresalgen. — Planta (Berl.) **16**: 413—432.

— (1932): Zur Frage der osmotischen Zustandsgrößen bei Meeresalgen. — Planta (Berl.) **17**: 805—809.

— (1933): Die Vegetation der Nord- und Ostsee. Tierwelt der Nord- und Ostsee. Teil 1 c.

— (1943): Der Salzgehalt des Seewassers als Lebensfaktor mariner Pflanzen. — Kieler Blätter, 3. H.: 160—176.

HOHENDORF, K. (1963): Der Einfluß der Temperatur auf die Salzgehaltstoleranz und Osmoregulation von *Nereis diversicolor* O. F. MUELL. — Kieler Meeresforsch. **19**: 196—218.

HOLLIDAY, F. G. T. & BLAXTER, J. H. (1960): The effects of salinity on the developing eggs and larvae of the herring. — J. Mar. Biol. Ass. U. K. **39**: 591—603.

HOLMES, R. M. (1960): Kidney function in the rainbow trout. — Ann. Rep. Challenger Soc. **3**: XIII.

HOLM-JENSEN, I. (1948): Osmotic regulation in *Daphnia magna* under physiological conditions and in the presence of heavy metals. — Biol. Medd. Kobenhavn 20, **11**: 1—64.

HOOP, M. (1940): Der Einfluß des Salzgehaltes des Wassers auf die Gewebe von euryhalinen Muscheln. — Zool. Jb. **73**: 391—442.

HOPKINS, A. E. (1936, 1938): Adaptation of the feeding mechanism of the oyster *Ostrea gigas,* to changes in salinity. — Bull. Bur. Fish, XLVIII, **21**: 345—364, 11 fig., Washington, D. C., J. du Cons., XIII, **1**: 127—128.

HOPKINS, D. L. (1946): The contractile vacuole and the adjustment to changing concentrations in fresh water amoebae. — Biol. Bull., Woods Hole **90**: 158—176.

HOPKINS, H. S. (1946): The influence of season, concentration of sea water and environmental temperature upon the oxygen consumption of tissues in *Venus mercenaria.* — J. Exp. Zool. **102**: 143—158.

— (1949): Metabolic reactions of clams tissues to change in salinity. I. Ciliary activity, narcotic and cyanide effects, and respiratory quotient. — Physiol. Zool. **22**: 295—308.

HOUSE, C. R. (1963): Osmotic regulation in the brackish water teleost. *Blennius pholis.* — J. Exp. Biol. **40**: 87—104.

HÜBEL, H. (1964): Die Primärproduktion des Phytoplanktons der nördlichen Rügenschen Boddengewässer unter Anwendung der C-Methode. — Diss. Greifswald.

— (1968): Die Bestimmung der Primärproduktion des Phytoplanktons der Nord-Rügenschen Boddengewässer unter Verwendung der Radiokohlenstoffmethode. — Int. Rev. ges. Hydrobiol. **53**: 601—633.

HUKADA, K. (1932): Changes of weight of marine animals in diluted media. — J. Exp. Biol. **9**: 61—68.

HYNES, H. B. N. (1954): The ecology of *Gammarus duebeni* LILLJEBORG and its occurence in fresh water in Western Britain. — J. Anim. Ecol. **23**: 38—84.

INMAN, O. L. (1921): Comparative studies in respiration. — J. Gen. Physiol. **3**: 663—666.

JANSSON, B.-O. (1962): Salinity resistance and salinity preference of two oligochaetes *Aktedrilus monospermaticus* KNÖLLNER and *Marionina preclitellochaeta* n. sp. from the interstitial fauna of marine sandy beaches. — Oikos **13**: 293—305.

JEUNIAUX, CH., BRICTEUX-GREGOIRE, S. & FLORKIN, M. (1962): Régulation osmotique intracellulaire chez *Asterias rubens.* Rôle du glycocolle et de la taurine. — Cah. biol. mar. **III**: 107—113.

JØRGENSEN, C. B. & DALES, R. P. (1957): The regulation of volume and osmotic regulation in some *Nereis* Polycaetes. — Physiol. compar. et oecol. **4**: 357—374, Amsterdam.

JØRGENSEN, C. B., LEVI, H. & ZERAHN, K. (1954): On active uptake of sodium and chloride ions in anurans. — Acta physiol. scand. (Stockh.) **30**: 178—190.

JØRGENSEN, C. B. & ROSENKILDE, P. (1956): On regulation of concentration and content of chloride in goldfish. — Biol. Bull. **110**: 300—303.

JØRGENSEN, E. G. (1960): The effects of salinity, temperature and light intensity on growth and chlorophyll formation of *Nitzschia ovalis.* — Yearb., Carnegie Inst. Wash. **59**: 348—349.

JONES, L. L. (1941): Osmotic pressure relations of nine species of crabs of the Pacific coast of N. America. — J. cell. comp. Physiol. **18**: 79—92.

JÜRGENS, O. (1935): Die Wechselbeziehungen von Blutkreislauf, Atmung, und Osmoregulation bei Polychaeten (*Nereis diversicolor* O. F. MÜLL.). — Zool. Jb. Physiol. **55**: 1—46.

KÄNDLER, R. (1953): Hydrographische Untersuchungen zum Abwasserproblem in den Buchten und Förden der Ostseeküste Schleswig-Holsteins. — Kieler Meeresforsch. **9**: 176—200.

— (1962): Die Fischereierträge der Meere als Ausdruck ihrer unterschiedlichen Produktionsleistungen. — Kieler Meeresforsch. **18**, 3. Sonderh.: 121—127.

KAIN, J. M. & FOGG, G. E. (1958): Studies on the growth of marine phytoplankton. I. *Asterionella japonica* GRAN. — J. Mar. Biol. Assoc. **37**: 397—413.

KALLE, K. (1943): Der Stoffhaushalt des Meeres. — Akad. Verlagsges. Becker & Erler Kom.-Ges., Leipzig.

— (1951): D. Hydrograph. Z. **4**: 13—17.

KAMADA, T. (1935): Contractile vacuole of *Paramecium.* — J. Sci. Tokyo Imp. Univ. **4**.

KAPLANSKY, S. & BOLDYREWA, N. (1934): K voprosu o reguliatsii mineral'noge obmena u gomoosmotiches kilsh ryb pri ismenenii mineral'noge sostava vody. 2. — Fiziol. Zhurn. **17**: 96—99.

KARPEVICH, A. F. (1958): Überlebensdauer, Fortpflanzung und Atmung von *Mesomysis kowalevskyi (Paramysis lacustris kowalevskyi* CZERN.*)* in Brackwasser der UDSSR. (Ökologisch-physiologische Grundlagen der Akklimatisierung von Mysideen im Aral-See, in der Ostsee und im Balchas-See). — Zool. Z. **37**, 1121—1135. (Russ. mit englischer Zusammenf.).

KESSELER, H. (1959): Mikrokryoskopische Untersuchungen zur Turgorregulation von *Chaetomorpha linum.* — Kieler Meeresforsch. **15**: 51—73.

— (1962): Beziehungen zwischen Atmung und Turgorregulation von *Chaetomorpha linum* in Abhängigkeit von Salzgehaltsänderungen und spezifischen Ionenwirkungen. — Helgoländer wiss. Meeresunters. **8**: 243—256.

— (1964): Die Bedeutung einiger anorganischer Komponenten des Seewassers für die Turgorregulation von *Chaetomorpha linum* (Cladophorales). — Helgoländer wiss. Meeresunters. **10**: 73—90.

— (1964): Zellsaftgewinnung, AFS (apparent free space) und Vakuolenkonzentration der osmotisch wichtigsten mineralischen Bestandteile einiger Helgoländer Meeresalgen. — Helgoländer wiss. Meeresunters. **11**, 3—4: 258—269.

KETCHUM, B. H. (1967): Phytoplankton nutrients in estuaries. — In "Estuaries" ed. by G. H. LAUFF, Publ. No. 83, 329—335. Amer. Assoc. Adv. Sci. Washington, D. C.

KEYS, A. B., (1931): A study of selective action of decreased salinity and of asphyxiation on the pacific killifish, *Fundulus parvipinnis.* — Bull. Scr. Inst. Oceanogr. Univ. Cal. Tech. Ser. **2**: 417—490.

— (1931): Chloride and water secretion and absorption by the gills of the eel. — Z. vergl. Physiol. **15**: 364—388.

— (1933): The mechanisms of adaptation to varying salinity in the commom eel the general problem of osmotic regulation in fishes. — Proc. Roy. Soc. (B) **112**: 184—199, London.

KING, E. N. (1965): The oxygen consumption of intact crabs and excised gills as a function of decreased salinity. — Comp. Biochem. Physiol. **15**: 93—102.

— (1966): Oxidative activity of crab gill mitochondria as a function of osmotic concentration. — Comp. Biochem. Physiol. **17**: 245—258.

KING, E. N. & SCHOFFENIELS, E. (1969): In vitro preparation of crab gill for use in ion transport studies. — Arch. Int. Physiol. Biochim. **77**: 105—111.

KINNE, O. (1952): Zur Biologie und Physiologie von *Gammarus duebeni* LILLJ., V: Untersuchungen über Blutkonzentration, Herzfrequenz und Atmung. — Kieler Meeresforsch. **9**: 134—150.

— (1954): Zur Biologie und Physiologie von *Gammarus duebeni* LILLJ. I. — Z. wiss. Zool. **157**: 427—491.

— (1954): Experimentelle Untersuchungen über den Einfluß des Salzgehaltes auf die Hitzeresistenz von Brackwassertieren. — Zool. Anz. **152**: 10—16.

— (1956): Über den Einfluß des Salzgehaltes und der Temperatur auf Wachstum, Form und Vermehrung bei dem Hydroidpolypen *Cordylophora caspia* (Pallas, Athecata, Clavidae). — Zool. Jb. (Allg. Zool. Physiol. Tiere) **66**: 565—638 u. **67**: 407—486, 1958.

— (1960): Growth, food intake, and food conversion in a euryplastic fish exposed to different temperatures and salinities. — Physiol. Zool. **33**: 288—317.

— (1963): Über den Einfluß des Salzgehaltes auf verschiedene Lebensprozesse des Knochenfisches *Cyprinodon macularius.* — Veröff. Inst. Meeresforsch. Bremerh., Sonderbd. **Y**, 3. Meeresbiol. Sympos.: 49—66.

— (1963): The effects of temperature and salinity on marine and brackish water animals. I. Temperature. — Oceanogr. Mar. Biol. Ann. Rev. **1**: 301—340.

— (1964): Physiologische und ökologische Aspekte des Lebens in Ästuarien. — Helgoländer wiss. Meeresuntersuch. **11**: 131—156.

— (1964): Non-genetic adaptation to temperature and salinity. — Helgoländer wiss. Meeresunters. **9**, 433—458.

— (1964): The effects of temperature and salinity on marine and brackish water animals. II. Salinity and temperature salinity combinations. — Oceanogr. Mar. Biol. Ann. Rev. **2**: 281—339.

— (1966): Physiological aspects of animal life in estuaries with special reference to salinity. — Netherlands J. Sea Res. **3**, 2: 222—244.

— (1967): Physiology of estuarine organisms with special reference to salinity and temperature: General aspects. — In "Estuaries" ed. by G. H. LAUFF, 525—540, Publ. No. **83**, Amer. Assoc. Adv. Sci., Washington, D. C.

KINNE, O. & KINNE, E. M. (1962): Rates of development in embryos of a cyprinodont fish exposed to different temperature-salinity-oxygen-combinations. — Canad. J. Zool. **40**: 231—253.

KINNE, O. & PFAFFENHÖFER, G.-A. (1966): Growth and reproduction as a function of temperature and salinity in *Clava multicornis* (Cnidaria, Hydrozoa). — Helgoländer wiss. Meeresunters. **13**: 62—72.

KINNE, O. & ROTTHAUWE, H.-W. (1952): Biologische Beobachtungen und Untersuchungen über die Blutkonzentration an *Heteropanope tridentatus* MAITLAND (Dekapoda). — Kieler Meeresforsch. **8**: 212—217.

KINNE-DIETRICH, E. M. (1955): Beiträge zur Kenntnis der Ernährungsphysiologie mariner Blaualgen. — Kieler Meeresforsch. **11**: 34—47.

KITCHING, J. A. (1934): The control of body volume in marine peritricha. J. Exp. Bio*l* **13**: 11—27.

— (1938): Contractile vacuoles. — Biol. Rev. Cambridge Phil. Soc. **13**: 403—444.

— (1948): The physiology of contractile vacuoles. V. The effects of short-term variations of temperature on an fresh-water peritrich ciliate. — J. Exp. Biol. **25**: 406—420.

— (1948): The physiology of contractile vacuoles. VI. Temperature and osmotic stress. — J. Exp. Biol. **25**: 421—436.

— (1951): The physiology of contractile vacuoles. VII. Osmotic relations in a suctorian, with special reference to the mechanism of control of vacuolar output. — J. Exp. Biol. **28**: 203—214.

— (1952): Contractile vacuoles. Symposia Soc. Exp. Biol. **6**: 145—146.

— (1954): Osmoregulation and ionic regulation in animals without kidneys. — Symposia Soc. Exp. Biol. **8**: 63—75.

KLEKOWSKI, R. Z. (1963): The influence of low salinity and desiccation on the survival, osmoregulation and water balance of *Littorina littorea* (L.) (Prosobranchia). — Pol. Arch. hydrobiol. **11**: 241—250.

KNIEP, H. (1907): Beiträge zur Keimungsphysiologie und -biologie von *Fucus*. — J. wiss. Bot. **44**: 635—725.

KNUDSEN, M. (1901): Hydrographical Tables. — Copenhagen: G. E. C. Gad.

— (1903): Gefrierpunkttabelle für Meerwasser. — Publ. Circ. Cons. Explor. Mer. No. 5: 11.

KOCH, H.-J. (1938): The absorption of chloride ions by the anal papillae of Diptera larvae. — J. Exp. Biol. **15**: 152—160.

KOCH, H.-J. & HEUTS, M. J. (1942): Influences de l'hormone thyroidienne sur la regulation osmotique *chez Gasterosteus aculeatus* L. forme *gymnurus* CUV. — Ann. Soc. Roy. Zool. Belg. **73**: 165—172.

KOCH, H.-J., EVANS, J. & SCHICKS, E. (1954): The active absorption of ions by the isolated gills of the crab *Eriocheir sinensis* (M. EDW.). — Meded. Vlaamse. Acad. Kl. Wet. **16**: 5, 1—16.

KOIZUMI, T. (1932, 1935): Studies on the exchange and the equilibrium of water and electrolytes in a holothurian, *Caudina chlensis*. — Sci. Rep. Tôhoku Univ. IV, **7**: 259—311 and **10**: 269—275.

KOLLER, G. (1939): Über die Nephridien von *Physcosoma japonicum*. — Verh. dt. Zool. Ges. **41**: 440—447.

KOSKE, P. H., KRUMM, H., RHEINHEIMER, G. & SZEKIELDA, K.-H. (1966): Untersuchungen über die Einwirkung der Tide auf Salzgehalt, Schwebestoffgehalt, Sedimentation und Bakteriengehalt in der Unterelbe. — Kieler Meeresforsch. **22**: 47—63.

KOTTE, H. (1914): Turgor und Membranquellung bei Meeresalgen. — Wiss. Meeresunters. Abt. Kiel, **17**: 119—167.

KOWALSKI, R. (1955): Untersuchungen zur Biologie des Seesternes *Asterias rubens* L. in Brackwasser. — Kieler Meeresforsch. **11**: 201—213.

KREPS, E. (1929): Untersuchungen über den respiratorischen Gaswechsel bei *Balanus crenatus* bei verschiedenem Salzgehalt des Außenmediums. — Pflügers Arch. ges. Physiol. **222**: 215—241.

KREY, J. (1953): Über die Fruchtbarkeit des Meeres. — Veröff. Inst. Meeresforsch. Bremerh. **2**: 1—13.

— (1956): Die Trophie küstennaher Meeresgebiete. — Kieler Meeresforsch. **12**: 46—71.

— (1957): Ergebnisse eines hydrographisch-produktionsbiologischen Längsschnitts durch die Ostsee im Sommer 1956. III. Die Verteilung des Gesamtphosphors. — Kieler Meeresforsch. **13**: 202—211.

— (1961): Beobachtungen über den Gehalt an Mikrobiomasse und Detritus in der Kieler Bucht 1958—1960. — Kieler Meeresforsch. **17**: 163—175.

— (1968): Unveröffentlichte Untersuchungen in der westlichen Ostsee.

KRIJGSMAN, B. J. & KRIJGSMAN, N. E. (1954): Osmorezeption in *Jasus lalandii*. — Z. vergl. Physiol. **37**: 78—81.

KROGH, A. (1938): The salt concentration in the tissues of some marine animals. — Skand. Arch. Physiol. **80**: 214—222.

— (1939): Osmotic regulation in aquatic animals. — Cambridge Univ. Press.

— (1946): The active and passive exchanges of inorganic ions through the surface of living cells and through membranes generally. — Proc. Roy. Soc. London, Ser. B., **133**: 140—200.

KRUMM, H. & RHEINHEIMER, G. (1966): Untersuchungen zur Hydrographie, Schwebestoffzusammensetzung und Bakteriologie des Nord-Ostsee-Kanals. Teil 2: Zur Schwebstoffzusammensetzung und Bakteriologie des Nord-Ostsee-Kanals. — Kieler Meeresforsch. **22**: 121—127.

KUENEN, D. J. (1939): Systematical and physiological notes on the brine shrimp, *Artemia*. — Arch. Néerland, Zool. **3**: 365—449.

KÜHL, H. & MANN, H. (1962): Modellversuche zum Stoffhaushalt in Aquarien bei verschiedenem Salzgehalt. — Kieler Meeresforsch. **18**, 3. Sonderh.: 89—92.

KYLIN, H. (1938): Über den osmotischen Druck und die osmotische Resistenz einiger Meeresalgen. — Svensk. Bot. Tidskr. (Stockholm) **32**: 238—248.

LABAT, R. & SERFATY, A. (1961): Modifications électrocardiographiques chez la carpe (*Cyprinus carpio* L.) au cours des changements des salinité. — Hydrobiologia (Den Haag) **18**: 185—191.

LAGERSPETZ, K. (1955): Physiological studies on the brackish water tolerance of some species of *Daphnia*. — Arch. Soc. Vanamo, **9**, suppl.: 138—143, Helsinki.

LAGERSPETZ, K. & MATTILA, M. (1961): Salinity reactions of some fresh- and brackish-water crustaceans. — Biol. Bull., Woods Hole **120**: 44—53.

LAGERSPETZ, K. & SIRKKA, A. (1995): Versuche über den Sauerstoffverbrauch von *Mytilus edulis* aus dem Brackwasser der finnischen Küste. — Kieler Meeresforsch. **15**: 89—96.

LANGE, J. (1964): The salinity tolerances of some estuarine planktonic crustaceans. — Biol. Bull. **127**: 108—118.

— (1965): Respiration and osmotic behaviour of the copepod *Acartia tonsa* in diluted sea water. — Comp. Biochem. Physiol. **14**: 155—165.

LANGE, R. (1963): The osmotic function of amino acids and taurine in the mussel, *Mytilus edulis*. — Comp. Biochem. Physiol. **10**: 173—179.

— (1964): The osmotic adjustment in the echinoderm *Strongylocentrotus droebachiensis*. — Comp. Biochem. Physiol. **13**: 205—216.

— (1965): The osmotic function of free ninhydrine positive substances in the sponge, *Halichondria panicea* (PALLAS), and the coelenterate, *Metridium senile* L. — Rep. Fourth Scand. Cell. Res. 40—41.

— (1968): The relation between the oxygen consumption of isolated gill tissue of the common mussel, *Mytilus edulis* L., and salinity. — J. exp. mar. Biol. Ecol. **2**: 37—45.

— (Ed. 1969): Chemical Oceanography. An Introduction. — Chapter 2, Marine biology and chemistry. pp. 35—46. — Universitetsforlaget Oslo.

LANGE, R. & FUGELLI, K. (1965): The osmotic adjustment in the euryhaline teleosts, the flounder, *Pleuronectes flesus* L., and the three-spined stickleback, *Gasterosteus aculeatus* L. — Comp. Biochem. Physiol. **15**: 283—292.

LARSEN, H. (1962): Halophilism. — In "The Bacteria" ed. by I. C. GUNSALUS & R. Y. STANIER, vol. IV: 297—342.

LEGENDRE, R. (1921): Influence de la salinité de l'eau de mer sur l'assimilation chlorophyllienne des algues. — C. r. Soc. Biol. **85**: 222—224.

LELOUP, J. (1948): Influence d'un abaissement de salinité sur la cuprémie de deux Téléostéens marins: *Muraena helena* L., *Labrus bergylta* Asc. — C. r. Soc. Biol. **142**: 178—179, Paris.

LERCHE, W. (1936/37): Untersuchungen über Entwicklung und Fortpflanzung in der Gattung *Dunaliella*. — Arch. Protistenkde. **88**: 236—268.

LEVRING, T. (1959): Some modern aspects of growth and reproduction in marine algae in different regions. — Int. Union Biol. Sci., Ser. B, Nr. 24.

LIENEMANN, L. J. (1938): The green glands as a mechanism for osmotic and ionic regulation in the crayfish (*Cambarus clarkii* GIRARD). — J. Cell. Comp. Physiol. **11**: 149—161.

LILLY, S. J. (1955): Osmoregulation and ionic regulation in Hydra. — J. Exp. Biol. **32**: 423—439.

LOCKWOOD, A. P. M. (1959): The regulation of the internal sodium concentration of *Asellus aquaticus* in the absence of sodium chloride in the medium. — J. Exp. Biol. **36**: 556—561.

— (1960): Some effects of temperature and concentration of the medium on the ionic regulation of the isopod *Asellus aquaticus* (L.). — J. Exp. Biol. **37**: 614—630.

— (1961): The urine of *Gammarus duebeni* and *G. pulex*. — J. Exp. Biol. **38**: 647—658.

— (1962): The osmoregulation of crustacea. — Biol. Rev. **37**: 257—305.

LOCKWOOD, A. P. M. & CROGHAN, P. C. (1957): The chloride regulation of the brackish and fresh-water races of *Mesidotea entomon* (L.). — J. Exp. Biol. **34**: 253—258.

LOEB, J. (1904): On the influence of sea water in the regeneration and growth of Tubulariens. — Univ. California Publ. Physiol. **2**: 139—147.

LOEB, J. & WASTENEYS, H. (1912): On the adaptation of fish *(Fundulus)* to higher temperatures. — J. Exp. Zool. **12**: 543—557.

LÖWENSTEIN, O. (1935): The respiratory rate of *Gammarus chevreuxi* in relation to difference in salinity. — J. Exp. Biol. **12**: 217—221.

LOFTS, B. (1956): The effect of salinity changes on the respiratory rate of the prawn *Palaemonetes varians* (LEACH). — J. Exp. Biol. **33**: 730—736.

LOVTRUP, S. & PIGON, A. (1951): Diffusion and active transport of water in the amoeba *Chaos chaos* L. — C. r. Lab. Carlsberg **28**: 1—28.

LOWENSTAM, H. A. (1954): Environmental relations of modification compositions of certain carbonate secreting marine invertebrates. — Proc. Nat. Acad. Sci. USA **40**: 39—48.

LUCKÉ, B. & McCUTCHEON, M. (1932): The living cell as an osmotic system and its per-permeability to water. — Physiol. Rev. **12**: 68—139.

LUMBYE, J. & LUMBYE, E. (1965): The oxygen consumption of *Potamopyrgus jenkinsi* (SMITH). — Hydrobiologia **25**, 3—4: 489—500.

LUQUE, O. & HUNTER, F. R. (1959): Osmotic studies of amphibian eggs. I. Preliminary survey of volume changes. — Biol. Bull. **117**: 458—467.

LYMAN, J. & FLEMING, R. H. (1940): Composition of sea water. — J. Mar. Res. **3**: 134.

MAC LEOD, R. A. (1965): The question of the existence of specific marine bacteria. — Bacteriol. Rev. **29**, 1: 9—23.

MAC LEOD, R. A. & ONOFREY, E. (1956): Nutrition and metabolism of marine bacteria. II. Observations on the relation of sea water to the growth of marine bacteria. — J. Bacteriol. **71**: 661—667.

MAC ROBBIE, E. A. & DAINTY, J. (1958): Ion transport in *Nitellopsis obtusa*. — J. gen. Physiol. **42**: 335—353.

MADANMOHANRAO, G. & PAMPAPATHI RAO, K. (1962): Oxygen consumption in a brackish water Crustacean, *Sesarma plicatum* (LATREILLE) and marine Crustacean, *Lepas anserifera*. — Crustaceana (Leiden) **4**: 75—81.

MALOEUF, N. S. R. (1938): Studies on the respiration (and osmoregulation) of animals. — Z. vergl. Physiol. **25**: 1—42.

MANIGGAULT, P. (1932): L'effet de variations expérimentales de salinité, de température de pH, sur *Littorina obtusata* L. subsp. *litoralis* L. — Bull. Inst. Océanogr. (Monaco) **605**: 1—8.

MARGARIA, R. (1931): The osmotic changes in some marine animals. — Proc. Roy. Soc. B, **107**: 606—624.

MARS, P. (1950): Euryhalinité de quelque Mollusques méditerrannéen. — Vie et Milieu **1**, 4: 441—448, Paris.

VON MARTENS, E. (1858): On the occurence of marine animal forms in fresh water. — Ann. Mag. Nat. Hist. (Ser. 3) **1**: 50, (cit. after PANIKAR 1940 b).

MARTRET, G. (1939): Variations de la concentration moléculaire et de la concentration en chlorures de l'urine des téléostéens sténohalins en fonction des variations de salinité du milieu extérieur. — Bull. Inst. Océanogr. (Monaco) **774**: 1—38.

MARX, W. & HENSCHEL, J. (1941): Die Befruchtung und Entwicklung von Plattfischen in verdünntem Nordseewasser im Vergleich zu den Befunden in der freien Ostsee. — Helgol. Wiss. Meeresunters. **2**: 226—243.

MC LEESE, D. W. (1956): Effects of temperature, salinity and oxygen on the survival of the American Lobster. — J. Fish. Res. Board of Canada **13**: 247—272.

MC LENNAN, H. (1955): The inorganic composition of the blood and eggs of a fresh-water crustacean *(Cambarus virilis)*. — Z. vergl. Physiol. **37**: 490—495.

MEDVEDEVA, N. B. (1927): Über den osmotischen Druck von *Artemia salina*. — Z. vergl. Physiol. **5**: 547—554.

MENZEL, D. W. & RYTHER, J. H. (1960, 1961): The annual cycle of primary production in the Sargasso Sea of Bermuda. — Deep-Sea Res. **6**: 351—367; **7**: 282—288.

MEYER, D. K. (1948): Physiological adjustement in chloride balance of the goldfish. — Sci. **108**: 305—307.

MEYER, H. (1935): Die Atmung von *Asterias rubens* und ihre Abhängigkeit von verschiedenen Außenfaktoren. — Zool. Jb. Physiol. **55**: 349—398.

MIYASAKI, I. (1933): Effects of temperature and salinity on the development of the eggs of a marine bivalve, *Mactra sulcatria* REEVE. — Bull. Jap. Soc. Fish, **II**, 4: 162—166. Tokyo. (Japan., Engl. Summary).

MIYAWAKI, M. (1951): Notes on the effect of low salinity on an actinian *Diadumene Luciae*. — J. Fac. Sci. Hokkaido Univ. Ser. IV, **10**, 2: 123—126, Sapporo.

MJAALAND, G. (1956): Some laboratory experiments on the coccolithophorid *Coccolithus hyxleyi*. — Oikos **7**, 2: 251—255.

MONTFORT, C. (1931): Assimilation und Stoffgewinn der Meeresalgen bei Aussüßung und Rückversalzung. — Ber. Dt. Bot. Ges. **49**: 49—66.

MORRIS, R. (1956): The osmoregulatory ability of the lampern (*Lampetra fluviatilis* L.) in sea water during the course of its spawning migration. — J. Exp. Biol. **33**: 235—248.

— (1958): The mechanism of marine osmoregulation in the lampern (*Lampetra fluviatilis* L.) and the causes of its breakdown the spawning migration. J. Exp. Biol. **35**: 649—665.

MOSEBACH, G. (1936): Kryoskopisch ermittelte Werte bei Meeresalgen. — Beitr. Biol. Pflanzen **24**.

MOTAIS, R. (1961): Sodium exchange in a euryhaline teleost, *Platichthys fiesus fiesus*. — Endocrinol. **70**: 724—726.

MOTAIS, R., GARCIA ROMEU, F. & MAETZ, J. (1966): Exchange diffusion effect and euryhalinity in teleosts. — J. gen. Physiol. **50**: 391—422.

MOTWANI, M. P. (1955): Experimental and ecological studies on the adaptation of *Mytilus edulis* L. to salinity fluctuations. — Proc. Nat. Inst. Sci. Indiana, Part B, **21**: 227—246.

MÜLLER, R. (1936): Die osmoregulatorische Bedeutung der kontraktilen Vakuolen von *Amoeba proteus*, *Zoothamnium hiketes* und *Frontonia marina*. — Arch. Protistenkde. **87**: 345—382.

MULLINS, L. J. (1950): Osmotic regulation in fish as studies with radioisotopes. — Acta physiol. scand. (Stockh.) **21**: 303—314.

MUNDAY, K. A. & THOMPSON, B. D. (1962): The effect of osmotic pressure on the activity of *Carcinus maenas* mitochondria. — Comp. Biochem. Physiol. **6**: 277—288.

MUUS, B. J. (1967): Some problems facing the ecologist races and subspecies of brackish--water animals. — In "Estuaries" ed. by G. H. LAUFF, 558—563, Publ. No. **83**, Amer. Assoc. Adv. Sci., Washington, D. C.

NAGEL, H. (1934): Die Aufgaben der Exkretionsorgane und der Kiemen bei der Osmoregulation von *Carcinus maenas*. — Z. vergl. Physiol. **21**: 468—491.

NAGABHUSHANAM, R. (1955): Tolerance of the marine wood borer, *Martesia striata* (L.), to waters of low salinity. — J. Zool. Soc. India **7**: 83—90.

NAKANISHI, M. & MONSI, M. (1965): Effect of variation in salinity on photosynthesis of phytoplankton growing in estuaries. — J. Fac. Sci. Univ. Tokyo, Sect. 3, **9**: 19—42.

NEB, K. E. (1952): Untersuchungen über Fortpflanzung und Wachstum an den Heringen der westlichen Ostsee mit besonderer Berücksichtigung der Kieler Förde als Laichgebiet und Fangplatz. — Inauguraldiss. Kiel.

NEEDHAM, J. (1930): On the penetration of marine organisms into fresh water. — Biol. Zbl. **50**: 504—509.

NELLEN, U. R. (1966): Über den Einfluß des Salzgehaltes auf die photosynthetische Leistung verschiedener Standortformen von *Delesseria sanguinea* und *Fucus serratus*. — Helgoländer wiss. Meeresunters. **13**: 288—313.

NELLEN, W. (1965): Beiträge zur Brackwasserökologie der Fische im Ostseeraum. — Kieler Meeresforsch. **21**: 192—198.

NEMENZ, H. (1960): On the osmotic regulation of the larvae of *Ephydra cinerea*. — J. Insect Physiol. **4**: 38—44.

NEUMANN, D. (1959): Morphologische und experimentelle Untersuchungen über die Variabilität des Farbmusters der Schale von *Theodoxus fluviatilis* L. — Z. Morph. Ökol. Tiere **48**: 349—411.

— (1960): Osmotische Resistenz und Osmoregulation der Flußdeckelschnecke *Theodoxus fluviatilis* L. — Biol. Zbl. **79**: 585—605.

— (1961): Osmotische Resistenz und Osmoregulation aquatischer Chironomidenlarven. — Biol. Zbl. **80**: 693—715.

— (1962): Die Analyse limitierender Ionenwirkungen bei Meeres- und Süßwassertieren mit Hilfe ökologischer, physiologischer und züchterischer Methoden. — Kieler Meeresforsch. **18**, 3: 38—54.

NORDL, E. (1953): Salinity and temperature as controlling factors for the distribution and mass occurrence of *Ceratia*. — Blyttia **11**.

NORDLI, E. (1957): Experimental studies on the ecology of *Ceratia*. — Oikos 8, 2: 201—265.

OBERTHÜR, K. (1937): Untersuchungen an *Frontonia marina* FARBEDOM, aus einer Binnenland-Salzquelle unter besonderer Berücksichtigung der pulsierenden Vakuole. — Arch. Protistenkde. **83**: 387—420.

OBUCHOWICZ, L. (1958): The influence of osmotic pressure of medium on oxygen consumption in the snail *Viviparus fasciatus* (O. F. MÜLLER), Streptoneura. — Bull. Soc. Sci. Poznan Sér. B, H. 14: 367—370.

ODUM, H. T. (1954): Factors controlling marine invasion into Florida fresh waters. — Bull. Mar. Sci. Gulf Caribb. 3: 134—156.

OGATA, E. & MATSUI, T. (1965): Photosynthesis in several marine plants of Japan as affected by salinity, drying and pH, with attention to their growth habitats. — Botanica Mar. **8**: 199—217.

OHLE, W. (1954): Sulfat als "Katalysator" des limnischen Stoffkreislaufe. — J. vom Wasser **21**: 13—32.

OLIVEREAU, M. (1948): Influence d'une diminution de salinité sur l'activité de la glande thyroide de deux Téléostéens marins: *Muraena helena* L., *Labrus bergylta* Asc. — C. r. Soc. Biol. **142**: 176—177, Paris.

OSTERHOUT, W. J. V. (1933): Permeability in large plant cells and in models. — Ergebn. Physiol. u. Exp. Pharm. **35**: 967—1021.

OSTWALD, W. (1929): Versuche über die Giftigkeit des Seewassers für Süßwassertiere (*Gammarus pulex* DE GEER). — Pflügers Arch. **106**: 568—598.

OTTO, J. P. (1934): Über den osmotischen Druck der Blutflüssigkeit von *Heteropanope tridentata* (MAITLAND). — Zool. Anz. **108**: 130—135.

— (1937): Über den Einfluß der Temperatur auf den osmotischen Wert der Blutflüssigkeit bei der Wollhandkrabbe *(Eriocheir sinensis)*. — Zool. Anz. **119**: 98—105.

OVERBECK, J. (1962): Das Nannoplankton (μ-Algen) der Rügenschen Brackwässer als Hauptproduzent in Abhängigkeit vom Salzgehalt. — Kieler Meeresforsch. **18**, 3. Sonderh.: 157—171.

— (1964): Der Fe/P-Quotient des Sediments als Merkmal des Stoffumsatzes in Brackwässern. — Helgol. wiss. Meeresunters. **10**: 430—447.

PALMHERT, H. W. (1933): Beiträge zum Problem der Osmoregulation einiger Hydroidpolypen. — Zool. Jb., Abt. Allg. Zool. u. Physiol. **53**: 212—260.

PANIKKAR, N. K. (1940): Influence of temperature on osmotic behaviour of some crustacea and its bearing on problems of animal distribution. — Nature **146**: 366—367.

— (1941): Osmoregulation on some palaemonid prawns. — J. Mar. Biol. Assoc. U. K. **25**: 317—359.

PANTIN, C. F. A. (1931): The adaptation of *Gunda ulvae* to salinity, I. The environment. III. The electrolyte exchange. — J. Exp. Biol. **8**: 63—72, 82—94.

PARRY, G. (1953): Osmotic and ionic regulation in the isopod crustacean *Ligia oceanica*. — J. Exp. Biol. **30**: 567—574.

— (1954): Ionic regulation in the palaemonid prawn *Palaemon* (= *Leander*) *serratus*. — J. Exp. Biol. **31**: 601—613.

— (1955): Urine production by the antennal glands of *Palaemonetes varians* (LEACH). — J. Exp. Biol. **32**: 408—422.

— (1966): Osmotic adaptation in fishes. — Biol. Rev. **41**: 392—444.

— (0000): Organ systems in adaptation the osmotic regulating. — Handb. Physiol. **14**: 245—257.

PEARSE, A. S. (1928): On the ability of certain marine invertebrates to live in diluted sea water. — Biol. Bull. **54**: 405—409.

PEARSE, A. S. & GUNTER, G. (1957): Salinity. In: Treatise on Marine Ecology. — Geol. Soc. Amer. Mem. **1**: 129—157.

PETERS, H. (1935): Über den Einfluß des Salzgehaltes im Außenmedium auf den Bau und die Funktion der Exkretionsorgane dekapoder Crustaceen (nach Untersuchungen an *Potamobius fluviatilis* und *Homarus vulgaris*). — Z. Morph. Ökol. Tiere **30**: 355—381.

PICHOTKA, J. & HÖFLER, W. (1954): Untersuchungen über die Wasserbindung in organischen Systemen. II. Die Wasserbindung in pflanzlichem Gewebe (Kartoffel). — Arch. Exp. Path. u. Pharmakol. **222**: 464—473.

PICKEN, L. E. R. (1936): The mechanism of urine formation in invertebrates. I. The excretion mechanism in certain Arthropoda. — J. Exp. Biol. **13**: 309—328.

— (1937): The mechanism of urine formation in invertebrates. II. The excretory mechanism in certain Mollusca. — J. Exp. Biol. **14**: 20—34.

PIEH, S. (1936): Über die Beziehungen zwischen Atmung, Osmoregulation und Hydratation der Gewebe bei euryhalinen Meeresevertebraten. — Zool. Jb., Allg. Zool. u. Physiol. **56**: 130—158.

PILGRIM, R. L. C. (1953): Osmotic relations in Molluscan contractile tissues. I. Isolated ventricle strip preparations from Lamellibranchs (*Mytilus edulis* L., *Ostrea edulis* L., *Anodonta cygnea* L.). — J. Exp. Biol. **30**: 297—317.

— (1953): Osmotic relations in Molluscan contractile tissues. II. Isolated gill preparations from Lamellibranchs (*Mytilus edulis* L., *Anodonta cygnea* L.). — J. Exp. Biol. **30**: 318—330.

PIRWITZ, W. (1954): Über die Fruchtbarkeit und das Wachstum der Plattfische im Nord-Ostsee-Raum. — Diss. Philos. Fak. Univ. Kiel.

PLATEAU, F. (1871): Recherches physico-chimiques sur les articulés aquatiques. Mém. Acad. Roy. Belg. **36**: 68—99.

Pora, E. A. (1937): La résistance aux salinitées des poissons d'eau douce sténohalins. — Bull. Soc. Sci. Cluj (Roumanie) 8: 612—614.
— (1938): Sur le comportment de *Palaemon squilla* aux variations de salinité. — Ann. Sci. Univ. Jassy (Roumanie) 24: 327—331.
— (1939): Sur l'adaptation d'un téléostéen dulcaquicole, *Carassius carassius* L., au milieu salin. — Bull. Soc. Sci. Cluj 9: 384—393.
— (1946): Problèmes de physiologie animale dans la Mer Noire. — Bull. Inst. Océanogr. (Monaco) 903: 1—43.
— (1949): Les réactions aux variations de salinité. 24. Influence du facteur salinité sur la vie aquatique, spécialement dans la Mer Noire. — Ann. Acad. Republ. popul. Rome 2, 10: 227—325.
Pora, E. A. & Acrivo, C. (1939): Considérations histophysiologiques sur les branchies des poissons téléostéens soumis aux variations de salinité du milieu ambiant. — Ann. Sci. Univ. Jassy (Roumanie) 25: 439—446, Fasc. 2.
Pora, A. E. & Bacescu, M. (1938): Sur la résistance du Mysidé *Gastrosaccus sanctus* de la Mer Noire aux variations de salinité du milieu ambiant. — Ann. Sci. Univ. Jassy (Roumanie) 25: 259—271.
Pora, A. E. & Carausu, S. (1938): Sur la résistance de l'amphipode *Pontogammarus maeoticus* de la Mer Noire aux variations de salinité du milieu ambiant. — Ann. Sci. Univ. Jassy (Roumanie) 25: 272.
Pora, A. E. & Rosca, D. I. (1944): La survie du vers polychaete *Nereis diversicolor* de la Mer Noire et du lac salé d'Eforia, dans des mileux de salinités différentes. — Ann. Sci. Univ. Jassy (Roumanie) 30: 1—17, Fasc. 2.
Pora, A. E. & Stoicovici, F. (1955): Cercetari asupra rolului sistemului nervos de la *Bufo viridis* in fenomenele de adaptare la salinitate. — Bull. stiint. Acad. române 7: 59—89.
Portier, P. (1938): Physiologie des animaux marins. — Paris.
Potts, W. T. W. (1952): Measurement of osmotic pressure in single cells. — Nature 169: 834.
— (1954): The inorganic composition of the blood of *Mytilus edulis* and *Anodonta cygnea*. — J. Exp. Biol. 31: 376—385.
— (1954): The energetics of osmotic regulation in brackish- and fresh-water animals. — J. Exp. Biol. 31: 618—630.
— (1954): The rate of urine production of *Anodonta cygnea*. — J. Exp. Biol. 31: 614—617.
— (1958): The inorganic and amino acid composition of some lamellibranch muscles. — J. Exp. Biol. 35: 749—764.
Potts, W. T. W. & Evans, D. H. (1967): Sodium and chloride balance in the killifish *Fundulus heteroclitus*. — Mar. Bull. 133: 411—425.
Potts, W. T. W. & Parry, G. (1964): Osmotic and ionic regulation in animals. Chapter IV. Osmotic regulation in brackish waters. pp. 119—163. — Pergamon Press, Oxford, 423 pp. (1964): Sodium and chloride Balance in the prawn, *Palaemonetes varians*. — J. Exp. Biol. 41: 591—601.
Precht, H. & Lindner, E. (1966): Reaktionen, Regulationen und Adaptationen der Tiere nach Veränderung von Temperatur und Salzgehalt. Versuche mit *Zoothamnium hiketes* (Ciliata, Peritricha). — Helgoländer wiss. Meeresunters. 13: 354—368.
Price, J. B. & Gunter, G. (1964): Studies of the chemistry of fresh and low salinity waters in Mississippi and the boundary between fresh and brackish water. — Internat. Rev. ges. Hydrobiol. 49: 629—636.
Prosser, C. L. (1955): Physiological variation in animals. — Biol. Rev. Cambridge 30: 229—262.
Prosser, C. L. & Brown. Jr., F. A. (1961): Comparative Animal Physiology. 2. ed. — W. B. Saunders Co. Philadelphia, London.
Prosser, C. L., Green, J. W. & Chow, T. J. (1955): Ionic and osmotic concentration in blood and urine of *Pachygrapsus crassipes* acclimated to different salinities. — Biol. Bull. 109: 99—107.
Raffy, A. (1932): Variations de la consommation d'oxygène dissous au cours de la mort de poissons marins stenohalins passant de l'eau de mer à l'eau douce. — C. r. Acad. Sci. 194: 1522—1524, Paris.

— (1932): Recherches physiologiques sur le mécanisme de la mort des poissons sténo-halins soumis à des variations de salinité. — Bull. Inst. Océanogr. (Monaco) **602**: 1—11.
— (1933): Recherches sur le métabolisme respiratoire des poikilothermes aquatiques. — Ann. Inst. Océanogr. (Monaco) **13**: 259—393.
— (1934): Influence des variations de salinité, sur l'intensité respiratoire de la Telphuse et de l'Ecrivisse. — C. r. Acad. Sci. Paris **198**: 680—681.
— (1949): L'euryhalinité de *Blennius pholis* L. — C. r. Soc. Biol. Paris. **153**: 23—24, 1575—1576.
— (1954): Influence des variations de la température sur l'osmoregulation de quelques Téléostéens marins. — C. r. Soc. Biol. Paris **148**: 1796—1798.
— (1957): Résistance de *Blennius gattorugine* L. à la dessalure. — C. r. Soc. Biol. Paris **150**: 2118—2120.
RAFFY, A. & FONTAINE, M. (1930): De l'influence des variations de salinité sur la respira-tion des civelles. — C. r. Soc. Biol. Paris. **104**: 466—468.
RAMSAY, J. A. (1949): A new method of freezing point determination for small quantities. — J. Exp. Biol. **26**: 57—64.
— (1950): Osmotic regulation in mosquito larvae. — J. Exp. Biol. **27**: 145—157.
— (1954): Movements of water and electrolytes in invertebrates. — Symposia Soc. Exp. Biol. **8**: 1—15.
RANADE, M. R. (1957): Observations on the resistance of *Tigriopus fulvus* (FISCHER) to changes in temperature and salinity. — J. Mar. Biol. Assoc. U. K. **36**: 115—119.
RANSON, G. (1948): Ecologie et répartion géographique des Ostréidés vivants. — XIII. Congrès internat. Zool. Paris: 455.
RAO, K. V. (1951): Observations on the probable effects of salinity on the spawning, develop-ment and setting of the Indian brackwater oyster *Ostrea madrasensis* PRESTON. — Proc. Ind. Acad. Sci. Sect. B. **33**: 231—256.
RASCHACK, M. (1968): Untersuchungen über Osmo- und Elektrolytregulation bei Knochen-fischen aus der Ostsee. — Int. Revue ges. Hydrobiol. **54**: 423—462.
RASQUIN, P. (1956): Cytological evidence for a role of the corpuscles of *Stannius* in the osmoregulation of teleosts. — Biol. Bull. **111**: 399—409.
RAYMONT, J. (1963): Plankton and Productivity in the Oceans. — Pergamon Press, Oxford, London.
REDEKE, H. C. (1933): Über den jetzigen Stand unserer Kenntnisse der Flora und Fauna des Brackwassers. — Verh. internat. Verein. Limnol. **6**: 46—61.
REMANE, A. (1934): Die Brackwasserfauna. — Verh. Dt. Zool. Ges. **36**: 34—74.
— (1940): Einführung in die zoologische Ökologie der Nord- und Ostsee. — In: Tier-welt der Nord- und Ostsee. **I a**: 1—238.
— (1950): Das Vordringen limnischer Tierarten in das Meeresgebiet der Nord- und Ostsee. — Kieler Meeresforsch. **7**, 2: 5—23.
— (1959): Regionale Verschiedenheiten der Lebewesen gegenüber dem Salzgehalt und ihre Bedeutung für die Brackwasser-Einteilung. — Arch. Oceanogr. e Limnol. (Vene-zia) **11**, (Suppl.): 35—46.
RESHÖFT, K. (1961): Untersuchungen zur zellulären osmotischen und thermischen Resistenz verschiedener Lamellibranchier der deutschen Küstengewässer. — Kieler Meeres-forsch. **17**: 65—84.
RESÜHR, B. (1935): Hydrations- und Permeabilitätsstudien an unbefruchteten *Fucus*-Eiern (*Fucus vesiculosus* L.). — Protoplasma **24**: 531—586.
RHEINHEIMER, G. (1966): Unpubl. observations on the salt requirements of the brackish water bacteria of the Schlei Estuary.
— (1966): Einige Beobachtungen über den Einfluß von Ostseewasser auf limnische Bakterienpopulationen. — Veröff. Inst. Meeresforsch. Bremerh., Sonderbd. II: 237—244.
RIEGEL, J. A. & LOCKWOOD, A. P. M. (1961): The role of the antennal gland in the osmotic and ionic regulation of *Carcinus maenas*. — J. Exp. Biol. **38**: 491—499.
ROBERTSON, J. D. (1939): The inorganic composition of the body fluids of three marine invertebrates. — J. Exp. Biol. **16**: 387—397.

344 Physiology of brackish water

— (1941): The function and metabolism of Calcium in the invertebrates. — Biol. Rev.
16: 106—133.
— (1949): Ionic regulation in some marine invertebrates. — J. Exp. Biol. 26: 182—200.
— (1953): Further studies on ionic regulation in marine invertebrates. — J. Exp. Biol.
30: 277—296.
— (1954): The chemical composition of the blood of some aquatic chordates, including
members of the Tunicata, Cyclostomata and Osteichthys. — J. Exp. Biol. 31: 424—442.
— (1957): Osmotic and ionic regulation in aquatic invertebrates. — Rec. Adv. Invertebrate Physiol., Univ. Oregon Publ.: 229—246.
— (1960): Ionic regulation in the Crab Carcinus maenas (L.) in relation to the moulting
cycle. — Comp. Biochem. Physiol. 1: 183—212.
ROBINSON, J. R. (1953): The active transport of water in living systems. — Biol. Rev. 28:
158—194.
— (1954): Secretion and transport of water. — Symposia Soc. Exp. Biol. 8: 42—62.
ROCH, F. (1924): Experimentelle Untersuchungen an Cordylophora caspia (PALLAS) (= lacustris ALLMAN) über die Abhängigkeit ihrer geographischen Verbreitung und ihrer
Wuchsformen von den physikalisch-chemischen Bedingungen des umgebenden Me-
diums. — Z. Morph. Ökol. Tiere 2: 350—426.
ROGENHOFER, A. (1905): Über das relative Größenverhältnis der Nierenorgane bei Meeres-
und Süßwassertieren. — Verh. Zool. Bot. Ges. Wien 55.
ROMANOVA, N. (1956): Variations of the biomass of higher Crustacea in the Northern
Caspian Sea, as observed in the course of several years. — DAN, USSR 109: 2 (R).
ROMEU, G. G. & MOTAIS, R. (1966): Mise en évidence d'échanges Na^+/NH_4^+ chez
l'Anguilla d'eau douce. — Comp. Biochem. Physiol. 17: 1201—1204.
ROTHSCHILD, LORD & BARNES, H. (1953): The inorganic constituents of the sea urchin egg.
— J. Exp. Biol. 30: 534—544.
RUDINSKA, M. A. & CHAMBERS, R. (1951): The activity of the contractile vacuole in a suc-
torian (Tokophrya infusionum). — Biol. Bull. Woods Hole 100: 49—58.
SAGATZ, K. (1931): Vergleichende Untersuchung der Assimilationsleistungen bei Süßwasser-
algen und Vaucheria aus einer Solquelle in abgestuften Salzlösungen. — Beitr. Biol.
Pfl. 19: 67—138.
SAVVATEEV, V. B. (1952): Über die Physiologie der Anpassung der Seepocken "Balanus
balanoides" an die Schwankungen des Salzgehaltes. — Zool. Z. 31: 861—865.
SCHACHTER, D. (1964): Etude comparative du métabolisme respiratoire de Sphaeroma
hookeri LEACH (Crustacé isopode) de la Durancole et des étangs méditerranéens. —
C. r. Acad. Sci. Paris 259: 2917—2919.
SCHELTEMA, R. S. (1965): The relationship of salinity to larval survival and development
in Nassarius obsoletus (Gastropoda). — Biol. Bull. 129, 2: 340—354.
SCHLESCH, H. (1937): Bemerkungen über die Verbreitung der Süßwasser- und Meeresmollusken im östlichen Ostseegebiet. — Tartu Ütikooli junres oleva Loodusuurijate
Seltsi aruanded 43: 37—64. (Tartu).
SCHLIEPER, C. (1929): Über die Einwirkung niederer Salzkonzentrationen auf marine Organismen. — Z. vergl. Physiol. 9: 478—514.
— (1929): Neue Versuche über die Osmoregulation wasserlebender Tiere. — Sitz. Ber.
Ges. Bef. Ges. Naturwiss. Marburg 64: 143—156.
— (1930): Die Osmoregulation wasserlebender Tiere. — Biol. Rev. 5: 309—356.
— (1931): Über das Eindringen mariner Tiere in das Süßwasser. — Biol. Zbl. 51:
401—412.
— (1932): Die Brackwassertiere und ihre Lebensbedingungen vom physiologischen
Standpunkt aus betrachtet. — Verh. internat. Verein. Limnol. 6: 113—146.
— (1934): Weitere Untersuchungen über die Beziehungen zwischen Bau und Funktion
bei den Excretionsorganen decapoder Crustaceen. — Z. vergl. Physiol. 20: 255—257.
— (1933): Über die osmoregulatorische Funktion der Aalkiemen. — Z. vergl. Physiol.
18: 682—695.
— (1935): Neuere Ergebnisse und Probleme aus dem Gebiet der Osmoregulation wasserlebender Tiere. — Biol. Rev. 10: 334—360.

— (1935): Über die osmoregulatorischen Funktionen der Nierenorgane und der Körperoberflächen wasserlebender Tiere. — Sitz. Ber. Ges. Bef. Ges. Naturwiss. **70**: 67—83.

— (1936): Die Abhängigkeit der Atmungsintensität der Organismen vom Wassergehalt und dem kolloidalen Zustand des Protoplasmas. — Biol. Zbl. **56**: 87—94.

— (1942): Die Osmotik des Tierkörpers. — Jenaische Z. Med. u. Naturwiss. **75**: 223—242.

— (1952): Versuch einer physiologischen Analyse der besonderen Eigenschaften einiger eurythermer Wassertiere. — Biol. Zbl. **71**: 450—461.

— (1955): Über die physiologischen Wirkungen des Brackwassers (Nach Versuchen an der Miesmuschel *Mytilus edulis*). — Kieler Meeresforsch. **11**: 22—33.

— (1955): Körperflüssigkeit und Lebensraum der Tiere. — Die Umschau, Frankfurt (Main): 653—655.

— (1957): Comparative study of *Asterias rubens* and *Mytilus edulis* from the North Sea (30⁰/₀₀ S) and the western Baltic Sea (15⁰/₀₀ S). — Internat. Conf. mar. Ecology. — Biol. Stat. Roscoff (France) 1956. — Biol. **33**: 117—127.

— (1956): Über die Physiologie der Brackwassertiere. — Verh. internat. Verein. Limnol. Helsinki.

— (1958): Sur l'adaptation des invertébrés marins à l'eau de mer diluée. Exposé fait au Laboratoire ARAGO de l'Université de Paris, Banyuls-sur-Mer, le 20 septembre 1957. — Vie et Milieu.

— (1960): Genotypische und phaenotypische Temperatur- und Salzgehalts-Adaptationen bei marinen Bodenevertebraten der Nord- und Ostsee. — Kieler Meeresforsch. **16**: 180—185.

— (1964): Ionale und osmotische Regulation bei ästuarlebenden Tieren. — Kieler Meeresforsch. **20**: 169—178.

— (1965): Praktikum der Zoophysiologie. 3. neubearb. u. erw. Aufl. — Gustav Fischer, Stuttgart.

— (1967): Genetic and nongenetic cellular resistance in marine invertebrates. — Helgoländer wiss. Meeresunters. **14**: 482—502.

— (Herausgeb.) (1968): Methoden der meeresbiologischen Forschung. — VEB Gustav Fischer, Jena, 322 pp.

SCHLIEPER, C., BLÄSING, J. & HALSBAND, E. (1952): Experimentelle Veränderungen der Temperaturtoleranz bei stenothermen und eurythermen Wassertieren. — Zool. Anz. **149**: 164—169.

SCHLIEPER, C., FLÜGEL, H. & RUDOLF, J. (1960): Temperature and salinity relationship in marine bottom invertebrates. — Experientia **16**: 470, 1—8.

SCHLIEPER, C., FLÜGEL, H. & THEEDE, H. (1967): Experimental investigations of the cellular resistance ranges of marine temperate and tropical bivalves: Results of the Indian Ocean Expedition of the German Research Association. — Physiol. Zool. **40**: 345 to 361.

SCHLIEPER, C. & HERRMANN, F. (1930): Beziehungen zwischen Bau und Funktion bei den Excretionsorganen decapoder Crustaceen. — Zool. Jb. Anat. **52**: 624—630.

SCHLIEPER, C. & KOWALSKI, R. (1956): Über den Einfluß des Mediums auf die thermische und osmotische Resistenz des Kiemengewebes der Miesmuschel *Mytilus edulis* L. — Kieler Meeresforsch. **12**: 37—45.

— — (1956): Quantitative Beobachtungen über physiologische Ionenwirkungen im Brackwasser. — Kieler Meeresforsch. **12**: 154—165.

— — (1957): Weitere Beobachtungen zur ökologischen Physiologie der Miesmuschel *Mytilus edulis* L. — Kieler Meeresforsch. **13**: 3—10.

SCHMIDT-NIELSEN, B. & LAWS, D. F. (1963): Invertebrate mechanisms for diluting and concentrating the urine. — Ann. Rev. Physiol. **25**: 631—658.

SCHMIDT-NIELSEN, K. (1941): Aktic ionoptagelse hos flodkrebs og strand krabbe. Med pavisning av ionoptagende celler. — Biol. Meed. Kobenhavn **16**, 6: 1—60.

— (1960): Comparative morphology and physiology of excretion. — In "Ideas in modern biology": 393—425. — Sutcliffe.

— (1963): Invertebrate mechanisms for diluting and concentrating the urine. — Ann. Rev. Physiol. **25**: 631—658.

SCHMITT, E. (1955): Über das Verhalten von Süßwasserplanarien (*Planaria gonocephala* DUGES und *Pl. lugubris* O. SCHMIDT) in Brackwasser. — Kieler Meeresforsch. **11**: 48—58.

SCHMITZ, W. (1956): Salzgehaltsschwankungen in der Werra und ihre fischereilichen Auswirkungen. — Jb. vom Wasser **23**.

— (1956): Ökologisch-physiologische Probleme der Besiedlung versalzener Binnengewässer. — Verh. internat. Verein. Limnol. Helsinki.

— (1959): Zur Frage der Klassifikation der binnenländischen Brackwässer. — Arch. Oceanograf. e Limnol. (Venezia) **11** (Suppl.): 179—226.

SCHOFFENIELS, E. (1962): Activité adénosine triphosphatasique des branchies *d'Eriocheir sinensis*. — Arch. Internat. Physiol. Bioch. **70**, 1: 160—161.

SCHOLANDER, P. F. (1966): Osmoregulation bei Mangrovebäumen. — Proc. Nat. Acad. Sci. USA **55**, 1407.

SCHOLANDER, P. F. et al. (1965): Osmoregulation bei Mangrovebäumen. — Sci. **148**: 339.

SCHOLANDER, P. F., VAN DAM, L., KANWISHER, J. W., HAMMEL, H. T., & GORDON, M. S. (1957): Supercooling and osmoregulation in arctic fish. — J. Cell. Comp. Physiol. **49**: 5—24.

SCHOLLES, W. (1933): Über die Mineralregulation wasserlebender Evertebraten. — Z. vergl. Physiol. **19**: 522—555.

SCHWABE, E. (1933): Über die Osmoregulation verschiedener Krebse (Malacostracen). — Z. vergl. Physiol. **19**: 183—236.

SCHWENKE, H. (1958): Über die Salzgehaltsresistenz einiger Rotalgen der Kieler Bucht. — Kieler Meeresforsch. **14**: 11—22.

— (1958): Über einige zellphysiologische Faktoren der Hypotonieresistenz mariner Rotalgen. — Kieler Meeresforsch. **14**: 130—150.

— (1959): Untersuchungen zur Temperaturresistenz mariner Algen der westlichen Ostsee. I. Das Resistenzverhalten von Tiefenrotalgen bei ökologischen und nichtökologischen Temperaturen. — Kieler Meeresforsch. **15**: 34—50.

— (1960a): Neuere Erkenntnisse über die Beziehungen zwischen den Lebensfunktionen mariner Pflanzen und dem Salzgehalt des Meer- und Brackwassers. — Kieler Meeresforsch. **16**: 28—47.

— (1960b): Vergleichende Resistenzuntersuchungen an marinen Rotalgen aus Nord- und Ostsee (Salzgehaltsresistenz). — Kieler Meeresforsch. **16**: 201—213.

SCOTT, T. G. & HAYWARD, R. (1955): Sodium and potassium regulation in *Ulva lactuca* and *Valonia macrophysa*. — Woods Hole Oceanograph. Inst. Coll. Reprints Part I, Contr. **597**: 35—64.

SECK, CH. (1958): Untersuchungen zur Frage der Ionenregulation bei in Brackwasser lebenden Evertebraten. — Diss. Philos. Fak. Univ. Kiel 1957, and Kieler Meeresforsch. **13**: 220—243.

SECONDAT, M. (1952): Influence des variations de salinité sur la consommation d'oxygène des alevins vésiculés de Saumon (*Salmo salar* L.). — Bull. Inst. Océanogr. (Monaco) **1013**: 1—8.

SEGAL, E. (1967): Physiological response of estuarine animals from different latitudes. — In "Estuaries" ed. by G. H. LAUFF, 548—553, Publ. No. **83**, Amer. Assoc. Adv. Sci., Washington, D. C.

SEGAL, E. & BURBANCK, W. D. (1963): Effects of salinity and temperature on osmoregulation in two latitudinally separated populations of an estuarine isopod, *Cyathura polita* (STIMPSON). — Physiol. Zool. (Chicago) **36**: 250—263.

SEGERSTRÅLE, S. G. (1949): The brackish-water fauna of Finland. — Oikos **1**: 127—141.

SHAW, J. (1955): A simple procedure for the study of ionic regulation in small animals. — J. Exp. Biol. **32**: 321—329.

— (1955): The permeability and structure of the cuticle of the aquatic larva of *Sialis lutaria*. — J. Exp. Biol. **32**: 330—352.

— (1955): Ionic regulation and water balance in the aquatic larva of *Sialis lutaria*. — J. Exp. Biol. **32**: 353—382.

— (1955): Ionic regulation in the muscle fibres of *Carcinus maenas* L. I. The electrolyte composition of single fibres. — J. Exp. Biol. **32**: 383—396.

— (1955): Ionic regulation in the muscle fibres of *Carcinus maenas* L. II. The effect of reduced blood concentration. — J. Exp. Biol. **32**: 664—680.

— (1959): The absorption of sodium ions by the crayfish, *Astacus pallipes* LEREBOULLET. I. The effect of external and internal sodium concentrations. — J. Exp. Biol. **36**: 126—144.

— (1959): Solute and water balance in the muscle fibres of the East African fresh-water crab, *Potamon niloticus* (M. EDW.). — J. Exp. Biol. **36**: 145—156.

— (1960): The mechanisms of osmoregulation. In "Comparative Biochemistry" ed. by M. FLORKIN & H. S. MASON, **2**: 479—518. — Acad. Press, New York.

— (1961): Studies on the ionic regulation in *Carcinus maenas* L. I. Sodium balance. — J. Exp. Biol. **38**: 135—153.

— (1961): Sodium balance in *Eriocheir sinensis* (M. EDW.). The adaptation of the crustacea to fresh water. — J. Exp. Biol. **38**: 153—162.

SHAW, J. & SUTCLIFFE, D. W. (1961): Studies on sodium balance in *Gammarus duebeni* LILLJEBORG and *G. pulex* L. — J. Exp. Biol. **38**: 1—15.

SHELBOURNE, J. E. (1957): Site of chloride regulation in marine fish larvae. — Nature **180**: 920—922, London.

SHOUP, C. S. (1932): Salinity of the medium and its effect on respiration in the Sea-Anemone. — Ecol. **8**: 81—85.

SMITH, C. W. (1956): The role of the endocrine organs in the salinity tolerance of trout. — Mem. Soc. Endocrinol. No. 5, Part II: 83—101, Cambridge.

SMITH, R. I. (1955): Comparison of the level of chloride regulation by *Nereis diversicolor* in different parts of its geographical range. — Biol. Bull. **109**: 453—474.

— (1959): Physiological and ecological problems of brackish-waters. — Mar. Biol., Proc. 20 th Ann. Biol. Coll., Oregon State Coll.: 59—69.

— (1963): A comparison of salt loss rate in three species of brackish-water nereid Polychaetes. — Biol. Bull. **125**, 2: 332—343.

— (1963): The reproduction of *Nereis diversicolor* (Polychaeta) on the south coast of Finland — some observations and problems. — Comment. Biol. **26**: 10, 1—12.

— (1964): D₂O uptake in two brackish-water nereid polychaetes. — Biol. Bull. **126**, 1: 142—149.

— (1964): On the early development of *Nereis diversicolor* in different salinities. — J. Morphol. **114**: 3.

— (1967): Osmotic regulation and adaptive reduction of water permeability in a brackish-water crab, *Rhithropanopeus harrisi* (Brachyura, Xanthidae). — Biol. Bull. **133**: 643—658.

STEEMANN-NIELSEN, E. (1958): A survey of recent Danish measurements of the organic productivity in the sea. — Rapp. Cons. Explor. Mer. **144**: 92—95.

— (1960): Productivity of the oceans. — Ann. Rev. Plant. Physiol. **11**: 341—362.

— (1964): Investigations of the rate of primary production at two Danish Light ships in the transition area between the North Sea and the Baltic. — Medd. Danmarks Fisk. og Havundersøg. **4**: 31—77.

STEEMANN-NIELSEN, E. & JENSEN, A. (1957): Primary oceanic production. The autotrophic production of organic matter in the oceans Galathea-Reports. — Sci. Res. of the Danish-Deep-Sea Exped. 1950—1952, **1**: 49—136.

STEINER, G. (1935): Der Einfluß der Salzkonzentration auf die Temperaturabhängigkeit verschiedener Lebensvorgänge. — Z. vergl. Physiol. **21**: 666—679.

STEPHENS, G. C. (1967): Dissolved organic material as nutritional source for marine and estuarine invertebrates. — In "Estuaries" ed. by G. H. LAUFF, 367—373, Publ. No. **83**, Am. Assoc. Adv. Sci., Washington D. C.

STEPHENS, G. C. & VIRKAR, R. A. (1966): Uptake of organic material by aquatic invertebrates. IV. The influence of salinity on the uptake of amino acids by the brittle star, *Ophiactis arenosa*. — Biol. Bull. **131**: 172—185.

STRODTMANN, S. (1906): Zur Biologie der Ostseesprotten. — Mitt. Dt. Seefisch 22: 12.
— (1906): Laichen und Wandern der Ostseefische. 2. — Ber. Wiss. Meeresunters. N. F., Abt. Helgoland 7: 132—216.
— (1918): Weitere Untersuchungen über Ostseefische. — Wiss. Meeresunters. N. F. Abt. Helgoland 14: 29—95.
SUOMALAINEN, P. (1956): Sauerstoffverbrauch finnischer Gammarus-Arten. — Verh. internat. Verein. Limnol. Helsinki.
SUTCLIFFE, D. W. (1960): Osmotic regulation in the larvae of some euryhaline Diptera. — Nature 187: 331—332.
SVERDRUP, H. U., JOHNSON, M. W. & FLEMING, R. H. (1942): The Oceans, their Physics, Chemistry and general Biology. — New York, 1087 pp.
Symposium in the classification of brackish-waters. 1959. Venice 8—14. April 1958. — Vol. XI, Suppl., Arch. Oceanograf. Limnol. Venezia, 248 pp.
TARUSSOV, B. (1927): Über den Einfluß der osmotischen Bedingungen auf die Oxydationsgeschwindigkeit. — Z. eksper. Biol. Med. (Russ.) 6: 229—240.
— (1929): Über Zellpermeabilität und Anpassungsfähigkeit bei Wassertieren. — Protoplasma 9: 97—105.
THEEDE, H. (1963): Experimentelle Untersuchungen über die Filtrationsleistung der Miesmuschel Mytilus edulis L. — Kieler Meeresforsch. 19: 20—41.
— (1964): Physiologische Unterschiede bei der Strandkrabbe Carcinides maenas L. aus der Nord- und Ostsee. — Kieler Meeresforsch. 20: 179—191.
— (1965a): Beziehungen zwischen dem Salzgehalt des Außenmediums und der Stoffwechselgröße bei wasserlebenden Tieren. — Bot. Gothoburgens. III, Proc. 5th Mar. Biol. Sympos. Göteborg: 221—231.
— (1965b): Vergleichende experimentelle Untersuchungen über die zelluläre Gefrierresistenz mariner Muscheln. — Kieler Meeresforsch. 21: 153—166.
— (1969): Einige neue Aspekte bei der Osmoregulation von Carcinus maenas. — Marine Biol. 2: 114—120.
THIENEMANN, A. (1928): Mysis relicta in sauerstoffarmem Tiefenwasser der Ostsee und das Problem der Atmung in Salzwasser und Süßwasser. — Zool. Jb., Abt. Zool. Physiol. 45: 371—384.
TODD, M. E. & DEHNEL, P. A. (1960): Effect of temperature and salinity on heat tolerance in two grapsoid carbs, Hemigrapsus nudus and Hemigrapsus oregonensis. — Biol. Bull. 118: 150—172.
TOPPING, F. L. & FULLER, J. L. (1942): The accomodation of some marine invertebrates to reduced osmotic pressures. — Biol. Bull. 82: 372—384.
TRAHMS, O. K. (1939): Die Größen- und Kalkreduktion bei Mytilus edulis L. in Rügenschen Binnengewässern. — Z. Morph. Ökol. Tiere 35, 2: 246—249.
TRAHMS, O. K. & STOLL, K. (1938): Hydrobiologische und hydrochemische Untersuchungen in den Rügenschen Boddengewässern während der Jahre 1936 und 1937. — Kieler Meeresforsch. 3: 61—98.
TREHERNE, J. E. (1954): The exchange of labelled sodium in the larva of Aedes aegypti L. — J. Exp. Biol. 31: 386—401.
TSCHANG-SI (1930): Action de l'eau de mer diluée sur le développement des Gastropodes, Opistobranches. — Congr. Int. de Océanogr. Hidrogr. marina e Hidrolcontinent., Sevilla, 1—7 Mayo 1929, 1: 252—255, Madrid.
VÄLIKANGAS, I. (1933): Über die Biologie der Ostsee als Brackwassergebiet. — Verh. internat. Verein. Limnol. 6: 62—112.
VENKATARAMAN, R. & SREENIVASAN, A. (1954): Salt tolerance of marine bacteria. — Food. Res. 19: 311—313, 1954 (cit. after Biol. Abstr. 28, Nr. 29027).
VERNBERG, J. F. (1967): Some future problems in the physiological ecology of estuarine animals. — In "Estuaries" ed. by G. H. LAUFF, pp. 554—557, Publ. No. 83, Amer. Assoc. Adv. Sci., Washington, D. C.
VERNBERG, J. F., SCHLIEPER, C. & SCHNEIDER, D. E. (1963): The influence of temperature and salinity on ciliary activity of excised gill tissue of molluscs activity from North Carolina. — Comp. Biochem. Physiol. 8: 271—285.

VERWEY, J. (1957): A plea for the study of temperature influence on osmotic regulation. — Ann. Biol. **33**: 129—149.

VESELOV, E. A. (1949): Effect of salinity of the environment on the rate of respiration in fish. — Zool. Zhurn. **28**: 85—98.

VOGEL, W. (1966): Über die Hitze- und Kälteresistenz von *Zoothamnium hiketes* PRECHT (Ciliata, Peritricha). — Z. wiss. Zool. **173**: 344—378.

WAEDE, M. (1954): Beobachtungen zur osmotischen, chemischen und thermischen Resistenz der Scholle *(Pleuronectes platessa)* und Flunder *(Pleuronectes flesus)*. — Kieler Meeresforsch. **10**: 58—67, and Diss. Kiel.

WALDES, V. (1939): Die chemische Beeinflussung der Cirren von *Balanus*. — Z. vergl. Physiol. **26**: 347—361.

WARREN, J. G. (1954): Osmotic responses in the Sipunculid *Dendrostomum zostericolum*. — J. Exp. Biol. **31**: 402—423.

WASMUND, C. (1939): Sedimendationsgeschichte des Großen Jasmunder Boddens. — Geologie-Meere-Binnengewässer **3**: 506—526.

WATTENBERG, H. & WITTIG, H. (1940): Über die Bestimmung der Titrationsalkalinität des Seewassers. — Kieler Meeresforsch. **3**: 258—262.

WEBB, D. A. (1940): Ionic regulation in *Carcinus maenas*. — Proc. Roy. Soc. London, Ser. B., **129**: 107—135.

WEEL, P. B. van (1957): Observations on the osmoregulation in *Aplysia juliana* PEASE (Aplysiidae, Mollusca). — Z. vergl. Physiol. **39**: 492—506.

WEIL, E. & PANTIN, C. F. A. (1931): The adaptation of *Gunda ulvae* to salinity. II. The water exchange. — J. Exp. Biol. **8**: 73—81.

WELLS, G. P. & LEDINGHAM, I.C. (1940): Physiological effects of a hypotonic environment. I. The action of hypotonic salines and isolated rhytmic preparations from polychaete worms *(Arenicola marina, Nereis diversicolor, Perinereis cultrifera)*. — J. Exp. Biol. **17**: 337—363.

WELLS, G. P., LEDINGHAM, I. C. & GREGORY, M. (1940): Physiological effects of a hypotonic environment. II. Shock effects and accomodation in cilia *(Pleurobrachia, Mytilus, Arenicola)*, following sudden salinity change. — J. Exp. Biol. **17**: 378—385.

WELLS, H. W. (1961): The fauna of oyster beds, with special reference to the salinity factor. — Ecol. Monogr. **31**: 239—266.

WIDMANN, E. (1936): Osmoregulation bei einheimischen Wasser- und Feuchtluft-Crustaceen. — Z. wiss. Zool. **147**: 132—169.

WIGGLESWORTH, B. (1933): The effects of salts on the anal gills of the mosquito larva. — J. Exp. Biol. **10**: 1—15.

— (1933): The adaptation of mosquito larvae to salt water. — J. Exp. Biol. **10**: 27—37.

WIKGREN, B. J. (1953): Osmotic regulation in some aquatic animals with special reference to the influence of temperature. — Acta zool. fenn. **71**: 3—102.

WILLIAMS, A. B. (1960): The influence of temperature on osmotic regulation in two species of estuarine shrimps *(Penaeus)*. — Biol. Bull. **119**: 560—571.

WILSON, K. G. & ARONS, A. B. (1955): Osmotic pressures of seawater solutions computed from experimental vapor pressure lowering. — J. Mar. Res. **14**: 195.

WINBERG, G. G. (1961): Der gegenwärtige Stand und die Aufgaben bei der Erforschung der Primärproduktion der Gewässer (Russ.). — Perwitschnaja produkzija morei i wnutrennich wod 11—24, Minsk.

WITTIG, H. (1940): Über die Verteilung des Kalzimus und der Alkalinität in der Ostsee. — Kieler Meeresforsch. **3**: 460—496.

WOHLSCHLAG, D. E. (1957): Differences in metabolic rates of migratory and resident freshwater forms of an arctic whitefish. — Ecol. **38**: 502—510.

WOODHEAD, P. M. J. & WOODHEAD, A. D. (1958): An effect of low temperature on the osmo-regulatory ability of the cod *(Gadus callarias)* in arctic waters. I. — Proc. Linn. Soc. London **169**: 63—66.

WÜST, G. (1957): Ergebnisse eines hydrographisch — produktionsbiologischen Längsschnittes durch die Ostsee im Sommer 1956. I. Die Verteilung von Temperatur, Salzgehalt und Dichte. — Kieler Meeresfrosch. **13**: 163—185.

YONGE, C. M. (1936): On the nature and permeability of chitin. 2. The permeability of the uncalcified chitin lining the foregut of *Homarus*. — Proc. Roy. Soc. London B **120**: 15—41.

ZAKS, M. G. & SOKOLOVA, M. M. (1960): The role of potassium in adaptation of the lugworm *(Arenicola)* tissue to hypotonic media. — Akad. Wiss. UDSSR, Zytol. **2**: 448—453.

ZEITZSCHEL, B. (1965): Zur Sedimentation von Seston, eine produktionsbiologische Untersuchung von Sinkstoffen und Sedimenten der westlichen und mittleren Ostsee. — Kieler Meeresforsch. **21**: 55—80.

ZELLER, A. (1931): Resistenzversuche an Rotalgen. — Acad. Wiss. Wien, Math.-Nat. Kl. Sitz.-Ber., Abt. I., Miner. Biol., Erdkde. **140**, 7: 543—552.

ZENKEVITCH, L. (1963): Biology of the seas of the U. S. S. R. — G. Allen & Unwin, London, 955 pp.

ZOBELL, C. E. (1941): Studies on marine bacteria. I. The cultural requirements of heterotrophic aerobes. — J. Mar. Res. **4**: 42—75.

— (1946): Marine Microbiology. — Waltham, Mass., USA.

ZWICKY, K. (1954): Osmoregulatorische Reaktionen der Larve von *Drosophila melanogaster*. — Z. vergl. Physiol. **36**: 367—390.

Addendum

ANDERSON, G. C. & BANSE, K. (1963): Hydrography and phytoplancton production. Proc. conf. on primary productivity measurements, 1961, pp. 61—71, (Editor M. S. DOTY), U. S. Atomic Energy Commiss., Ref. TID-7633, Biol. and Med.

BERGE, G. (1958): The primary production in the Norwegian Sea, June 1954, as measured by an adapted ^{14}C-technique. — Rapp. Cons. Int. Explor. Mer **144**: 85—91.

FLÜGEL, H. (1963): Elektrolytregulation u. Temperatur bei *Crangon crangon* L. und *Carcinus maenas* L. — Kieler Meeresforsch. **19**: 189—195.

LANGE, R. (1970): Isosmotic intracellular regulation and euryhalinity in marine bivalves. — J. exp. mar. Biol. Ecol. **5**: 170—179.

LAUFF, G. H. (Ed.) (1967): Estuaries — Amer. Assoc. Adv. Sci. Publ. No. 83, 757 pp.

LUMBYE, J. (1958): The oxygen consumption of *Theodoxus fluviatilis* (L.) and *Potamopyrgus jenkinsi* (SMITH) in brackish and fresh water. — Hydrobiologica **10**: 245—262.

SMITH, R. I. (1970a): Chloride regulation at low salinities by *Nereis diversicolor*. — J. Exp. Biol. **53**: 75—92.

— (1970b): Hypo-osmotic urine in *Nereis diversicolor*. — J. Exp. Biol. **53**: 101—108.

STEEMANN-NIELSEN, E. (1951): The marine vegetation of the Isefjord. A study on ecology and production. — Medd. Kommiss. Danmarks Fisk. og Havundersogelser, Ser. Plankton V, 1—114.

WEBER, R. E. & SPAARGAREN, D. H. (1970): On the influence of temperature on the osmoregulation of *Crangon crangon* and its significance under estuarine conditions. — Netherlands J. Sea Res. **5**: 108—120.

WINKLE, JR., WEBSTER VAN (1968): The effects of season, temperature and salinity on the oxygen consumption of bivalve gill tissue. — Comp. Biochem. Physiol. **26**: 69—80.

Subject Index

adaptation, genetic 55
, individual 220
, nongenetic 55, 56
Agigea (Black Sea) 221
Algae, Brown 95
, Green 96
, Red 95, 308
alkalinity 307
ambulacral fluid, starfish 265, 266
Amphipoda from brackish water 91
Azov (USSR), sea of 131, 162

Baltic Sea 17, 18, 71, 135, 313 ff
Baltic Sea, arctic species 141, 142
, hydrographic sections 11, 317
, temperature of bottom water 139
Bankflunder (Baltic Sea) 220
bathymetric distribution of species 23
Belt Sea 17
BERGMANN's law 25
Black Sea 20, 102, 130
, hydrographic sections 146
, isobaths 144
, mud community 159
, Zostera-region 158
Blyth estuary (England) 214
Bornholm basin, bottom fauna Figs. 61, 62
Bosphorus 24
brachyhaline 6
brackish water-sterility 44
breeding experiments 56

calcareous content of Mytilus shells 236
calcium-content of brackish water 237, 307
calcium influence on Delesseria 308
Gunda 307
Mytilus gills 310
permeability 292
Caspian Sea (USSR) 75
cellular salinity ranges 223, 224
chemical composition of starfish 237
chloride transport 256, 274
ciliary activity 235
coelomic fluid, starfish 265, 266
competition 59
composition of ovaries, starfish 243
Concarneau (France) 213
contractile vacuoles as volume regulators 240, 241
Copepoda (Calanoidea), euryhaline marine 92
Copepoda-Harpacticoidea, euryhaline 93, 94
cultures with graded salinities 100

definition of brackish water 4, 211
Desmidiaceae 83
Diatomaceae 99, 115, 216
Diptera, halobiont 82
, secondary marine 82
discontinuity layer 14
Dybsöfjord (Denmark) 161

ecological tolerance 56
Ekenäs region (Finnland) 52
electrical conductivity 250, 304, 306
Enteromorpha zone 9
euhaline 5
euryhaline 6, 87
euryhalinity, degree of 213
euhalobia 8

F¹ — hybrids 221
facultatively holeuryhaline species 59
Farbstreifenwatt 12
filtration rates 233, 234
Finland, higher plants of coast 84, 85
Finngrundet (Sweden) 13
Finnish coast 24
fishes, euryhaline 82
fishes of the Baltic 28, 115
food conversion 238
free amino acids 262, 263 270, 276
freezing point of body fluids 251 ff
of brackish water 250
of single cells 251
freezing resistance, bivalves 301, 302

Gotland Basin, section 317
growth curves, Clupea 29
, Gadus 30
, Gammarus 231
, Leuciscus 38
, Lyngbya 231
, Perca 36
, Pleuronectes 34

haline stratification 10
halobionts 87
halolenitobionts 103
halophilons 6
halotolerant 2
heart beat rate 235
heat resistance, Idus 301
, Mytilus 298
, Planaria 301
, Pleuronectes 300
, Tigriopus 299
, Zoothamnium 299
homoiosmotic behavior 254

Index of Genera

Triarthra 99, 100
Tribus 149
Trichocerca 87, 163
Trichocladius 81, 82
Trichodina 65
Trigonostomum 155
Trilobodridus 176
Trilobus 68
Trinectes 95
Triops 73
Tripyla 69, 108
Tripyloides 91
Trochammina 88, 89, Table 5
Trochus 143
Troglochaetus 51, 97, 180
Troglomysis 207
Tubifex 70, 110, 165
Turbanella 91, 108, 176
Turbinaria 119
Tvaerminnea 103, 106, 107
Typha 124
Typhlopharoma 207

Uca 166
Udotea 119
Ulothrix 61, 86, 170
Ulva 96, 119, 164, 277, 289, 346
Uncinais 70
Uncinorhynchus 91
Unio 37, 71
Upogebia 159
Uronema 61
Uronychia 90
Urospora 32, 33
Uteriporus 179
Utricularia 43, 85, 124

Vallisneria 19
Valonia 276, 277, 346

Valvata 72
Vaucheria 43, 86, 120—123, 344
Vejdovskya 51, 103, 106, 107, 116, 178, 179
Venerupis 245
Venus 127, 159, 334
Verruca 49
Victorella 114, 197, 208
Viscosia 60, 91
Viviparus 72, 340
Volgocuma 149
Volsella 130
Vorticella 64

Waerniella 95
Westbladiella 179
Wierzeiskiella 68
Wierzejskia 177
Wigrella 68

Xenotrichula 176, 177
Xestoleberis 33, 92, 93, Table 5, 112, 167

Yoldia 135

Zagrabica 152, 153
Zannichellia 84, 86, 116, 171
Zaus 116, 117
Zelinkiella 68
Zippora 50, 88, 94, 136
Zoarces 94
Zonorhynchus 107
Zoothamnium 215, 240, 241, 298, 299, 339, 342, 349
Zostera 8, 43, 50, 52, 69, 74, 80, 81, 84, 86, 96, 115, 116, 119, 123, 154, 158, 161, 164, 165, 197, 257
Zygnema 43, 86, 123

Author Index